The TAB Electronics Yellow Pages
Equipment, Components, and Supplies

Andrew Yoder

McGraw-Hill

New York San Francisco Washington, D.C. Auckland Bogota
Caracas Lisbon London Madrid Mexico City Milan
Montreal New Delhi San Juan Singapore
Sydney Tokyo Toronto

McGraw-Hill
A Division of The McGraw-Hill Companies

Copyright © 1997 by The McGraw-Hill Companies, Inc. All rights reserved. Printed in the United States of America. Except as permitted under the United States Copyright Act of 1976, no part of this publication may be reproduced or distributed in any form or by any means, or stored in a data base or retrieval system, without prior written permission of the publisher.

1 2 3 4 5 6 7 8 9 0   AGM/AGM   9 0 1 0 9 8 7 6

ISBN   0-07-076510-3 (PBK)
       0-07-076512-X (HC)

The sponsoring editors for this book were Roland Phelps and Scott Grillo, the editing supervisors were Lori Flaherty and Sally Glover, and the production supervisor was Clare Stanley. It was set in Humanist 777 light.

Printed and bound by Quebecor / Martinsburg.

McGraw-Hill books are available at special quantity discounts to use as premiums and sales promotions, or for use in corporate training programs. For more information, please write to the Director of Special Sales, McGraw-Hill, 11 West 19th Street, New York, NY 10011. Or contact your local bookstore.

Information contained in this work has been obtained by The McGraw-Hill Companies, Inc. ("McGraw-Hill") from sources believed to be reliable. However, neither McGraw-Hill nor its authors guarantee the accuracy or completeness of any information published herein and neither McGraw-Hill nor its authors shall be responsible for any errors, omissions, or damages arising out of the use of this information. This work is published with the understanding that McGraw-Hill and its authors are supplying information but are not attempting to render engineering or other professional services. If such services are required, the assistance of an appropriate professional should be sought.

Dedication

This book is dedicated to the memory of TAB Books' office in Blue Ridge Summit, Pennsylvania. For over 30 years, we published some excellent books and had a great time doing it. It was a privilege to have worked with the fine staff. I will miss everyone.

Acknowledgments

This book could not have been published in its entirety without the help of: Roland Phelps, Stacey Spurlock, Dawn Bowling, Stephen Moore (who ressurrected a scrambled disk file), Lori Flaherty, and Jen Priest of TAB Books; Steve Chapman, Scott Grillo, Sally Glover, and Clare Stanley of McGraw-Hill; Alan Wimstanley of *Everyday Practical Electronics*; Bob Grove of *Monitoring Times*; Thomas E. Black of Digital Products Company; Susan Roe of Magellan's; John Bachellier of MBI USA; Robert E. Sokol of Two-Bit Computing; Betty Cho and Clare Charnley of PMC-Sierra, Inc., Christopher A. Nielsen of Zorin, Leigh Goldstein of Leigh's Computers, David Kranser of EZ Systems, George Varvitsiotes of Ham Radio Outlet, and Larry Sribnick of SR Batteries.

Introduction

I have been interested in radio, audio, and computers for years. The hardware companies all seem to have my name duplicated on their mailing lists. Even if they don't all know who I am, I try to keep track of who they are. I track rare components for antique radio equipment, find the best prices for audio components, and look for the best ratio of price to service for computer equipment. So, I have a long section of shelves that is dedicated to catalogs from all of these suppliers and manufacturers.

But sometimes, I am working on a tight project and I don't have the time to merge in my freshly received catalogs. A pile forms under the bed. Then, the answering machine dies and I have to hunt for parts through old catalogs. A pile forms on the floor of the TV room. Then, someone from work asks me where they can get a good deal on some car stereo equipment. A pile forms across my desk at work. By this time, my entire natural habitat is covered with catalogs, magazines, newsletters, and forms. What a mess!

In a worst-case scenario, I will lose the catalog and waste an hour stomping around the house looking for it. Later on, I'll discover that I could've saved 30% if I would have just looked in catalog X. Argh!

After a few years of this torture, I thought "Wouldn't it be great if a book covered electronics and computer contact information?" Light bulb! So, I tediously worked through my shelves of catalogs and nearly 100 different magazines. I contacted dozens of Internet forums and collected advertising from hamfests and computerfests around the United States.

Now that this book is finished, my floors and desk are almost tidy. The space under the bed isn't cleared out, but it's improving. And some of the former catalog space on the bookshelves is now filled with books and magazines.

Not only is my house cleaner, but I'm finding information much quicker and easier. While I was working on this book, some friends asked me what companies sold different products. Easy! In a few seconds, I had a dozen or so phone numbers to call for price information. Too cool.

I'm convinced that this book is a must for anyone who needs to buy, repair, or upgrade anything electronic. In most cases, it will save more than the price of the book in time and money. To help you get more out of the book, I also worked with a number of manufacturers and dealers, and money-saving coupons are included in the back of the book. Be sure to check them out before you buy anything!

I hope that you find the book and the coupons useful. If you notice any changes in contact information or new companies, please feel free to drop me a line:

Andrew Yoder
P.O. Box 840
Mont Alto, PA 17237
ayoder@cvn.net (Internet e-mail address)

By the time you read this, I hope to have a Web page that will include updated information on the listings in this book. I don't have a URL yet, so if you're interested, search for *The TAB Electronics Yellow Pages*.

How to use this book

This book contains contact information (addresses, telephone numbers, fax numbers, e-mail addresses, and Internet Web sites), some warranty or catalog information, and a brief description of what the company sells or manufactures.

Most of the listings here are for companies in North America, although a number of companies are included from Europe, Australia, Taiwan, and New Zealand, as well. This book isn't comprehensive; it doesn't have a listing for that little electric nose flute shop in Blagoveshchensk or the quaint electric yak milker company in Thimphu, Bhutan. Still, I think you'll find that there's a wealth of information here.

A few rules and general information guidelines concerning the listings:

* *Most known information is listed.* Some companies might stock products that they no longer advertise (some of this information might not be listed). Or some companies might accept credit cards, but not advertise that they do. If you can only buy via credit card and the listing has no credit cards listed, ask the company if it accepts credit cards anyway. It might.

* *Companies constantly change.* They go out of business, get into business, and change product lines. Although I used up-to-date information to compile this book (and I updated it through the course of the project), some companies will have changed by the time you read this. If you borrowed this book from a library or from a friend, check the date and make sure that it is an up-to-date edition (it will probably be updated every two or three years).

* *Company status is not listed in the name header.* For example, "Pico Technology Ltd." is listed as "Pico Technology," and "B&L Communications, Inc." is listed as "B&L Communications." It's simpler and less confusing that way.

* *Some companies have the same name.* This is in part because the company status (Ltd., Co., Inc., etc.) is not included and because companies from around the world are listed.

* *American telephone numbers that begin with 800 are toll-free numbers.* These numbers have restrictions as to where they can be called from; some work only in the United States, some work only in a particular state, some work only in the continental United States. If you want to call a company

that has an 800 number, give it a shot; it might work for you.

If you have problems calling an overseas telephone number, try looking up the international calling section of your phone book. There might be some particular rules about dialing internationally in your country. For additional help, try calling the telphone operator for assistance.

The Notes portion of each listing might not entirely cover the number of products that a company sells or manufactures. For surplus dealers, for example, the inventory is constantly changing and is almost impossible to pin down.

Credit card information, COD costs, money-back guarantees, etc., are listed where they were known. Warranties vary so significantly in coverage that they were not listed at all. If you have any questions about this information, be sure to ask the company before ordering an item.

Whether or not a company manufactures or just sells a product is a grey area. Most of the computer "manufacturers" buy parts from overseas manufacturers and merely assemble the systems. Other companies design a piece of equipment, then farm out the work to an outside manufacturer. Are these companies manufacturers or merely sellers? I took a liberal approach; if the company's name was on the item, then I counted the company as a manufacturer.

Some of the companies only manufacture products, and they don't deal with the public. You will not be able to buy directly from a number of those in this book. However, the listings in this book should be helpful for finding warranty or technical information on a product.

8th St. Music
1023 Arch St., Philadelphia, PA 19107 USA
(800) 878-8882 (tel)
Sells instruments for professional and personal musical applications.

111 Systems
423 S. Lynnhaven Rd., #107, Virginia Beach, VA 23452 USA
(800) 809-1319 (tel)
(800) 771-4049 (tel)
(804) 486-8389 (tel)
(804) 498-2432 (purchase tel)
(804) 486-8225 (tech support tel)
(804) 498-1211 (fax)
(804) 486-5916 (BBS)
Visa, MasterCard, Discover, American Express
Manufactures and sells personal desktop computers and computer parts and boards, such as floppy drives, hard drives, monitors, motherboards, modems, cases, etc. Offers a 30-day money-back guarantee.

168 Club
9852 W. Katella Ave., Ste. 289, Anaheim, CA 92804 USA
(800) 688-8588 (tel)
Visa, MasterCard, COD
Sells computer parts and boards, such as hard drives, floppy drives, motherboards, modems, sound cards, keyboards, etc.

3M
P.O. Box 140407, Austin, TX 78754 USA
(800) 745-7459 (tel)
(512) 984-5811 (fax)
Manufactures test instruments for professional cable TV applications, and many other electronics and nonelectronics products.

AA Computertech
28170 Crocker Ave., #105, Valencia, CA 91355 USA
(800) 360-6801 (tel)
(805) 257-6801 (tel)
(805) 257-6804 (tech support tel)
(805) 257-6805 (fax)

Visa, MasterCard, Discover, American Express, COD
Sells computer floppy drives and hard drives.

A&A Engineering
2521 W. LaPalma, #K, Anaheim, CA 92801 USA
(714) 952-2114 (tel)
Sells spectrum analyzers assembled or in kit form. Offers more information for an SASE.

AAMP of America
13160 56th Ct., Ste. 502-508, Clearwater, FL 34620 USA
(813) 572-9255 (tel)
(813) 573-9326 (fax)
Manufactures car batteries and capacitors for mobile audio applications.

AAVID Thermal Technologies
1 Kool Path, P.O. Box 400, Laconia, NH 03247 USA
(603) 528-3400 (tel)
(603) 528-1478 (fax)
Manufactures heatsinks for a wide variety of electronics applications.

Abate Electronics
P.O. Box 143, Hillsdale, NJ 07642 USA
Sells an in-line electronic blocker to prevent cable descramblers from being effective.

Abbott Electronics
2727 S. La Cienga Blvd., Los Angeles, CA 90034 USA
(310) 202-8820 (tel)
Manufactures dc power converters for a wide variety of electronics applications.

ABC (American Buyer's Club)
130 Hwy. 33, Englishtown, NJ 07726 USA
(800) 354-1324 (tel)
(908) 780-6600 (tel)
(908) 294-7480 (tel)
Visa, MasterCard, Discover, American Express, COD
Sells home and amateur audio and video equipment, such as camcorders, VCRs, speakers, cassette decks, TVs, home theater equipment, etc.

ABC Cable Products
6950 S. Tuscon Way, Ste. C, Englewood, CO 80112 USA
(800) 777-2259 (tel)
(303) 792-2552 (tel)
(303) 792-2642 (fax)
Manufactures optical links and remote-control units for professional cable TV applications.

ABC Communications
17550 15th Ave. N.E., Seattle, WA 98155 USA
(206) 364-8300 (tel)
Sells ICOM amateur radio equipment.

ABC Drives
8717 Darby Ave., Northridge, CA 91325 USA
(818) 885-7157 (tel)
(818) 885-6846 (fax)
Visa, MasterCard, Discover, American Express
Sells data storage systems, such as hard drives and tape drives, for a wide variety of computer applications.

AB Computer
117 N. Lake Ave., Pasadena, CA 91101 USA
(818) 666-1628 (tel)
(818) 666-1622 (tel)
Manufactures and sells personal desktop computers; also sells computer parts and boards, such as motherboards, video boards, hard drives, etc. Offers a free price list.

Aberdeen
9728 Alburtis Ave., Santa Fe Springs, CA 90670 USA
(800) 552-6868 (tel)
(310) 801-5626 (tel)
(310) 801-5629 (fax)
Visa, MasterCard
Manufactures and sells personal desktop computers; also sells computer parts and boards, such as tape drives, hard drives, monitors, modems, video adapters, motherboards, etc. Offers a 30-day money-back guarantee.

Abe's of Maine
1957 Coney Island Ave., Brooklyn, NY 11223 USA
(800) 545-2237 (tel)
(800) 992-2379 (service tel)
(718) 645-1818 (tel)
(718) 998-5216 (order fax)
(718) 382-8596 (service fax)
Visa, MasterCard, Discover, American Express, Diner's Club, Canon, COD
Sells high-end and home-photographic audio and video equipment, such as cameras, camcorders, CD players, cassette decks, video editors, monitors, VCRs, cables, etc. Offers a free catalog.

Abbott Electronics
2727 S. La Cienga Blvd., Los Angeles, CA 90034 USA
(310) 202-8820 (tel)
(310) 836-1027 (fax)
Manufactures dc-to-dc converters for a wide variety of electronics applications.

ABLE Communications
2823 McGaw, Irvine, CA 92714 USA
(800) 654-1223 (tel)
(714) 553-8825 (tel)
(714) 553-1320 (fax)
Manufactures routers and multiplexers for professional computer network applications.

ABS Computer Technologies
1295 Johnson Dr., City of Industry, CA 91745 USA
(800) 876-8088 (tel)
(800) 685-3471 (tech support tel)
(818) 937-2300 (tel)
(818) 937-2322 (fax)
Visa, MasterCard, Discover, American Express
Manufactures and sells personal desktop and laptop computers; also sells computer parts, such as floppy drives, hard drives, monitors, motherboards, video cards, modems, keyboards, etc.

Absopulse Electronics
110 Walgreen Rd., Carp, ON K0A 1L0 Canada
(613) 836-3511 (tel)
(613) 836-7488 (fax)
Manufactures high-voltage rectifier/battery chargers for a wide variety of electronics applications.

ABTECH
10211 Pacific Mesa Blvd., Ste. 409, San Diego, CA 92121 USA
(800) 474-7397 (tel)
(619) 450-6992 (tel)
(619) 622-0350 (fax)
33-1-4723-5022 (Europe tel)
abtech@powergrid.electriciti.com
Sells new and refurbished Hewlett-Packard workstations, peripherals, and upgrades for a wide variety of computer applications.

ACCO USA
770 S. ACCO Plz., Wheeling, IL 60090 USA
(708) 541-9500 (tel)
(800) 962-0576 (fax)
Manufactures paper shredders for a wide variety of office applications.

AC Components
901 S. 4th St., St. La Crosse, WI 54601 USA
(608) 784-4579 (tel)
(608) 784-6367 (fax)
Sells a wide variety of components for high-end audio systems, including speakers, power resistors, capacitors, etc.

Accurate Audio Video
2301 N. Central, Ste. 182, Plano, TX 75075 USA
(800) 414-1849 (tel)
Visa, MasterCard, Discover, American Express
Sells a wide variety of high-end home audio and video equipment.

Accurite Technologies
231 Charcot Ave., San Jose, CA 95121 USA
(408) 433-1980 (tel)
(408) 433-1716 (fax)
Manufactures diagnostics cards for computer testing applications.

Accusound
13802 N. Scottsdale Rd. N.E., Ste. 104, Phoenix, AZ 85254 USA
(603) 483-5398 (tel)
Manufactures amplifiers and loudspeakers for car stereo applications.

Accutech International
22837 Ventura Blvd., #304, Woodland Hill, CA 91364 USA
(800) 456-4224 (tel)
(818) 888-0086 (tel)
(818) 888-0085 (fax)
Sells telephones, power adapters, data ports, battery chargers, etc. for cellular telephones.

ACE Audio
RD #3 Box 351, Homer City, PA 15748 USA
Manufactures and sells electronic crossovers for high-end audio applications. Offers a free flier.

Ace Car Audio
71-14 Bang-Yee-Dong, Songpa-Gu, Seoul 138-150 Korea
Sells head units, speakers, amplifiers, receivers, cassette decks, CD players, etc. for home audio and car audio applications.

Acecad
791 Foam St., Ste. 200, Monterey, CA 93940 USA
(800) 676-4ACE (tel)
(408) 655-1900 (tel)
(408) 655-1988 (BBS)
(408) 655-1919 (fax)
Manufactures a pen-type pointing device for a wide variety of computer applications.

ACE Communications
10707 E. 106th St., Fishers, IN 46038 USA
(800) 445-7717 (tel)
(800) 448-1084 (fax)
(317) 849-8683 (fax facts)
(317) 579-2045 (BBS)
(317) 845-7436 (tech support)
(317) 849-8794 (international fax)
acescans@aol.com
Visa, MasterCard, Discover, American Express
Sells a wide variety of handheld, mobile, and tabletop scanners, shortwave radios, two-way radios, and accessories. Offers a one-week trial.

ACE Hardware Hobbies
1863 El Camino Real, Burlingame, CA 94010 USA
Visa, MasterCard, Discover, COD
Sells batteries, etc. for radio-control models. Offers a 50-page catalog for $5.

ACE Micro
7258 Clairemont Mesa Blvd., San Diego, CA 92111 USA
(800) 455-9555 (tel)
(619) 569-8380 (tel)
(619) 569-1319 (fax)
Visa, MasterCard, COD
Sells memory, hard drives, video cards, cases, etc. for personal laptop computers and laser printers. Offers a 30-day money-back guarantee.

Ace Music Center
13630 W. Dixie Hwy. N., Miami, FL 33161 USA
(305) 892-2733 (tel)
(305) 893-7934 (fax)
Visa, MasterCard, Discover, American Express
Sells a wide variety of audio and video equipment for professional broadcasting and professional recording.

AC Enterprises
4325 W. Beltline Hwy., Addison, WI 53711 USA
(800) 588-3554 (tel)
(608) 831-6004 (tel)
(608) 831-1944 (fax)
Sells used and refurbished Hewlett-Packard laser printers for a wide variety of computer applications.

Acer
(800) 529-ACER (tel)
(800) 239-ACER (tel)
(408) 428-0140 (BBS)
Visa, MasterCard, American Express
tsup@smtplink.altos.com
http://www.acer.com/aac/
Manufactures and sells personal desktop computers. Offers a 30-day money-back guarantee.

AcerOpen
2641 Orchard Pkwy., San Jose, CA 95134 USA
(800) 369-OPEN (tel)
(408) 922-2935 (fax)
Manufactures and sells personal desktop and laptop computer components such as monitors, add-on cards, motherboards, keyboards, etc.

Ace R/C
116 W. 19th St., P.O. Box 472, Higginsville, MO 64037 USA
(800) 322-7121 (tel)
(816) 584-7766 (fax)
Manufactures sport and computer radios, battery chargers, and servos for radio-control model airplanes.

Ace Systems
RD#1 Box 83, Wilcox, PA 15870 USA
(814) 965-5937 (tel)
Manufactures an automatic Morse code keyer for amateur radio operators.

ACI
P.O. Box 700081, Dallas, TX 75370 USA
(800) 652-7876 (tel)
(214) 243-5774 (customer service tel)
(214) 243-3144 (fax)
Sells memory and CPU chips and motherboards for a wide variety of computer applications.

A.C.I.
15111 N. Hayden Rd., 160-151, Scottsdale, AZ 85260 USA
(603) 971-8102 (tel)
(603) 404-3149 (fax)
Manufactures plug blocks, crossover blocks, etc. for car stereo applications.

Ackerman Computer Sciences (ACS)
4276 Lago Way, Sarasota, FL 34241 USA
(941) 377-5775 (tel)
(941) 378-4226 (fax)
(941) 379-9643 (BBS)
acsscott@gate.net
Manufactures CPU boards for a wide variety of computer applications.

Ack Radio Supply
3101 4th Ave., S. Birmingham, AL 35233 USA
(800) 338-4218 (tel)
(205) 322-0588 (tel)
(205) 322-0580 (fax)
554 Deering Rd., Atlanta, GA 30367 USA
(404) 351-6340 (tel)
(404) 351-1879 (fax)
Sells a variety of amateur radio equipment, including shortwave and VHF/UHF transceivers, Morse code keyers, antennas, power supplies, DSPs, antenna tuners, etc.

ACK Technology
35 Eastman St., Easton, MA 02334 USA
(508) 230-9251 (tel)
(508) 230-9257 (fax)
Manufactures thermoelectric cooler modules for a wide variety of electronics applications.

Acma
47988 Fremont Blvd., Fremont, CA 94538 USA
(800) 786-6888 (tel)
(800) 786-8998 (tech support)
(510) 651-1606 (BBS)
(510) 629-0629 (fax)
Manufactures and sells personal desktop computers.

Acom
46600 Landing Pkwy., Fremont, CA 94538 USA
(800) 898-2665 (tel)
(510) 353-9200 (fax)
http://www.sinanet.com/bay/acom
Manufactures and sells personal laptop computers. Offers a 30-day money-back guarantee.

A-Comm Electronics
7198 S. Quince St., Englewood, CO 80112 USA
(303) 290-8012 (tel)
(303) 290-8133 (fax)
Sells test equipment, including resistance bridges, spectrum analyzers, RF bridges, D/A converters, etc., for a wide variety of electronics testing applications.

Acopian
P.O. Box 638, Easton, PA 18044 USA
(610) 258-5441 (tel)
(610) 258-2842 (fax)
Manufactures and sells power supplies for a wide variety of electronics applications. Offers a free 60-page catalog.

Acoustic Research
535 Getty Ct., Bldg. A, Benecia, CA 94510 USA
(800) 969-AR4U (tel)
330 Turnpike St., Canton, MA 02021 USA
(617) 821-2300 (tel)
(617) 784-4102 (fax)
Manufactures amplifiers and loudspeakers for high-end home audio and home theater applications.

Acoustic Sciences Corp. (ASC)
P.O. Box 1189, Eugene, OR 97440 USA
(800) 272-8823 (tel)
bdcrow@continent.com
Manufactures portable sound deadeners and tubes for professional studio recording.

ACPEAM
P.O. Box 92215, 2900 Warden Ave., Scarborough, ON M1W 3Y9 Canada
(416) 498-7915 (tel)
Manufactures power-line conditioners for high-end home audio systems. Offers a 30-day money-back guarantee.

Acqutek
1549 S. 1100 E., Ste. B, Salt Lake City, UT 84105 USA
(801) 485-4594 (tel)
(801) 485-4555 (fax)
Manufactures single-board computers and other computer boards for professional and industrial computer applications.

Acromag
30765 Wixom Rd., P.O. Box 437, Wixom, MI 48393 USA
(800) 336-8655 (tel)
(810) 624-1541 (tel)
(810) 624-9234 (fax)
Manufactures I/O boards for professional and industrial computer applications.

Acropolis Systems
1575 McCandless Dr., Milpitas, CA 95035 USA
(800) 735-4311 (tel)
(408) 946-6947 (tel)
(408) 946-8715 (fax)
Manufactures workstations and data storage systems for a wide variety of professional computer network applications.

ACR Systems
12960 84th Ave., Unit 210, Surrey, BC V3W 1K7 Canada
(800) 663-7845 (tel)
(604) 591-1128 (fax)
Manufactures data and temperature loggers for industrial applications.

ACS
2630 Walnut Ave., #A, Tustin, CA 92680 USA
(800) 774-7014 (tel)
Sells computer parts and boards, such as hard drives, floppy drives, multimedia upgrades, monitors, etc.

ACS Equipment
10696 Haddington, Ste. 130, Houston, TX 77043 USA
Sells personal desktop and network computer systems; also sells computer parts and boards, such as hard drives, tape drives, etc.

AC Technology
8201 Greensboro Dr., Ste. 220, McLean, VA 22102 USA
(800) 228-4365 (tel)
(703) 448-5581 (tel)
(703) 847-4223 (fax)
Manufactures data storage systems for a wide variety of professional computer network applications.

Action Electronics
198 Chung Yuan Rd., Chung Li Industrial Zone, Tao-yuan, Taiwan
(886) 3-451-5494 (tel)
(886) 3-452-0697 (fax)
(886) 596-5955 (tel)
(886) 2-5928138 (fax)
(909) 444-1300 (US tel)
(909) 444-1308 (US fax)
49-2166-9527-0 (Europe tel)
49-2166-9527-27 (Europe fax)
Manufactures monitors for personal computer applications.

Action Instruments
8601 Aero Dr., San Diego, CA 92123 USA
(800) 767-5726 (tel)
(619) 279-5726 (tel)
(619) 279-6290 (fax)
actionsales@actionio.com
Manufactures signal conditioners for professional industrial applications.

Adams-Smith
34 Tower St., Hudson, MA 01749 USA
(508) 562-3801 (tel)
Manufactures an audio/video control system for professional studio recording.

Adaptec
(800) 804-8886 (tel)
(800) 655-3977 (tel)
(408) 945-2570 (fax)
http://www.adaptec.com
Manufactures SCSI adapter cards for a wide variety of computer applications.

Adastra Systems
26232 Executive Place, Hayward, CA 94545 USA
(510) 732-6900 (tel)
(510) 732-7655 (fax)
Manufactures single-board computers for professional and industrial computer applications.

ADAX
614 Bancroft Way, Berkeley, CA 94710 USA
(510) 548-7047 (tel)
(510) 548-5526 (fax)
Manufactures protocol controllers for professional computer network applications.

ADC Video Systems
999 Research Parkway, Meriden, CT 06450 USA

(800) 504-4443 (tel)
(203) 630-5701 (fax)
Manufactures complete systems for professional video and telephone network services.

Adcom
11 Elkins Rd., E. Brunswick, NJ 08816 USA
(908) 390-1130 (tel)
(908) 390-9152 (fax)
Manufactures and sells high-end stereo preamplifiers and power amplifiers for home audio, car stereo, and home theater applications.

A.D.C.S.
1830 Vernon St., Ste. 5, Roseville, CA 95678 USA
(800) 783-8979 (tel)
(916) 781-6952 (tel)
(916) 781-6962 (fax)
Buys and sells network equipment for a wide variety of computer applications.

Addtron Technology
97968 Fremont Blvd., Fremont, CA 94538 USA
(800) 998-4638 (tel)
(800) 998-4646 (tel)
(510) 770-0120 (tel)
(510) 770-0272 (BBS)
(510) 770-0171 (fax)
Manufactures concentrators for professional computer network applications.

A/D Electronics
10421 Burnham Dr. NW, Bldg. 4, Gig Harbor, WA 98332 USA
(206) 851-8005 (tel)
(206) 851-8090 (fax)
Manufactures connectors for a wide variety of computer and electronics applications.

A&D Electronics
P.O. Box 601, Monsey, NY 10952 USA
(914) 356-7541 (tel)
(914) 356-7505 (fax)
Visa, MasterCard
Sells surveillance equipment and kits, including Part 15 FM transmitters, miniature video cameras, bug detectors, video transmitters, night vision equipment, etc. Offers a free catalog.

ADI Systems
2115 Ringwood Ave., San Jose, CA 95131 USA
(800) 228-0530 (tel)
(408) 944-0100 (tel)
Manufactures monitors for personal desktop computers.

a/d/s
One Progress Way, Wilmington, MA 01887 USA
(800) 522-4434 (tel)
(617) 729-1140 (tel)
(508) 658-8498 (fax)
Manufactures loudspeakers for high-end home audio and home theater applications.

ADS International
434 Cloverleaf Dr., Baldwin Park, CA 91706 USA
(818) 369-2332 (tel)
(818) 855-5059 (fax)
Manufactures data storage systems for a wide variety of professional computer network applications.

AD Systems
34 Nelson St., Oakville, ON L6L 3H6 Canada
(800) 300-4756 (US/Canada tel)
+44-0181-943-4949 (UK tel)
+44-0181-943-5155 (UK fax)
Manufactures a digital computerized audio workstation for professional studio recording.

Advance
9 Rothwell Ave., North Harbour Industrial Estate, Albany, New Zealand
09-415-8310 (tel)
09-415-8382 (fax)
364 Cashel St., Christchurch, New Zealand
03-366-1687 (tel)
03-366-8890 (fax)
Unit 2, 7 McCormack Pl., Ngauranga, New Zealand
04-473-3165 (tel)
04-473-3184 (fax)
Manufactures desktop personal computer systems.

Advanced Composite Audio (A.C.A.)
97 Lafayette Rd., Hampton Falls, NH 03844 USA
(603) 929-4700 (tel)
(603) 929-0445 (fax)
Manufactures loudspeakers for car stereo applications.

Advanced Computer Products
1310 E. Edinger, Santa Ana, CA 92705 USA
(800) FON.E.-ACP
(714) 558-8813 (sales tel)
(714) 558-8849 (fax)
Sells a variety of semiconductors.

Advanced Computer Services
P.O. Box 21, Seaforth, Liverpool L21 5LE UK
0151-286-0370 (tel)
0151-286-0361 (fax)
Visa, Access, Switch, Delta
Sells computer parts, including CPUs, video cards, hard drives, monitors, motherboards, cases, etc., for personal computer applications.

Advanced Digital Industrial
195 Riviera Dr., Unit 2, Markham, ON L3R 5J6 Canada
(905) 940-9414 (tel)
(905) 940-9417 (fax)
Flat 26, 6/F, Block B, Local Industrial Centre, 21 Man Lok St., Hunghom, Kowloon, Hong Kong
(852) 2362-2298 (tel)
(852) 2330-5768 (fax)
(852) 2330-5770 (fax)
Manufactures motherboards and video cards for a wide variety of computer applications.

Advanced Digital Information
10201 Willows Rd., Redmond, WA 98073 USA
(800) 366-1233 (tel)
(206) 881-8004 (tel)
(206) 881-2296 (fax)
Manufactures data storage systems for a wide variety of professional and personal computer applications.

Advanced Educational Systems (AES)
970 W. 17th St., Santa Ana, CA 92706 USA
(800) 730-3232 (tel)
(714) 550-8094 (tel)
(714) 550-9941 (fax)
sales@aesmicro.com
Manufactures microcontroller boards for a wide variety of educational computer applications.

Advanced Equipment Products
9245 Lazy Lane, Tampa, FL 33614 USA
(813) 935-3614 (tel)
(813) 935-4048 (fax)
Sells discounted personal desktop computers and printers.

Advanced Linear Devices
415 Tasman Dr., Sunnyvale, CA 94089 USA
(408) 747-1155 (tel)
(408) 747-1286 (fax)
Manufactures integrated circuits for a wide variety of electronics applications.

Advanced Logic Research
9401 Jeronimo, Irvine, CA 92718 USA
(800) 257-1230 (tel)
(714) 581-9240 (tel)
(714) 458-0532 (fax)
(714) 581-3332 (fax back)
(714) 581-6770 (fax)
(714) 458-6834 (BBS)
http://www.alr.com
Manufactures and sells desktop computers for personal and professional applications.

Advanced Micro Devices
P.O. Box 3453, Sunnyvale, CA 94088 USA
(800) 222-9323 (tel)
(408) 749-5703 (fax)
Manufactures integrated circuits for a wide variety of computer applications.

Advanced Micro Solutions
4630 Campus Dr., Ste. 109, Newport Beach, CA 92660 USA
(800) 838-0860 (tel)
Visa, MasterCard, Discover, American Express
Sells computer parts and boards, such as hard drives,

Sells computer parts and boards, such as hard drives, motherboards, memory chips, monitors, CD-ROM drives, video boards, etc.

Advanced Portable Technologies
Tech Rentals House, Level 1, 2009 Victoria St., P.O. Box 27-431, Upper Willis St., Wellington, New Zealand
04-385-2838 (tel)
04-385-2845 (fax)
Sells accessories for laptop computers.

Advanced Radar Technologies
3355 Lenox Rd., Ste. 750, Atlanta, GA 30326 USA
(800) 688-1200 (tel)
Visa, MasterCard
Manufactures and sells electronic radar scramblers for personal auto applications.

Advanced Receiver Research
P.O. Box 1242, Burlington, CT 06013 USA
(203) 582-9409 (tel)
Visa, MasterCard
Manufactures 10-GHz and 24-GHz transceivers for amateur radio applications.

Advanced Recovery
3 Montgomery St., Belleville, NJ 07109 USA
(201) 450-9797 (tel)
Buys and sells memory chips for a wide variety of computer applications.

Advanced RF Design
39 Maple Stream Rd., E. Windsor, NJ 08520 USA
(800) 669-2733 (tel)
(609) 448-0910 (tel)
(609) 448-6689 (fax)
x703jko@chelsea.ios.com
Visa, MasterCard
Manufactures and sells VHF/UHF low-noise/high-gain preamplifiers for amateur radio applications.

Advanced Specialties
P.O. Box 1099R, Lodi, NJ 07644 USA
(201) 843-2067 (tel/fax)

Imports and distributes wattmeters, switch boxes, tuner/matchboxes, etc. for citizens band and amateur radio applications.

Advanced System Products
1150 Ringwood Ct., San Jose, CA 95131 USA
(800) 525-7443 (tel)
(408) 383-9400 (tel)
(408) 383-9612 (fax)
Manufactures disk controller boards for a wide variety of professional computer network applications.

Advanced Testing Technologies (ATTI)
5036 Jericho Tnpke., Commack, NY 11725 USA
(800) ATTI-VXI (tel)
(516) 462-1900 (tel)
(516) 462-2671 (fax)
Manufactures automatic test systems for professional electronics applications.

Advance Power
32111 Aurora Rd., Solon, OH 44139 USA
(216) 349-0755 (tel)
(216) 349-0142 (fax)
Manufactures power supplies for a wide variety of electronics applications.

Advantage Electronics
1125 Riverwood Dr., Burnsville, MN 55337 USA
(800) 952-3916 (tel)
Visa, MasterCard, Discover, American Express, COD
Sells cable TV converters and accessories.

Advantage Instrument Corp.
3817 S. Carson St., #818, Carson City, NV 89701 USA
(702) 885-0234 (tel)
(702) 885-7600 (fax)
Visa, MasterCard, Discover, American Express
Sells test equipment, including spectrum analyzers and oscilloscopes, for a wide variety of electronics applications.

Advantage Microsystems International
22 Baker Terrace, Buffalo, NY 14150 USA

Advantage Microsystems International
(800) 252-3856 (tel)
(716) 743-0831 (tel)
(716) 743-0654 (international order tel)
(716) 743-0914 (fax)
Visa, MasterCard, Discover, American Express, COD
Manufactures and sells personal desktop computers. Offers a 90-day, money-back guarantee.

Advantech
750 E. Arques Ave., Sunnyvale, CA 94086 USA
(800) 800-6889 (tel)
(408) 205-6678 (tel)
(408) 245-8268 (fax)
http://www.advantek.com
info@advantek.com
Manufactures remote data acquisition and control modules, personal computers, plug-in I/O cards, etc. for professional industrial applications. Offers a free, 100-page Solution Guide.

Advent
25 Tri-State Office Ctr., #400, Lincolnshire, IL 60069 USA
(800) 477-3257 (tel)
(800) 284-7234 (tel)
(708) 317-3826 (fax)
Manufactures loudspeakers for high-end home audio, home theater, and personal computer multimedia applications.

Advent Electronics
1865 Minor St., Des Plaines, IL 60016 USA
(800) 323-1270 (tel)
(800) 255-4771 (tel)
Sells a wide variety of electronics, including equipment for audio and video applications.

Advin Systems
1050-L E. Duane Ave., Sunnyvale, CA 94086 USA
(800) 627-2456 (tel)
(408) 243-7000 (tel)
(408) 736-2503 (fax)
advinsales@aol.com
Manufactures universal chip programmers for professional electronics applications.

AEA (Advanced Electronic Applications)
P.O. Box C2160, Lynnwood, WA 98036 USA
(800) 432-8873 (literature tel)
(206) 774-5554 (tel)
(206) 775-2340 (fax)
(206) 775-7373 (tech support)
Manufactures packet controller boards and antenna analyzers for amateur radio applications.

AEL Industries
305 Richardson Rd., Lansdale, PA 19446 USA
(215) 822-2929 (tel)
(215) 822-3654 (fax)
Manufactures laser transmission products for professional cable TV applications.

AEMC
99 Chauncy St., Boston, MA 02111 USA
(617) 451-0227 (tel)
(617) 423-2952 (fax)
Manufactures RMS power meters for professional electronics testing applications.

AEQ America
2211 48th St. S., Ste. II, Tempe, AZ 85282 USA
(603) 431-0334 (tel)
(603) 431-0497 (fax)
Manufactures digital audio codecs, digital telephone hybrids, mixers, and audio monitors for professional broadcasting.

AER Energy Resources
1500 Wilson Way, Ste. 250, Smyrna, GA 30082 USA
(404) 431-2084 (tel)
Manufactures batteries for a wide variety of electronics applications.

Aero Product Supply
1400B Miamisburg-Centerville Rd., Dayton, OH 45459 USA
(800) 590-7222 (tel)
Visa, MasterCard, Discover, American Express
Sells GPS and LORAN systems for professional aviation applications.

Aero Scientific
P.O. Box 292, Grayslake, IL 60030 USA
(708) 223-9066 (tel)
(708) 223-9719 (fax)
Manufactures and sells controls, servos, receivers, etc. for model radio-control airplanes.

Aero Spectra
P.O. Box 3021, Boulder, CO 80307 USA
(303) 499-2584 (tel/fax)
Manufactures a handheld radio channel analyzer for interference control with radio-control model clubs.

Aero Sport
370 Laurelwood Rd., #113, Santa Clara, CA 95054 USA
(408) 980-0800 (tel)
Manufactures engine data loggers for radio-control model airplanes.

AEROTECH
101 Zeta Dr., Pittsburgh, PA 15238 USA
(412) 963-7470 (tel)
(412) 963-7459 (fax)
01734-817274 (UK tel)
01734-815022 (UK fax)
0911-52-10-31 (Germany tel)
0911-521-52-35 (Germany fax)
Manufactures motion-control systems for professional industrial applications.

Aerotech Models
2740 31st Ave., S. Minneapolis, MN 55406 USA
(612) 721-1285 (tel)
Manufactures and sells a circuit to slow the servo reaction time for radio-control model airplanes.

Advanced Educational Systems (AES)
1407 N. Batavia St., Orange, CA 92667 USA
(800) 730-3232 (tel)
(714) 744-0981 (tel)
(714) 744-2693 (fax)
Visa, MasterCard
Sells a "complete learning system" computer/microcontroller board.

AES
285 Newbury St., Peabody, MA 01960 USA
(508) 535-7310 (tel)
Manufactures batteries for a wide variety of electronics applications.

AET
P.O. Box 2541, Toluca Lake, CA 91610 USA
(310) 572-6309 (tel)
Manufactures digital music production systems for professional audio applications.

Affordable Portables
23819 A El Toro Rd., Lake Forest, CA 92630 USA
(714) 580-3548 (tel)
Sells receivers, antennas, and accessories for shortwave listening applications.

AFIC Technologies
160 Broadway, Ste. 600, New York, NY 10038 USA
(212) 406-2503 (tel)
(212) 406-2415 (fax)
Manufactures workstations for a wide variety of professional computer network applications.

Agrelo Engineering
1145 Catalyn St., Schenectady, NY 12303 USA
(800) 588-4300 (tel)
(518) 381-1057 (tech tel)
(518) 381-1058 (fax)
Visa, MasterCard
Manufactures and sells electronics kits, including a Part 15 FM transmitter, a voice scrambler/descrambler, and two digital voice recorders.

A.I. Credit
160 Water St., 19th Fl., New York, NY 10038 USA
(212) 428-5470 (tel)
(212) 428-5518 (fax)
Sells new, used, and refurbished personal computer systems and printers; also sells computer parts and boards, such as CPUs, hard drives, tape drives, etc.

Air Comm
4614 E. E. McDowell Rd., Phoenix, AZ 85008 USA

Air Comm

(603) 275-4505 (tel)
(603) 275-4555 (fax)
Sells base stations, repeaters, pagers, etc. for cellular microwave radio and other two-way applications.

Aircraft Spruce & Specialty

201 W. Truslow Ave., P.O. Box 424, Fullerton, CA 92632 USA
(800) 824-1930 (tel)
(714) 994-2221 (tel)
(800) 329-3020 (fax)
Visa, MasterCard, Discover, American Express
Sells GPS and LORAN systems, transceivers, batteries, etc. for professional aviation applications. Offers a 448-page catalog for $5.

Airtronics

11 Autry, Irvine, CA 92718 USA
(714) 830-8769 (tel)
Sells sport and computer radios and servos for radio-control model airplanes.

AIS

Rt. 4, Box 664, Spiro, OK 74959 USA
(918) 962-3349 (tel)
(918) 962-2148 (fax)
Sells radar guns for a wide variety of sports applications.

AITech

47971 Fremont Blvd., Fremont, CA 94538 USA
(800) 882-8184 (tel)
(510) 226-9169 (tel)
(510) 226-8996 (fax)
Manufactures scan converter boards to display computer video on televisions.

Aiwa

800 Corporate Dr., Mahwah, NJ 07430 USA
(201) 512-3600 (tel)
(201) 512-3705 (fax)
14821 E. Northam St., La Mirada, CA 90368 USA
(714) 522-2492 (tel)
Manufactures a variety of home and car stereo equipment, including DAT decks, portable stereos, component systems, etc.

AJ's Power Source

4105 Cox Dr., Land O'Lakes, FL 34639 USA
(813) 996-1922 (tel)
Manufactures dc-to-dc power converter modules for a wide variety of electronics applications.

Akai

1316 E. Lancaster Ave., Ft. Worth, TX 76113 USA
(817) 336-5114 (tel)
(817) 870-1271 (fax)
Manufactures digital samplers, cassette decks, etc. for a wide variety of professional and home audio applications.

AKG Acoustics

8500 Balboa Blvd., Northridge, CA 91329 USA
(818) 893-8411 (tel)
(818) 787-0788 (fax)
(818) 909-4576 (fast fax)
Manufactures microphones for professional studio and broadcasting applications.

Alantec

70 Pluneria Dr., San Jose, CA 95134 USA
(800) ALANTEC (tel)
(408) 955-9000 (tel)
(408) 955-9500 (fax)
Manufactures routers for professional computer network applications.

Alcatel Network Systems

1225 N. Alma Rd., Richardson, TX 75081 USA
(800) 252-2835 (tel)
Manufactures fiberoptic cable, transmitter links, etc. for professional cellular microwave radio applications.

Alcoa Fujikura

150 Ridgeview Cir., Duncan, SC 29334 USA
(800) 552-5688 (tel)
Manufactures fiberoptic equipment for a wide variety of data and communications applications.

Alden Electronics

40 Washington St., Westborough, MA 01581 USA
(508) 366-8851 (tel)

(508) 898-2427 (fax)
http://www.alden.com
Manufactures and sells equipment for shortwave radio fax applications.

Alesis
3630 Holdrege Ave., Los Angeles, CA 90016 USA
(800) 5-ALESIS (tel)
(310) 841-2272 (tel)
alecorp@alesis1.usa.com
Manufactures several digital multieffects processors and studio reference monitor speakers for professional studio recording.

Alexander Aeroplane Co.
P.O. Box 909, Griffin, GA 30224 USA
(800) 831-2949 (tel)
(404) 229-2329 (fax)
Sells aluminum tubing and other items for airplanes. The tubing can also be used for a variety of antennas, including shortwave, amateur radio, CB, TV, and UHF/VHF antennas. There is a minimum $10 order. Offers a free catalog.

Alexander Batteries
P.O. Box 1508, Mason City, IA 50401 USA
(800) 247-1821 (tel)
(800) 526-ALEX (tel)
(515) 423-8955 (tel)
(515) 423-1644 (fax)
Manufactures battery chargers, batteries for two-way radio applications, two-way radio antennas, etc.

Alfa Electronics
741 Alexander Rd., Princeton, NJ 08540 USA
(800) 526-2532 (tel)
(609) 520-2002 (tel)
(609) 520-2007 (fax)
Sells a variety of electronic test equipment.

Alinco Electronics
438 Amapola Ave., #130, Torrance, CA 90501 USA
(310) 618-8616 (tel)
(310) 618-8758 (fax)
Sells handheld VHF amateur transceivers.

Alis Technologies
3410 Griffith St., Montreal, PQ H4T 1A7 Canada
(514) 738-9171 (tel)
(514) 342-0318 (fax)
Manufactures terminals for a wide variety of professional computer network applications.

Aljon Technologies
3892 Del Amo Blvd., #703, Torrance, CA 90503 USA
(310) 793-1788 (tel)
Visa, MasterCard
Manufactures microwave audio/video transmitters and receivers for hobby and experimental applications.

Allegheny Paper Shredders
Old William Penn Hwy E., Delmont, PA 15626 USA
(800) 245-2497 (tel)
(412) 468-4300 (fax)
Manufactures paper shredders for a wide variety of office applications.

All Electronics
P.O. Box 567, Van Nuys, CA 91408 USA
(800) 826-5432 (tel)
(818) 904-0524 (tel)
(818) 781-2653 (fax)
905 S. Vermont Ave., Los Angeles, CA 90006 USA
(213) 380-8000 (tel)
14928 Oxnard St., Van Nuys, CA 91411 USA
(818) 997-1806 (tel)
Visa, MasterCard, Discover
Sells surplus electronic and computer equipment, parts, and assemblies. No mininum order. Offers a free, 64-page catalog.

Allegro
1900 N. Austin, Chicago, IL 60639 USA
(312) 745-5140 (tel)
Manufactures lighting systems for amateur and professional video production.

Allen Systems
2346 Brandon Rd., Columbus, OH 43221 USA

Allen Systems
(614) 488-7122 (tel)
Manufactures I/O boards for a wide variety of computer applications.

Allen Telecom Group
30500 Bruce Industrial Pkwy., Cleveland, OH 44139
(216) 349-8695 (tel)
(216) 349-8692 (fax)
http://www.indra.com/unicom/company/allentelgroup.html
Manufactures two-way mobile and base antennas for VHF/UHF, cellular, citizens band, marine, amateur radio, etc.

Alliance Rotators
Baldwin Green, #110, Woburn, MA 01801 USA
(617) 932-9070 (tel)
(617) 932-3553 (fax)
Manufactures and sells antenna rotators.

Alliant Techsystems
104 Rock Rd., Horsham, PA 19044 USA
(215) 674-3800 (tel)
Manufactures batteries for a wide variety of electronics applications.

Allied Electronics
7410 Pebble Dr., Ft. Worth, TX 76118 USA
(800) 433-5700 (tel)
http://www.allied.avnet.com
Sells a wide variety of electronics components, instruments, and tools. Offers an almost 900-page catalog!

Allied Resources
108 Dodd Ct., American Canyon, CA 94559 USA
(800) 843-5432 (tel)
(800) 443-5432 (Calif tel)
headsetco@aol.com
Sells headsets for telecommunications applications.

AlliedSignal Aerospace
400 N. Rogers Rd., Olathe, KS 66062 USA
(913) 768-3000 (tel)
(913) 791-1302 (fax)
Manufactures a variety of GPS systems for professional aviation applications.

Allied Telesyn International
190155 N. Creek Pkwy, Ste. 200 Bothell, WA 98011 USA
(800) 424-4284 (tel)
Manufactures hubs for professional computer network applications.

Alligator Technologies
2900 Bristol St., #E-101, Costa Mesa, CA 92626 USA
(714) 850-9984 (tel)
(714) 850-9987 (fax)
Manufactures low-pass filter computer boards for professional and industrial computer applications.

Allison Acoustics
478 Stanford Ave., Danville, KY 40422 USA
(606) 236-8298 (tel)
http://www.hifi.com/allison.html
Manufactures loudspeakers for home audio and home theater applications.

Allison Technology
8343 Carvel, Houston, TX 77036 USA
(800) 980-9806 (tel)
(713) 777-0401 (tech tel)
(713) 777-4746 (fax/BBS)
Manufactures an adapter with software so that a PC can be used as a digital storage oscilloscope.

AllMicro
18820 U.S. Hwy. 19N, #215, Clearwater, FL 34624 USA
(800) 653-4933 (tel)
(813) 539-7283 (tel)
(813) 531-0200 (fax)
(813) 535-9042 (BBS)
allmicro@ix.netcom.com
Visa, MasterCard, American Express
Manufactures diagnostic cards to troubleshoot personal computer systems.

Allpass Technologies
2844 Charmont Dr., Apopka, FL 32703 USA
Manufactures and sells crossover networks for high-end home audio applications. Offers a design guide for $2.

All Phase Video-Security
70 Cain Dr., Brentwood, NY 11717 USA
(800) 945-0909 (tel)
(516) 435-7494 (fax)
Visa, MasterCard, Discover, American Express
Sells cable TV converters, microwave testers, oscilloscopes, miniature cameras, etc. for a wide variety of electronics applications. Offers a free catalog.

Alltech Electronics
2618 Temple Heights, Oceanside, CA 92056 USA
(619) 724-2404 (tel)
(619) 724-8808 (fax)
Visa, MasterCard
Sells surplus computer parts, including power supplies, cash drawers, monitors, modems, RAM chips, video cards, etc. for a wide variety of computer applications.

Alltronics
2300 Zanker Rd., San Jose, CA 95131 USA
(408) 943-9773 (tel)
(408) 943-9776 (fax)
(408) 943-0622 (BBS)
Sells surplus components, grab bags, tubes, etc. Minimum order: $15.

Alpha Delta Communications
P.O. Box 620, Manchester, KY 40962 USA
(606) 598-2029 (tel)
(606) 598-4413 (fax)
Manufactures and sells amateur radio and shortwave radio antennas.

Alpha Electronics
741 Alexander Rd., Princeton, NJ 08540 USA
(800) 526-2532 (tel)
(609) 520-2002 (tel)
(609) 520-2007 (fax)
Visa, MasterCard, Discover
Sells test equipment, including oscilloscopes, audio generators, bench power supplies, DMMs, LCR meters, etc. Offers a free catalog and a 15-day, money-back guarantee.

Alpha International Business
129 E. Arrow Hwy., San Dinas, CA 91773 USA
(800) 933-5455 (tel)
(909) 599-3567 (tel)
(909) 599-7584 (fax)
alpha1@primenet.com
http://www.primenet.com/_7Ealpha1
Manufactures and sells personal desktop computers. Offers a 30-day money-back guarantee.

AlphaLab
1280 S. 300 W., Salt Lake City, UT 84101 USA
(800) 769-3754 (tel)
(801) 487-9492 (fax)
Manufactures electromagnetic meters for a wide variety of electronics testing applications.

Alphanumeric Keyboards
1640 Fifth St., Ste. 224, Santa Monica, CA 90401 USA
(800) 200-9496 (tel)
Manufactures ruggedized keyboards for professional computer applications.

Alpha Semiconductor
5776 Sonoma Dr., Pleasanton, CA 94566 USA
(510) 417-1391 (tel)
Manufactures integrated circuits for a wide variety of electronics applications.

Alphasonik
701 Heinz Ave., Berkeley, CA 94710 USA
(510) 548-4005 (tel)
(510) 548-1478 (fax)
Manufactures line drivers and preamplifiers for car stereo applications.

Alpha Stereo
57 Smithfield Blvd., Plattsburgh, NY 12901 USA
(518) 561-2822 (tel)
(518) 561-2961 (fax)
Visa, MasterCard, Discover, American Express
Sells high-end home and amateur audio and video equipment, such as camcorders, VCRs, speakers, cassette decks, TVs, home theater equipment, etc.

Alpha Technologies
3767 Alpha Way., Bellingham, WA 98225 USA
(800) 421-8089 (tel)
(360) 647-2360 (tel)
http://www.alpha_us.com
Manufactures broadband power systems, batteries, UPS systems, etc. for professional cable TV broadcasting applications.

Alphatronix
P.O. Box 13798, 4022 Stirrup Creek Rd., Research Triangle Park, NC 27709 USA
(800) 849-2611 (tel)
(919) 544-0011 (tel)
(919) 544-4079 (fax)
Manufactures data storage systems for a wide variety of professional computer network applications.

Alpine
19145 Gramercy Pl., Torrance, CA 90501 USA
(800) 421-2284 (tel)
(310) 326-8000 (tel)
(310) 782-0726 (fax)
http://www.alpine1.com
Manufactures head units, CD changers, amplifiers, and loudspeakers for mobile audio applications.

Alpine Computer Sales
605 Tennant Ave. Ste. H, Morgan Hill, CA 95037 USA
(800) 655-1628 (tel)
(702) 782-5208 (tel)
(702) 782-5244 (fax)
Manufactures UNIX workstations for a wide variety of professional computer network applications.

ALS
P.O. Box 11, 1 Farm Rd., Ravena, NY 12143 USA
(518) 756-3797 (tel)
(518) 756-2910 (fax)
Manufactures loudspeakers for car stereo applications.

Al-Salah Trading
C.R. 10 10 187, P.O. Box 2387, SA-Alkhobar 31952 Saudi Arabia
Sells head units, speakers, amplifiers, receivers, cassette decks, CD players, etc. for home audio and car audio applications.

Alta Engineering
58 Cedar Ln., New Hartford, CT 06057 USA
(203) 489-8003 (tel/fax)
Manufactures PLD programmers for a wide variety of professional computer applications.

Altec Lansing
P.O. Box 277, Milford, PA 18337 USA
(800) 258-3288 (tel)
(800) 648-6663 (multimedia div tel)
(717) 296-4434 (tel)
(717) 296-2213 (fax)
Manufactures loudspeakers for high-end home audio and home theater applications.

Altera
2610 Orchard Pkwy., San Jose, CA 95134 USA
(800) 925-8372 (tel)
Manufactures semiconductors for a wide variety of electronics applications.

Alternative Computer Products (ACPC)
1120 Holland Dr., Ste. 5, Boca Raton, FL 33487 USA
(407) 994-9899 (tel)
(407) 994-9855 (fax)
Visa, MasterCard, Discover
Sells IBM personal desktop computers and parts and accessories, including memory chips, system boards, hard drives, adapters, network cards, etc.

Alternative Computer Solutions
2341 Lawrenceville Hwy., Ste. B, Decatur, GA 30033 USA
(404) 294-7806 (tel)
(404) 294-0565 (fax)
raviator@atlmindspring.com
Buys and sells MacIntosh computer equipment and memory chips, etc.

Alternative Marketing
51 McAndrews, #111, Burnsville, MN 55337 USA
(800) 200-8897 (tel)

Sells cable TV converters for a variety of video applications.

Altronic Research
P.O. Box 249, Yellville, AR 72687 USA
(501) 449-4093 (tel)
Manufactures water- and air-cooled dummy loads for broadcast and amateur radio transmitters.

Aluma Tower Co.
P.O. Box 2806, Vero Beach, FL 32961 USA
(407) 567-3423 (tel)
(407) 567-3432 (fax)
Manufactures and sells crank-up aluminum towers for amateur radio applications.

Amateur and Advanced Communications
3208 Concord Pike, Rt. 202, Wilmington, DE 19803 USA
(302) 478-2757 (tel)
Sells a variety of amateur radio equipment, including shortwave and VHF/UHF transceivers, Morse code keyers, antennas, power supplies, DSPs, antenna tuners, etc.

Amateur Communications
263 Mink, San Antonio, TX 78213 USA
Sells a variety of amateur radio equipment, including shortwave and VHF/UHF transceivers, Morse code keyers, antennas, power supplies, DSPs, antenna tuners, etc.

Amateur Electronic Supply (AES)
5710 W. Good Hope Rd., Milwaukee, WI 53223 USA
(800) 558-0411 (tel)
(414) 358-0333 (tel)
(414) 358-3337 (fax)
(414) 358-3472 (BBS)
aesham@execpc.com
28940 Euclid Ave., Wickliffe, OH 44092 USA
(800) 321-3594 (tel)
(216) 585-7388 (tel)
(216) 585-1024 (fax)
621 Commonwealth Ave., Orlando, FL 32803 USA
(800) 327-1917 (tel)

(407) 894-3238 (tel)
(407) 894-7553 (fax)
1898 Drew St., Clearwater, FL 34625 USA
(800) 557-4237 (tel)
(813) 461-4267 (tel)
(813) 443-7893 (fax)
1072 N. Rancho Dr., Las Vegas, NV 89106 USA
(800) 634-6227 (tel)
(702) 647-3114 (tel)
(702) 647-3412 (fax)
http://www.execpc.com/_7Eaesham
Visa, MasterCard, Discover, American Express
Sells a a variety of amateur radio equipment, including transceivers, receivers, amateur satellite equipment, antenna systems, accessories, etc. Offers a free catalog.

Amateur Radio Supply
5963 Corson Ave. S., Ste. #140, Seattle, WA 98108 USA
(206) 767-3222 (tel)
(206) 763-8176 (fax)
Sells a variety of amateur radio equipment, including shortwave and VHF/UHF transceivers, Morse code keyers, antennas, power supplies, DSPs, antenna tuners, etc.

Amateur Radio Team of Spokane
S. 25 Girard, Spokane, WA 99212 USA
(509) 928-3073 (tel)
Sells a variety of amateur radio equipment, including shortwave and VHF/UHF transceivers, Morse code keyers, antennas, power supplies, DSPs, antenna tuners, etc.

A Matter of Fax
65 Worth St., New York, NY 10013 USA
(800) 433-3FAX (tel)
(212) 941-8877 (tel)
Visa, MasterCard, Discover, COD
Sells fax machines, printers, personal laptop computers, monitors, copiers, and scanners. Offers a 30-day money-back guarantee.

AmberWave Systems
42 Nagog Park, Acton, MA 01720 USA

AmberWave Systems
(508) 266-2900 (tel)
(508) 266-1159 (fax)
info@amberwave.com
http://www.amberwave.com
Manufactures Ethernet switches for professional computer network applications.

Ambiance Acoustics
P.O. Box 27115, San Diego, CA 92198 USA
(619) 485-7514 (tel)
Manufactures monitor loudspeakers for professional studio recording applications.

Ambrosia Microcomputer Products
98 W. 63rd St., Ste. 371, Willowbrook, IL 60514 USA
(708) 655-0610 (tel)
Manufactures a computerized radio-control model flight simulator.

AMCC MarCom
6195 Lusk Blvd., San Diego, CA 92121 USA
(800) 755-2622 (tel)
(619) 450-9885 (fax)
Manufactures interface boards for professional network computer applications. Offers a free Network Products Databook.

AM Communications
1900 AM Dr., Quakertown, PA 18951 USA
(800) 248-9004 (tel)
(215) 536-1354 (tel)
(215) 536-1475 (fax)
Manufactures monitors for professional cable TV broadcasting applications.

AMC Sales
193 Vaquero Dr., Boulder, CO 80303 USA
(800) 926-2488 (tel)
(303) 499-5405 (tel)
(303) 494-4924 (fax)
Visa, MasterCard
Sells surveillance and security kits and assembled devices, such as a call forwarder, caller ID, voice-activated switch (VOX), phone scrambler, voice changer, etc. Offers a free catalog.

AMD (Advanced Micro Devices)
P.O. Box 3453, Sunnyvale, CA 94088 USA
(800) 222-9323 (tel)
http://www.amd.com
Manufactures microprocessors for personal desktop computers.

AMDEV Communications
20975 Woodside Way, Groveland, CA 95321 USA
(209) 962-5900 (tel)
(209) 962-5800 (fax)
Sells new and reconditioned voice mail systems for a wide variety of computer applications.

Amega Technology
Loddon Business Centre, Roentgen Rd., Daneshill East, Basingstoke, Hampshire, RG24 8NG UK
01256 330301 (tel)
01256 330302 (fax)
Manufactures integrated circuits for a wide variety of electronics applications.

American Advantech
750 E. Arques Ave., Sunnyvale, CA 94086 USA
(800) 800-6889 (tel)
(408) 245-6678 (tel)
(408) 245-8268 (fax)
Manufactures single-board computers for a wide variety of professional and industrial computer applications.

American Antenna
1500 Executive Dr., Elgin, IL 60123 USA
(800) 323-6768 (tel)
(708) 888-7200 (tel)
(708) 888-7094 (fax)
Manufactures and sells shortwave and VHF amateur radio antennas and accessories.

American Audio & Video
754 Franklin Ave., Franklin Lakes, NJ 07417 USA
(800) 695-6590 (tel)
Sells high-end home, amateur, and professional audio and video equipment, such as camcorders, VCRs, video editors, monitors, laser disk players, etc.

American Bass
5242 Warrensville Center Rd., Maple Heights, OH 44137 USA
(216) 475-9311 (tel)
(717) 296-2213 (fax)
Manufactures loudspeakers for car stereo applications.

American Computer Hardware
2205 S. Wright St., Santa Ana, CA 92705 USA
(800) 447-1237 (tel)
(714) 549-2688 (tel)
(714) 662-0491 (fax)
Visa, MasterCard
Sells enclosures for computer data storage systems.

American Electronics
P.O. Box 301, 173 E. Broadway, Greenwood, IN 46142 USA
(800) 872-1373 (tel)
(317) 888-7265 (tel)
(317) 888-7368 (fax)
Sells a variety of citizens band, two-way, and scanner radio equipment, including transceivers, receivers, accessories, tools, etc.

American Eltec
101 College Rd. E., Princeton, NJ 08540 USA
(609) 279-0111 (tel)
Manufactures CPU boards for a wide variety of computer applications.

American Hobby Center
146 M W. 22nd St., New York, NY 10011 USA
(800) 989-7951 (tel)
(212) 675-8922 (tel)
(800) 323-2992 (fax)
Visa, MasterCard, Discover, COD
Sells sport and computer radios, batteries, and servos for radio-control model airplanes. Offers a free catalog.

American Innovations
119 Rockland Center, Ste. 315, Nanuet, NY 10954 USA
(914) 735-6127 (tel)
(914) 735-3560 (fax)
Visa, MasterCard, Discover, American Express
Manufactures phone tap detectors, transmitter detectors, etc. for countersurveillance applications.

American Megatrends
6145 F Northbelt Pkwy., Norcross, GA 30071 USA
(800) 828-9264 (tel)
joel@american.megatrends.com
http://www.megatrends.com
Manufactures motherboards for a wide variety of computer applications.

American Micro Computer Center (AMCC)
6073 NW 167th St., Unit C-27, Miami, FL 33015 USA
(305) 825-5565 (tel)
(305) 825-7774 (fax)
Sells IBM personal desktop computers.

American Micro Products Technology
5351 Naiman Pkwy., Solon, OH 44139 USA
(800) 619-0508 (tel)
(216) 498-9499 (international sales tel)
(216) 349-6169 (tech support tel)
(216) 349-6160 (fax)
Visa, MasterCard, Discover, American Express, COD
Sells computer parts and boards, such as floppy drives, hard drives, monitors, video cards, modems, motherboards, memory chips, keyboards, etc.

American Microsystems
2190A Regal Pkwy., Euless, TX 76040 USA
(800) 648-4452 (tel)
(817) 571-9015 (tel)
(817) 685-6232 (fax)
Visa, MasterCard, American Express
Sells bar code printers, UPC scanners, card scanners, etc. for computerized business applications. Offers a 30-day money-back guarantee.

American Power Conversion (APC)
132 Fairgrounds Rd., W. Kingston, RI 02892 USA
(800) 800-4272 (tel)
(401) 789-5735 (fax)

American Power Conversion
+33-1-64-62-59-00 (France tel)
+49-89-958-23-5 (Germany tel)
+44-753-511022 (UK tel)
+35-391-702000 (Ireland tel)
(401) 789-5735 (Latin America tel)
+83-5295-1988 (Japan tel)
apctech@apcc.com
http://www.apcc.com
Manufactures uninterruptible power supplies for a wide variety of personal computer applications.

American Precision Instruments
262 Quince Dr., San Jacinto, CA 92582 USA
(909) 487-0195 (tel)
(909) 487-9276 (fax)
(909) 487-1365 (BBS)
Sells cash drawers, UPC scanners, etc. for computerized business applications. Offers a free catalog.

American Pro Audio
6026 Blue Circle Dr., Minnetonka, MN 55343 USA
(612) 938-7777 (tel)
Sells digital workstations, mixing consoles, recorders, DATs, etc. for professional studio recording and live music performances.

American Recorder
4545-6A Industrial St., Simi Valley, CA 93063 USA
(800) 527-9580 (tel)
(800) 527-1433 (fax)
Manufactures interconnects for car stereo applications.

American Reliance
11801 Goldring Rd., Arcadia, CA 91006 USA
(800) AMREL-44 (tel)
(818) 303-5838 (fax)
Manufactures benchtop dc power supplies for a variety of professional industrial applications. Offers a free catalog.

American Ribbon & Toner
2895 W. Prospect Rd., Ft. Lauderdale, FL 33309 USA
(800) 327-1013 (tel)
Visa, MasterCard, Discover, American Express
Sells ribbons, inkjet refills, toner cartridges, etc. for personal computer printers.

American Science and Surplus
P.O. Box 48838, Niles, IL 60714 USA
(708) 475-8440 (tel)
(708) 864-1589 (fax)
Offers nearly anything imaginable in surplus materials, including a number of electronics components, assemblies, and products. The $12.50 minimum order includes a $4 shipping charge.

American Terminal
48925 West Rd., Wixom, MI 48393 USA
(800) 826-4697 (tel)
(313) 380-8890 (fax)
Manufactures audio cables, fuse panels, distribution blocks, etc. for car stereo applications.

American Trap
(800) 684-0527 (tel)
http://www.gocable.com
Visa, MasterCard, Discover
Manufactures negative traps for cable TV applications.

American Video Tape Warehouse
3201 Nicholsolville Rd., Ste. 420, Lexington, KY 40503 USA
(800) 598-8273 (tel)
Sells a variety of bulk video tapes.

American Wholesale Center
817 New Churchman's Rd., New Castle, DE 19720 USA
(800) 845-4962 (tel)
(302) 323-1814 (fax)
Sells personal laptop computers, printers, and accessories.

America's Hobby Center
146 CA W. 22nd St., New York, NY 10011 USA
(800) 989-7950 (tel)
(212) 675-8922 (tel)
(800) 323-2992 (fax)
Visa, MasterCard, Discover, COD ($6)

Sells radios, batteries, battery chargers, speed controls, etc. for radio-control models.

Americomp
5380 E Naiman Pkwy., Solon, OH 44139 USA
(800) 217-2667 (tel)
(216) 498-9620 (tel)
(216) 498-9630 (fax)
Visa, MasterCard, Discover, American Express, COD
Manufactures and sells personal desktop computers; also sells computer parts and boards, such as floppy drives, hard drives, monitors, motherboards, modems, keyboards, controllers, etc. Offers a 30-day money-back guarantee.

AmeriData
9421 Science Center Dr., New Hope, MN 55428
(800) 555-8393 (tel)
(612) 531-3808 (fax)
http://www.ameridata.com
Sells used personal computer systems and printers.

Ameri-Shred
P.O. Box 46130, Monroeville, PA 15146 USA
(412) 823-4222 (tel)
(412) 798-7329 (fax)
Manufactures paper shredders for a wide variety of office applications.

Ameritron
116 Willow Rd., Starkville, MS 39759 USA
(800) 647-1800 (tel)
(601) 323-8211 (tel)
(601) 323-6551 (fax)
Manufactures and sells a variety of RF linear amplifiers for amateur radio use. Offers a free catalog.

AMEX Computers
37 W. 37th St., 5th Fl., New York, NY 10018 USA
(212) 944-7860 (tel)
(212) 944-7868 (fax)
Sells personal desktop and laptop computer systems.

AMI (American Microsystems) Semiconductors
2300 Buckskin Rd., Pocatello, ID 83201 USA
(800) 639-7264 (tel)
Manufactures semiconductors for a wide variety of electronics applications.

Amkotron Electronics
14821 Spring Ave., Santa Fe Springs, CA 90670 USA
(800) 344-3882 (tel)
(310) 802-0102 (tel)
(310) 802-0061 (fax)
Sells replacement parts and accessories for audio equipment.

Amot Controls
401 First St., Richmond, CA 94801 USA
(510) 236-8300 (tel)
(510) 234-9950 (fax)
Western Way, Bury, St. Edmunds, 1P33 3SZ UK
0284-76-2222 (tel)
0284-76-0256 (fax)
230 Orchard Rd., 09-230 Faber House, Singapore 0923
65-235-8187 (tel)
65-235-8869 (fax)
Manufactures electronic pressure controls for industrial applications.

AMP
P.O. Box 3608, Harrisburg, PA 17105 USA
(800) 522-6752 (tel)
(717) 564-0100 (tel)
(717) 986-7575 (fax)
Manufactures electrical environment monitors and edge connectors for a wide variety of electronics applications.

AMPAC Computers & Software
5750 Bintliff, 7302 Harwin Dr., Ste. 206, Houston, TX 77036 USA
(800) 37-AMPAC (tel)
(713) 977-4692 (tel)
(713) 977-9172 (fax)
Sells personal desktop and laptop computer systems; also sells computer parts and boards, such as hard drives, cases, floppy drives, CD-ROM drives, monitors, etc.

Ampeg
1400 Ferguson Ave., St. Louis, MO 63133 USA
(314) 727-4512 (tel)
Manufactures amplifiers for professional and personal musical applications.

Ampex
401 Broadway, MS 22-02, Redwood City, CA 94063 USA
(800) 227-8443 (tel)
Manufactures a variety of analog and digital tape formats for consumer and professional studio recordings.

Amphenol
1 Kennedy Ave., Danbury, CT 06810 USA
(800) 881-9913 (tel)
(203) 796-2072 (tel)
(203) 796-2091 (fax)
Manufactures connectors for a wide variety of RF and audio signal applications.

Amplifier Research
160 School House Rd., Souderton, PA 18964 USA
(215) 723-8181 (tel)
(215) 723-5688 (fax)
Manufactures microwave amplifiers for a variety of professional industrial applications.

Amprobe Instrument
P.O. Box 329, Lynbrook, NY 11563 USA
(800) 477-8658 (tel)
(516) 593-5682 (fax)
Manufactures high-voltage insulation testers for professional electronics applications.

AmPro
1301 Armstrong Dr., Titusville, FL 32780 USA
(407) 269-6680 (tel)
5 Wheeling Ave., Woburn, MA 01801 USA
(617) 932-4800 (tel)
(617) 932-3434 (service tel)
(617) 932-8756 (fax)
Manufactures large-screen projectors for a wide variety of multimedia applications.

Ampro Computers
990 Almanor Ave., Sunnyvale, CA 94086 USA
(408) 522-2100 (tel)
Manufactures CPU interfaces for a wide variety of computer applications.

Amptech
2305 Montgomery St., Ft. Worth, TX 76107 USA
(800) 364-9966 (tel)
(800) 324-4405 (fax)
Manufactures high-output alternators for a wide variety of electronics applications.

Amprotector
943 Wymore Rd., Altamonte Springs, FL 32714 USA
(800) 788-9321 (tel)
(407) 862-3665 (fax)
Manufactures power amplifier cooling systems for car stereo applications.

AmRam
2059 Camden Ave., #330, San Jose, CA 95124 USA
(408) 975-2440 (tel)
(408) 975-2446 (fax)
Visa, MasterCard, Discover, COD
Sells computer parts and boards, such as memory chips, hard drives, CD-ROM drives, etc.

Amroh
P.O. Box 370, NL-1380 AJ Weesp, Netherlands
Sells head units, speakers, amplifiers, receivers, cassette decks, CD players, etc. for home audio and car audio applications.

AMS Direct
12881 Ramona Blvd., Irwindale, CA 91705 USA
(800) 868-8148 (tel)
(818) (814) 8851 (tel)
(818) (814) 0782 (fax)
Visa, MasterCard, Discover, American Express
Manufactures and sells personal laptop computers.

Am-Search Tower
833 Denmark Dr., Mesquite, TX 75149 USA
(800) 410-5988 (tel)

(214) 288-3388 (tel)
Manufactures crank up towers for amateur radio applications.

AMS Neve
6357 W. Sunset Blvd. 402, Hollywood, CA 90028 USA
(213) 461-6383 (Calif tel)
(213) 461-1620 (Calif fax)
(212) 949-2324 (NY tel)
(212) 450-7339 (NY fax)
(416) 365-3363 (Canada tel)
(416) 365-1044 (Canada fax)
+44-1282-475011 (UK tel)
+44-1282-39542 (UK fax)
ams-neve.com@aol.com
Manufactures digital audio mixing consoles for professional studio recording.

Amtel
2125 O'Neal Dr., San Jose, CA 95131 USA
(800) 365-3375 (tel)
(408) 441-0311 (tel)
(408) 436-4300 (fax)
Manufactures semiconductor chips for a wide variety of electronics applications.

Analog Devices
1 Technology Way, P.O. Box 9106, Norwood, MA 02062 USA
(800) 262-5643 (tel)
(617) 461-3881 (tel)
(617) 461-3392 (tel)
(800) 446-6212 (fax data)
(617) 329-1241 (fax)
http://www.analog.com
Manufactures semiconductors for a wide variety of electronics circuits. Offers free application information.

Analogics
5261 Maple Ave. E., Geneva, OH 44041 USA
(216) 466-6911 (tel)
Sells MIDI analog synthesizer kits for professional and personal musical applications.

Analog Modular Systems
(213) 850-5216 (tel)
(213) 850-1059 (fax)
polyfusion@aol.com
Buys, sells, and trades analog music synthesizers for personal and professional musical applications.

Anatek
3938 N. Fraser Way, Burnaby, BC V5J 5H6 Canada
(800) 736-1271 (tel)
Manufactures MIDI-to-XLR line drivers for professional music recording and performance applications.

Anbonn
Lok Sing Centre, Flat B-6, 16/Fl., 19-31 Yee Wo St., Causeway Bay, Hong Kong
(852) 2576-2969 (tel)
(852) 2576-2247 (fax)
5th Fl., No. 787, Chung Ming South Rd., Taichung, Taiwan
(886) 4-263-8176 (tel)
(886) 4-263-7714 (fax)
Manufactures monitors for a wide variety of computer applications.

Ancot
115 Constitution Dr., Menlo Park, CA 94025 USA
(415) 322-5322 (tel)
(415) 322-0455 (fax)
sales@ancot.com
Manufactures SCSI switches for professional computer network applications.

Andataco
10140 Mesa Rim Rd., San Diego, CA 92121 USA
(800) 453-9191 (tel)
(619) 453-9191 (tel)
(619) 453-9294 (fax)
Manufactures data storage systems for a wide variety of professional computer network applications.

Klay Anderson Audio
1856 Grover Ln., Salt Lake City, UT 84124
(801) 942-8346 (tel)
(801) 942-3136 (fax)

Klay Anderson Audio

klay@klay.com
Sells audio equipment for recording studios, professional broadcasting, and home theater applications.

Anderson Electronics
18603 N. Hwy. 1, Ste. 74, Ft. Bragg, CA 95437 USA
Manufactures and sells a variety of kits. Offers free information.

ANDO
290 N. Wolfe Rd., Sunnyvale, CA 94086 USA
(408) 991-6700 (tel)
(408) 991-6701 (fax)
7617 Standish Place, Rockville, MD 20855 USA
(301) 294-3365 (tel)
(301) 294-3359 (fax)
Manufactures test equipment for professional cable TV applications.

Andol Audio Products
4212 14th Ave., Brooklyn, NY 11219 USA
(800) 221-6578 (tel)
(718) 435-7322 (tel)
(718) 853-2589 (fax)
Visa, MasterCard, Discover, American Express
Sells blank recording cassettes, DATs, video tapes, MiniDiscs, etc. Offers a free catalog.

Andrew
10500 W. 153rd St., Orland Park, IL 60462 USA
(800) 255-1479 (tel)
(800) 349-5444 (fax)
Manufactures cable and connectors for a variety of transmitting and receiving applications.

Andrew Electronics
25158 Stanford Ave., Santa Clarita, CA 91355 USA
(800) 289-0300 (servicer tel)
(800) 274-4666 (consumer tel)
(800) 289-0301 (fax)
Sells replacement parts and accessories for audio equipment.

Andromeda Research
P.O. Box 222, Milford, OH 45150 USA

(513) 831-9708 (tel)
(513) 831-7562 (fax)
Visa, MasterCard, COD ($5)
Sells EPROM programming systems. Offers a one-year warranty.

Angstrom
5273 Commerce Ave., Ste. 1, Moorpark, CA 93201 USA
(805) 523-7864 (tel)
(805) 523-2791 (fax)
Manufactures switcher/sound processors for high-end home theater applications.

ANLI
20277 Valley Blvd., #J, Walnut, CA 91789 USA
(800) 666-2654 (tel)
(909) 869-5711 (tel)
(909) 869-5710 (fax)
Manufactures and sells antennas and 3-, 6-, 9-, and 12-Vdc battery power supplies.

Anritsu America
365 W. Passaic St., Rochelle Park, NJ 07662 USA
(201) 843-2690 (tel)
(201) 843-2665 (fax)
Manufactures displacement meters for a variety of professional manufacturing applications.

Antares Microsystems
160B Albright Way, Los Gatos, CA 95030 USA
(800) 726-SBUS (tel)
(408) 370-7287 (tel)
(408) 370-7649 (fax)
Manufactures video accelerator boards for a wide variety of computer applications.

Antec
2859 Bayview Dr., Fremont, CA 94538 USA
(510) 770-1200 (tel)
(510) 770-1288 (fax)
Manufactures card readers, power supplies, file server cases, etc. for a wide variety of personal computer applications.

Antenna Research
11317 Fredrick Ave., Beltsville, MD 20705 USA

(301) 937-8888 (tel)
(301) 937-2796 (fax)
Manufactures broadband antennas for a variety of professional communications applications. Also manufactures fiberoptic links and camera systems for professional industrial applications.

Antennas America
4880 Robb St., Unit 6, Wheat Ridge, CO 80033 USA
(800) 508-6532 (tel)
(303) 421-4063 (tel)
(303) 424-5085 (fax)
Manufactures and sells disguised antennas for professional two-way radio applications.

Antennas Are Us
P.O. Box 2958 W., Columbia, SC 29171 USA
(800) 835-5228 (tel)
(803) 951-9156 (tel)
(803) 957-7718 (fax)
Visa, MasterCard, Discover
Sells a variety of amateur radio antennas and accessories.

Antennas Etc.
P.O. Box 4215, Andover, MA 01810 USA
(508) 475-7831 (tel)
(508) 975-2711 (tel)
(508) 474-8949 (fax)
Sells Unadilla antennas and equipment from James Millen.

Antenna Mart
P.O. Box 699, Loganville, GA 30249 USA
(404) 466-4353 (tel)
Manufactures and sells antenna switches and mounts for amateur and citizens band radio applications.

Antenna Supermarket
P.O. Box 563, Palantine, IL 60078 USA
(708) 359-7092 (tel)
(708) 359-8161 (fax)
Visa, MasterCard
Manufactures and sells shortwave receiver antennas and lightning arrestors.

Antennas West
P.O. Box 50062, Provo, UT 84605 USA
(800) 926-7373 (tel)
(801) 373-8425 (tel)
Manufactures and sells a wide variety of shortwave and scanner antenna kits, including a yagi in a walking stick.

Antique Audio
5555 N. Lamar, Bldg. H-105, Austin, TX 78751 USA
(512) 467-0304 (tel)
Sells electron tubes, electronics components, books, and kits for antique audio applications.

Antique Electronic Supply
6221 S. Maple Ave., Tempe, AZ 85283 USA
(603) 820-5411 (tel)
(800) 706-6789 (fax)
Sells a wide variety of components for and materials to repair antique radios, including tubes, variable capacitors, paint, decals, cleaners, etc.

Antique Radio Components
1065 Faith Dr., Meadow Vista, CA 95722 USA
(800) 649-6550 (tel)
(916) 878-9421 (fax)
Buys and sells antique audio, industrial, and collector electron tubes.

Antique Triode
P.O. Box A2, Farnham, NY 14061 USA
(716) 549-5379 (tel/fax)
Sells surplus electron tubes for a wide variety of electronics applications.

Antonio Precise Products Manufactory
Rm. 1314-16, 13/F, Block B, Yau Tong Industrial City, 17 Ko Fai Rd., Yau Tong, Kowloon, Hong Kong
(852) 2772-3101 (tel)
(852) 2775-5178 (fax)
(852) 2775-7861 (fax)
Manufactures loudspeakers, headphones, and microphones for computer multimedia applications.

AOC International
311 Sinclair Frontage Rd., Milpitas, CA 95035 USA

AOC International

(408) 956-1070 (tel)
(408) 956-1516 (fax)
Sells monitors for a wide variety of computer applications.

AOR
2-6-4 Misuji, Taito-Ku, Tokyo 111 Japan
Adam Bede Centre, Derbe Rd., Wirksworth, Derbys DE4 4BG UK
Manufactures and sells VHF/UHF scanners and shortwave receivers.

AOS
534 N. Derr Dr., Ste. 305, Lewisburg, PA 17837 USA
Sells amplifiers, mixers, equalizers crossovers, loudspeakers, electric guitars, etc. for professional and personal musical applications. Offers a free catalog.

APEM
134 Water St., P.O. Box 544, Wakefield, MA 01880 USA
(617) 246-1007 (tel)
Manufactures toggle switches for a wide variety of electronics applications.

Apex Data
4305 Cushing Pkwy., Fremont, CA 94538 USA
(800) 841-APEX (tel)
sales@smartm.com
http://www.apexdata.com
http://www.smartm.com
Manufactures computer modems.

Apex PC Solutions
20031 142nd Ave. N.E., Woodinville, WA 98072 USA
(800) 861-5858 (tel)
(206) 402-9393 (tel)
(206) 402-9494 (fax)
info@pcsol.com
Manufactures screen-based concentrator switches for a wide variety of computer applications.

Apex Technology
B1, No. 28, Ln. 80, Sec. 3, Nan-Kang Rd., Taipei, Taiwan
(886) 2-783-8331 (tel)

(886) 2-785-2554 (fax)
Manufactures touch mouse controllers for a wide variety of computer applications.

Aphex
11068 Randall St., Sun Valley, CA 91352 USA
(818) 767-2929 (tel)
(818) 767-2641 (fax)
Manufactures sound processors, digital coders, and microphone preamps for broadcast stations.

api
7655-G Fullerton Rd., Springfield, VA 22153 USA
(703) 455-8188 (tel)
(703) 455-4240 (fax)
Manufactures modular compressors, noise gates, filters, and microphone preamplifiers for professional studio recording.

A-Plus Computer
21038 Commerce Pointe Dr., Walnut, CA 91789 USA
(800) 745-0880 (tel)
(909) 468-3723 (tel)
(909) 468-3728 (fax)
Visa, MasterCard, Discover, American Express, COD
Manufactures and sells personal desktop computers.

Apollo Display Technologies
194-22 Morris Ave., Holtsville, NY 11742 USA
(516) 654-1143 (tel)
(516) 654-1496 (fax)
Manufactures LCD modules for a wide variety of electronics applications.

Apollo Presentation Products
60 Trade Zone Ct., Ronkonkoma, NH 11779 USA
(800) 777-3750 (tel)
Manufactures LCD projection panels for computer multimedia applications.

Apple Computer
1 Infinite Loop, Cupertino, CA 95104 USA
(800) 795-1000 (tel)
(800) 538-9696 (tel)

(408) 996-1010 (tel)
(800) 600-7808 (computers/disability tel)
(800) 755-0601 (computers/disability TTY)
(408) 996-0275 (fax)
http://www.apple.com/
http://www.info.apple.com/gomobile/
Manufactures and sells personal desktop computers.

Appleseed Electronics
P.O. Box 8228, Fort Wayne, IN 46898 USA
(219) 489-9802 (tel)
Sells components for a wide variety of electronics applications.

Applied Digital Systems
30 LiftBridge Ln. E., Fairport, NY 14450 USA
(716) 377-7000 (tel)
(716) 377-5544 (fax)
Manufactures CD-ROM drives, optical drives, and recorders for a wide variety of professional computer network applications.

Applied Kilovolts
38/40 Eastbrook Rd., Portslade, Sussex BN41 1PB UK
+44-1273-439440 (tel)
Manufactures filament power supplies for a wide variety of electronics applications.

Applied Microsystems
P.O. Box 97002, Redmond, WA 98073 USA
(800) 426-3925 (tel)
info@amc.com
http://www.amc.com
Manufactures debuggers for professional computer troubleshooting applications.

Applied Research & Technology
215 Tremont St., Rochester, NY 14608 USA
(716) 436-2720 (tel)
Manufactures microphone preamplifiers for professional and personal musical applications.

APPRO International
2032 Bering Dr., San Jose, CA 95131 USA
(800) 927-5464 (tel)
(408) 452-9200 (tel)
(408) 452-9210 (fax)
Visa, MasterCard, American Express
Manufactures ruggedized, rack-mount computer equipment for professional and industrial computer applications.

Apogee
3145 Donald Douglas Loop S., Santa Monica, CA 90405 USA
(310) 915-1000 (tel)
(310) 391-6262 (tel)
Manufactures digital-to-analog converters for car stereo applications.

Apogee Acoustics
35 York Ave., Randolph, MA 02368 USA
(508) 988-0124 (tel)
Manufactures loudspeakers for home audio and home theater applications.

APS Technologies
6131 Deramus, Kansas City, MO 64120 USA
(800) 418-6401 (tel)
(800) 374-5802 (info fax)
Visa, MasterCard, Discover, American Express
Manufactures and sells hard disk drives and tape drives for personal desktop computers. Offers a 30-day money-back guarantee.

APTEC
P.O. Box 101, El Segundo, CA 90245 USA
(310) 640-7262 (tel)
Manufactures instrument simulators for aviation applications.

The ARC
1014 Central Ave., Tracy, CA 95376 USA
(800) 753-0114 (tel)
(209) 832-4300 (tel)
(209) 832-3270 (fax)
Buys and sells used Apple and MacIntosh computer systems and parts for a wide variety of computer applications.

ARC
190 Munsonhurst Rd., Ste. 13, Franklin, NJ 07416 USA

ARC

(201) 827-8055 (tel)
(201) 827-0591 (fax)
fastscsi@aurora.planet.net
Visa, MasterCard
Sells data storage systems, including tape drives and disk drives, for a wide variety of computer applications.

ARCAL
2732 Bay Rd., Redwood City, CA 94063 USA
(800) 272-2591 (tel)
(415) 369-7446 (fax)
Sells a variety of recording supplies, including bulk audio and video cassettes, DAT, reel-to-reel tape, data storage media, etc., for professional recording studios.

ARCOM
P.O. Box 6729, Syracuse, NY 13217 USA
(315) 422-1230 (tel)
(315) 422-2963 (fax)
Manufactures traps and filters for professional cable TV applications.

Arcom Control Systems
Clifton Rd., Units 8-10, Cambridge, CB1 4WH UK
Manufactures I/O boards for a wide variety of computer applications.

Ar_FD Communications Products
P.O. Box 1242, Burlington, CT 06013 USA
(203) 485-0310 (tel)
Visa, MasterCard, COD
Manufactures telephone taps for a wide variety of surveillance applications.

A.R.E. Electronic Surplus
15272 St. Rt. 12E, Findlay, OH 45840 USA
(419) 422-1558 (tel)
Sells surplus components, tools, and equipment for a wide variety of electronics applications.

ARE Communications
6 Royal Parade, Hanger Lane, Ealing, London W5A 1ET UK
44-081-997-4476 (tel)
Sells kits to modify Icom receivers for greater frequency range and SSB reception.

Arena Services
700 W. Washington St., Norristown, PA 19401 USA
(800) WT-ARENA (tel)
(610) 630-0320 (tel)
(610) 630-8202 (fax)
Sells and reconditions a wide variety of equipment for professional cable TV applications.

Argaph
111 Asia Pl., Carlstadt, NJ 07072 USA
(201) 939-7722 (tel)
Manufactures batteries for a wide variety of electronics applications.

Argus Technologies
3767 Alpha Way, Bellingham, WA 98226 USA
(206) 647-2360 (tel)
7033 Antrim Ave., Burnaby, BC V5J 4M5 Canada
(604) 436-5900 (tel)
Manufactures power systems for professional cellular microwave radio applications.

Aria
9244 Commerce Hwy., Pennsauken, NJ 08110 USA
(800) 524-0441 (tel)
Manufactures electric guitars for professional and personal musical applications.

Aria Technologies
5309 Randall Pl., Fremont, CA 94538 USA
(510) 226-6222 (tel)
(510) 226-6242 (fax)
Manufactures fiberoptic assemblies for professional cable TV applications.

Aries Electronics
P.O. Box 130, Frenchtown, NJ 08825 USA
(908) 996-6841 (tel)
(908) 996-3891 (fax)
Manufactures specialty correction circuit boards for professional electronics applications.

Aries Manufacturing
1181 Tower Rd., Schaumburg, IL 60173 USA
(708) 885-0021 (tel)
(708) 885-0913 (fax)
Manufactures battery kits, call connectors, power adapters, etc. for cellular telephone applications.

Arista
780 Montague Expressway, #702, San Jose, CA 95131 USA
(408) 526-1820 (tel)
(408) 526-1823 (fax)
Manufactures CPU boards for a wide variety of computer applications.

Aristo Computers
6700 SW 105th Ave., Ste. 307, Beaverton, OR 97005 USA
(800) 327-4786 (tel)
(503) 626-6333 (tel)
(503) 626-6492 (fax)
Manufactures memory chip testers for a variety of professional computer applications.

ARK Logic
1737 N. 1st St., Ste. 680, San Jose, CA 95112 USA
(408) 467-1988 (tel)
Manufactures semiconductors for a wide variety of electronics applications.

Arlen Supply
7409 W. Chester Pike, Upper Darby, PA 19082 USA
(610) 352-9311 (tel)
(610) 352-9388 (fax)
Sells a variety of electron tubes--more than 1 million stocked. Minimum order: $150.

Arlington Computer Products
851 Commerce Ct., Buffalo Grove, IL 60089 USA
(800) 548-5105 (tel)
(708) 541-6595 (shipp tel)
(708) 541-6333 (cust. service tel)
(708) 541-6694 (corp. sales tel)
(708) 541-6583 (tech tel)
(708) 541-6881 (fax)
Visa, MasterCard, Discover, American Express
Sells personal desktop and laptop computers. Also sells computer parts, such as floppy drives, hard drives, monitors, motherboards, video cards, modems, keyboards, etc.

Armadillo Enterprises
923 McMullen Booth Rd., Clearwater, FL 34619 USA
(813) 796-8888 (tel)
(813) 797-9448 (fax)
Manufactures keyboard synthesizers for a wide variety of musical applications.

ARM Computer
2149-A O'Toole Ave., San Jose, CA 95131 USA
(800) 765-1767 (tel)
(408) 428-9800 (tel)
(408) 428-9528 (fax)
Visa, MasterCard, Discover, American Express, COD
Manufactures and sells personal laptop computers. Offers a 30-day money-back guarantee.

Arnav Systems
P.O. Box 73730, Puyallup, WA 98373 USA
(206) 848-6060 (tel)
Manufactures GPS and Loran systems for professional aviation applications.

Arnes Radio
Hovedvejen 102, 2600 Glostrup, Denmark
42-45-45-44 (tel)
Sells high-end home audio equipment, such as speakers, cassette decks, amplifiers, preamplifiers, etc.

Arnet
618 Grassmere Park Dr., Nashville, TN 37211 USA
(800) 453-5395 (tel)
(615) 834-8000 (tel)
(615) 834-5399 (fax)
Manufactures single-board computers and terminal servers for professional and industrial computer applications.

Yakov Aronov Audio Lab
7418 Beverley Blvd., Los Angeles, CA 90036 USA
(213) 653-3045 (tel)
(213) 937-6905 (fax)

Yakov Aronov Audio Lab

Manufactures high-end preamplifiers and amplifiers for audio and home theater applications.

Arrakis Systems
2619 Midpoint Dr., Ft. Collins, CO 80525 USA
(303) 224-2248 (tel)
(303) 493-1076 (fax)
Manufactures digital audio workstations, audio consoles, studio furniture, and prewired system packages for broadcast stations.

Arriflex
600 N. Victory Blvd., Burbank, CA 91502 USA
(818) 841-7070 (tel)
Manufactures lighting systems for amateur and professional video production.

Arrow Electronics
1860 Smithtown Ave., Ronkonkoma, NY 11779 USA
(800) 93-ARROW (tel)
Sells a variety of electronics components. Mininum order: $25.

ART
215 Tremont St., Rochester, NY 14608 USA
(716) 436-2720 (tel)
(716) 436-3942 (fax)
Manufactures distortion pedals for personal and professional music applications. Offers a full-color catalog for $2.

Artais
4660 Kenny Rd., Columbus, OH 43220 USA
(800) 327-AWOS (tel)
(614) 451-8388 (tel)
(614) 451-0229 (fax)
Manufactures automated weather systems for professional aviation applications.

Artecon
P.O. Box 9000, Carlsbad, CA 92018 USA
(800) 872-2783 (tel)
(619) 931-5500 (tel)
(619) 931-5527 (fax)
Manufactures data storage systems for a wide variety of professional computer network applications.

Arteq Sound & Light
Vendsysselgade 9, 9000 Ålborg, Denmark
98-10-12-66 (tel)
Sells lighting systems for live music performances and discos.

ASA
P.O. Box 3461, Myrtle Beach, SC 29578 USA
(803) 293-7888 (tel)
Visa, MasterCard
Manufactures 2-meter antennas for amateur radio applications.

Asahi Kasei Microsystems
2001 Gateway Pl., San Jose, CA 95110 USA
(408) 436-8580 (tel)
Manufactures integrated circuits for a wide variety of electronics applications.

Asanté Technologies
821 Fox Ln., San Jose, CA 95131 USA
(800) 662-9686 (fax)
(408) 435-8388 (tel)
(800) 741-8607 (fax)
sales@asante.com
Manufactures hubs for professional computer network applications.

ASAPc Direct
3847 W. Oakton St., Skokie, IL 60076 USA
(800) 404-7228 (tel)
(708) 673-2300 (fax)
http://www.asapc2u.com
Visa, MasterCard, Discover, American Express
Sells personal home and laptop computers and parts and accessories, including memory chips, tablets, plotters, modems, monitors, etc. Offers a 30-day money-back guarantee.

ASCOR
47790 Westinghouse Dr., Fremont, CA 94539 USA
(510) 490-8819 (tel)
(510) 490-8493 (fax)
Manufactures VXI switching and digital modules, chassis, and ITAs for a wide variety of professional electronics applications.

Aserton
2923 Saturn St., Ste. 3B, Berea, CA 92621 USA
(800) 247-9329 (tel)
(714) 577-0325 (tel)
(714) 577-8237 (fax)
Visa, MasterCard, Discover, American Express
Manufactures and sells personal desktop computers. Offers a 30-day money-back guarantee.

Ashcroft
250 E. Main St., Stratford, CT 06497 USA
(800) 328-8258 (tel)
(203) 378-8281 (tel)
(203) 385-0476 (fax)
Manufactures pressure and temperature transducers for a wide variety of professional electronics applications.

Ashly Audio
847 Holt Rd., Webster, NY 14580 USA
(800) 828-6308 (tel)
(716) 872-0010 (tel)
(716) 872-0739 (fax)
(714) 440-0760 (international exports tel)
(416) 696-2779 (Canada tel)
http://www.ashly.com
info@ashly.com
Manufactures power audio amplifiers for professional studio recording and live music performances.

Ashtech
1170 Kifer Rd., Sunnyvale, CA 94086 USA
(408) 524-1400 (tel)
Manufactures GPS systems for professional aviation applications.

Ashtek
2600-B Walnut Ave., Tustin, CA 92680 USA
(800) 801-9400 (tel)
(714) 505-3157 (tel)
(714) 505-3831 (tel)
(714) 505-2693 (fax)
ashtek@aol.com
Visa, MasterCard, Discover, American Express, COD
Sells customized computers and computer parts and boards, such as hard drives, motherboards, memory chips, keyboards, etc.

Ashton Systems
Guanajuato 224-5, Col. Roma, Mexico
(525) 264-1210 (tel)
(525) 264-3087 (fax)
Manufactures systems for professional computer network applications.

ASI
452 Crompton St., Charlotte, NC 28273 USA
(800) 681-SCAN (tel)
Sells UPC scanners, etc. for business applications.

ASKA Communication
3540 W. 256 St., Ste. 206, Ft. Lauderdale, FL 33309 USA
(800) 317-6625 (tel)
(305) 486-0039 (tel)
(305) 486-0202 (fax)
Manufactures amplifiers, splitters, taps, combiners, band separators, matching transformers, etc. for professional cable TV applications.

ASK LCD
1099 Wall St. W., Ste. 396, Lyndhurst, NJ 07071 USA
(800) 275-5231 (tel)
Manufactures projection panels, projectors, and monitors for use with computers.

ASK Technology
Unit 1, 4/F, Henley Industrial Centre, 9-15 Bute St., Mongkok, Kowloon, Hong Kong
(852) 2398-3223 (tel)
(852) 2391-1003 (fax)
askkenho@hk.net
Block 16, Pao Chun Industrial Zone A, Shajing, Baoan, Shenzhen, China
86-755-720-2818 (tel)
86-755-720-2819 (tel)
86-755-720-2820 (fax)
Manufactures system boards, Windows accelerator boards, and MPEG boards for a wide variety of computer applications.

Aspen Computer Products
12780 S. Redwood Rd., Riverton, UT 84065 USA

Aspen Computer Products
(800) 275-SIMM (tel)
(801) 254-4282 (tel)
(801) 253-2558 (fax)
Visa, MasterCard, Discover, American Express
Sells memory chips and computers for personal computer applications.

A.S. & S.
P.O. Box 25, 6061 Western Australia, Australia
00-61-9344-5905 (tel)
00-61-9349-9413 (fax)
Sells more than 30 different types of miniature video cameras for home video and surveillance applications.

Associated Electrics
3585 Cadillac Ave., Costa Mesa, CA 92626 USA
(714) 850-9342 (tel)
(714) 850-1744 (fax)
Sells speed controls for radio-control models.

Associated Radio
8012 Conser, Box 4327, Overland Park, KS 66204 USA
(800) 497-1457 (tel)
(913) 381-5900 (tel)
(913) 648-3020 (fax)
Specializes in selling vintage amateur radio transmitters, transceivers, and receivers. Offers a catalog for $3.

Associates Computer Supply
275 W. 123rd St., Riverdale, NY 10463 USA
(718) 543-8686 (tel)
(718) 548-0343 (fax)
Visa, MasterCard, Discover, American Express, COD
Sells computer parts and boards, such as motherboards, video cards, hard drives, memory chips, etc.

A&S Speakers
3170 23rd St., San Francisco, CA 94110 USA
(415) 641-4573 (tel)
(415) 648-5306 (fax)
Manufactures complete loudspeakers and loudspeaker kits for home audio and home theater applications.

Astatic
P.O. Box 120, Harbor & Jackson, Conneaut, OH 44030 USA
(216) 593-1111 (tel)
(216) 693-5395 (fax)
Manufactures microphones for a wide variety of two-way radio applications.

Astec
6339 Paseo Del Lago, Carlsbad, CA 92009 USA
401 Jones Rd., Oceanside, CA 92054 USA
(800) 451-2677 (tel)
(619) 757-1880 (tel)
(619) 930-0881 (fax)
Manufactures power supplies, power converter modules, etc. for a wide variety of electronics applications.

AST Research
16215 Alton Pkwy., P.O. B 57005, Irvine, CA 92619 USA
(800) 876-4278 (tel)
(800) 727-1278 (tech support tel)
(714) 727-4141 (tel)
http://www.ast.com
Manufactures and sells personal desktop computers.

AstroFlight
13311 Beach Ave., Marina Del Rey, CA 90292 USA
(310) 821-6242 (tel)
Manufactures battery chargers for radio-control model airplanes.

Astron
9 Autry, Irvine, CA 92718 USA
(714) 458-7277 (tel)
(714) 458-0826 (fax)
Manufactures and sells inverters, power supplies, and converters for amateur and professional two-way radio applications.

A2Z Computers
701 Beta Dr., Ste. 19, Cleveland, OH 44143 USA
(800) 983-8889 (tel)
(216) 442-9028 (tech support tel)

Atlas/Soundolier

(216) 442-8891 (fax)
http://www.a2z.com
Manufactures and sells personal desktop computers; also sells computer parts and boards, such as floppy drives, hard drives, monitors, video cards, modems, tape drives, etc. Offers a 30-day money-back guarantee.

AT
550 Lakeside Dr., Sunnyvale, CA 94086 USA
(800) 858-2173 (tel)
(408) 774-9010 (tech support tel)
(408) 774-9011 (fax)
Visa, MasterCard, Discover, American Express
Manufactures and sells personal desktop computers. Also sells computer parts and boards, such as hard drives, video cards, motherboards, modems, memory chips, etc. Offers a 30-day money-back guarantee.

Atari
1196 Borregas Ave., Sunnyvale, CA 94089 USA
http://www.atari.com
Manufactures video game systems, software, and personal computers.

ATC Power Systems
45 Depot St., Merrimack, NH 03054 USA
(603) 429-0391 (tel)
(603) 429-0795 (fax)
Manufactures power supply modules for a wide variety of electronics applications.

A-TECH Electronics
2210 W. Magnolia Blvd., Burbank, CA 91506 USA
(818) 845-9203 (tel)
(818) 846-2298 (fax)
Sells a variety of amateur radio equipment, including shortwave and VHF/UHF transceivers, antennas, amplifiers, power supplies, antenna tuners, etc.

ATI Technologies
33 Commerce Valley Dr. E., Thornhill, ON L3T 7N6 Canada
(905) 882-2600 (tel)
(905) 882-2620 (fax)
http://www.atitech.ca
Manufactures graphics accelerator cards for personal desktop computers.

Atlanta Signal Processors (ASPI)
1375 Peachtree St. N.E., Ste. 690, Atlanta, GA 30309 USA
(404) 892-7265 (tel)
(404) 892-2512 (fax)
info@aspi.com
Manufactures DSP boards for a wide variety of professional computer applications.

Atlantic Ham Radio
368 Wilson Ave., Downsview, ON M3H 1S9 Canada
(416) 222-2506 (tel)
(416) 636-3636 (tel)
(416) 631-0747 (fax)
Sells a wide variety of amateur radio equipment and accessories, including amateur radio transceivers, receivers, antenna tuners, antennas, amplifiers, etc.

Atlantic Logic South
(407) 253-0304 (tel)
(407) 253-0307 (fax)
Sells printers for a wide variety of computer applications.

Atlantic Select
812 Warsaw Ave., Blackwood, NJ 08012 USA
(609) 228-2070 (tel)
(609) 228-2711 (fax)
Sells used IBM personal computer systems; also sells IBM computer parts and boards, such as processor cards, system boards, network equipment, etc.

Atlantic Technology
343 Vanderbilt Ave., Norwood, MA 02062 USA
(617) 762-6300 (tel)
Manufactures loudspeakers for high-end home audio and home theater applications.

Atlas/Soundolier
1859 Intertech Dr., Fenton, MO 63026 USA
Manufactures loudspeakers for home audio and professional audio applications.

ATL Products
240 E. Palais Rd., Anaheim, CA 92805 USA
(714) 774-6900 (tel)
(714) 774-5909 (fax)
Manufactures automated back-up and archive systems for professional and industrial computer applications.

A-Trend Technology
1257 Reamwood Ave., Sunnyvale, CA 94089 USA
(800) 866-0188 (tel)
(408) 745-0400 (tel)
(408) 745-0922 (fax)
10F, No. 75 Hsin Tai Wu Rd., Sec. 1, Hsi-Chih, Taipei Hsien, Taiwan 221
(886) 2-698-2199 (tel)
(886) 2-698-2369 (fax)
Energiestraat 36, 1411 AT Naarden, Netherlands
31-2159-44444 (tel)
31-2159-43345 (fax)
1-12-10 Shiba, Minato Ku, Tokyo 105, Japan
81-3-3769-1054 (tel)
81-3-3769-1060 (fax)
Manufactures motherboards for a wide variety of computer applications.

AT&T
Microelectronics Division
Room 21Q-133BA, 555 Union Blvd., Allentown, PA 18103 USA
(800) 372-2447 (tel)
(610) 712-4106 (fax)
(800) 447-1124 (tel)
(800) 233-2650 (tel)
(800) 553-2448 (Canada tel)
(800) 526-7819 (power systems tech info)
(800) 325-7466 (global business communications systems tel)
http://www.att.com/nsg/
Manufactures semiconductors, personal and professional computers, CD-ROM WORM drives, CD-ROM recorders, telephones, power systems, and computer-compatible telephones.

ATTO Technology
40 Hazelwood Dr., Ste. 106, Amherst, NY 14228 USA
(716) 691-1999 (tel)
(716) 691-9353 (fax)
Manufactures disk controller boards for a wide variety of professional computer network applications.

ATx Telecom Systems
1251 Frontenac Rd., Napierville, IL 60563 USA
(708) 778-2900 (tel)
(708) 369-4299 (fax)
Manufactures fiberoptic links for professional cable TV broadcasting applications.

Audax Direct
100 Produce Ave., S. San Francisco, CA 94080 USA
(800) 589-8709 (tel)
(415) 589-8700 (tel)
(415) 589-9100 (fax)
Visa, MasterCard, American Express
Sells adapters, converters, print servers, hubs, routers, modems, connectors, repeaters, etc. for professional computer network applications. Offers a free, on-disk catalog.

Audio Art
152 S. Brent Circle, City of Industry, CA 91789 USA
(909) 598-0515 (tel)
(909) 595-4694 (fax)
Manufactures loudspeakers for car stereo applications.

Audio Artistry
8312 Salem Dr., Apex, NC 27502 USA
(919) 319-1375 (tel)
(919) 319-1416 (fax)
Manufactures high-end speakers for audio and home theater applications.

Audioarts Engineering
7305 Performance Dr., Syracuse, NY 13212 USA
(315) 452-5000 (tel)
(315) 452-0160 (fax)
Manufactures mixing consoles for broadcast stations.

Audio Broadcast Group (ABG)
2342 S. Division Ave., Grand Rapids, MI 49507 USA

(800) 999-9281 (tel)
(616) 452-1652 (fax)
75371.144@compuserve.com
support@abg.com
Manufactures a digital instant replay for broadcast stations.

Audio by Van Alstine
2022 River Hills Dr., Burnsville, MN 55337 USA
(612) 890-3517 (tel)
(612) 894-3675 (fax)
Manufactures power amplifiers, preamplifiers, and phase inverters (in complete and kit form). Offers a free catalog.

Audio Car
915 Broadway, Chula Vista, CA 91511 USA
(619) 422-1449 (tel)
(525) 531-8033 (Mexico tel)
Sells head units, amplifiers, loudspeakers, etc. for mobile audio applications.

Audio Classics
P.O. Box 176, Walton, NY 13856 USA
(607) 865-7200 (tel)
(607) 865-7222 (fax)
Sells a variety of new and used high-end audio components, including loudspeakers, amplifiers, CD players, processors, cassette decks, crossovers, headphones, equalizers, etc. Offers a free catalog.

AudioControl
22410 70th Ave. W., Mountlake Terrace, WA 98043 USA
(206) 775-8461 (tel)
(206) 778-3166 (fax)
Manufactures time spectrum analyzers for professional audio applications. Offers a five-year parts and labor warranty.

Audio Crafters Guild
5102 E. 38 Pl., Tulsa, OK 74135 USA
Manufactures a digital-to-analog converter kit for high-end audio applications.

Audio Design Innovations
P.O. Box 402, Osseo, MN 55369 USA
(612) 424-8737 (tel/fax)
Manufactures loudspeakers for car stereo applications.

Audio Excellence
143 W. 26th St., New York, NY 10001 USA
(212) 229-1622 (tel)
Visa, MasterCard, Discover, American Express
Sells a wide variety of high-end home audio equipment.

Audio Gods
5101 Lankershim Blvd., N. Hollywood, CA 91601 USA
(818) 509-8879 (tel)
(818) 760-8484 (fax)
Manufactures head units, amplifiers, crossovers, loudspeakers, etc. for car stereo applications.

Audio Influx
P.O. Box 381, Highland Lakes, NJ 07422 USA
(201) 764-8958 (tel)
Manufactures loudspeakers for high-end home audio and home theater applications.

Audio Intervisual Design
1155 N. La Brea, Los Angeles, CA 90038 USA
(213) 845-1155 (tel)
(213) 845-1170 (fax)
Distributes Sanken microphones and dB Technologies professional digital audio equipment in the United States.

AudioLink
23241 La Palma Ave., Yorba Linda, CA 92687 USA
(800) 327-5905 (tel)
(714) 692-0068 (fax)
Manufactures crossovers for car stereo applications.

Audio Logic
9837 Owensmouth G., Chatsworth, CA 91311 USA
(800) 700-9750 (tel)
(818) 700-9768 (fax)
Manufactures amplifiers for car stereo applications.

Audio Nexus
33 Union Pl., Union, NJ 07901 USA
(908) 277-0333 (tel)
Sells high-end home audio equipment, such as speakers, cassette decks, amplifiers, preamplifiers, etc.

Audio Outlet
69 S. Moger Ave., Mt. Kisco, NY 10549 USA
(914) 666-0550 (tel)
(914) 666-0554 (fax)
Visa, MasterCard, Discover, American Express
Sells a wide variety of high-end home audio equipment. Offers a 94-page catalog for $4.

audiophile
7416 Washington Ave. S., Eden Prarie, MN 55344 USA
(800) 727-6863 (tel)
(612) 944-8335 (fax)
Manufactures loudspeakers for car stereo applications.

Audio Precision
P.O. Box 2209, Beaverton, OR 97075 USA
(800) 231-7350 (tel)
(503) 627-0832 (tel)
(503) 641-8906 (fax)
Manufactures audio analyzers for professional studio recording.

Audio Products
Rt. 6, Box 149, Donthan, AL 36303 USA
Manufactures loudspeakers for mobile audio applications.

Audioquest
915 Calle Amanecer #O, San Clemente, CA 92673 USA
(714) 498-2770 (tel)
(714) 498-5112 (fax)
Manufactures audio and power cables for high-end, amateur, and professional audio and video applications.

audio research
5740 Green Circle Dr., Minnetonka, MN 55343 USA
(612) 939-0600 (tel)
(612) 939-0604 (fax)
Manufactures high-end audio power amplifiers for audio and home theater applications.

Audioscan
Fredriksborggade 50, 1360 Kobenhavn, Denmark
33-32-37-11 (tel)
Lyngby Hovedgade 37, 2800 Lyngby, Denmark
45-87-78-87 (tel)
Norregade 30, 9000 Aalborg, Denmark
98-16-34-00 (tel)
Sells high-end home audio equipment, such as speakers, cassette decks, amplifiers, preamplifiers, etc.

Audio Solutions
5576 Chamblee Dunwoody Rd., Atlanta, GA 30338 USA
(404) 804-8977 (tel)
Sells high-end home audio equipment, such as speakers, cassette decks, amplifiers, preamplifiers, etc.

Audio Sound
4900 Nakskov, Denmark
54-95-62-05 (tel)
Sells high-end home audio equipment, such as speakers, cassette decks, amplifiers, preamplifiers, etc.

AudioSource
1327 N. Carolan Ave., Burlingame, CA 94010 USA
(800) 227-5087 (tel)
(415) 348-8114 (tel)
(415) 348-8083 (fax)
Manufactures loudspeakers and sound processors for high-end home audio and home theater applications.

audio studio
303 Newbury St., Boston, MA 02115 USA
(617) 267-1001 (tel)
414 Harvard St., Brookline, MA 02146 USA
(617) 277-0111 (tel)

(617) 277-2415 (fax)
Visa, MasterCard, Discover, American Express
Sells home and amateur audio and video equipment, such as camcorders, VCRs, speakers, cassette decks, TVs, home theater equipment, etc.

Audio-Technica
1221 Commerce Dr., Stow, OH 44224 USA
(216) 686-2600 (tel)
(216) 688-3752 (fax)
pro@atus.com
Old Ln., Leeds, LS11 8AG UK
+44-0113-274-1441 (tel)
+44-0113-270-4836 (fax)
Manufactures microphones for professional studio recording applications.

Audio Technologies
328 W. Maple Ave., Horsham, PA 19044 USA
(800) 959-0307 (tel)
(215) 443-0330 (tel)
(215) 443-0394 (fax)
Manufactures audio amplifiers and preamplifiers, mixing consoles, and studio metering systems for broadcast stations.

Audio Toys (ATI)
99017-C Mendenhall Ct., Columbia, MD 21045 USA
(410) 381-7879 (tel)
(410) 381-5025 (fax)
Manufactures several full-audio mixing consoles and rack-mount mixing consoles for professional studio recording.

Audio Unlimited
2341 W. Yale Ave., Englewood, CO 80110 USA
(303) 691-3407 (tel/fax)
Visa, MasterCard
Sells high-end home audio equipment, such as speakers, cassette decks, amplifiers, preamplifiers, etc.

Audio Upgrades
6410 Matilija Ave., Van Nuys, CA 91401 USA
(818) 780-1222 (tel)
(818) 346-2308 (fax)
Makes component-level upgrades for professional studio recording mixing consoles, tape machines, microphones, etc.

Audio Video Parts
P.O. Box 19670, 1071 S. La Brea Ave., Los Angeles, CA 90019 USA
(213) 933-8141 (tel)
(213) 933-7008 (fax)
Sells replacement parts and accessories for audio equipment.

Audio Village
P.O. Box 2902, Palm Springs, CA 92262 USA
(619) 320-0728 (tel)
Sells audio patch cords and patch bays for professional recording studios and live music performances.

Audio Visual Assistance
565 Sherwood Rd., Shorewood, MN 55126 USA
(612) 481-9715 (tel/fax)
Visa, MasterCard
Sells microphone "pop" guards for professional studio recording.

Audiovox
150 Marcus Blvd., Hauppauge, NY 11788 USA
(800) 645-7750 (tel)
(516) 233-3300 (tel)
(516) 231-7750 (tel)
(516) 434-3995 (fax)
Manufactures a variety of car stereo equipment, including head units, speakers, amplifiers, etc. Also manufactures telephones for cellular telephone applications.

Audio World
4858 Queen Florence Ln., Woodland Hills, CA 91364 USA
(310) 275-4277 (tel)
(818) 340-4331 (fax)
Sells a variety of audio equipment, including mixing consoles, microphones, compressors, preamplifiers, etc. for professional studio recording and live music performances.

Audison USA
99 Tulip Ave., Floral Park, NY 11001 USA

Audison USA
(516) 231-7750 (tel)
(516) 231-7897 (fax)
Manufactures amplifiers and loudspeakers for car stereo applications.

Auditronics
3750 Old Getwell Rd., Memphis, TN 37116 USA
(901) 362-1350 (tel)
(901) 365-8629 (fax)
Manufactures mixing consoles for professional broadcasting applications.

August Mark
636 Scalp Ave., Ste. 213-215, Johnstown, PA 15904 USA
(800) 636-6373 (tel)
(814) 269-1398 (tel)
(814) 269-9221 (fax)
Sells cash drawers, UPC scanners, receipt printers, etc. for computerized business applications.

Ault
7300 Boone Ave. N., Minneapolis, MN 55428 USA
(612) 483-1900 (tel)
(612) 483-1911 (fax)
Manufactures battery chargers for a wide variety of electronics applications.

Aura
2335 Alaska Ave., El Segundo, CA 90245 USA
(800) 909-AURA (tel)
Manufactures cushioned audio accessories so that you can feel the audio from a computer.

Aurora Computech
4570 Enterprise St., Fremont, CA 94538 USA
(800) 852-3344 (tel)
(510) 440-1628 (tel)
(510) 490-3147 (fax)
Visa, MasterCard, COD ($5)
Sells computer parts and boards, such as cables, switch boxes, power supplies, CD-ROMs, audio cards, etc.

Auskits
Amblecote Crescent, Mulgrave, VIC 3170, Australia
Sells amateur radio kits, including QRP transmitters, etc.

Auspex Systems
5200 Great America Pkwy., Santa Clara, CA 95054 USA
(800) 735-3177 (tel)
(408) 986-2000 (tel)
(408) 986-2020 (fax)
Manufactures file servers for a wide variety of professional computer network applications.

Austin Amateur Radio Supply
5325 N., I-35, Austin, TX 78723 USA
(800) 423-2604 (tel)
(512) 454-2994 (tel)
(512) 454-3069 (fax)
Visa, MasterCard, Discover
Sells a wide variety of amateur radio equipment and accessories, including amateur radio transceivers, receivers, antenna tuners, towers, antennas, power supplies, amplifiers, etc.

Austin Antenna
10 Main St., Gonic, NH 03839 USA
(603) 335-6339 (tel)
(603) 335-1756 (fax)
Manufactures and sells antennas for VHF/UHF and microwave applications.

Austin Direct
2121 Energy Dr., Austin, TX 78758 USA
(512) 339-3500 (tel)
(512) 454-1357 (fax)
(800) 874-7182 (tel)
(800) 833-4472 (gov't sales)
(800) 622-5506 (Fortune 1000 sales)
(800) 341-7500 (fax info)
Visa, MasterCard, Discover, American Express
Manufactures and sells laptop and desktop systems. Offers 24-hour technical support and a 30-day money-back guarantee.

Autech
4-14-3-Minami-Azabu, Minato-ku, Japan
Sells head units, speakers, amplifiers, receivers, cassette decks, CD players, etc. for home audio and car audio applications.

Autek Research
P.O. Box 8772, Madeira Beach, FL 33738 USA
(813) 886-9515 (tel)
Visa, MasterCard
Manufactures and sells an RF analyzer, SWR/ wattmeter, and an SSB/CW/AM filter for amateur radio applications.

Authorized Audio/Video/Car
2898 N. University Dr., Ste. #35, Coral Springs, FL 33065 USA
(800) 348-7799 (tel)
Visa, MasterCard
Sells a wide variety of high-end home audio, auto audio, and home video equipment, such as speakers, cassette decks, amplifiers, preamplifiers, etc.

Authorized Parts Company
208 Berg St., Algonquin, IL 60102 USA
(800) 654-6464 (tel)
(708) 658-0582 (fax)
Manufactures high-value capacitors for car stereo applications.

Auto-ID Products
150 Newmark Rd., West Grove, PA 19390 USA
(800) 876-0743 (tel)
(610) 869-4388 (fax)
Sells UPC scanners, etc. for computerized business applications.

Automated Control Concepts
3535 Rt. 66, Neptune, NJ 07753 USA
(908) 922-6611 (tel)
(908) 922-9611 (fax)
Manufactures ruggedized computer workstations for industrial applications.

Automated Tech Tools
8515 Freeway Dr., Unit B, Macedonia, OH 44056 USA
(800) 413-0767 (tel)
(216) 468-0771 (tel)
(216) 468-3313 (fax)
Visa, MasterCard
Specializes in the sales of computer design products, including printers, plotters, monitors, and digitizing tablets.

Automation & Electronics Engineering (AEE)
13667 Floyd Cir., Dallas, TX 75243 USA
(800) 527-4596 (tel)
Manufactures telephone controls for a wide variety of communications applications.

Autogram
1500 Capital Ave., Plano, TX 75074 USA
(800) 327-6901 (tel)
(214) 423-6334 (fax)
Manufactures signal routing and control products and audio mixing consoles for professional broadcasting.

Autotek
855 Cowan Rd., Burlingame, CA 94010 USA
(415) 692-2444 (tel)
(415) 692-2448 (fax)
Manufactures amplifiers for mobile audio applications.

Autotime
6605 SW Macadam, Portland, OR 97201 USA
(503) 452-8577 (tel)
(503) 452-0208 (fax)
(503) 452-8495 (tel)
info@autotime.com
http://www.teleport.com/_7Eautotime.com
Sells computer memory chip adapter boards and converters.

Autronic
Schörli-Hus, CH-8600 Dübendorf, Switzerland
Sells head units, speakers, amplifiers, receivers, cassette decks, CD players, etc. for home audio and car audio applications.

Avalanche
2630 Nova Dr., Dallas, TX 75229 USA
(214) 484-8881 (tel)
(214) 484-8819 (fax)

Avalanche

Manufactures amplifiers and loudspeakers for car stereo applications.

Avalon Design
P.O. Box 5976, San Clemente, CA 92673 USA
(714) 492-2000 (tel)
(714) 492-4284 (fax)
Manufactures class-A microphone preamplifiers, preamplifier remotes, and class-A direct boxes for professional studio recording.

Avalon Micro Liquidators
688 #D, Wells Rd., Boulder City, NV 89005 USA
(800) (610) 1012 (tel)
(702) 293-2300 (tel)
(702) 293-4453 (fax)
Visa, MasterCard, Discover, American Express, COD ($5)
Sells computer parts and boards, such as memory, math coprocessor, and CPU chips.

Avcom of Virginia
500 Southlake Blvd., Richmond, VA 22236 USA
(804) 794-2500 (tel)
(804) 794-8284 (fax)
(804) 379-0500 (fax)
Manufactures test equipment, including spectrum analyzers, for a wide variety of electronics testing applications.

AVEL Transformers
47 S. End Plaza, New Milford, CT 06776 USA
(203) 355-4711 (tel)
(203) 354-8597 (fax)
Manufactures toroid line transformers--especially for high-end audio applications.

Aventel OY
Kuoppatie 4, 00730 Helsinki, Finland
358-0-346-3415 (tel)
358-0-346-3540 (fax)
Sells receivers, antennas, and accessories for shortwave radio-listening applications.

AVerMedia
47923A Warm Springs Blvd., Fremont, CA 94539 USA
(510) 770-9899 (tel)
(510) 770-9901 (fax)
(604) 278-3224 (Canada tel)
(604) 278-2909 (Canada fax)
Manufactures PC/Mac to TV converters for desktop video applications.

AVG
Birostraße 8-10, A-1232 Vienna, Austria
Sells head units, speakers, amplifiers, receivers, cassette decks, CD players, etc. for home audio and car audio applications.

AVI
2328 Durfee Ave., Unit J, El Monte, CA 91732 USA
(818) 442-8755 (tel)
Sells a wide variety of blank video tapes, including VHS, S-VHS, Hi-8, etc.

Aviation Surplus
2603 Oakcrest, Waukesha, WI 53188 USA
(800) SKY-KING (tel)
(414) 549-6709 (tel)
Sells aviation surplus for professional aviation applications.

Avital Auto Security
1535 Barclay Blvd., Buffalo Grove, IL 60089 USA
(800) 9-AVITAL (tel)
(800) 361-3444 (tel)
(525) 605-0382 (Mexico tel)
(582) 237-9074 (South America tel)
+48-2-679-4449 (Poland tel)
+42-5-43-21-61-01 (Czech Republic tel)
+42-7-328-164 (Slovakia tel)
+36-1-270-0402 (Hungary tel)
+62-21-625-5333 (Indonesia tel)
+65-744-7866 (Singapore tel)
+63-241-9601 (Phillipines tel)
avisales@avitalautosecurity.com
http://avitalautosecurity.com
Manufactures systems for car security applications.

AVR Systems
372 Main St., Watertown, MA 02172 USA

(617) 924-0660 (Boston tel)
(203) 289-9475 (CT tel)
Sells a wide variety of new and used audio equipment for professional recording studios and live music performances.

AVT
1310 S. Dixie Hwy., Ste. 18W, Pompano Beach, FL 33060 USA
(800) 741-2490 (tel)
(305) 785-2490 (service tel)
(305) 785-3202 (tel)
Sells refurbished monitors for a wide variety of computer applications.

Avtec
4335 Augusta Hwy., Gilbert, SC 29054 USA
(803) 892-2181 (tel)
(803) 892-3715 (fax)
Manufactures digital switching systems for professional voice communications.

AVX
P.O. Box 867, Myrtle Beach, SC 29578 USA
(803) 946-0414 (tel)
Manufactures surface-mount capacitors for a wide variety of electronics applications.

Axil Computers
3151 Coronado Dr., Santa Clara, CA 95054 USA
(408) 486-5700 (tel)
(408) 654-5720 (fax)
Manufactures workstations for a wide variety of professional computer network applications.

Axiom
Clifton, NJ 07014 USA
(800) 526-7452 (tel)
Manufactures battery chargers, antennas, etc. for cellular telephone applications.

Axiom Technology
13791 Roswell Ave., Unit D, Chino, CA 91710 USA
(909) 464-1881 (tel)
(909) 464-1882 (fax)
Manufactures single-board computers for professional and industrial computer applications.

Axionix
844 S. 200 E., Salt Lake City, UT 84111 USA
(800) 866-9797 (tel)
Manufactures portable multimedia systems with CD-ROM players for portable computer applications.

Axis Communications
99 Rosewood Dr., Ste. 170, Danvers, MA 01923 USA
(800) 444-AXIS (tel)
(508) 777-7957 (tel)
(508) 777-9905 (fax)
Manufactures print servers for professional computer network applications.

AXM
11791 Loara St., Ste. B, Garden Grove, CA 92640 USA
(800) 755-7169 (tel)
(714) 638-8807 (tel)
(714) 638-9556 (fax)
Sells a variety of citizens band radio equipment.

Axsys Comunicaciones
Planes 1057, 1405 Buenos Aires, Argentina
(541) 582-2915 (tel)
(541) 582-1372 (fax)
Sells a wide variety of equipment for professional cable TV applications.

Azden
147 New Hyde Park Rd., Franklin Square, NY 11010 USA
(800) 643-7655 (tel)
(516) 328-7501 (tel)
(516) 328-7506 (fax)
Visa, MasterCard
Manufactures amateur and commercial radio VHF and UHF transceivers and accessories and VHS wireless microphones for professional videography.

Azonix
900 Middlesex Tnpk., Bldg. 6, Billerica, MA 01821 USA

Azonix

(800) 365-1663 (tel)
(508) 670-6300 (tel)
(508) 670-8855 (fax)
Manufactures data acquisition, analysis, control systems, and plug-in boards for professional industrial applications.

Babcock Display Products Group

14930 E. Alondra Blvd., La Mirada, CA 90638 USA
(714) 944-6500 (tel)
(714) 994-3013 (fax)
Manufactures dc-plasma displays for a wide variety of electronics applications.

BAF Communications

314 Northstar Ct., Sanford, FL 32771 USA
(800) 633-8223 (tel)
(407) 324-8250 (tel)
(407) 324-7860 (fax)
Manufactures satellite news vans for professional broadcasting.

Bagnall Electronics

179 May, Fairfield, CT 06430 USA
Sells digital multimeter kits.

Ballard Technology

3229A Pine St., Everett, WA 98201 USA
(800) 829-1553 (tel)
(206) 339-0281 (tel)
(206) 339-0915 (fax)
Manufactures databus boards for computerized avionics applications.

Bang & Olufsen

1200 Business Center Dr., #100, Mt. Prospect, IL 60056 USA
(800) 323-0378 (tel)
Manufactures complete systems, amplifiers, and loudspeakers for home stereo applications; also manufactures telephones and answering machines.

Barbetta

(805) 529-3607 (tel)
(818) 992-4633 (fax)
barbetta1@aol.com
http://www.magicisland.com/barbetta/
Manufactures loudspeakers for a wide variety of professional audio and musical applications.

Barco Chromatics

2558 Mountain Industrial Blvd., Tucker, GA 30084 USA
(404) 493-7000 (tel)
Manufactures ruggedized flat-panel displays for a wide variety of industrial computer applications.

Barco Communications Systems

Th. Sevenslaan 106, B-8500 Kortrijk, Belgium
+32-56-233-388 (tel)
+32-56-233-334 (fax)
Manufactures TV modulators for professional TV broadcasting applications.

Bar Code Discount Warehouse

6862 Engle Rd. #212
Middleburg Hts., OH 44130 USA
(800) 888-2239 (tel)
(216) 826-0283 (tel)
(216) 826-0653 (fax)
Sells cash drawers, UPC scanners, receipt printers, etc. for computerized business applications.

Barcus-Berry

5381 Production Dr., Huntington Beach, CA 92649 USA
(800) 854-6481 (tel)
(714) 897-6766 (tel)
(714) 898-9211 (tel)
(714) 896-0736 (fax)
Manufactures piano pickups for a wide variety of musical applications.

N. R. Bardwell

288 Abbeydale Rd., Sheffield S7 1FL UK
0114-2552886 (tel)
0114-2500689 (fax)
Sells a wide variety of electronics components, including switches, capacitors, diodes, transistors, voltage regulators, etc. Offers a list for a stamp.

Barfield

4101 NW 29th St., Miami, FL 33142 USA

(800) 321-1039 (tel)
(305) 871-3900 (tel)
(800) 241-6769 (tel)
(404) 761-0114 (tel)
Manufactures test equipment for professional aviation applications.

Barker & Williamson
10 Canal St., Bristol, PA 19007 USA
(215) 788-5581 (tel)
Manufactures electronics components for amateur radio use, antenna items, etc.

Barksdale
3211 Fruitland Ave., Los Angeles, CA 90058 USA
(800) 835-1060 (tel)
(213) 589-3463 (fax)
Manufactures pressure sensors, monitors, and controls for industrial applications.

Barnett's Computers
475 Fifth Ave., New York, NY 10017 USA
(800) 931-7070 (tel)
(212) 447-1407 (tel)
(212) 447-7055 (fax)
Sells personal laptop computers, scanners, printers, portable data drives, etc. for portable personal computer applications.

Barry Electronics
540 Broadway, New York, NY 10012 USA
(800) 990-2929 (tel)
(212) 925-7000 (tel)
(212) 925-7001 (fax)
Sells a variety of shortwave receivers, amateur radio equipment, antennas, accessories, and literature.

Bartek Electronics
P.O. Box 410344, Charlotte, NC 28241 USA
(800) 960-8378 (tel)
(704) 588-9883 (tel)
(704) 588-5991 (fax)
Buys, sells, and trades equipment for a wide variety of electronics testing applications.

Bartolini
P.O. Box 934, Livermore, CA 94551 USA
(510) 443-1037 (tel)
Manufactures guitar pickups for professional and personal musical applications.

Base Station
1839 E. St., Concord, CA 94520 USA
(510) 685-7388 (tel)
Sells a variety of citizens band radio equipment, including base, mobile, and handheld transceivers, antennas, and accessories.

BASF
9 Oak Park Dr., Bedford, MA 01730 USA
(800) 225-4350 (US tel)
(800) 446-BASF (fax)
(800) 661-8273 (Canada tel)
Manufactures a variety of analog and digital tape formats for consumer and professional studio recordings.

Basic Electrical Supply & Warehousing
P.O. Box 8180, Bartlett, IL 60103 USA
(800) 577-8775 (tel)
(708) 582-2099 (tel)
Sells cable TV equipment.

Bason Hard Drive Warehouse
20130 Plummer St., Chatsworth, CA 91311 USA
(800) 238-4453 (tel)
Visa, MasterCard
Sells computer parts and boards, such as hard drives, memory chips, and SCSI adapters. Offers a 30-day money-back guarantee.

Basslines
5427 Hollister Ave., Santa Barbara, CA 93111 USA
(805) 964-9610 (tel)
Manufactures in-bass tone circuits for professional and personal musical applications.

Batima
118 Rue du Marechal Foch, F-67380 Lingolsheim, France

Batima

088-780012 (tel)
Sells equipment for shortwave radio receiving applications.

Batteries, etc.

P.O. Box 4800, Wayne, NJ 07474 USA
(800) 697-9900 (tel)
Visa, MasterCard, Discover, American Express
Sells rechargeable batteries for computers and cellular and cordless telephones, camcorders, etc.

Battery-Biz

31352 Via Colinas, #104, Westlake Village, CA 91362 USA
(800) 848-6782 (tel)
(818) 706-3234 (fax)
Visa, MasterCard
Sells batteries for laptop computers and uninterruptible power systems.

Battery Engineering

1636 Hyde Park Ave., Hyde Park, MA 02136 USA
(617) 361-7555 (tel)
(617) 361-1835 (fax)
Manufactures batteries for a wide variety of electronics applications.

Battery Express

713 Gladstone St., Parkersburg, WV 26101 USA
(800) 666-2296 (tel)
(304) 428-2296 (tel)
(304) 428-2297 (fax)
Visa, MasterCard, Discover, American Express, COD
Sells batteries for a wide variety of electronics applications, including laptop computers, camcorders, cordless phones, radio-control models, etc.

Battery-Tech

28-25 215 Pl., Bayside, NY 11360 USA
(800) 442-4275 (tel)
(718) 631-4275 (tel)
(718) 631-5117 (fax)
Visa, MasterCard, Discover
Sells a wide variety of batteries and chargers for amateur radio and camcorder applications. Offers a free catalog.

Baumer Electric

122 Spring St., C-6, Southington, CT 06489 USA
(800) 937-9336 (tel)
Manufactures proximity switches, encoders, ultrasonic sensors, etc. for a wide variety of professional industrial applications.

Baycom

400 Daily Ln., P.O. Box 5210, Grants, OR 97527 USA
(800) 822-9722 (tel)
Manufactures packet/modems for amateur radio use.

Bay Networks

4401 Great American Pkwy., Santa Clara, CA 95054 USA
(800) 8-BAYN.E.T (tel)
http://www.baynetworks.com
Manufactures hubs, switches, and routers for professional network computer applications.

B and B Enterprises

208 S. Pulaski St., Baltimore, MD 21223 USA
(410) 566-5388 (tel)
Sells discounted Radio Shack equipment for a wide variety of electronics applications.

BBE Sound Inc.

5500 Bolsa Ave., Ste. 245, Huntington Beach, CA 92649 USA
(714) 897-6766 (tel)
(714) 895-6728 (fax)
Manufactures audio processors for a wide variety of professional audio and musical applications.

BC Communications

The 211 Bldg.-Depot Rd., Huntington Stn., NY 11746 USA
(800) 924-9884 (tel)
(516) 549-8833 (tel)
(516) 549-1277 (tel)
(516) 549-8820 (fax)
Sells a variety of amateur radio equipment, including shortwave and VHF/UHF transceivers, Morse code keyers, antennas, power supplies, DSPs, antenna tuners, etc.

Beach
203 Rt. 22 E., Greenbrook, NJ 08812 USA
(800) 634-1811 (tel)
(908) 424-1101 (info tel)
(908) 424-1105 (fax)
(908) 424-1103 (service tel)
Visa, MasterCard, Discover, American Express, COD
Sells home, amateur, and professional audio and video equipment, such as camcorders, video editors, karaoke players, VCRs, cables, etc.

Beals Brothers
2165 Lehigh Station Rd., Pittsford, NY 14534 USA
(716) 334-3815 (tel)
(716) 334-8215 (fax)
Manufactures amplifiers for car stereo applications.

Bear Labs
4 Jennings Rd., Hannacroix, NY 12087 USA
(518) 756-9894 (tel)
(518) 756-3092 (fax)
Manufactures and sells high-end audio amplifier kits.

Bear Mountain Audio
12700 Yukon Ave., Hawthorne, CA 90250 USA
(310) 639-1183 (tel)
(310) 639-1187 (fax)
Manufactures amplifiers for mobile audio applications.

Bear Rock Technologies
4140 Mother Lode Dr., Ste. 100, Shingle Springs, CA 95682 USA
(800) 232-7625 (tel)
(916) 672-1103 (tel)
(916) 672-0244 (fax)
Sells UPC scanners, etc. for computerized business applications.

Becker Autoradio
16 Park Way, Upper Saddle River, NJ 07458 USA
(201) 327-3434 (tel)
(201) 327-9061 (tel)
(201) 327-2084 (fax)
Manufactures car audio systems.

Bedrock Amplification
1600 Concord St., Framingham, MA 01701 USA
(508) 877-4055 (tel)
(508) 877-4125 (fax)
Manufactures amplifiers for professional and personal musical applications.

BEI Sensors & Systems
2700 Systron Dr., Concord, CA 94518 USA
(800) 227-1625 (tel)
(510) 671-6601 (tel)
(510) 671-6464 (cust serv tel)
(510) 671-6647 (fax)
44-1304216-281 (Europe tel)
44-1304214-638 (Europe fax)
Manufactures rotation sensors for professional industrial applications.

Belden Wire and Cable
P.O. Box 19880, Richmond, IN 47375 USA
(800) BELDEN-1 (tel)
(800) BELDEN-4 (tel)
(317) 983-5200 (tech)
11611 N. Meridian St., Ste. 300, Carmel, IN 46032 USA
(800) 246-2673 (tel)
Manufactures cable and wire for a wide variety of electronics applications.

Belkin Components
1303 Walnut Park Way, Compton, CA 90220 USA
(310) 898-1100 (tel)
(310) 898-1111 (fax)
Manufactures SCSI adapters for a wide variety of computer applications.

John Bell
1381 Saratoga St., Minden, NV 89423 USA
(702) 267-2704 (tel)
Manufactures and sells a repeater controller with autopatch for amateur radio applications.

Floyd Bell
897 Higgs Ave., Columbus, OH 43212 USA
(614) 294-4000 (tel)
(614) 291-0823 (fax)
Manufactures piezoelectric alarms for a wide variety of electronics applications.

F.W. Bell
6120 Hanging Moss Rd., Orlando, FL 32807 USA
(407) 678-6900 (tel)
(407) 677-5765 (fax)
Manufactures noncontact milliameters for a wide variety of electronics applications.

Bellari
5143 S. Main St., Salt Lake City, UT 84107 USA
(801) 263-9053 (tel)
(801) 263-9068 (fax)
Manufactures tube microphone preamplifiers, compressor/limiters, and direct boxes for a wide variety of professional audio and musical applications.

Bel Merit
26242 Dimension Dr., Ste. 110, Lake Forest, CA 92630 USA
(714) 586-3700 (tel)
(714) 586-3399 (fax)
Manufactures test instruments for a wide variety of electronics testing applications.

Bencher
831 N. Central Ave., Wood Dale, IL 60191 USA
(708) 238-1183 (tel)
(708) 238-1186 (fax)
Manufactures and sells low-pass filters for amateur radio applications. Offers a free brochure.

Benchmark Computer Systems
10513 W. 84th Terr., Lenexa, KS 66214 USA
(800) 279-9192 (tel)
(913) 894-4860 (tel)
(913) 894-4824 (fax)
Manufactures data storage systems for a wide variety of professional computer network applications.

Benchmark Media Systems
5925 Court Street Rd., Syracuse, NY 13206 USA
(800) 262-4675 (tel)
(315) 437-8119 (fax)
info@benchmarkmedia.com
Manufactures assignable mixers for professional broadcast applications.

Benchmarq Microelectronics
17919 Waterview Pkwy., Dallas, TX 75252 USA
(800) 966-0011 (tel)
(214) 437-9195 (tel)
(214) 437-9198 (fax)
benchmarq@benchmarq.com
http://www.benchmarq.com
Manufactures semiconductors for a wide variety of electronics applications.

Bendix/King
400 N. Rodgers Rd., Olathe, KS 66062 USA
(913) 782-0400 (tel)
Manufactures GPS and IFR systems for professional aviation applications.

Berger Brothers
209 Broadway (Rt. 110), Amityville, NY 11701 USA
(800) 262-4160 (tel)
(516) 264-4160 (tel)
(516) 264-1007 (fax)
Sells a wide variety of photographic and video equipment, including cameras, camcorders, VCRs, etc. Offers a full money-back guarantee and a 240-page catalog.

Berger Electronics
(800) 588-KITS (tel)
(510) 229-3441 (fax)
Manufactures and sells electronics kits, including FM transmitters, thermometers, audio links, logic probes, etc.

Bergquist
5300 Edina Industrial Blvd., Minneapolis, MN 55439 USA
(800) 347-4572 (tel)
(612) 835-2322 (tel)
(612) 835-0430 (fax)
Manufactures heatsinks for a wide variety of electronics applications.

Bering Technology
1357 Dell Ave., Campbell, CA 95008 USA
(800) 237-4641 (tel)
(408) 379-6900 (tel)
(408) 374-8309 (fax)

Manufactures CD-ROM WORM drives for a wide variety of professional computer network applications.

Berler Pro-Audio
4069 Joseph Dr., Waukegan, IL 60087 USA
(847) 263-6400 (tel)
(847) 263-6455 (fax)
Sells a variety of audio equipment, such as microphones, MiniDisk recorders, samplers, mixing consoles, etc., for professional studio recording and live music performances.

Bertan High Voltage
121 New South Rd., Hicksville, NY 11801 USA
(800) 966-2776 (tel)
(516) 433-3110 (tel)
(516) 935-1766 (fax)
Manufactures high-voltage power supplies for professional electronics applications.

Best Data Products
21800 Nordhoff St., Chatsworth, CA 91311 USA
(800) 632-2378 (tel)
(818) 773-9600 (tel)
(818) 773-9619 (fax)
Manufactures modems for a wide variety of computer applications.

Best Power
P.O. Box 280, Necedah, WI 54646 USA
(800) 356-5794 (tel)
(608) 565-7200 (tel)
(608) 565-2221 (fax)
Manufactures uninterruptible power supplies for personal desktop computer applications.

Better Business Systems
7949 Woodley Ave., Van Nuys, CA 91406 USA
(800) 697-9993 (tel)
(818) 373-7250 (tel)
(818) 373-7525 (tech support)
(818) 376-1581 (fax)
Visa, MasterCard, Discover, American Express
Sells CD-ROM drives, CD recorders, CD-Rs, etc. for professional and personal computer applications.

Better Reception
2471 Montpelier Rd., Columbia, KY 42728 USA

Distributes Alpermann + Velte Video editing equipment to North America.

BEXT
1045 10th Ave., San Diego, CA 92101 USA
(619) 239-8462 (tel)
(619) 239-8474 (fax)
Manufactures programmable FM and TV exciters, translators, and amplifiers for professional broadcasting.

beyerdynamic
56 Central Ave., Farmingdale, NY 11735 USA
(800) 293-4463 (tel)
Manufactures a variety of microphones for professional studio recording.

Beyond Oil
P.O. Box 99, Mill Valley, CA 94942 USA
(415) 388-0838 (tel)
(415) 388-9178 (fax)
73144.237@compuserve.com
Sells test equipment, including oscilloscopes, signal generators, power supplies, etc., for a wide variety of electronics testing applications.

BHA Trading
15 Oakwell Close, Stevenage, Herts SG2 8UG UK
01438-816474 (tel)
01438-812569 (fax)
Sells a wide variety of electronics components, including semiconductors, relays, resistors, connectors, etc. Offers a list for a SASE.

B&H Audio-Video
119 W. 17th St., New York, NY 10011 USA
(800) 947-5508 (order tel)
(212) 444-6698 (order tel)
(800) 947-9003 (order fax)
(212) 444-5001 (order fax)
(800) 221-5743 (order info tel)
(212) 807-7479 (order info tel)
(212) 366-3738 (order info fax)
76623.570@compuserve.com
Visa, MasterCard, Discover
Sells a wide variety of audio and video equipment for professional broadcasting and professional recording,

B&H Audio-Video

including monitor speakers, mixers, power amplifiers, DAT recorders, digital audio workstations, microphones, audio processors, etc.

B&H Sales and Service
707 N. Baltimore, Derby, KS 67037 USA
(316) 788-4225 (tel)
Sells a variety of amateur radio equipment, including antenna tuners, shortwave transceivers, DSPs, antenna switches, antennas, packet equipment, etc.

B&K Components
2100 Old Union Rd., Buffalo, NY 14227 USA
(716) 656-0026 (tel)
Manufactures tuner/preamplifiers for high-end home theater applications.

B+K Precision
6470 W. Cortland St., Chicago, IL 60635 USA
(312) 889-1448 (tel)
Manufactures a variety of test equipment for professional video applications.

BIC America
883 E. Hampshire, Stow, OH 44224 USA
(216) 928-2011 (tel)
(216) 928-8392 (fax)
Manufactures loudspeakers for home stereo and home theater applications.

Big Blue Products
586 New York Ave., Huntington, NY 11743 USA
(516) 351-3600 (tel)
(516) 351-3670 (tel)
Buys and sells IBM, Compaq, and Apple personal computer parts.

Big Briar
554-C Riverside St., Asheville, NC 28801 USA
(800) 948-1990 (tel)
(704) 251-0090 (tel)
(704) 254-6223 (fax)
Visa, MasterCard
Manufactures theremin kits for a wide variety of home and professional audio applications.

Bilal
137 Manchester Dr., Florissant, CO 80816 USA
(719) 687-0650 (tel)
Manufactures and sells shortwave antennas.

Billingsley Magnetics
2600 Brighton Dam Rd., Brookeville, MD 20833 USA
(301) 774-7707 (tel)
(301) 774-0745 (fax)
magnetic@access.digex.net
Manufactures altitude control sensors, magnetometers, gradiometers, and heading error-correction systems for professional aviation applications. Offers a free brochure.

Bi-Link Computer
11606 E. Washington Blvd., Whittier, CA 90606 USA
(800) 888-5369 (tel)
(310) 692-5345 (tel)
(310) 695-5166 (tech support tel)
(310) 695-9623 (fax)
Visa, MasterCard
Manufactures rack-mount and portable workstations for professional and industrial computer applications.

Binary Technology
P.O. Box 541, Carlisle, MA 01741 USA
(508) 369-9556 (tel)
(508) 369-9549 (fax)
Visa, MasterCard
Manufactures microcontroller boards for a wide variety of professional computer applications.

Bird Electronic
30303 Aurora Rd., Solon, OH 44139 USA
(216) 248-1200 (tel)
(216) 248-5426 (fax)
(805) 646-7255 (tel)
(805) 646-0275 (fax)
353-61-360583 (Europe tel)
353-61-360585 (Europe fax)
65-2992537 (Pan-Asia tel)
65-2998509 (Pan-Asia fax)
Manufactures antenna testers, wattmeters, etc. for a wide variety of radio testing applications.

Bisme Computer
1298 Reamwood Ave., Sunnyvale, CA 94089 USA
(800) 899-6430 (tel)
(408) 745-7206 (service tel)
(408) 745-7203 (tel)
(800) 745-1670 (fax)
Sells computer parts and boards, such as floppy drives, hard drives, monitors, motherboards, video cards, modems, keyboards, etc. Offers a 30-day money-back guarantee.

BIT Computer
21068 Commerce Pointe Dr., Walnut, CA 91789 USA
(800) 935-0209 (tel)
(909) 598-0086 (tel)
(909) 598-1391 (tech support/info tel)
(909) 598-1544 (fax)
bit@intelisys.com
http://www.bit.l1.com
Visa, MasterCard
Manufactures and sells personal desktop computers. Offers a 30-day money-back guarantee.

Bits & Byte Technologies
146 W. 29th St., 10th Fl., New York, NY 10001 USA
(212) 967-1616 (tel)
(212) 967-0672 (fax)
Visa, MasterCard
Sells personal desktop and laptop computers.

Bittware Research Systems
33 N. Main St., Concord, NH 03301 USA
(800) 848-0436 (tel)
(603) 226-6667 (fax)
bittware@bittware.com
Manufactures DSP computer boards and modules for professional and industrial computer applications.

B.K. Electronics
Unit 1, Comet Way, Southend-on-Sea, Essex SS3 9NR UK
01702-527572 (tel)
01702-420243 (fax)
Visa, Access
Manufactures loudspeakers for a wide variety of professional and home audio applications.

Black Cat Products
5930 E. Royal Ln., Ste. 291, Dallas, TX 75230 USA
(800) 929-5889 (tel)
Manufactures theremins for personal and professional audio and musical applications.

Black Feather Electronics
645 Temple 7A, Long Beach, CA 90814 USA
(800) 526-3717 (tel)
(310) 434-5641 (tel)
(310) 434-9142 (fax)
Manufactures and sells video stabilizers to eliminate video problems caused by copyguards on video tapes.

Blackrock Hardware Store
2501 Blichmann Ave., Grand Junction, CO 81505 USA
(800) 262-2002 (tel)
Manufactures and sells personal desktop computers.

Blade
3641 McNicoll Ave., Scarborough, ON M1X 1G5 Canada
(416) 321-1800 (tel)
(416) 321-1500 (fax)
Manufactures DSPs, equalizers, and power amplifiers for car stereo applications.

Blaupunkt
P.O. Box 4601 N., Suburban, IL 60197 USA
(708) 865-5200 (tel)
(708) 865-5209 (fax)
Manufactures head units, amplifiers, loudspeakers, etc. for car stereo applications.

Bliley Electric
2545 W. Grandview Blvd., P.O. Box 3428, Erie, PA 16508 USA
(814) 838-3571 (tel)
(814) 833-2712 (fax)
Manufactures crystal oscillators for a wide variety of telecommunications applications.

Blitz-Safe
260 Columbia Ave., #18, Ft. Lee, NJ 07024 USA

Blitz-Safe
(800) 587-7233 (tel)
(201) 224-0101 (fax)
Manufactures systems for car security applications.

Blonder Tongue Laboratories
1605 E. Iola, Broken Arrow, OK 74012 USA
(800) 331-5997 (tel)
Sells receivers, modulators, traps, connectors, etc. for a professional cable TV applications.

BLR Communications
P.O. Box 976, Ellicott City, MD 21041 USA
(800) 442-9199 (tel)
(410) 750-1400 (tel)
(410) 750-0052 (fax)
Buys and sells a wide variety of new and used equipment for professional cable and satellite TV applications.

Blue Line Communications Warehouse
808 W. Vermont, Anaheim, CA 92805 USA
(800) 258-7810 (tel)
(714) 999-2441 (tech support tel)
(714) 999-2442 (fax)
(714) 999-1775 (BBS)
blueline@bluecomm.com
Visa, MasterCard, Discover, American Express
Sells computer accessories and network products, such as modems, acoustic couplers, etc.

Blue Moon Gadgets
6059 Allentown Blvd., #608, Harrisburg, PA 17112 USA
(800) 671-8117 (tel)
(717) 671-8650 (tel)
Visa, MasterCard
Manufactures radar jammers for mobile electronics applications.

Blue Star Computer
8 Tech Circle, Natick, MA 01760 USA
(508) 647-0800 (tel)
(508) 647-0899 (fax)
Sells personal computer parts and accessories, including system boards, hard drives, power supplies, network equipment, etc.

BMC
13700 Alton Pkwy., Ste. 154-282, Irvine, CA 92718 USA
Visa, MasterCard
(800) 532-3221 (tel)
(714) 586-2310 (tel)
(714) 586-3399 (fax)
Sells test equipment, including function generators, frequency counters, power supplies, audio generators, oscilloscopes, multimeters, etc. Offers a 1-year warranty.

BMC Communications
141-43 71st. Ave., Flushing, NY 11367 USA
(718) 575-1936 (tel)
(718) 544-3628 (fax)
Manufactures interfaces for a wide variety of computer applications.

BMG Engineering
9935 Garibaldi, Temple City, CA 91780 USA
Manufactures and sells a UHF/VHF direction-finding antenna.

BMI (Benjamin Michael Industries)
9445 Seven Mile Rd., Caledonia, WI 53108 USA
(414) 835-4299 (tel)
(414) 835-4298 (fax)
Manufactures and sells a cassette recorder activator and 24-hour time clocks.

BNFE
134R Newbury St., Rt. 1 S., Peabody, MA 01960 USA
(508) 536-2000 (tel)
(508) 536-9090 (fax)
Manufactures and sells personal desktop computers; also sells computer parts and boards, such as floppy drives, hard drives, monitors, video cards, modems, keyboards, etc.

Bob's CB
72 Eastern Ave., Malden, MA 02148 USA
(800) 473-9708 (tel)
Visa, MasterCard, Discover, American Express
Manufactures and sells audio compressors; also sells a wide variety of CB transceivers, antennas, and accessories.

Boca Computer Technology (BCT)
7400 N. Federal Hwy., Boca Raton, FL 33487 USA
(800) 689-3475 (tel)
(407) 994-8333 (tel)
(407) 994-0479 (tech support tel)
(407) 994-6001 (fax)
Visa, MasterCard, Discover, American Express
Manufactures personal desktop computers; also sells computer parts and boards, such as floppy drives, hard drives, video cards, modems, etc.

BOCA Research
6413 Congress Ave., Boca Raton, FL 33487 USA
(407) 997-6227 (tel)
(407) 997-0918 (fax)
Manufactures modems for a wide variety of computer telecommunications applications.

A. Böck
Mollardgasse 30, A-1060 Wien, Austria
Sells equipment for shortwave radio receiving applications.

Bodden Model Products
P.O. Box 8095, Redlands, CA 92375 USA
(909) 793-2514 (tel)
Manufactures and sells landing gear/door sequencers, servo motion controllers, etc. for radio-control model airplanes.

Boffin
2500 W. County Rd. 42, Ste. #5, Burnsville, MN 55337 USA
(603) 894-0595 (tel)
(612) 894-6175 (fax)
sales@boffin.com
http://www.boffin.com
Manufactures CD-ROM drive arrays for professional computer applications.

Bogen Communications
50 Spring St., Ramsey, NJ 07446 USA
(800) 646-1447 (tel)
Sells voice mail/fax mail storage systems for professional and personal computer applications.

Bogen Photo
565 E. Crescent Ave., Ramsey, NJ 07446 USA
(201) 818-9500 (tel)
Manufactures tripods for amateur and professional video systems.

Boger-funk
Grundesch 15, W-88326, Aulendorf-Steinenbach, Germany
Sells equipment for shortwave radio receiving applications.

Bonigor USA
8339 NW 66 St., Miami, FL 33166 USA
(305) 592-0428 (tel)
(305) 592-6841 (fax)
Sells battery chargers, power supplies, antennas, etc. for cellular telephone applications.

Boomerang Musical Products
P.O. Box 541595, Dallas, TX 75354 USA
(800) 530-4699 (tel)
Manufactures and sells audio samplers for professional and personal musical applications.

Borbely Audio
Melchior Fanger Strasse 34A, 82205 Gilching, Germany
+49-8105-5291 (tel)
+49-8105-24605 (fax)
Visa, MasterCard
Sells high-end audio components and kits, including preamplifiers, power supplies, crossovers, power amplifiers, etc.

Bose
The Mountain, Framingham, MA 01701 USA
(800) 845-BOS.E. (tel)
(508) 879-7330 (tel)
Manufactures loudspeakers and radios for high-end home audio and home theater applications.

Boss/AVA
3628 E. Olympic Blvd., Los Angeles, CA 90023 USA
(800) 999-1236 (tel)
(213) 881-0838 (tel)
(213) 881-0835 (fax)
Manufactures amplifiers, head units, crossovers, and loudspeakers for mobile audio applications.

Boston Acoustics
70 Broadway, Lynnfield, MA 01940 USA
(617) 592-9000 (tel)
(617) 592-6148 (fax)
Manufactures loudspeakers for high-end home audio and home theater applications.

Bostwick
125 Old Monroe Rd., Ste. 200, Bogart, GA 30622 USA
(706) 543-9494 (tel)
(706) 369-9519 (fax)
Manufactures loudspeakers for car stereo applications.

Bourns
1200 Colombia Ave, Riverside, CA 92507 USA
(909) 781-5500 (tel)
(909) 781-5700 (fax)
Manufactures miniature switches for a wide variety of electronics applications.

Box Hill Systems
161 Ave. of the Americas, Ste. 901, New York, NY 10013 USA
(800) 727-3863 (tel)
(212) 989-4455 (tel)
(212) 989-6817 (fax)
Manufactures CD-ROM WORM drives, CD-ROM recorders, and optical disk drives, etc. for a wide variety of professional computer network applications.

Boxlight
17771 Fjord Dr. N.E., Poulsbo, WA 98370 USA
(800) 762-5757 (tel)
(800) 497-4009 (government tel)
(800) 736-6956 (resellers tel)
(360) 779-7901 (international tel)
(800) 736-6954 (rental tel)
(360) 779-3299 (fax)
Visa, MasterCard, American Express
Manufactures LCD projection panels for computer multimedia applications. Offers a free catalog and a 30-day money-back guarantee.

Bozak Antenna
100 Church Hill, Waterford, NY 12188 USA
(518) 373-8069 (tel)
(518) 373-0701 (fax)
Manufactures FM antennas for amateur radio applications.

BPB Associates
1977 High St., Longwood, FL 32750 USA
(407) 330-3408 (tel)
(407) 330-3498 (fax)
Sells amplifiers for cellular telephone applications.

Bill Bradley Microphones
4911 Hamilton Rd., Medina, OH 44256 USA
(216) 723-6494 (tel)
(216) 723-6595 (fax)
Visa, MasterCard
bradleyb@bright.net
Buys and sells tube microphones for professional studio recording and live music performances.

Audio Intervisual Design
1155 N. LaBrea, W. Hollywood, CA 90038 USA
(213) 845-1155 (tel)
(213) 845-1170 (fax)
Sells a wide variety of professional studio recording equipment, including mixing consoles, editing systems, etc.

Brandess-Kalt
5441 N. Kadzi Ave., Chicago, IL 60625 USA
(312) 588-8601 (tel)
Manufactures lighting systems for amateur and professional video production.

Brant Computer
1611 S. Walter Reed Dr., Arlington, VA 22204 USA
(703) 521-2870 (tel)
(703) 521-2872 (fax)
Manufactures and sells personal desktop computers; also sells computer parts and boards, such as floppy drives, hard drives, monitors, motherboards, modems, cases, etc.

Brent Averill Enterprises
14300 Hortense St., Sherman Oaks, CA 91423 USA
(818) 784-2046 (tel)

(818) 784-0750 (fax)
Manufactures preamplifiers for professional studio audio applications.

Brigar Electronics
7-9 Alice St., Binghamton, NY 13904 USA
(607) 723-3111 (tel)
(607) 723-5202 (fax)
brigar2@aol.com
http://www.binghamton.com/brigar/
Sells surplus electronics, including meters, power supplies, answering machines, EEPROM chips, etc. for a wide variety of electronics applications. Minimum order: $30. Offers a free catalog.

Bright Creation International
Rm. 407, Kodak House II, 321 Java Rd., Hong Kong
(852) 2516-6922 (tel)
(852) 2561-7069 (fax)
Manufactures sound cards, MPEG cards, and multimedia upgrade kits for computer multimedia applications.

Bri-Tech A/V
3900 Veterans Hwy., Ste. 120, Bohemia, NY 11716 USA
(800) 467-7707 (tel)
Visa, MasterCard, Discover, American Express
Sells a wide variety of new and used high-end home audio and video equipment.

Brix
565 Display Way, Sacramento, CA 95838 USA
(800) 359-2749 (tel)
414 W. Bedford Ave., Ste. 101, Fresno, CA 93711 USA
(800) 733-2749 (tel)
Sells two-way radios.

Broadcast Devices
5 Crestview Ave., Peekskill, NY 10566 USA
(914) 737-5032 (tel)
(914) 736-6916 (fax)
Manufactures composite D/As and switchers for broadcast stations.

Broadcast Electronics
4100 N. 24th St., P.O. Box 3606, Quincy, IL 62305 USA
(217) 224-9600 (tel)
(217) 224-9607 (fax)
Manufactures a variety of AM and FM exciters and transmitters, cart machines, STLs, and digital studio systems for broadcast stations.

Broadcom
10924 Wilshire Blvd., Los Angeles, CA 90024 USA
(310) 443-4490 (tel)
Manufactures semiconductors for a wide variety of electronics applications.

Brookline Technologies
2035 Carriage Hill Rd., Allison Park, PA 15101 USA
(800) 366-9290 (tel)
Manufactures several volume control boxes for high-end home theater applications.

Brooktrout Technology
144 Gould St., Needham, MA 02194 USA
(617) 449-4100 (tel)
(617) 449-9009 (fax)
Manufactures fax boards for a wide variety of computer telecommunications applications.

Brother International
200 Cottontail Ln., Somerset, NJ 08875 USA
Manufactures printers for personal computers.

Bruckner Hobbies
2908 Bruckner Blvd., Bronx, NY 10465 USA
(800) 288-8185 (tel)
(718) 863-3434 (tel)
Sells sport and computer radios, batteries, and servos for radio-control model airplanes. Offers a free catalog.

Thomas J. Bruckner
P.O. Box 151, Livingston, NJ 07039 USA
(201) 227-1085 (tel)
(201) 227-8469 (fax)
Sells a variety of electron tubes, especially for radio transmitting applications.

Bryston
106 W. Lime Ave., Ste. 207, Monrovia, CA 91016

Bryston
USA
(800) 673-7899 (tel)
(818) 359-9672 (fax)
P.O. Box 2170, 677 Neal Dr., Peterborough, ON K9J 7Y4 Canada
(705) 742-5325 (tel)
(705) 742-0882 (fax)
57 Westmore Dr., Rexdale, ON M9V 3Y6 Canada
(416) 746-1800 (tel)
(416) 746-0308 (fax)
Manufactures high-end power amplifiers for audio, home theater, and professional recording applications.

BSI
9440 Telstar Ave., #4, El Monte, CA 91731 USA
(800) 872-4547 (tel)
(818) 442-7038 (information tel)
(818) 442-4527 (fax)
Manufactures and sells personal laptop, portable, and rack-mount computers.

B&S Sales
51756 Van Dyke St., #330, Shelby Twp, MI 48316 USA
(819) 566-7248 (tel)
(810) 566-7258 (fax)
Sells converters, converter/descramblers, and add-on descramblers for use with video. Offers a free calculator with every order.

B & T RC Products
508 Lake Winds Trail, Rougemont, NC 27572 USA
(909) 471-2060 (tel/fax)
Sells batteries for radio-control models. Offers a free catalog.

Buchla and Associates
P.O. Box 10205, Berkeley, CA 94709 USA
(510) 528-4446 (tel)
(510) 526-1955 (fax)
Manufactures electronic instruments for a wide variety of musical applications.

Buffalo
2805 19th St. S.E., Salem, OR 97302 USA
(800) 345-2356 (tel)
(503) 585-4505 (fax)
Manufactures and sells printer interfaces and memory boards for computer printers.

Bulldog Computer Products
3241-E Washington Rd., Martinez, GA 30907 USA
(800) 438-6039 (tel)
(708) 541-2394 (tel)
(708) 541-2394 (shipping tel)
(708) 541-6988 (fax)
Visa, MasterCard, Discover, American Express
Sells personal and laptop computers. Also sells computer parts, such as hard drives, monitors, tape drives, modems, network hardware, etc.

Bull Electrical
250 Portland Rd., Hove, Sussex BN3 5QT UK
01273-203500 (tel)
01273-323077 (fax)
Visa, American Express, Access, Switch
Manufactures and sells a wide variety of electronics kits, including transmitters, surveillance kits, timers, function generators, etc. Offers a 100-page catalog for a 45p stamp.

Burk Electronics
35 N. Kensington, LaGrange, IL 60525 USA
(708) 482-9310 (tel)
Sells a wide variety of amateur radio equipment and accessories, including amateur radio transceivers, receivers, antenna tuners, antennas, amplifiers, etc.

Burk Technology
7 Beaver Rd., Littleton, MA 01460 USA
(800) 255-8090 (tel)
(508) 486-0086 (tel)
(508) 486-0081 (tel)
danrau@burk.com
Manufactures wireless RF data links for broadcast stations.

Burlington A/V Recording Media
106 Mott St., Oceanside, NY 11572 USA
(800) 331-3191 (tel)
(516) 678-4414 (tel)
(516) 678-8959 (fax)

Visa, MasterCard, Discover, American Express
Sells a variety of professional audio/video tapes, data media, equipment, and accessories.

Burqhardt Amateur Center
182 N. Maple, P.O. Box 73, Watertown, SD 57201 USA
(800) 927-4261 (tel)
(605) 886-7314 (tel)
(605) 886-3444 (fax)
(605) 886-6914 (product info fax)
Visa, MasterCard, Discover
Sells a wide variety of amateur radio equipment and accessories, including amateur radio transceivers, receivers, antenna tuners, antennas, amplifiers, etc.

Burr-Brown
P.O. Box 11400, Tucson, AZ 85734 USA
(800) 548-6132 (tel)
(603) 746-1111 (tel)
(800) 548-6133 (fax)
Manufactures semiconductors for a wide variety of electronics applications. Offers free brochures.

Business Information Zygoun (BIZI)
144 Rangeway Rd., N. Billerica, MA 01862 USA
(508) 667-4926 (tel)
(508) 663-0607 (fax)
Sells new and used network equipment for professional computer applications.

Business System Solutions
981 Mt. Eden Rd., Three Kings, Auckland, New Zealand
P.O. Box 24-196, Royal Oak, Auckland, New Zealand
09-625-6091 (tel)
09-625-6671 (fax)
Sells laptop systems for a wide variety of personal computer applications.

Bustronic
44350 Grimmer Blvd., Fremont, CA 94538 USA
(510) 490-7388 (tel)
(510) 490-1853 (fax)
Manufactures backplane computer boards for professional and industrial computer applications.

Butternut Manufacturing
831 N. Central Ave., Wood Dale, IL 60191 USA
(708) 238-1854 (tel)
(708) 238-1186 (fax)
Manufactures and sells vertical antennas for amateur radio applications.

Buy-Comm
29669 N. 45th St., Cave Creek, AZ 85331 USA
(603) 585-3900 (tel)
(603) 585-6900 (fax)
Sells conventional and trunking equipment, paging transmitters, site equipment, etc. for cellular microwave telephone and other two-way radio communications.

BWC
8850 Rivers Ave., Ste. D, Charleston, SC 29406 USA
(803) 764-0474 (tel)
(803) 824-0278 (fax)
Sells modems, hard drives, floppy drives, motherboards, keyboards, etc. for a wide variety of personal computer applications. Offers a free price list.

B/W Controls
1080 N. Crooks Rd., Clawson, MI 48017 USA
(810) 435-0700 (tel)
(810) 435-8120 (fax)
Manufactures tank gauging transducers for a wide variety of electronics applications.

B&W Loudspeakers of America
54 Concord St., N. Reading, MA 01864 USA
(800) 370-3740 (tel)
(508) 664-2870 (tel)
(508) 664-4109 (fax)
Manufactures loudspeakers for high-end home audio and home theater applications.

Byers Chassis Kits
5120 Harmony Grove Rd., Dover, PA 17315 USA
(717) 292-4901 (tel)
Manufactures aluminum cabinet and chassis kits, antenna grounding kits, rack equipment cabinets, etc. for amateur radio applications.

Bytek
543 NW 77th St., Boca Raton, FL 33487 USA
(407) 994-3520 (tel)
(407) 994-3615 (fax)
Manufactures EPROM programmers for professional electronics applications.

C & A Audio
193 Bellevue Ave., Upper Montclair, NJ 07043 USA
(201) 744-0600 (tel)
Visa, MasterCard, Discover, American Express
Sells a wide variety of high-end home audio and video equipment.

Cable AML
(702) 363-5660 (tel)
(702) 363-2960 (fax)
Manufactures transmitters, receivers, and repeaters for professional cable TV applications.

Cable Link
280 Cozzins St., Columbus, OH 43215 USA
(614) 221-3131 (tel)
(614) 222-0581 (fax)
Sells remanufactured converters, line gear, and headend equipment for professional cable TV applications.

Cable Linx
(800) 501-0095 (tel)
Visa, MasterCard, Discover, American Express
Sells cable TV converters and descramblers.

Cablemaster
20710 Lassen St., Chatsworth, CA 91311 USA
(818) 341-9200 (tel)
(818) 341-3002 (fax)
Manufactures high-voltage cable testers for a variety of professional industrial applications.

Cables America
824 Leo St., Dayton, OH 45404 USA
(800) 348-USA4 (tel)
(800) FAX-USA5 (fax)
Visa, MasterCard, American Express
Manufactures cables, adapters, switch boxes, etc. for a wide variety of computer applications. Offers a free minicatalog and a 30-day money-back guarantee.

Cablescan
145 E. Emerson Ave., Orange, CA 92665 USA
(800) 898-5783 (tel)
(714) 998-1961 (tel)
(714) 921-2493 (fax)
Manufactures wire harness testers for professional computer applications.

Cables & Connector Technologies
P.O. Box 621, Dayton, OH 45405 USA
(800) 452-8770 (tel)
(513) 277-4304 (tel)
(513) 277-4369 (fax)
Visa, MasterCard, COD
Sells a wide variety of cables, connectors, and adapters for computer applications.

Cable Source International
3528 N. High St., #2, Columbus, OH 43214 USA
(614) 268-3700 (tel)
(614) 268-6818 (fax)
Buys and sells new, used, and refurbished equipment for cable television, satellite, and microwave communications applications.

Cables to Go
1501 Webster St., Dayton, OH 45404 USA
(800) 225-8646 (tel)
(800) 331-2841 (fax)
Visa, MasterCard, Discover, COD
Manufactures cables, adapters, switch boxes, etc. for a wide variety of computer applications. Offers a free minicatalog.

Cable Technologies International
2500 Office Center, #300, Willow Grove, PA 19090 USA
(215) 657-3300 (tel)
(215) 657-9578 (fax)
Sells a wide variety of equipment for professional cable TV applications.

Cabletron Systems
35 Industrial Way, P.O. Box 5005, Rochester, NH 03867 USA
(603) 332-9400 (tel)
(603) 332-7386 (fax)
Manufactures routers for professional computer network applications.

Cable Warehouse
10117 W. Oakland Park Blvd., #315, Sunrise, FL 33351 USA
(800) 284-8432 (tel)
(305) 749-3122 (tel)
Visa, MasterCard, American Express
Sells a variety of cable TV descramblers. Offers a free catalog, 24-hour shipments, and a 30-day money-back guarantee.

Cable X-Perts
113 McHenry Rd., #240, Buffalo Grove, IL 60089 USA
(800) 828-3340 (tel)
(708) 506-1886 (tech support)
Manufactures and sells coaxial cable, antenna wire, antennas, baluns, connectors, etc.

Cactus Logic
14120 E. Live Oak Ave., #A2, Bladwin Park, CA 91706 USA
(800) 847-1998 (tel)
(818) 337-4547 (tel)
(818) 337-0689 (fax)
Visa, MasterCard, American Express
Manufactures in-circuit emulators for a wide variety of computer applications.

Cadence
6519 Hwy. 9 N., Howell, NJ 07731 USA
(908) 370-5400 (tel)
(908) 370-5553 (fax)
Manufactures loudspeakers for mobile audio applications.

Caddock Electronics
1717 Chicago Ave., Riverside, CA 92507 USA
(909) 788-1700 (tel)
(909) 369-1151 (fax)
17271 N. Umpqua Hwy., Roseburg, OR 97470 USA
(503) 496-0700 (tel)
(503) 496-0408 (fax)
Manufactures a variety of electronics components, including power film resistors. Offers a free catalog.

Cadex Electronics
111-7400 MacPherson Ave., Burnaby, BC V5J 5B6 Canada
(800) 565-5228 (tel)
(604) 451-7900 (tel)
(604) 451-7991 (fax)
Manufactures programmable battery analyzers for professional two-way radio applications.

CAD & Graphics
1175 Chess Dr., #C, Foster City, CA 94404 USA
(800) 288-1611 (tel)
(800) 387-9945 (tel)
(415) 378-6400 (tel)
(415) 378-6414 (fax)
Visa, MasterCard, Discover
Specializes in the sales of CAD design products, including printers, plotters, monitors, digitizing tablets, and scanners. Offers a 30-day money-back guarantee.

CAD Warehouse
8515-D Freeway Dr., Macedonia, OH 44056 USA
(800) 487-0485 (tel)
(216) 468-1468 (cust service tel)
(216) 468-1937 (fax)
Sells computer systems, large-format scanners and printers, digitizers, etc. for professional CAD computer applications.

C & A Electronics
P.O. Box 25070, Athens, 10026 Greece
52-42-867 (tel)
52-42-537 (tel)
Manufactures and sells radio and audio kits. Offers a 53-page catalog.

CAL Crystal Lab
1142 N. Gilbert, Anaheim, CA 92801 USA
(800) 333-9825 (tel)

CAL Crystal Lab

(714) 491-9825 (fax)
Manufactures crystals for a variety of radio applications.

Calculated Industries
4840 Hytech Dr., Carson City, NV 89706 USA
(800) 854-8075 (tel)
(702) 885-4900 (tel)
(702) 885-4949 (fax)
Manufactures dimension calculators for proportional scale modeling applications.

Calex Manufacturing
2401 Stanwell Dr., Concord, CA 94520 USA
(800) 542-3355 (tel)
(510) 687-3333 (fax)
Manufactures dc-to-dc power converter modules for a wide variety of electronics applications.

California Amplifier
460 Calle San Pablo, Camarillo, CA 93012 USA
(805) 987-9000 (tel)
(805) 987-8359 (fax)
33-1-48-64-52-52 (Paris tel)
33-1-48-64-52-55 (Paris fax)
55-11-884-6411 (Sao Paulo tel/fax)
66-2-870-8333 (Bangkok tel)
Manufactures broadband scrambling systems for professional cable TV applications.

California Instruments
9025 Balbola Ave., San Diego CA 92123 USA
(800) 4AC-POWER (tel)
http://www.calinst.com
Manufactures harmonic generation and analysis systems for professional industrial applications.

California Microwave
855 Mission Ct., Fremont, CA 94539 USA
(800) 999-1920 (tel)
Manufactures two-way radios for cellular microwave applications.

California PC Products
205 Apollo Way, Hollister, CA 95023 USA
(800) 394-4122 (tel)
Visa, MasterCard
Manufactures and sells cases for personal desktop computers. Offers a free catalog.

CalTex Computer
10635 Tower Oaks Blvd., Ste. G, Houston, TX 77070 USA
(800) 959-5507 (tel)
(713) 894-5496 (tel)
(713) 894-9469 (fax)
Visa, MasterCard, Discover, American Express
Sells new and remanufactured computer systems, printers, etc. for a wide variety of computer applications.

Cambex
360 Second Ave., Waltham, MA 02154 USA
(800) 292-RISC (tel)
(617) 890-6000 (tel)
(617) 890-1038 (fax)
Manufactures data storage systems for a wide variety of professional computer network applications.

Cambridge AccuSense
1000 Mount Laurel Circle, Shirley, MA 01464 USA
(800) 313-9271 (tel)
(508) 425-4062 (fax)
Manufactures pressure sensors for a wide variety of electronics applications.

Cambridge Microprocessor Systems
Unit 17-18 Zone D, Chelmsford Rd. Ind. Est.
Great Dunmow, Essex CM6 1XG UK
Manufactures and sells micro module kits for a wide variety of experimental computer applications.

Cambridge Soundworks
154 California St., #104, Newton, MA 02158 USA
(800) 367-4434 (tel)
(617) 332-9229 (fax)
(617) 332-5936 (outside North America)
Visa, MasterCard, Discover, American Express
Manufactures a variety of hi-fi speakers and sells audio components and entire home stereo systems. Offers a free catalog.

Cameleon Technology
2081 Bering Dr., Unit L, San Jose, CA 95131 USA
(800) 440-7466 (tel)
(408) 436-8681 (tel)
(408) 436-8685 (fax)
Manufactures memory chip adapters for computer applications.

Camera Sound
Rt. 252 S., Berwyn, PA 19312 USA
(800) 477-0022 (tel)
(215) 949-2124 (tel)
(215) 949-1085 (fax)
1104 Chestnut St., Philadelphia, PA 19312 USA
110 Lincoln Hwy., Fairless Hills, PA 19030 USA
(215) 949-1086 (tel)
Visa, MasterCard, Discover, American Express, Diner's Club, Canon, COD
Sells high-end home, amateur, and professional audio and video equipment, such as camcorders, video editors, cassette decks, VCRs, receivers, etc.

Camera World of Oregon
500 S.W. 5th Ave., Portland, OR 97204 USA
(800) 729-8937 (tel)
(800) 729-8940 (free catalog tel)
(503) 227-6008 (tel)
(503) 222-7070 (fax)
Visa, MasterCard, Discover, American Express
Sells a wide variety of photographic, audio, video, and office equipment, including cameras, lenses, camcorders, video editors, telephones, TVs, receivers, karaoke decks, ministereos, etc. Offers a free catalog.

Camintonn
22 Morgan, Irvine, CA 92718 USA
(800) 368-4726 (tel)
(714) 454-1500 (tel)
(714) 830-4726 (fax)
+44-1-865-784747 (UK tel)
+44-1-865-784750 (UK fax)
Manufactures RAM chips and boards for professional industrial applications.

Campbell Scientific
815 W. 1800 N., Logan, UT 84321 USA
(801) 753-2342 (tel)
(801) 750-9540 (fax)
Manufactures acceleration/vibration measurement and control systems for professional industrial applications.

Campbell Technical
1190 Dell Ave., Unit N, Campbell, CA 95008 USA
(800) 474-8399 (tel)
(408) 364-1789 (tel)
(408) 364-0689 (fax)
Sells data storage systems, such as hard drives and tape drives, for a wide variety of computer applications.

Cannon R/C Systems
2756 N. Green Valley Pkwy., #405, Henderson, NV 89014 USA
(702) 896-7203 (tel)
(702) 896-7206 (fax)
Visa, MasterCard, COD (20% deposit required)
Manufactures and sells receivers and transmitters for radio-control model airplanes.

Canon
30-2, Shimomaruko 3-chome, Ohta-ku, Tokyo 146 Japan
1 Canon Plaza, Lake Success, NY 11042 USA
(800) 828-4040 (tel)
(800) 848-4123 (tel)
(800) 221-3333 (tel)
(516) 488-6700 (tel)
(516) 328-5960 (tech support tel)
http://www.usa.canon.com
Manufactures audio and photographic equipment, such as loudspeakers, copiers, computer printers, and cameras.

Canseda
F.O. Petersons Gata 24 B, S-421 31 Västra Frölunda, Sweden
Sells head units, speakers, amplifiers, receivers, cassette decks, CD players, etc. for home audio and car audio applications.

Canton
915 Washington Ave., Minneapolis, MN 55415 USA
(800) 328-8040 (tel)
(612) 333-1150 (tel)
(612) 338-8129 (fax)
(905) 567-1920 (Canada tel)
cantonusa@aol.com
Neugasse 21-23, D-61276 Weilrod, Germany
06083-287-0 (tel)
06083-28113 (fax)
Manufactures audio amplifiers and loudspeakers for home audio and mobile audio applications.

Capital Computer Group
2627 Humboldt Ave. S., Minneapolis, MN 55408 USA
(612) 377-9294 (tel)
(612) 377-9094 (fax)
Buys and sells new and used personal home and laptop computers and parts and accessories.

Capital Electronics
852 Foster Ave., Bensenville, IL 60106 USA
(708) 350-9510 (tel)
(708) 350-9760 (fax)
(708) 350-9761 (BBS)
info@capital-elec.com (info)
order@capital-elec.com (ordering)
quote@capital-elec.com (price quotes)
http://www.capital-elec.com
Sells microcontrollers and EEPROMs for a wide variety of electronics applications.

Capital Resource Recovery
200 N. 2nd St., Ste. 300, Minneapolis, MN 55401 USA
(800) 452-6670 (tel)
(612) 376-0426 (tel)
(612) 376-0458 (fax)
Sells personal desktop computer parts and accessories, including motherboards, hard drives, power supplies, monitors, etc.

Capricorn Capital Group
2938 Waterview Dr., Rochester Hills, MI 48309 USA
(800) 853-7980 (tel)
(800) 899-4666 (tel)
(810) 853-0526 (fax)
Buys, sells, and leases refurbished personal desktop computers and parts and accessories, including memory chips, video boards, hard drives, power supplies, controllers, etc.

Capteur Sensors & Analysers
66 Milton Pk., Abingdon, Oxon OX14 4RY UK
+44-0-1235-821323 (tel)
+44-0-1235-820632 (fax)
Manufactures a variety of gas sensors for professional industrial applications.

Cardservice International
26072 Merrot Circle, Ste. 3118, Laguna Hills, CA 92653 USA
(800) 748-6318 (tel)
Visa, MasterCard, Discover, American Express
Manufactures and sells credit card processors for professional computerized business applications.

Carlingswitch
60 Johnson Ave., Plainville, CT 06062 USA
(860) 793-9281 (tel)
(860) 793-9231 (fax)
Manufactures circuit breakers for a wide variety of electronics applications. Offers a free catalog.

Carlson Computer International
11956 Bernardo Plaza Dr., #503, San Diego, CA 92128 USA
(800) 334-7073 (tel)
(619) 675-3377 (tel)
(619) 675-3379 (fax)
Buys, sells, trades, and leases new and used computers and peripherals for professional computer applications.

Carlson-Strand
P.O. Box 3761, San Clemente, CA 92672 USA
(714) 492-8978 (tel)
(714) 492-9638 (fax)
Modifies equipment for professional video applications. Offers a free catalog.

Carolina Computer Concepts
1380 Shearers Rd., P.O. Box 1261, Mooresville, NC 28115 USA
(704) 662-0669 (tel)
(704) 662-0670 (fax)
Buys and sells mainframes, lightpens, and parts, etc. for a wide variety of computer applications.

Carolina Tubeworks
20 Ranchor Dr., Thomasville, NC 27360 USA
(910) 476-8309 (tel)
Sells electron tubes for a wide variety of electronics applications.

Carrot Computer
4911 Sepulveda Blvd., Culver City, CA 90230 USA
(310) 313-4540 (tel)
(310) 313-4542 (fax)
(310) 313-4543 (BBS)
Visa, MasterCard, COD
Sells portable computer hard drives.

Caruso Music
94 State St., New London, CT 06320 USA
(203) 442-9600 (tel)
(203) 442-0463 (fax)
sales@caruso.net
Visa, MasterCard, Discover, American Express
Sells new and used audio recording keyboard and outboard equipment for professional recording studios and live music performances.

Carver
P.O. Box 1237, Lynnwood, WA 98046 USA
(206) 775-1202 (tel)
(416) 847-8888 (Canada tel)
Manufactures a full line of home theater, home audio, and car stereo equipment. Offers free brochures.

Carvin
12340 World Trade Dr., San Diego, CA 92128 USA
(800) 854-2235 (tel)
(619) 487-8700 (tel)
(619) 487-8160 (fax)
7414 Sunset Blvd., Hollywood, CA 90068 USA
(213) 851-4200 (tel)
1907 N. Main St., Santa Ana, CA 92708 USA
(714) 558-0655 (tel)
Manufactures electric guitars and amplifiers for professional and personal musical applications. Offers a free catalog.

Cascade Audio
19135 Kiowa Rd., Bend, OR 07702 USA
(541) 389-6821 (tel)
(541) 389-5273 (fax)
Manufactures power supplies and battery chargers for car stereo applications.

The Case Depot
14102 E. Firestone Blvd., Santa Fe Springs, CA 90670 USA
(800) 200-6118 (tel)
(310) 404-9725 (tel)
(310) 404-8525 (fax)
Visa, MasterCard, Discover, American Express
Sells a wide variety of cases, power supplies, and complete home computer systems.

Casio
570 Mt. Pleasant Ave., Dover, NJ 07801 USA
(201) 361-5400 (tel)
(201) 361-3819 (fax)
Manufactures watches, clocks, musical keyboards, cordless telephones, etc.

Cassette House
219 Merrylog Ln., Kingston Springs, TN 37082 USA
(800) 321-5738 (tel)
artmuns@tape.com
http://www.tape.com
Sells blank recording cassettes, DATs, CD-Rs, etc.

CastleRock
20 S. Santa Cruz Ave., #310, Los Gatos, CA 95030 USA
(800) 695-8206 (tel)
(408) 395-8206 (tel)
(408) 395-8236 (fax)
Manufactures workstations for a wide variety of professional computer network applications.

C & A Systems
3814 N. 1st Blvd., B, Abilene, TX 79603 USA
(800) 675-5702 (tel)
(915) 675-5700 (tech support tel)
(915) 675-0731 (fax)
Visa, MasterCard
Sells new, refurbished, and used personal desktop and laptop computers.

CATS
7368 S.R., #105, Pemberville, OH 43450 USA
Sells a variety of antennas, rotators, wire, cable, etc.

CATV Services
4014 Chase Ave., Ste. 211, Miami, FL 33140 USA
(800) 227-1200 (tel)
(305) 535-3033 (tel)
(305) 535-2528 (fax)
catvserv@aol.com
Sells a wide variety of surplus equipment for professional cable TV applications.

C.B. City International (CBCI)
P.O. Box 31500, Phoenix, AZ 85046 USA
Sells kits and modifcation kits for citizens band (CB) transceivers.

CB Shop
410 E. Fayette Ave., Effingham, IL 62401 USA
(217) 342-3054 (tel)
Sells equipment for citizens band radio applications.

C.B.S. of Tidewater
375 Independence Blvd., Virginia Beach, VA 23462 USA
(804) 473-0354 (tel)
(804) 473-9406 (fax)
Sells personal desktop computers and parts, such as hard drives, motherboards, cases, controller boards, etc.

CCA Electronics
P.O. Box 426, Fairburn, GA 30213 USA
(404) 964-2222 (fax)
(404) 964-3530 (tel)
Manufactures a variety of FM exciters and transmitters for professional broadcasting.

CCNS Group
3194 De La Cruz Blvd., #6, Santa Clara, CA 95054 USA
(800) 367-2267 (tel)
(408) 970-8673 (tel)
(408) 970-0639 (fax)
Manufactures file servers for a wide variety of professional computer network applications.

C-Comm
6115 15th Ave., NW, Seattle, WA 98107 USA
(800) 426-6528 (tel)
(206) 784-7337 (tel)
(206) 784-0541 (fax)
Sells a wide variety of amateur radio equipment and accessories, including amateur radio transceivers, receivers, antenna tuners, antennas, amplifiers, etc.

C-COR Electronics
60 Decibel Rd., State College, PA 16901 USA
(800) 233-2267 (tel)
(814) 238-2461 (tel)
(814) 238-4065 (fax)
47323 Warm Springs Blvd., Fremont, CA 94539 USA
(510) 440-0330 (tel)
(510) 440-0218 (fax)
12742 E. Caley Ave., #A, Englewood, CO 80111 USA
(303) 799-1100 (tel)
(303) 643-1743 (fax)
P.O. Box 10.265, 1301 AG Almere, Netherlands
31-36-536-4199 (tel)
31-36-536-4255 (fax)
377-MacKenzie Ave., Unit 5, Ajax, ON L1S 2G2 Canada
(800) 427-2559 (tel)
(905) 427-0366 (tel)
(905) 428-0927 (fax)
Manufactures fiberoptics and RF products for professional cable TV applications.

CCS Audio Products
670 N. Beers St., Bldg. 4, Holmdel, NJ 07733 USA
(908) 739-5600 (tel)
(908) 739-1818 (fax)
Manufactures ISDN decoders for professional studio recording.

CD-Rom Direct
383 California St., Newton, MA 02158 USA
(800) 332-2404 (tel)
(617) 332-1783 (fax)
Visa, MasterCard
Sells CD-ROM recorders and towers for professional computer applications.

CD-R Solutions
2355 N. Steves Blvd. #C, Flagstaff, AZ 86004 USA
(800) 278-3480 (tel)
(602) 396-3616 (tel)
(602) 396-0483 (fax)
Visa, MasterCard
Sells CD-ROM recorders, duplicators, and towers and tape drives, CD-Rs, DAT tapes, etc. for professional and personal computer applications.

C & E Communications & Energy
7395 Taft Park Dr. E., Syracuse, NY 13057 USA
(800) 882-1587 (tel)
(315) 452-0709 (tel)
(315) 452-0732 (fax)
Manufactures filters for professional cable TV applications.

CED
2500 NW 39th St., Miami, FL 33142 USA
(305) 633-8020 (tel)
(305) 635-5445 (fax)
Manufactures amplifiers, modulators, upconverters, and exciters for professional satellite and UHF TV broadcasting applications.

CED
4209 Evergreen Ln., Annandale, VA 22003 USA
(703) 354-2545 (tel)
Manufactures oscilloscope frequency divider modules for a wide variety of electronics testing applications.

Cedko Electronics
3002 S. Oak St., Santa Ana, CA 92707 USA
(714) 540-8454 (tel)
(714) 540-1299 (fax)
(714) 546-4810 (BBS)
Manufactures PC boards for a wide variety of electronics applications.

CEH/Kirby Enterprises
4120 Kirby Rd., Cincinnati, OH 45223 USA
(800) 237-9654 (tel)
(513) 542-8870 (fax)
Manufactures crystals for a variety of radio applications.

Celestion
89 Doug Brown Way, Holliston, MA 01746 USA
(800) 235-7757 (tel)
(508) 429-6706 (tel)
Manufactures loudspeakers for high-end home audio and home theater applications.

Cellcast
2002 Ford Ave., Springdale, AR 72764 USA
(501) 750-0277 (tel)
(501) 751-3761 (fax)
Manufactures remote broadcast mixers with cellular transceivers for professional broadcasting.

Cell-Tel International
9454 Phillips Hwy., #8, Jacksonville, FL 32256 USA
(904) 268-3003 (tel)
(904) 262-0067 (fax)
Buys and sells refurbished cellular systems and ancillary parts for professional cellular microwave radio applications.

Cellular Products
5909 Clinton Dr., Houston, TX 77020 USA
(800) 542-5542 (tel)
(713) 672-8305 (tel)
(713) 676-2977 (fax)
Sells telephones, batteries, etc. for cellular telephone applications.

Cellular Security Group
4 Gerring Rd., Gloucester, MA 01930 USA
(800) 487-7539 (tel)
(508) 281-8892 (tel)
(508) 768-7486 (fax)
Manufactures and sells antennas for VHF/UHF applications.

Cellular Wholesalers
5151 W. Church St., Skokie, IL 60077 USA
(800) 395-0505 (tel)
(708) 965-2300 (tel)
(708) 676-8960 (fax)
3100 NW 72nd Ave., Unit 102, Miami, FL 33122 USA
(305) 592-1100 (tel)
(305) 592-4949 (fax)
Sells pagers, etc. for cellular telephone applications.

Celwave
2 Ryan Rd., Marlboro, NJ 07746 USA
(800) CELWAVE (tel)
(908) 462-1880 (tel)
(908) 462-6919 (fax)
(908) 431-8388 (fax)
Manufactures repeaters and amplifiers for professional cellular microwave radio applications.

Centari Technologies
146 Londonderry Tnpke., Hooksett, NH 03106 USA
(603) 645-8510 (tel)
(603) 645-8520 (fax)
Visa, MasterCard
Sells personal desktop computer parts and accessories, including memory chips, hard drives, monitors, printers, etc.

Centaur Electronics
3720 S. Park Ave., #604, Tucson, AZ 85713 USA
(520) 622-6672 (tel)
Manufactures a variable duty-cycle station tuning instrument for amateur radio applications.

Central Data
1602 Newton Dr., Champaign, IL 61812 USA
(800) 482-0315 (tel)
(217) 359-8010 (tel)
(217) 359-6904 (fax)
info@cd.com
http://www.cd.com
ftp://cd.com
Manufactures SCSI serial ports and print servers for professional computer applications.

Central Model Marketing
P.O. Box 17291, Denver, CO 80217 USA
(800) 962-2010 (tel)
Visa, MasterCard, Discover, COD ($4)
Sells sport and computer radios for radio-control model boats.

Central Semiconductor
145 Adams Ave., Hauppauge, NY 11788 USA
(516) 435-1110 (tel)
(516) 435-1824 (fax)
Manufactures semiconductors for a wide variety of electronics applications.

Central Technologies
387 Zachary Ave., Bldg. 103, Moorpark, CA 93021 USA
(800) 838-6423 (tel)
(805) 532-9165 (intrnl tel)
(805) 523-1329 (fax)
(805) 532-9174 (fax)
(805) 532-9182 (voice mail tel)
cencomp@aol.com
Manufactures and sells voice mail systems for a wide variety of computer business applications.

Centronic
2088 Anchor Ct., Newbury Park, CA 91320 USA
(800) 700-2088 (tel)
(805) 499-7770 (tel)
(805) 499-5902 (fax)
Manufactures photodiodes for a wide variety of electronics applications.

Centurion
P.O. Box 82846, Lincoln, NE 68501 USA
(800) 228-4563 (tel)
(402) 467-4491 (tel)
(800) 848-3825 (fax)
(402) 467-4528 (fax)
Manufactures cellular microwave telephones.

Centurian Surplus
375 Tennant Ave., Morgan Hill, CA 95037 USA
(408) 778-2001 (tel)

(408) 778-9420 (fax)
Sells new, used, and refurbished WYSE terminals, printers, CPU boards, keyboards, and parts for professional computer applications.

Centurion Technology Systems (CTS)
43-26, 52 St., Woodside, NY 11377 USA
(718) 565-2535 (tel)
(718) 565-2346 (fax)
Sells personal home and laptop computers, and printers.

Cermark
107 Edward Ave., Fullerton, CA 92633 USA
(800) 704-6229 (tel)
(714) 680-5888 (tel)
(714) 680-5880 (fax)
815 Oakwood #D, Lake Zurich, IL 60047 USA
(800) 416-6299 (tel)
(708) 438-2233 (tel)
(708) 438-2898 (fax)
Visa, MasterCard
Manufactures autopilot devices for radio-control model airplanes. Offers more information for an SASE. Mininum order: $35.

The Cerplex Group
10 Jeanne Dr., Newburgh, NY 12550 USA
(800) 434-3880 (tel)
(914) 566-7071 (tel)
(914) 566-7075 (fax)
Visa, MasterCard
Sells personal home computer and printer parts and accessories.

Cerwin-Vega
55 E. Easy St., Simi Valley, CA 93065 USA
(805) 584-9332 (tel)
Manufactures loudspeakers for high-end home audio and home theater applications.

CFS Electronics
3321 E. Oakland Park Blvd., Ste. 214, Ft. Lauderdale, FL 33308 USA
(800) 995-1749 (tel)
Visa, MasterCard, American Express, COD
Sells cable TV equipment.

C & H
2176 E. Colorado Blvd., Pasadena, CA 91107 USA
P. O. Box 5356, Pasadena, CA 91117 USA
(800) 325-9465 (tel)
(213) 681-4925 (tel)
(818) 796-2628 (tel)
(818) 796-4875 (fax)
Sells surplus electronics, especially power supplies, transformers, motors, solenoids, etc. Offers a free catalog.

Chainford Electronics
P.O. Box 7-058, Taipei, Taiwan
No. 171, Pao Hsin St., Hsi Chih County, Taipei, Taiwan
886-2-7542515 (tel)
886-2-7085800 (fax)
Manufactures amplifiers for professional cable TV applications.

Champion Technologies
2553 N. Edgington St., Franklin Park, IL 60131 USA
(800) 888-1499 (tel)
(708) 451-1000 (tel)
(708) 451-7585 (fax)
Manufactures oscillator modules for a wide variety of electronics applications.

Channell
P.O. Box 9022, Temecula, CA 92589 USA
(800) 423-1863 (US tel)
(909) 694-9160 (US tel)
(909) 694-9170 (US fax)
(800) 387-8332 (Canada tel)
(905) 567-6751 (Canada tel)
(905) 567-6756 (Canada fax)
+44-171-589-3304 (UK tel)
+44-171-589-1021 (UK fax)
972-4-341329 (Israel tel)
972-4-341387 (Israel fax)
Manufactures broadband and telecom cabinets for professional cable TV applications.

Channel Master
P.O. Box 1416, Industrial Park Dr., Smithfield, NC 27577 USA

Channel Master

(919) 989-2205 (tel)
(919) 989-2212 (tel)
(919) 989-2215 (fax)
(919) 989-2200 (fax)
Manufactures baluns, antenna couplers, preamplifiers, distribution amplifiers, attenuators, filters, modulators, matching transformers, connectors, switches, cables, splitters, wall plates, signal meters, power dividers, etc. for a wide variety of satellite and terrestrial TV applications.

Chaparral Communications

2450 N. 1st St., San Jose, CA 95131 USA
(408) 435-1530 (tel)
(408) 435-1429 (fax)
(408) 435-3088 (fax back)
Industrielstaat 1
1704 AA Heehugowaard, Netherlands
31-072-5721171 (tel)
31-072-5721585 (fax)
http://www.chaparral.net
Manufactures feedhorns, receivers, controllers, etc. for home and professional satellite communications.

Charge Guard

400 Highland Ave., Altoona, PA 16602 USA
(800) 458-3410 (tel)
Manufactures battery-charging timers for cellular microwave telephone applications.

Charles Radio

6836 W. Pearl City Rd., Freeport, IL 61032 USA
(815) 235-7733 (tel)
Manufactures and sells antennas, transceivers, power boosters, connectors, coaxial switches, etc. for citizens band radio applications.

Charleswater

90 Hudson Rd., Canton, MA 02021 USA
(617) 821-8370 (tel)
(617) 575-0172 (fax)
Manufactures digital surface resistance meters for professional electronics testing applications.

Charlie's

2955-A3 Cochran St., Simi Valley, CA 93065 USA
(805) 584-0125 (tel)
(805) 584-0792 (fax)
Visa, MasterCard, Discover, American Express
Sells speed controls and battery chargers for radio-control model airplanes.

Charlie's Trains

P.O. Box 158, Hubertus, WI 53033 USA
(414) 628-1544 (tel)
(414) 628-2651 (fax)
Sells miniature light bulbs for model railroads.

Charlie Stringer's Strings

P.O. Box 4241, Warren, NJ 07059 USA
(908) 469-2828 (tel)
(908) 469-2882 (fax)
Manufactures guitar chord computers for personal musical applications.

Chase Research

545 Marriot Dr., #100, Nashville, TN 37122 USA
(800) CHASE-US (tel)
(615) 872-0770 (tel)
(615) 872-0771 (fax)
Manufactures routers for professional computer network applications.

Chase Technologies

111 2nd Ave. NE, # 700A, St. Petersburg, FL 33701 USA
(800) 531-0631 (tel)
(813) 896-7899 (fax)
Manufactures and sells high-end equipment for audio and home theater applications, including a home theater decoder and a remote line controller.

C & H Distributing

215 S. George St., York, PA 17403 USA
(717) 843-7881 (tel)
(717) 843-3875 (fax)
Sells surplus electronics, including semiconductors, electron tubes, etc., for a wide variety of electronics applications.

Chelco Electronics

61 Water St., Mayville, NY 14757 USA

(800) 388-8521 (tel)
(716) 753-3220 (fax)
Manufactures printed circuit boards. Offers free shipping.

Cherokee Data
1221 N. Classen Blvd, Oklahoma City, OK 73106 USA
(405) 528-5270 (tel)
(405) 528-5303 (fax)
Sells hard drives and tape drives for a wide variety of computer applications.

Cherry Electrical Products
3600 Sunset Ave.
Waukegan, IL 60087 USA
(708) 360-3434 (tel)
Manufactures plasma displays for a wide variety of electronics applications.

Cherry Semiconductor
2000 S. County Trail, E. Greenwich, RI 02818 USA
(800) 272-3601 (tel)
Manufactures semiconductors for a wide variety of electronics applications.

CHI
31200 Carter St., Solon, OH 44139 USA
(800) 828-0599 (tel)
(216) 349-8605 (tel)
(216) 349-8609 (fax)
(800) 928-9099 (West Coast tel)
(602) 548-9010 (West Coast tel)
(602) 548-9041 (West Coast fax)
Sells recordable CD-ROM drives, optical hard drives, optical jukeboxes, etc.

Chief Aircraft
1301 Brookside Blvd., Grants Pass, OR 97526 USA
(800) 447-3408 (tel)
(503) 476-1869 (tel)
(503) 476-6605 (customer service tel)
(503) 479-4431 (fax)
Sells instruments, intercoms, gyros, monitors, GPS systems, and transceivers for professional aviation applications.

Chilton Pacific
5632 Van Nuys Blvd. #222, Van Nuys, CA 91401 USA
(800) 717-7780 (tel)
Visa, MasterCard
Sells shortwave receivers and accessories. Offers a catalog.

Chimera Lighting
1812 Valtec Ln., Boulder, CO 80301 USA
(800) 424-4075 (tel)
Manufactures lighting systems for amateur and professional video production.

Chinon America
Information Equipment Division
615 Hawaii Ave., Torrance, CA 90503 USA
(800) 441-0222 (tel)
(310) 533-0274 (tel)
(310) 533-1727 (fax)
Electronic Imaging Division
1065 Bristol Rd., Mountainside, NJ 07092 USA
(800) 932-0374 (tel)
(908) 654-6656 (fax)
Waldstrasse 23/B4, 63128 Dietzenbach, Germany
49-6074-82230 (tel)
49-6074-31406 (fax)
http://www.chinon.com
Manufactures digital cameras and CD-ROM drives for a wide variety of computer applications.

Chipcom
Southborough Office Park, 118 Turnpike Rd.
Southborough, MA 01772 USA
(800) 666-2447 (tel)
(508) 460-8900 (tel)
(508) 460-8950 (fax)
Manufactures routers, switches, and hubs for professional computer network applications.

The Chip Merchant
4870 Viewridge Ave., San Diego, CA 92123 USA
(619) 654-2700 (tel)
(619) 268-4774 (tel)
(619) 268-0874 (fax)
Buys and sells memory chips for a wide variety of computer applications.

The Chip Ship
40 Exchange Pl., #1318, New York, NY 10035 USA
(212) 785-4500 (tel)
(212) 785-4800 (fax)
Visa, MasterCard, American Express
Sells computer parts and boards, such as memory chips, hard drives, floppy drives, fans, etc.

Chrislin Industries
31312 Via Colinas, #108, Westlake Village, CA 91362 USA
(800) 468-0736 (tel)
(818) 991-2254 (tel)
(818) 991-3490 (fax)
Manufactures memory boards for a wide variety of computer applications.

C & H Technologies
P.O. Box 14765, Austin, TX 78761 USA
(512) 251-1171 (tel)
(512) 251-1963 (fax)
Manufactures test equipment, such as pulse generators, for a wide variety of professional electronics applications.

Chyron
5 Hub Dr., Melville, NY 11747 USA
(516) 845-2026 (tel)
(516) 845-3895 (fax)
Manufactures and sells a character generator board for home, amateur, and professional video editing on computer.

C3I
406 N. Pitt St., Alexandria, VA 22314 USA
(800) 224-5137 (tel)
(703) 684-6980 (tel)
(703) 684-1382 (fax)
Manufactures audio delay modules, DTMF modules, repeater controllers, remote base switches, etc. for a wide variety of amateur radio repeater applications.

Cimarron Technologies
934 S. Andreasen Dr., #G, Escondido, CA 92029 USA
(800) 487-7184 (tel)
(619) 738-3282 (fax)
Manufactures RF modems for vehicular tracking applications.

CIMCO
2257 N. Penn Rd., Hatfield, PA 19440 USA
(215) 822-2171 (tel)
(215) 822-6401 (customer service tel)
(215) 822-3795 (fax)
Manufactures overhead ionizers to prevent electrostatic shocks for a variety of professional electronics work areas.

Cincinnati Electronics
7500 Innovation Way, Mason, OH 45040 USA
(800) 852-5105 (tel)
(513) 573-6290 (fax)
Manufactures infrared imaging systems for a variety of professional industrial applications.

Ciprico
2800 Campus Dr., Plymouth, MN 55441 USA
(800) 727-4669 (tel)
(612) 551-4000 (tel)
(612) 551-4002 (fax)
Manufactures data storage systems for a wide variety of professional computer network applications.

Circo Technology
148 8th Ave. #D, City of Industry, CA 91746 USA
(800) 678-1688 (tel)
(818) 369-5779 (tel)
(818) 369-2769 (fax)
73547.1345@compuserve.com
Visa, MasterCard, Discover
Sells cases, power supplies, and other components for personal computer applications.

Circuit Research Labs
2522 W. Geneva Dr., Tempe, AZ 85252 USA
(800) 535-7648 (tel)
(602) 438-0888 (tel)
(602) 438-8227 (fax)
Manufactures stereo processor/generators, RBDS generators, and a variety of audio processors for broadcast stations.

Circuit Specialists
P.O. Box 3047, Scottsdale, AZ 85271 USA
(800) 528-1417 (tel)
Sells electronic components, including RF parts for radio transmitting applications.

Cirkit Distribution
Park Ln., Broxbourne, Hertfordshire EN10 7NQ UK
01992-448899 (tel)
01992-471314 (fax)
Visa, Access
Sells electronic and computer components, including motherboards, sound cards, video cards, I/O cards, power supplies, etc. Offers a 280-page catalog for £1.95.

Cirris Systems
1991 Parkway Blvd., Salt Lake City, UT 84119 USA
(800) 441-9910 (tel)
(801) 973-4600 (tel)
(801) 973-4609 (fax)
Manufactures wire harness testers for professional computer applications.

Cirrus Logic
3100 W. Warren Ave., Fremont, CA 94538 USA
(800) 858-0487 (tel)
(510) 623-8300 (tel)
(510) 252-6020 (fax)
http://www.cirrus.com
Manufactures integrated circuits for a wide variety of personal computer applications.

Citel
1111 Parkcentre Blvd., #474, Miami, FL 33169 USA
(305) 621-0022 (tel)
(305) 621-0766 (fax)
Manufactures surge protectors for a wide variety of computer applications.

Cititronix
1641 Dielman Rd., St. Louis, MO 63132 USA
(800) 846-2484 (tel)
(314) 427-3420 (tel)
(800) 397-8587 (fax)
Sells replacement parts and accessories for audio equipment.

Citizen
2450 Broadway, #600, Santa Monica, CA 90411 USA
(800) 421-6516 (tel)
(310) 453-0614 (tel)
(310) 453-2814 (fax)
Manufactures color LCD camcorder monitors for professional, amateur, and home video recording applications.

C&K Components
57 Stanley Ave., Watertown, MA 02172 USA
(800) 635-5936 (tel)
(800) 334-7729 (tel)
(617) 926-6846 (fax)
Manufactures switches, torroids, etc. for a wide variety of electronics applications. Offers a free catalog.

Clarion
661 W. Redondo Beach Blvd., Gardena, CA 90247 USA
(310) 327-9100 (tel)
(310) 327-9377 (fax)
(310) 327-1999 (fax)
(800) 347-8667 (dealer fax)
jzak@clarionmultimedia.com
http://www.clarionmultimedia.com
Visa, MasterCard
Manufactures head units, amplifiers, security systems, and loudspeakers for mobile audio applications.

David Clark
360 Franklin St., Box 15054, Wercester, MA 01615 USA
(508) 751-5800 (tel)
(508) 753-5827 (fax)
Manufactures radio VOX headsets for two-way radio applications. Offers free information and a demonstration.

Clarke-Hess Communication Research
220 W. 19th St., New York, NY 10011 USA

(212) 255-2940 (tel)
(212) 691-8158 (fax)
Manufactures V-A-W meters for professional electronics testing applications.

Clary
1960 S. Walker Ave., Monrovia, CA 91016 USA
(800) 44-CLARY (tel)
75702.324@compuserve.com
Manufactures uninterruptible power supplies for a wide variety of personal computer applications. Offers a free catalog.

Classic Air Supply
801 S. Lincoln, Marshfield, WI 54449 USA
(715) 387-3057 (tel)
Manufactures intercoms for professional aviation applications.

CLE
P.O. Box 1913, Sarasota, FL 34230 USA
(813) 922-2633 (tel)
(813) 922-2633 (fax)
Visa, MasterCard
Sells quick-mount acoustic guitar pickups that can also be used as contact microphones.

ClearTek
P.O. Box 1123, Crystal Beach, FL 34681 USA
(615) 954-9221 (tel)
Manufactures and sells antennas for shortwave-listening applications.

Clif Designs
1602 Babcock St., Costa Mesa, CA 92627 USA
(800) 845-7002 (tel)
(714) 645-8740 (fax)
Manufactures loudspeakers for car stereo applications.

Clifford Electronics
20750 Lassen St., Chatsworth, CA 91311 USA
(800) CLIFFORD (brochure tel)
(800) 824-3208 (customer service tel)
askclifford@clifford.com (info)
techsupport@clifford.com (tech support)
http://www.clifford.com
Manufactures systems for mobile security applications.

Clinton Electronics
6701 Clinton Rd., Rockford, IL 61111 USA
(815) 633-1444 (tel)
(815) 633-8712 (fax)
Manufactures monitors for a wide variety of computer applications.

CLM Technologies
1139 McConnell Dr., Decatur, GA 30033 USA
(404) 266-8315 (USA tel)
(819) 778-2053 (Canada tel)
http://www.magi.com
Manufactures emergency radio dispatch systems for professional communication applications.

Closed Circuit Products
6395 Gumpark Dr., Ste. C, Boulder, CO 80301 USA
(800) 999-3130 (tel)
Sells a variety of products for closed-circuit TV/ security systems, such as video cameras.

CMD Technology
1 Vanderbilt, Irvine, CA 92178 USA
(800) 426-3832 (tel)
(714) 454-0800 (tel)
Manufactures PCI-to-SCSI caching adapter cards for professional network applications.

C&M Engineering
P.O. Box 701-353, West Valley City, UT 84170 USA
(801) 974-5757 (tel)
(801) 974-0869 (fax)
Manufactures battery discharge systems, etc. for radio-control models.

CMO
101 Reighard Ave., Williamsport, PA 17701 USA
(800) 233-8950 (tel)
(717) 327-9200 (tel)
(717) 327-1217 (fax)

http://www.newmmi.com
Visa, MasterCard, American Express
Manufactures complete computer systems and sells a variety of products, including notebook computers, upgrade chips, printers, hard drives, multimedia packages, modems, monitors, scanners, CD-ROM drives, etc. Offers a free catalog.

C & N Electronics
6104 Egg Lake Rd., Hugo, MN 55038 USA
(800) 421-9397 (tel)
(612) 429-9397 (tel)
(612) 429-0292 (fax)
Sells electron tubes for a wide variety of vintage electronics applications.

Coactive Aesthetics
4000 Bridgeway, #303, Sausilito, CA 94965 USA
(415) 289-1722 (tel)
(415) 289-1320 (fax)
coactive@coactive.com
Manufactures microcontroller cards for a wide variety of computer applications.

Coast to Coast
2570 86th St., Brooklyn, NY 11214 USA
(800) 788-5555 (tel)
(718) 265-1723 (service tel)
(718) 265-6800 (info tel)
(718) 265-1754 (fax)
Visa, MasterCard, Discover, Federal Express
Sells high-end home and amateur audio and video equipment, such as camcorders, video editors, monitors, VCRs, cassette decks, car audio equipment, etc. Offers extended warranties.

Cobalt Industrial
Flat M, 7/F, Yue Cheung Centre, 1-3 Wong Chuk Yeung St.
Fo Tan, Shatin, NT, Hong Kong
852-2693-0360 (tel)
852-2(601) 5865 (fax)
Manufactures loudspeakers for computer multimedia applications.

Cobra Electronics
6500 W. Cortland St., Chicago, IL 60635 USA
(800) 262-7222 (tel)
(312) 889-8870 (tel)
(800) 595-2433 (fax)
(312) 794-1930 (fax)
Visa, MasterCard, Discover, American Express
Manufactures and sells transceivers and antennas for citizens band radio applications, cordless telephones, answering machines, and radar and laser detectors.

Cogent Data Technologies
175 West St., P.O. Box 920, Friday Harbor, WA 98250 USA
(800) 4-COGENT (tel)
(206) 603-0333 (tel)
Manufactures PCI adapter boards for professional computer network applications.

Coherent Communications Systems
44084 Riverside Pkwy., Lansdowne, VA 22075 USA
(800) 443-0726 (tel)
Manufactures digital signal-processing systems for professional cellular microwave radio applications.

Coilcraft
1102 Silver Lake Rd., Cary, IL 60013 USA
(800) 322-2645 (tel)
(708) 639-1469 (fax)
Manufactures coils, filters, transformers, etc. for a wide variety of electronics applications.

Collins
400 Collins Rd. NE, Cedar Rapids, IA 52498 USA
(319) 395-4064 (tel)
http://www.rockwell.com
Manufactures satellite-based communication and navigation systems for professional aviation applications, and also equipment for amateur radio applications.

Collins Sun Electronics
1000 Brioso Dr., Costa Mesa, CA 92627 USA
(800) 678-1115 (tel)
(714) 631-3709 (fax)
Manufactures loudspeakers for car audio applications.

Colonel Video & Audio
16451 Space Center Blvd., Houston, TX 77058 USA
(800) 423-VCRS (orders tel)
(800) Video-97 (repeat callers tel)
(713) 486-8866 (customer service tel)
(713) 486-8300 (fax)
Visa, MasterCard, Discover, TeleCheck, Colonel Video & Audio, COD
Sells high-end home, amateur, and professional audio and video equipment, such as camcorders, VCRs, video editors, monitors, TVs, home theater equipment, etc.

Colorgraphic Communications
5980 Peachtree Rd., Atlanta, GA 30341 USA
(770) 455-3921 (tel)
(770) 458-0616 (fax)
sales@colorgfx.com
http://www.colorgfx.com
Manufactures graphics adapters for a wide variety of computer applications.

Columbia Data Products
P.O. Box 142584, Altamonte Springs, FL 32714 USA
(407) 869-6700 (tel)
http://www.cdp.com
Manufactures data-storage products for professional computer applications

Columbia Research Laboratories
1925 Macdade Blvd., Woodlyn, PA 19094 USA
(800) 813-8471 (tel)
(610) 872-3900 (tel)
(610) 872-3882 (fax)
Manufactures vibration, acceleration, displacement, and velocity measuring systems for professional testing applications.

Combinet
333 W. El Camino Real, Sunnyvale, CA 94087 USA
(800) 967-6651 (tel)
(408) 522-9020 (tel)
(408) 522-5479 (fax)
http://www.combinet.com/
Manufactures routers for professional computer network applications.

Comco DatNet
P.O. Box 349, Bettendorf, IA 52722 USA
(800) 4-DATNET (tel)
(319) 355-1212 (tel)
(319) 355-4055 (fax)
Sells digital linear tape drives, etc.

COMDAC (Communications Data Corp.)
1051 Main St., St. Joseph, MI 49085 USA
(800) 382-2562 (tel)
(616) 982-0404 (tel)
Visa, MasterCard, Discover, COD
Sells a wide variety of amateur radio equipment and accessories, including amateur radio transceivers, receivers, antenna tuners, antennas, amplifiers, etc.

Comer Communications
609 Washingtonia Dr., San Marcos, CA 92069 USA
(619) 744-7266 (tel)
(619) 744-4745 (fax)
(619) 744-4032 (BBS)
Manufactures general-coverage shortwave receivers with DSP card for use with personal computers.

Comet North America
89 Taylor Ave., Norwalk, CT 06854 USA
(203) 852-1231 (tel)
(203) 838-3827 (fax)
Manufactures vacuum capacitors for broadcast stations.

Comlinear
4800 Wheaton Dr., Ft. Collins, CO 80525 USA
(800) 776-0500 (tel)
(303) 226-0564 (fax)
44-1(203) 422958 (Europe tel)
44-1(203) 422961 (Europe fax)
Manufactures semiconductors for a wide variety of electronics applications.

Command Technologies
1207 W. High St., P.O. Box 7082, Bryan, OH 43506 USA
(800) 736-0443 (tel)
(419) 636-0443 (tel)
(419) 636-2269 (fax)

Visa, MasterCard, Discover, American Express
Manufactures and sells linear amplifiers for amateur radio applications.

Comm-Net
(800) 283-5158 (tel)
(800) 337-6475 (fax)
Manufactures and sells timer boards for professional two-way radio applications.

Comm-Pute
1057 E. 2100 S., Salt Lake City, UT 84106 USA
(800) 942-8873 (tel)
(801) 484-7388 (tel)
(801) 467-8873 (tel)
Sells a wide variety of amateur radio equipment and accessories, including amateur radio transceivers, receivers, antenna tuners, antennas, amplifiers, etc.

Communication Headquarters
3832 Oleander Dr., Wilmington, NC 28403 USA
(800) 688-0073 (tel)
(910) 791-8885 (service & tech tel)
(910) 452-3891 (fax)
chq@wilmington.net
http://www.wilmington.net/chq
Visa, MasterCard, Discover
Sells a wide variety of amateur radio equipment and accessories, including amateur radio transceivers, receivers, antenna tuners, antennas, amplifiers, etc.

Communications Associates
P.O. Box 2399, Joliet, IL 60434 USA
(800) 435-9313 (tel)
(800) 284-4934 (fax)
Sells base-station antennas, cavity filters, multicouplers, duplexers, etc. for professional cellular microwave radio applications.

Communication Concepts
508 Millstone Dr., Xenia, OH 45385 USA
(513) 426-8600 (tel)
(513) 220-9677 (tel)
(513) 429-3811 (fax)
Manufactures and sells HF linear amplifier kits for amateur radio use.

Communications Consultants
16128 Cohasset St., Van Nuys, CA 91406 USA
(818) 901-9711 (tel)
(818) 901-0549 (fax)
Sells replacement parts and accessories for cellular telephone applications.

Communications Electronics
P.O. Box 1045, Ann Arbor, MI 48106 USA
(800) USA-SCAN (tel)
(313) 996-8888 (tel)
(313) 663-8888 (fax)
(900) 555-SCAN ($2/min tech support)
Visa, MasterCard, Discover, American Express
Sells a variety of scanners, CB/GMRS radios, weather stations, shortwave receivers, amateur transceivers, and accessories.

Communications Specialists
426 W. Taft Ave., Orange, CA 92665 USA
(800) 854-0547 (tel)
(714) 998-3021 (tel)
(800) 850-0547 (fax)
(714) 974-3420 (fax)
Visa, MasterCard
Manufactures encoders, decoders, video modems, tone panels, CW stations identifiers, etc. for a wide variety of professional tone signalling applications. Offers a free catalog.

Communications Supply Group
680 Industrial Circle S., Shakopee, MN 55379 USA
(800) 451-9032 (tel)
(612) 445-8423 (fax)
Sells equipment for professional cable TV applications.

Communications Systems Laboratories
(800) 529-9483 (tel)
Visa, MasterCard
Manufactures surveillance camera switcher/sequencers for a wide variety of electronics security applications.

Communication Task Group
463 Amherst St., Buffalo, NY 14207 USA

Communication Task Group

(716) 873-4205 (tel)
(716) 875-0758 (fax)
ctg@usa.pipeline.com
Visa, MasterCard, Discover, American Express
Sells audio equipment, including mixing consoles, recorders, amplifiers, microphones, sound processors, etc. for professional studio recording and live music performances.

Compac
1320-12 Lincoln Ave., Holbrook, NY 11741 USA
(516) 585-3400 (tel)
(510) 227-1064 (twx)
Manufactures RFI enclosures for a variety of applications.

Compac Electronics
3L Weylond Rd., Dagenham, Essex RM8 3AB UK
+44-0181-984-0831 (tel/fax)
Sells remote controls, integrated circuits, connectors, switch boxes, keyboards, CD-ROM drives, etc. for a wide variety of electronics applications.

Compaq Computer
P.O. Box 692000, Houston, TX 77269 USA
(800) 345-1518 (tel)
(800) 888-5925 (tel)
(713) 370-0670 (tel)
(713) 374-1740 (fax)
www.compaq.com
careerpaq@compaq.com (Internet, available Compaq jobs)
http://www.monster.com
Manufactures a variety of personal computers. Offers a free catalog.

Compaq Works
10251 N. Freeway, Houston, TX 77037 USA
(800) 318-6919 (tel)
Visa, MasterCard, Discover, American Express
Sells refurbished Compaq personal computers.

Compass Design Automation
1865 Lundy Ave., San Jose, CA 95131 USA
(800) 433-4880 (tel)
(800) 200-8435 (tel)
submicron@compass-da.com
Manufactures integrated circuits etc. for computer applications.

Compatible Systems
P.O. Box 17220, Boulder, CO 80308 USA
(800) 356-0283 (tel)
(303) 444-9532 (tel)
(303) 444-9595 (fax)
info@compatible.com
http://www.compatible.com
Manufactures Internet routers for professional network computer applications.

CompCo Engineering
611 Pkwy. D4, Gatlinburg, TN 37738 USA
(423) 436-5189 (tel)
(423) 436-6333 (BBS)
http://www.hometeam.com/worthing/sniffer.htm
Manufactures infrared, two-way, and hardwired product controllers for home remote-control applications.

Compelec
14 Constable Rd.
St. Ives, Huntingdon, Cambs PE17 6EQ UK
01480-300819 (tel/fax)
Sells a wide variety of electronics components, including switches, relays, diodes, crystals, tubes, etc.

Competition Electronics
3469 Precision Dr., Rockford, IL 61109 USA
(815) 874-8001 (tel)
(815) 874-8181 (fax)
Sells battery chargers for radio-control models.

Competition R/C Imports
109-13060, 80th Ave., Surrey, BC V3W 3B2 Canada
(604) 543-9328 (tel)
(604) 543-8890 (fax)
Sells batteries for radio-control models.

Comp Express
3381 Walnut Ave., Fremont, CA 94638 USA
(800) 381-4701 (tel)
(510) 794-9311 (tel)
(510) 794-9314 (fax)

Visa, MasterCard
Sells personal desktop and laptop computers. Offers a 30-day money-back guarantee.

Complete Systems-N-More
1106 Commerce Dr., Richardson, TX 75081 USA
(800) 705-9596 (tel)
(214) 705-9595 (tel)
(214) 705-9594 (fax)
Visa, MasterCard, Discover, American Express
Manufactures and sells personal desktop computers; also sells computer parts and boards, such as floppy drives, hard drives, monitors, motherboards, modems, controllers, etc.

Component Distributors
5950 Crooked Creek Rd., Ste. 215, Norcross, GA 30092 USA
(800) 777-7334 (tel)
cdisales@compdist.com
http://www.comp.dist.com/cdi
Sells equipment for professional computer network applications.

Component Parts
c/o Franklin Lee, 230 Hampton Pkwy., Kenmore, NY 14217 USA
Sells Van de Graaff generators, kits, etc. for experimental electrical applications.

Component Technology
4950 Allison Pkwy. #C, Vacaville, CA 95688 USA
(800) 878-0540 (tel)
(707) 452-8037 (tel)
(707) 452-8060 (fax)
Visa, MasterCard, Discover
Sells flyback transformers for a wide variety of electronics applications. Offers a free catalog.

Com-Power
114 Olinda Dr., Brea, CA 92621 USA
(714) 528-8800 (tel)
Manufactures spectrum analyzers, preamplifiers, antennas, comb generators, RF amplifiers, etc. for a variety of professional radio and industrial applications.

Comptek Computer
16560 Harbor Blvd. #D, Fountain Valley, CA 92708 USA
(800) 88-FIXPC (tel)
(714) 839-7274 (tel)
(714) 839-3787 (fax)
Sells monitors, terminals, and printers for a wide variety of computer applications.

Compu America
295 Ocean Pkwy., Ste. A7, Brooklyn, NY 11218 USA
(800) 974-2392 (tel)
(800) 794-2392 (wholesale tel)
(718) 871-3368 (tel)
(718) 436-4952 (fax)
Visa, MasterCard, Discover
Sells computer printers, parts, and boards, such as motherboards, memory chips, etc.

Compucon Computers
09-443-8088 (tel)
09-443-8222 (fax)
Manufactures personal computers.

Compu.D International
6741 Van Nuys Blvd., Van Nuys, CA 91405 USA
(800) 929-9333 (tel)
(818) 787-3282 (sales tel)
(818) 787-4548 (tel)
(818) 780-0816 (export tel)
(818) 780-9457 (dealer tel)
(818) 787-5555 (fax)
(818) 787-2111 (fax)
74673.1503@compuserve.com
Visa, MasterCard, American Express
Sells personal desktop and laptop computers, printers, and scanners; also sells computer parts and boards, such as hard drives, monitors, processor upgrades, graphics cards, etc.

Compudoc
6225 S. Harrison Ave., #5, Las Vegas, NV 89120 USA
(800) 895-2575 (tel)
(702) 878-7993 (tech support tel)
(702) 878-9209 (fax)

Compudoc

Visa, MasterCard, Discover, American Express
Sells computer parts and boards, such as hard drives, motherboards, memory chips, CPUs, monitors, etc.

Comp-U-Plus
20 Robert Pitt Dr., Monsey, NY 10952 USA
(800) 287-2323 (tel)
(914) 352-8100 (tel)
Visa, MasterCard, Discover, American Express, COD
Sells computer parts and boards, such as tape drives, hard drives, CD-ROM drives, video cards, modems, memory chips, etc.

CompUSA Direct
15167 Business Ave., Dallas, TX 75244 USA
(800) 266-7872 (tel)
(214) 888-5770 (tel)
(800) FAX-2212 (fax)
(214) 888-5760 (fax)
(900) 896-9393 (tech support tel)
Visa, MasterCard, Discover, American Express, Diner's Club, CompUSA
Sells personal desktop and laptop computers, printers, and scanners; also sells computer parts and boards, such as monitors, memory upgrades, modems, etc. Offers a free catalog with more than 5,000 different products.

CompuShack
903 N. Bowser Rd. #120, Richardson, TX 75081 USA
(800) 500-2896 (tel)
(214) 644-7211 (tel)
(214) 644-7351 (fax)
Visa, MasterCard, American Express, COD
Manufactures and sells personal desktop computers; also sells computer parts and boards, such as tape drives, hard drives, monitors, motherboards, modems, drive controllers, etc.

Compustar Computers
61A Church Ave., Brooklyn, NY 11218 USA
(800) 369-5411 (tel)
(718) 853-7888 (tel)
(718) 854-1820 (fax)
compustar9@aol.com

http://www.prismdesign.com/cstar
Sells computer parts and boards, such as hard drives, motherboards, memory chips, sound cards, monitors, multimedia upgrades, modems, etc.

ComputAbility Consumer Electronics
P.O. Box 17882, Milwaukee, WI 53217 USA
(800) 554-9954 (tel)
(414) 357-8181 (tech tel)
(414) 357-7814 (fax)
http://www.compuability.com
Visa, MasterCard, Discover, American Express
Sells home, amateur, and professional audio video equipment, such as camcorders, VCRs, video editors, microphones, TVs, home theater equipment, etc.

Computek Systems
10703 Plano Rd., #300, Dallas, TX 75238 USA
(214) 503-6500 (tel)
(214) 340-9482 (fax)
Sells Zenith spare parts for a wide variety of computer applications.

Computer Add-Ons
143-16 45th Ave., Flushing, NY 11355 USA
(800) 989-2264 (tel)
(718) 939-7976 (tel)
(718) 886-2025 (tel)
(718) 939-7193 (fax)
Visa, MasterCard, Discover
Sells personal computers, scanners, and printers for personal computer applications.

Computer Aided Technologies
P.O. Box 18285, Shreveport, LA 71138 USA
(318) 636-1234 (tel)
(318) 686-0449 (fax)
(318) 687-2555 (tech support tel)
(318) 631-3082 (BBS)
j.springer7@geis.com
Sells scanner antennas and accessories.

Computer and Software Sales
501 Mecham Ave., Elmont, NY 11003 USA
(800) COMP-440 (tel)
(516) 775-7800 (tel)

(516) 775-7822 (fax)
Sells personal desktop and laptop computer systems; also sells computer parts and boards, such as hard drives, processor chips, monitors, floppy drives, system boards, etc.

Computer Automated Technology
6301 Zanes Ct., Plano, TX 75023 USA
(214) 517-6152 (tel)
Visa, MasterCard, Discover
Sells blueprint scanners for professional desktop architectural applications.

ComputerBoards
125 High St., Mansfield, MA 02048 USA
(508) 261-1123 (tel)
(508) 261-1094 (fax)
pcmcia@comp-boards.com
Manufactures analog input, D/A voltage output, serial network interface, GPIB interface, and computer boards for a variety of portable computer data acquisitions applications.

Computer Brokerage Service (CBS)
6313 Ft. Worth Ave., Ft. Worth, TX 76112 USA
(800) 657-9555 (tel)
(817) 496-3904 (tel)
(817) 496-3924 (fax)
Buys and sells refurbished computer equipment.

Computer Classics
Rt. 1 Box 117, Cabool, MO 65689 USA
(417) 469-4571 (tel)
(417) 467-4571 (fax)
Sells IBM parts for a wide variety of personal computer applications.

Computer Commodity (CCI)
1405 SW 6th Ct. #B, Pompano Beach, FL 33069 USA
(305) 942-6616 (tel)
(305) 942-7815 (fax)
Buys and sells new and used personal computer systems and printers; also sells computer parts and boards, such as hard drives, memory chips, system boards, network equipment, etc.

Computer Component Source
P.O. Box 9022, 135 Eileen Way, Syosset, NY 11791 USA
(800) 356-1227 (tel)
(800) 926-2062 (fax)
Sells components for a wide variety of personal computer applications.

Computer Connection
1101 W. 80th St., Minneapolis, MN 55420 USA
(612) 884-0758 (tel)
(612) 884-2206 (fax)
Sells NCR computer systems, terminals, and parts.

Computer Discounters
10543 Ewing Rd., Beltsville, MD 20705 USA
(301) 595-0500 (tel)
(301) 595-5112 (fax)
Visa, MasterCard, COD
Sells surplus personal desktop and laptop computers, printers, monitors, memory chips, etc.

Computer Discount Warehouse
1020 E. Lake Cook Rd., Buffalo Grove, IL 60089 USA
(800) 726-4CDW (tel)
(708) 465-6800 (fax)
info@cdw.com
http://www.cdw.com
Visa, MasterCard, Discover, American Express
Sells personal desktop computers and parts and accessories, including system boards, hard drives, CPU boards, software, printers, video cards, etc. Offers a free catalog.

Computer Disk Service
Newbury Park, CA 91320 USA
(805) 499-6355 (tel)
(805) 499-0346 (fax)
Visa, MasterCard, Discover
Sells data storage devices, including hard drives, tape drives, optical drives, IDE drives, etc. for a wide variety of computer applications.

Computer Dynamics
7640 Pelham Rd., Greenville, SC 29615 USA

Computer Dynamics

(803) 627-8800 (tel)
(803) 675-0106 (fax)
sales@cdynamics.com
Manufactures a variety of flat-panel displays and monitors for professional computer applications. Offers a free 84-page Designer's Guide.

Computer Friends

14250 NW Science Park Dr., Portland, OR 97229 USA
(800) 547-3303 (tel)
(503) 626-2291 (tel)
(503) 643-5379 (fax)
cfriends@teleport.com
Sells computer scanners, printers, VGA-to-TV encoders, etc. for professional and personal desktop artwork, publishing, and video applications.

Computer Gate International

2960 Gordon Ave., Santa Clara, CA 95051 USA
(408) 730-0673 (tel)
(408) 730-0735 (fax)
http://www.aimnet.com/_7Ecgate/cgate.htm
Visa, MasterCard, Discover, COD
Sells computer parts and boards, such as cables, adapters, video cards, speakers, mice, surge protectors, etc. Offers a free catalog.

Computer Goldmine

4890 Pearl St., Boulder, CO 80301 USA
(800) 878-5554 (tel)
(303) 443-5554 (tel)
(303) 443-5557 (fax)
fmills@csn.net
http://www.csn.net/~fmills
Sells computer systems, peripherals, plotters, printers, image scanners, etc. for a wide variety of computer applications.

Computer Innovations

1129 Broad St., Shrewsbury, NJ 07702 USA
(800) 922-0169 (tel)
(908) 542-5920 (tel)
(908) 542-6121 (fax)
Manufactures data storage systems for a wide variety of professional and personal computer applications.

Computer Junction

1730 N. Greenville, Richardson, TX 75081 USA
(800) 783-9399 (tel)
(214) 783-9393 (tel)
(214) 234-6160 (fax)
Visa, MasterCard, Discover, American Express
Sells new and used computer parts and boards, such as motherboards, video cards, monitors, cases, power supplies, etc.

Computerlane

7500 Topanga Cyn Blvd., Canoga Park, CA 91303 USA
(800) 526-3482 (tel)
(818) 884-8644 (tel)
(818) 884-8253 (fax)
ssolim01@interserve.com
Visa, MasterCard
Sells personal desktop and laptop computer systems and printers.

Computer Leasing Exchange

1003 Wirt, Ste. 100, Houston, TX 70055 USA
(800) 428-2532 (tel)
(713) 467-9384 (tel)
atiwa@iamerica.net
Sells personal desktop and laptop computers, printers, and monitors.

Computer Marketing International

85 Flagship Dr., N. Andover, MA 01845 USA
(800) 497-4264 (tel)
(508) 687-3700 (tel)
(508) 689-2031 (fax)
(508) 687-4395 (international fax)
Sells new and refurbished workstations, peripherals, and upgrades for a wide variety of computer applications.

Computer Marketing Investments

4941 Allison St., Unit 2, Arvada, CO 80002 USA
(313) 422-3231 (tel)
(313) 422-3932 (fax)
Sells computer systems and terminals for a wide variety of computer applications.

Computer Recyclers

The Computer Memory Outlet
13628 Beta Rd., Dallas TX 75244 USA
(800) 617-2220 (tel)
(214) 385-0001 (tel)
(214) 385-0004 (fax)
Sells memory chips and computers for personal computer applications.

Computer Monitor Service
2414 Hwy. 80 E., #203, Mesquite, TX 75149 USA
(214) 285-9935 (tel)
(214) 285-5330 (fax)
Buys and sells used and reconditioned computer monitors.

Computer Network Integration
111 Wood St., Pittsburgh, PA 15222 USA
(800) 873-0303 (tel)
(412) 391-7804 (tel)
(412) 391-9512 (fax)
Manufactures gateways and servers for professional computer network applications.

Computer Network Systems
2212 Superior Ave., Cleveland, OH 44114 USA
(800) 552-7888 (tel)
(216) 241-2002 (tel)
(216) 241-4440 (tech support tel)
(216) 241-1628 (fax)
Visa, MasterCard, Discover, American Express, COD
Manufactures and sells personal desktop computers; also sells computer parts and boards, such as CD-ROM drives, hard drives, monitors, video cards, modems, tape drives, etc. Offers a 30-day money-back guarantee.

The Computer Outlet
1500 Beville Rd., Daytona Beach, FL 32114 USA
(800) 900-0695 (tel)
(904) 947-4865 (fax)
Visa, MasterCard, Discover, American Express
Sells new and used personal computer systems; also sells computer parts and boards, such as hard drives, memory chips, cases, floppy drives, keyboards, etc.

Computer Parts & Pieces
301 W. Abram St., Arlington, TX 76010 USA
(800) 235-0096 (tel)
(817) 274-4716 (tech support tel)
(817) 274-2388 (fax)
Visa, MasterCard
http://www.parts-n-pcs.com
Sells personal computer parts and accessories, including network adapters, system boards, hard drives, power supplies, controllers, etc.

Computer Parts Unlimited
5069 Maureen Ln., Moorpark, CA 93021 USA
(805) 532-2500 (tel)
(805) 532-2599 (fax)
Visa, MasterCard, Discover, American Express
Buys and sells used and refurbished personal desktop and laptop computers, monitors, and parts.

Computer Products
1431 S. Cherryvale Rd., Boulder, CO 80303 USA
(800) 338-4273 (tel)
(303) 442-3685 (tel)
(303) 442-0501 (fax)
Specializes in selling computer hard drives, but also sells CD-ROM drives, modems, tape drives, etc. for computer applications.

Computer Products
7 Elkins St., S. Boston, MA 02127 USA
(617) 268-1170 (tel)
Manufactures dc-to-dc power converters for a wide variety of electronics applications.

Computer Products
Power Conversion Europe, Springfield Industrial Estate, Youghal, Co. Cork, Ireland
+353-24-93130 (tel)
Manufactures dc-to-dc power converters for a wide variety of electronics applications.

Computer Recyclers
3784 Realty Rd., Dallas, TX 75244 USA
(214) 484-6447 (tel)
(214) 406-1505 (fax)
Sells new and used computer parts and boards, such as floppy drives, hard drives, video cards, modems,

Computer Recyclers

motherboards, memory chips, network cards, etc. Offers a free catalog.

Computer Recyclers
1010 N. State St., Orem, UT 84057 USA
(801) 226-1892 (tel)
(801) 226-2129 (fax)
Sells personal computer parts and accessories, including hard drives, power supplies, modems, network equipment, etc.

Computer Reset
P.O. Box 461782, Garland, TX 75046 USA
(214) 276-8072 (tel)
(214) 272-7920 (fax)
sales@c-reset.com
http://www.unicomp.net/c-reset
Visa, MasterCard, American Express, COD
Sells used personal desktop and laptop computers, monitors, printers, etc. Offers a free catalog.

Computer Salvage Discount
7944-A Gateway E., El Paso, TX 79915 USA
(915) 598-6000 (tel)
(915) 598-6016 (fax)
Buys and sells used personal desktop computers, and parts and accessories, including memory chips, motherboards, etc.

Computers Direct
3613 LaFayette Rd., Portsmouth, NH 03801 USA
(800) 222-4070 (tel)
Visa, MasterCard, Discover
http://www.ssdonline.com
Sells personal desktop and laptop computers, scanners, printers, etc. Offers a 30-day money-back guarantee on desktop computer systems.

Computer Service Center
1310 E. Edinger Ave.; #E, Santa Ana, CA 92705 USA
(714) 543-3034 (tel)
(714) 543-7607 (fax)
Sells computer printers and printer parts.

Computers, Parts, and Commodities (CPAC)
22349 La Palma Ave., Ste. 114, Yorba Linda, CA 92687 USA
(800) 778-2722 (tel)
(714) 692-5044 (tel)
(714) 692-6680 (fax)
Buys and sells new and refurbished IBM personal desktop and laptop computer systems; also sells computer parts and boards, such as network equipment, monitors, adapters, etc.

Computer Terminal Repair
8971 Quite Canyon Rd., Placerville, CA 95667 USA
(800) 675-2880 (tel)
(916) 621-2888 (fax)
Sells refurbished printheads for a wide variety of computer applications.

Computer Trade Exchange
3A Great Meadows Ln., E. Hanover, NJ 07936 USA
(201) 887-9300 (tel)
(201) 887-9303 (fax)
Sells personal home and laptop computer and printers, parts and accessories, including memory chips, system boards, hard drives, power supplies, floppy drives, etc.

Computer Trend
104 Rainbow Industrial Blvd., Rainbow City, AL 35906 USA
(205) 442-6376 (tel)
(205) 442-4615 (fax)
Visa, MasterCard
Buys and sells IBM computer parts for a wide variety of computer applications.

Computer Ventures
619 W. Estes Ave., Schaumberg, IL 60193 USA
(708) 894-8857 (tel)
(708) 894-8859 (fax)
Manufactures workstations for a wide variety of professional computer network applications.

Computer Warehouse
2819 Blystone Ln., Dallas, TX 75220 USA

(214) 351-4993 (tel)
(214) 350-4350 (fax)
Visa, MasterCard
Sells used and reconditioned printers and terminals for a wide variety of computer applications.

Computerware Plus

93 Wordsworth St., Christchurch, New Zealand
03-365-4195 (tel)
03-366-3212 (fax)
BCS House, 190 The Parade, Island Bay, New Zealand
04-383-9822 (tel)
04-383-5203 (fax)
14 Teed St., Newmarket, New Zealand
09-523-3058 (tel)
09-523-3059 (fax)
Manufactures personal computers.

Computer Wholesalers

3246 Marjan Dr., Doraville, GA 30340 USA
(800) 229-2897 (tel)
(404) 457-5841 (fax)
Sells Data General computer systems for professional computer applications.

Computerwise

302 N. Winchester, Olathe, KS 66062 USA
(800) 255-3739 (tel)
(913) 829-0600 (tel)
(913) 829-0810 (fax)
Manufactures bar-coded attendance clock systems for a wide variety of professional computing applications.

Computone

1100 Northmeadow Pkwy., #150, Roswell, GA 30076 USA
(800) 241-3946 (tel)
(404) 475-2725 (tel)
(404) 475-2707 (fax)
Manufactures terminal servers for professional computer network applications.

CompuTrend (CTI)

3115 Gateway Dr., #F, Norcross, GA 30071 USA
(770) 734-9220 (tel)
(770) 734-9244 (fax)
Sells data storage systems, such as CD-ROM drives and optical drives, for a wide variety of computer applications.

CompuWorld

24441 Miles Rd., Unit A-1, Cleveland, OH 44128 USA
(800) 666-6294 (tel)
(216) 595-6500 (tel)
(800) 99-TECH-9 (tech support tel)
(216) 595-6565 (fax)
Visa, MasterCard, Discover, American Express
Manufactures and sells personal desktop computers; also sells computer parts and boards, such as floppy drives, hard drives, monitors, motherboards, modems, video cards, etc. Offers a 30-day money-back guarantee.

Comp View

13405 NW Cornell Rd., Portland, OR 97229 USA
(800) 448-8439 (tel)
(503) 641-8439 (tel)
(503) 643-9461 (fax)
Manufactures projection panels for a wide variety of professional computer applications.

Com-Rad Industries

P.O. Box 88, Wilson, NY 14172 USA
(716) 751-9945 (tel)
(716) 751-9879 (fax)
Manufactures and sells low-profile antennas for cellular radio applications.

Comrex

65 Nonset Path, Acton, MA 01720 USA
(508) 263-1800 (tel)
(508) 635-0401 (fax)
Manufactures compact mixing consoles and a variety of digital audio codecs for remote broadcasting.

Comroad

Bruckmanring 32, D-85764 Oberschleissheim, Germany
(800) 784-7243 (US tel)
+49-89-31-57-19-0 (tel)

Comroad

+49-89-3-15-91-46 (fax)
http://www.solidinfo.com
Manufactures and sells computer-based mobile systems that combine computers, GPS, RDS, etc.

ComSonics
1350 Port Republic Rd.,
P.O. Box 1106, Harrisonburg, VA 22801 USA
(800) 336-9681 (tel)
(540) 434-5965 (tel)
(540) 432-9794 (fax)
(540) 434-9847 (fax)
Manufactures leakage detectors for professional cable TV applications.

Comstar
5250 W. 74th St., Minneapolis, MN 55439 USA
(612) 835-5502 (tel)
(612) 835-1927 (fax)
Buys and sells network equipment for professional computer applications.

Comtec Automated Solutions
10000 Old Katy Rd., #150, Houston, TX 77055 USA
(713) 935-3666 (tel)
(713) 935-3650 (fax)
Manufactures data storage systems for a wide variety of professional computer network applications.

ComTek Systems
P.O. Box 470565, Charlotte, NC 28247 USA
(714) 542-4808 (tel)
(704) 542-9652 (fax)
Manufactures and sells a hybrid phased-array antenna network for amateur radio applications.

Comtelco Industries
501 Mitchell Rd., Glendale Heights, IL 60139 USA
(800) 634-4622 (tel)
(708) 790-9799 (fax)
Manufactures base antennas for professional two-way radio applications.

Comtrade
1215 Bixby Dr., City of Industry, CA 91745 USA
(800) 969-2123 (tel)
(800) 868-5588 (corp orders tel)
(818) 961-6688 (tel)
(818) 369-1479 (fax)
(818) 961-6098 (BBS)
Manufactures and sells personal desktop and laptop computers. Offers a 30-day money-back guarantee.

Comtrol
2675 Patton Rd., St. Paul, MN 55113 USA
(800) 926-6876 (tel)
(612) 631-7654 (tel)
(612) 631-8117 (fax)
Manufactures serial boards for professional computer network applications.

Com-West Radio Systems
8179 Main St., Vancouver, BC V5X 3L2 Canada
(604) 321-3200 (tel)
(604) 321-6560 (fax)
Sells a wide variety of amateur radio equipment and accessories, including amateur radio transceivers, receivers, antenna tuners, antennas, amplifiers, etc.

Comtrad Industries
2820 Waterford Lake Dr., #106, Midlothian, VA 23113 USA
(800) 992-2966 (tel)
Sells wireless video and audio transmitter/receivers to couple VCRs with TVs in other locations, etc.

Concept Seating
W227 N6193 Sussex Rd., Sussex, WI 53089-3969 USA
(414) 246-0900 (tel)
(414) 246-0909 (fax)
Randall Marketing & Consulting
18 Twickenham Ct., Thornhill, ON L3T 5T7 Canada
(905) 709-4452 (tel)
(905) 709-4460 (fax)
Manufactures vehicle consoles for professional data/communications equipment storage.

Concord
501 Center St., Huntington, IN 46750 USA
(219) 356-1200 (tel)

(219) 356-2830 (fax)
Manufactures loudspeakers and amplifiers for car stereo applications.

Concorde Technologies
9770 Carroll Center Rd., Ste. F, San Diego, CA 92126 USA
(800) 359-0282 (tel)
Manufactures CD-ROM recorder systems for professional computer applications.

Condor D.C. Power Supplies
2311 Stratham Pkwy., Oxnard, CA 93033 USA
(800) 235-5929 (tel)
(805) 486-4565 (tel)
(805) 487-8911 (fax)
Manufactures dc power supplies for a wide variety of computer and electronics applications.

Conec
72 Devon Rd., Unit 1, Brampton, ON L6T 5B4 USA
(905) 790-2200 (tel)
(905) 790-2201 (fax)
Manufactures connectors and filter adapters for a wide variety of computer applications.

Conifer
1400 N. Roosevelt, P.O. Box 1025, Burlington, IA 52601 USA
(800) 843-5419 (tel)
(319) 752-3607 (tel)
(319) 753-5508 (fax)
Manufactures wireless receivers for professional cable TV applications.

Conley
420 Lexington Ave., 10th Fl., New York, NY 10017 USA
(212) 682-0162 (tel)
(212) 682-0071 (fax)
Manufactures data storage systems for a wide variety of professional computer network applications.

Conneaut Audio Devices (CAD)
P.O. Box 120, Conneaut, OH 44030 USA
(216) 593-1111 (tel)
Manufactures microphones for professional studio recording.

Connecticut Microcomputer
568 Danbury Rd., New Milford, CT 06776 USA
(800) 426-2872 (tel)
(203) 354-9395 (tel)
(203) 355-8258 (fax)
Sells equipment for personal and professional computer applications.

Connectix
2655 Campus Dr., San Mateo, CA 94403 USA
(800) 839-3628 (tel)
(415) 571-5100 (tel)
(415) 571-5195 (fax)
info@connectix.com
Manufactures tiny video cameras for a wide variety of computer applications, including video conferencing, movie making, etc.

Connectors Unlimited
P.O. Box 5983, Manchester, NH 03108 USA
(800) 549-5955 (tel)
(603) 641-1179 (fax)
Sells a wide variety of connectors for audio, video, cellular, CATV, data, transmitting, etc.

Connect Systems
2259 Portola Rd., Ventura, CA 93003 USA
(800) 545-1349 (tel)
(805) 642-7184 (tel)
(805) 642-7271 (fax)
Manufactures decoders for professional two-way radio applications.

Connectware
1301 E. Arapaho, Richardson, TX 75081 USA
(214) 907-1093 (tel)
(214) 907-1594 (fax)
Manufactures and sells radiation protection/antiglare filters for computer monitors.

Conner Peripherals
3081 Zanker Rd., San Jose, CA 95134 USA

Conner Peripherals

(800) 626-6637 (tel)
(408) 456-4500 (tel)
(407) 262-4755 (East Coast tel)
(408) 456-4903 (fax)
Manufactures hard disk drives for personal and professional computer applications.

Conrad-Johnson Design

2733 Merilee Dr., Fairfax, VA 22031 USA
(703) 698-8581 (tel)
(703) 560-5360 (fax)
Manufactures amplifiers and preamplifiers for high-end home audio and home theater applications.

Continental Computers

10524 S. La Cienega Blvd., Inglewood, CA 90304 USA
(310) 410-0133 (tel)
(310) 410-4942 (fax)
Visa, MasterCard
Sells personal desktop computers and parts and accessories, including system boards, hard drives, CPU boards, etc.

Continental Electronics

P.O. Box 270879, Dallas, TX 75227 USA
(800) 733-5011 (tel)
(214) 381-7161 (tel)
(214) 381-4949 (fax)
Manufactures AM, FM, and shortwave transmitters for professional broadcasting applications.

Control Signal (CSC)

1985 S. Depew, #7, Denver, CO 80227 USA
(800) 521-2203 (tel)
(303) 989-8000 (tel)
Manufactures Morse code station identifiers for professional radio applications.

Conversion Devices

15 Jonathan Dr., Brockton, MA 02401 USA
(508) 559-0880 (tel)
(508) 559-9288 (fax)
Manufactures dc-to-dc power converters for a wide variety of electronics applications.

Conversion Equipment

330 W. Taft Ave., Orange, CA 92665 USA
(614) 637-2970 (tel)
Manufactures power supplies for telecommunications applications.

Converters

77 W. Main St., Smithtown, NY 11787 USA
(800) 322-9690 (tel)
Visa, MasterCard
Sells converters for home cable TV applications.

Cooke International

Unit 4, Fordingbridge Site, Main Rd.
Barnham, Bognor Regis, W. Sussex PO22 0EB UK
44-01243-545111 (tel)
44-01243-542457 (fax)
Buys and sells used electronics test equipment, including voltmeters, spectrum analyzers, oscilloscopes, power meters, etc.

Cool-Lux

409 Colle San Pablo, Ste. 105, Camarillo, CA 93012 USA
(800) 223-2589 (tel)
(805) 482-4820 (tel)
(805) 482-0736 (fax)
Manufactures and sells a variety of lights for professional and amateur video recording and batteries and battery belts, especially for camcorders. Offers a free catalog.

Cooper Instruments & Systems

P.O. Box 3048, Warrenton, VA 22186 USA
(800) 344-3921 (tel)
(703) 347-4755 (fax)
Manufactures force cells, pressure transducers, pressure gauges, etc. for a wide variety of professional industrial applications.

R.J. Cooper & Associates

(800) 752-6673 (tel)
(714) 240-4852 (tel)
rjcoop@aol.com
Manufactures and sells joysticks so that people with

disabilities can use Windows on personal desktop computers.

Copley Controls
410 University Ave., Westwood, MA 02090 USA
Manufactures servo-amplifiers for professional electronics applications.

Copper Electronics
3315 Gilmore Industrial Blvd., Louisville, KY 40213 USA
(800) 626-6343 (tel)
(502) 968-8500 (tel)
(502) 968-0449 (fax)
Sells a variety of CBs, scanners, shortwave radios, antennas, accessories, etc. Offers a free catalog.

Cord-Lox
1658 Precision Park Ln., San Yfidro, CA 92173 USA
(800) CORD-LOX (tel)
Manufactures Velcro cable ties for audio and radio cables. Offers a free sample and brochure.

Cornell Dubilier
1605 E. Rodney French Blvd., New Bedford, MA 02744 USA
(508) 996-8561 (tel)
(508) 996-3830 (fax)
Manufactures capacitors for a wide variety of electronics applications.

Corning
P.O. Box 7429, Endicott, NY 13760 USA
(800) 525-2524 (tel)
http://usa.net/corning-fiber
Manufactures fiberoptic cable for a wide variety of electronics applications. Offers a free brochure.

Corollary
2802 Kelvin, Irvine, CA 92714 USA
(800) 338-4020 (tel)
(714) 250-4040 (tel)
(714) 250-4043 (fax)
Manufactures file servers and concentrators for a wide variety of professional computer network applications.

Corporate Raider
1449 39th St., Brooklyn, NY 11218 USA
(800) 453-3555 (tel)
(718) 633-7916 (customer service tel)
(718) 633-9100 (wholesale order tel)
(718) 633-7841 (fax)
http://www.rothnet.com/corpraid/corpraid.html
Visa, MasterCard, Discover, American Express
Sells printers, scanners, fax machines, etc. for a wide variety of personal computer and office applications.

Cort
3451 W. Commercial Ave., Northbrook, IL 60062 USA
(708) 498-6491 (tel)
(708) 498-5370 (fax)
cortguitar@aol.com
http://www.cort.com
Manufactures electric guitars for professional and personal musical applications. Offers a catalog for $2.

COSMOSIC
P.O. Box 112-468, Taipei, Taiwan
886-2-752-8485 (tel)
886-2-752-7477 (fax)
COSMOSIC is the Asian representative for Borbely Audio (Germany).

CoStar
599 W. Putnam Ave., Greenwich, CT 06830 USA
(203) 661-9700 (tel)
(203) 661-1540 (fax)
(203) 661-6292 (BBS)
75300.2225@compuserve.com
http://www.costar.com
Manufactures printers, address writers, etc. for personal and professional computer applications.

Costello's Music
10 W. Main St., Fredonia, NY 14063 USA
(716) 672-3429 (tel)
(716) 672-5115 (fax)
Sells equipment for professional and personal musical applications. Offers a free catalog.

Cotronics
2250 SE Federal Hwy., Stuart, FL 34994 USA
(800) 848-3004 (tel)
Sells Radio Shack scanners and shortwave receivers.

Coustic
4260 Charter St., Vernon, CA 90058 USA
(213) 582-2832 (tel)
(213) 582-4328 (fax)
Manufactures loudspeakers, crossovers, amplifiers, and head units for car stereo applications.

Covia Technologies
9700 W. Higgins Rd., Rosemont, IL 60018 USA
(800) 566-1969 (tel)
(708) 518-4166 (tel)
(708) 518-4850 (fax)
Manufactures communications integrators for professional computer network applications.

Hal Cox
164 Tamalpais Ave., Mill Valley, CA 94941 USA
(415) 388-5711 (tel)
(415) 388-3359 (fax)
Visa, MasterCard
Buys and sells loudspeakers for home audio applications.

CPI
1186 Commerce Dr., Richardson, TX 75801 USA
(800) 869-9128 (tel)
(214) 437-5320 (tel)
Manufactures remote-control systems for Midland two-way radios.

C.P.I.
7946 S. State St., Midvale, UT 84047 USA
(800) 942-8873 (tel)
(801) 567-9494 (tel)
Sells a wide variety of amateur radio equipment and accessories, including amateur radio transceivers, receivers, antenna tuners, towers, antennas, radio teletype equipment, amplifiers, etc.

Craig Consumer Electronics
13845 Artesia Blvd., Cerritos, CA 90701 USA
(310) 926-9944 (tel)
(416) 673-3307 (Canada tel)
(310) 926-9269 (fax)
Manufactures head units, amplifiers, equalizers, loudspeakers, etc. for car stereo applications.

C. Crane
558 10th St., Fortuna, CA 95540 USA
(800) 522-8863 (tel)
(707) 725-9000 (tel)
(707) 725-9060 (fax)
Sells shortwave radios, scanners, antennas, and accessories. Offers a free catalog.

Cranel
8999 Gemeni Pkwy., Columbus, OH 43240 USA
(800) 288-3475 (tel)
(614) 431-8000 (tel)
(614) 431-8388 (fax)
Manufactures file servers for professional computer network applications.

Crate
1400 Fergusin Ave., St. Louis, MO 63133 USA
Manufactures amplifiers for professional and personal musical applications.

Creative Labs
1523 Cimarron Plaza, Stillwater, OK 74075 USA
(800) 998-5227 X110 (info tel)
(800) 998-5227 (tel)
(405) 742-6622 (tech support)
(405) 742-6633 (fax support)
(405) 742-6660 (BBS)
65-773-0233 (int'l tel, Singapore)
65-773-0353 (int'l fax, Singapore)
http://www.creativelabs.com
Sells a wide variety of multimedia packages that include sound cards, CD-ROMs, games, multimedia software, speakers, microphones, etc.

Creative Micro Electronics (CME)
P.O. Box 4477, Englewood, CO 80155 USA
(800) 771-1295 (tel)
(303) 771-1288 (tel)

(303) 771-1136 (fax)
Visa, MasterCard, Discover
Manufactures microwave audio/video transmitters and receivers for hobby and experimental applications.

Crest Audio
100 Eisenhower Dr., Paramus, NJ 07652 USA
(201) 909-8700 (tel)
(201) 909-8744 (fax)
+44-0-1273-693513 (UK tel)
+44-0-1273-692894 (UK fax)
+49-0-2173-915450 (Germany tel)
+49-0-2173-168247 (Germany fax)
(201) 909-8700 (Asia tel)
(201) 909-8744 (Asia fax)
Manufactures audio consoles for professional studio recording and live music performances.

Crestron
101 Broadway, Cresskill, NJ 07626 USA
(800) 237-2041 (tel)
(201) 894-0660 (tel)
(201) 894-1192 (fax)
+32-15-730-974 (Europe tel)
+32-15-233-579 (Europe fax)
http://www.crestron.com
Manufactures touch-screen home control systems.

Cricklewood Electronics
40-42 Cricklewood Broadway, London NW2 3ET UK
Sells a wide variety of electronics components; professional, home, and personal audio equipment; computer accessories; and test equipment. Offers a catalog for £2.50.

Critical Mass
1647 Acme St., Orlando, FL 32805 USA
(800) 545-6277 (tel)
(407) 872-5772 (fax)
Manufactures loudspeakers for car stereo applications.

The CR Supply
127 Chesterfield Rd., Sheffield S8 0RN UK
0114-2557771 (tel)

Sells a wide variety of electronics components, including capacitors, transistors, diodes, voltage regulators, etc. Offers a list for a stamp.

Crown International
1718 W. Mishawaka Rd., Elkhart, IN 46517 USA
(800) 342-6939 (tel)
(219) 294-8050 (tel)
(219) 294-8302 (fax)
http://www.crownintl.com
Manufactures transmitters, translators, and hard drive storage systems for professional broadcasting applications; also manufactures studio amplifiers for professional studio recording applications.

Cruising Equipment
6315 Seaview Ave. NW, Seattle, WA 98107 USA
(206) 782-5869 (tel)
(206) 782-4336 (fax)
Manufactures battery monitors for a wide variety of electronics applications.

Crunch
RR #6, Box 149, Dothan, AL 36303 USA
(334) 983-6542 (tel)
(334) 983-6205 (fax)
Manufactures loudspeakers and amplifiers for car stereo applications.

Crutchfield
1 Crutchfield Pk., Charlottesville, VA 22906 USA
(800) 890-4302 (tel)
(804) 973-1862 (tel)
catalog@crutchfield.com
administration@crutchfield.com
http://www.crutchfield.com
Visa, MasterCard, American Express
Sells a wide variety of car audio and home theater equipment, including speakers, head units, amplifiers, crossovers, receivers, equalizers, etc. Offers a 100+ page catalog.

Crydom
411 N. Central Ave., Glendale, CA 91203 USA
(818) 956-3900 (tel)

Crydom

Manufactures liquid-level sensors for industrial applications.

Crystal Group
1165 Industrial Ave., Hiawatha, IA 52233 USA
(800) 378-1636 (tel)
(319) 378-1636 (tel)
(319) 393-2338 (fax)
Manufactures rack-mount computers for professional and industrial computer applications.

Crystal-Line
2075 Byberry Rd., #10, Bensalem, PA 19020 USA
(215) 638-1600 (tel)
(215) 638-4515 (fax)
Manufactures signal processors and crossovers for car stereo applications.

Crystaloid
5282 Hudson Dr., Hudson, OH 44036 USA
(216) 655-2429 (tel)
(216) 655-2176 (fax)
Manufactures LCD touch screens for a wide variety of electronics applications.

Crystek Crystals
2351/2371 Crystal Dr., Ft. Myers, FL 33907 USA
(800) 237-3061 (tel)
Manufactures a variety of electronics crystals.

Crystronics Pager Parts Unlimited
(305) 776-0031 (tel)
(305) 776-0109 (fax)
Sells crystals and pager components for cellular telephone applications.

CS Electronics
1342 Bell Ave., Tustin, CA 92680 USA
(714) 259-9100 (tel)
Manufactures cabling for professional computer network applications.

C & S Electronics
P.O. Box 2142, Norwalk, CT 06852
(203) 866-3208 (tel)

Sells specialty semiconductors, including RF chips, etc.

C&S Sales
1245 Rosewood, Deerfield, IL 60015 USA
(800) 292-7711 (USA tel)
(800) 445-3201 (Canada tel)
(708) 541-0710 (tel)
(708) 520-0085 (fax)
Visa, MasterCard
Sells a variety of new test equipment, including oscilloscopes, probes, scopemeters, meters, etc. Offers a free catalog and a 15-day money-back warranty.

C.S. Source
15338 E. Valley Blvd., City of Industry, CA 91746 USA
(800) 967-5292 (tel)
(800) 967-2269 (tech serv tel)
(818) 855-2700 (tel)
(818) 330-6807 (fax)
Visa, MasterCard
Manufactures and sells personal desktop and laptop computers.

CSS Power
8905 Kelso Dr., Baltimore, MD 21221 USA
(800) 780-3120 (tel)
(410) 780-0241 (fax)
Manufactures power breakers for professional cellular microwave radio applications.

CST
2336 Lu Field Rd., Dallas, TX 75229 USA
(214) 241-2662 (tel)
(214) 241-2661 (fax)
(214) 241-3782 (BBS)
Manufactures memory chip testers for a variety of professional computer applications.

C.S.U. Industries
207 Rockaway Tnpke., Lawrence, NY 11559 USA
(516) 239-4310 (tel)
(516) 239-8374 (fax)
Sells Hewlett-Packard workstations for professional computer applications.

C Sys Labs
1430 Koll Cir., Ste. 103, San Jose, CA 95112 USA
(408) 453-5380 (tel)
(408) 453-5382 (fax)
http://www.sbusiness.com
Manufactures LCD touch panels for a wide variety of electronics applications.

CT2M
56 rue de Londres, Paris 75008 France
+33-43-87-89-29 (tel)
+33-43-87-28-71 (fax)
Manufactures file servers for a wide variety of professional computer network applications.

CTP
517 Lower Terr., Huntington, WV 25705 USA
(304) 525-6372 (tel)
Sells analog S-meters for use with scanners. Offers a free catalog.

C-Tronics
P.O. Box 192, Ramsey, NJ 07446 USA
(201) 818-4289 (tel)
Manufactures and sells flight alarms for radio-control model airplanes.

CTX International
20530 Earlgate St., Walnut, CA 91789 USA
(800) 289-8808 (National sales tel)
(800) 888-2012 (tel)
(201) 646-0707 (East sales tel)
(404) 729-8909 (South sales tel)
(909) 598-8094 (West sales tel)
(818) 837-4341 (fax back)
Manufactures monitors for personal desktop computer systems and manufactures LCD projection panels for computer multimedia applications.

Cubix
2800 Lockheed Way, Carson City, NV 89706 USA
(800) 829-0550 (tel)
(702) 888-1000 (tel)
(702) 888-1001 (fax)
1 Hunter Rd., Kirkton South, Livingston, Scotland EH54 9DH UK
+44-1506-465065 (tel)
+44-1506-465430 (fax)
Manufactures servers for professional computer network applications.

CUI Stack
9640 SW Sunshine Ct., #700, Beaverton, OR 97005 USA
(503) 643-4899 (tel)
(503) 643-6129 (fax)
Manufactures plug-in power supplies for a wide variety of electronics applications.

Cummins-Allison
891 Feehanville St., Mt. Prospect, IL 60056 USA
(800) 786-5528 (tel)
(708) 299-4939 (fax)
Manufactures paper shredders for a wide variety of office applications.

Cunard Associates
RD#6, Box 104, Bedford, PA 15522 USA
(814) 623-7000 (tel/fax)
Visa, MasterCard
Sells rebuilt batteries for a wide variety of electronics applications. Offers a free catalog.

Curcio Audio Engineering
P.O. Box 643, Trexlertown, PA 18087 USA
(215) 391-8821 (tel)
Manufactures high-resolution vacuum-tube audio modules, including digital processors, preamplifiers/crossovers, and power amplifiers.

Curtis Electro Devices
4345 Pacific St., Rocklin, CA 95677 USA
(800) 332-2790 (tel)
(916) 632-0600 (tel)
(916) 632-0636 (fax)
Manufactures system analyzers for professional cellular microwave radio applications.

Curtis Manufacturing
30 Fitzgerald Dr., Jaffrey, NH 03452 USA

Curtis Manufacturing

(800) 955-5544 (tel)
Manufactures surge protectors for a wide variety of electronics applications.

Cushcraft/Signals
P.O. Box 4680, 48 Perimeter Rd.
Manchester, NH 03103 USA
(603) 627-7877 (tel)
(603) 627-1764 (fax)
Manufactures vertical and mobile antennas for the shortwave (HF) amateur radio bands.

Custom Autosound
808 W. Vermont Ave., Anaheim, CA 92805 USA
(714) 535-1091 (tel)
(714) 533-0361 (fax)
Manufactures custom cassette head units for antique car stereo applications.

Custom Computers
1174 Sweetwater Circle, Lawrenceville, GA 30244 USA
(800) 239-4467 (tel)
(404) 381-9114 (fax)
Visa, MasterCard
Sells computer parts and boards, such as hard drives, motherboards, memory chips, floppy drives, multimedia upgrades, video cards, etc.

Cutting Edge
2101 Superior Ave., Cleveland, OH 44114 USA
(216) 241-3343 (tel)
(216) 241-4103 (fax)
info@cuttingedge.zephyr.com
Manufactures the Unity series of audio processors for AM and FM broadcast stations.

CW Crystals
570 N. Buffalo St., Marshfield, MO 65706
Manufactures a variety of electronics crystals. Offers a catalog for $1.

CyberMax Computer
133 N. 5th St., Allentown, PA 18102 USA
(800) 443-9868 (tel)

(610) 770-1808 (tel)
Visa, MasterCard, Discover, American Express
Manufactures and sells personal desktop computers. Offers a 30-day, money-back guarantee.

Cybernetics
Tera One, Yorktown, VA 23693 USA
(804) 833-9000 (tel)
(804) 833-9300 (fax)
Sells data storage systems, such as tape drives, for a wide variety of computer applications.

Cybex
4912 Research Dr., Huntsville, AL 35805 USA
(205) 430-4000 (tel)
(205) 430-4030 (fax)
http://www.cybex.com/
Manufactures switching centers for professional computer network applications.

Cyclades
44140 Old Warm Springs Blvd., Fremont, CA 94538 USA
(800) 347-6601 (tel)
(510) 770-9727 (tel)
(510) 770-0355 (fax)
Manufactures Ethernet cards for professional computer network applications.

Cylix
2637 Townsgate Rd., #100, Westlake Village, CA 91361 USA
(800) 877-3735 (tel)
Manufactures surge protectors for a wide variety of computer applications.

Cytec
2555 Baird Rd., Penfield, NY 14526 USA
(800) 346-3117 (tel)
Manufactures VXI bus switch modules for a wide variety of professional electronics applications.

DAD
168 Main St., P.O. Box 711, Chadron, NE 69337 USA
(800) 669-4548 (tel)

(308) 432-2122 (tel)
(308) 432-2002 (fax)
Manufactures and sells receivers for radio-control models.

Daddy's Junky Music
P.O. Box 1018, Salem, NH 03079 USA
(603) 894-6492 (tel)
(603) 893-7023 (fax)
usedbymail@aol.com
http://www.daddys.com
Sells used electric guitars, synthesizer keyboards, amplifiers, sound effects, accessories, etc. for professional and personal audio and musical applications.

Daewoo Electronics
120 Chubb Ave., Lyndhurst, NJ 07071 USA
(800) DAEWOO8 (tel)
(800) 258-2088 (order tel)
(201) 460-2000 (tel)
(201) 935-5284 (fax)
100 Daewoo Pl., Carstadt, NJ 07062 USA
(201) 935-8700 (tel)
(201) 935-6491 (fax)
Manufactures and sells personal computer monitors.

Peter Dahl Co.
5869 Waycross Ave., El Paso, TX 79924 USA
(915) 751-0768 (tel)
(915) 751-2300 (tel)
Sells a wide variety of transformers, chokes, capacitors, etc.--especially hard-to-find transformers for antique amateur radio equipment.

Daily Electronics
10914 N.E. 39th St., Vancouver, WA 98682 USA
(800) 346-6667 (tel)
(206) 896-8856 (tel)
(206) 896-5476 (fax)
Sells a wide variety of used and new/old stock tubes.

DakTech
4900 Ritter Rd., Mechanicsburg, PA 17055 USA
(800) 325-3238 (tel)
(717) 795-9544 (tel)
(717) 795-9420 (fax)

daktech@ix.netcom.com
Specializes in IBM and Compaq replacement parts, printers, and complete systems for personal and professional computer applications.

Dalco Electronics
275 Pioneer Blvd., Springboro, OH 45066 USA
(800) 445-5342 (tel)
(800) 543-2526 (tech support tel)
(800) 228-9859 (dealer sales tel)
(800) 445-5342 (custom cables tel)
(800) 449-8487 (customer service tel)
(513) 743-8042 (business office tel)
(513) 743-9251 (sales fax)
(513) 743-2244 (BBS)
Visa, MasterCard, Discover, American Express
Sells computer parts and boards, such as modems, controller cards, headphones, speakers, keyboards, power supplies, cases, cables, etc.

Dale Pro Audio
7 East 20th St., New York, NY 10003 USA
(212) 475-1124 (tel)
(212) 475-1963 (fax)
Sells a variety of components and instruments for professional studio recording and high-end consumer audio applications.

Da-Lite Screen Company
3100 N. Detroit St., P.O. Box 137, Warsaw, IN 46581 USA
(800) 622-3737 (tel)
(219) 267-8101 (tel)
(219) 267-7804 (fax)
Manufactures projection screens for high-end home theater applications.

Dallas Computer Parts
1135 E. Plano Pkwy., #2, Plano, TX 75074 USA
(214) 422-1580 (tel)
(214) 422-5193 (fax)
Sells personal desktop and laptop computer systems, printers, terminals, monitors, etc.

Dallas Semiconductor
4401 S. Beltwood Pkwy., Dallas, TX 75244 USA

Dallas Semiconductor
(214) 450-0448 (tel)
(214) 450-3715 (fax)
Manufactures semiconductors for a variety of electronics applications.

Dallas Solar Power
P.O. Box 611927, San Jose, CA 95161 USA
(800) 345-4913 (tel)
Sells photovoltaic panels, solar cells, and chargers.

Damark
7101 Winnetka Ave. N., P.O. Box 29900
Minneapolis, MN 55429 USA
(800) 827-6767 (tel)
Visa, MasterCard
Sells a wide variety of new and refurbished electronics, including car and home stereo equipment, home and semiprofessional video equipment, personal computers, office equipment, etc.

Danbar Sales
14455 N. 79th St., Unit #C, Scottsdale, AZ 85260 USA
(602) 483-6202 (tel)
(602) 483-6403 (fax)
Sells test equipment, including oscilloscopes, ac calibrators, voltage dividers, spectrum analyzers, etc. for a wide variety of electronics testing applications. Offers a free catalog.

Dan Discolight
Bellahojvej 129, 2720 Vanlose, Denmark
38-33-52-33 (tel)
Sells disc jockey equipment, such as speakers, double CD deck, lighting systems, etc.

Dandys
120 N. Washington, Wellington, KS 67152 USA
(316) 326-6314 (tel)
Sells a wide variety of amateur radio equipment and accessories, including amateur radio transceivers, receivers, antenna tuners, antennas, amplifiers, etc.

Danitas Radio
Hammershusgade 11, 2100 Kobenhavn O, Denmark
31-20-00-31 (tel)
Manufactures handheld transceivers for CB radio applications. Offers a free catalog.

Danka
9715 Burnet Rd., Austin, TX 78758 USA
(800) 221-8330 (tel)
Manufactures fax machines for a wide variety of office applications.

Dansk Audio Teknik
Fredriksberg Alle 6, 1800 Fredriksberg C, Denmark
3131-3736 (tel)
Manufactures high-end loudspeakers for home audio applications.

Darkhorse Systems
12201 Technology Blvd., #135, Austin, TX 78727 USA
(512) 258-5721 (tel)
(512) 257-0296 (fax)
Manufactures memory chip testers for professional computer applications.

Ted Dasher & Associates
4117 Second Ave. S., Birmingham, AL 35222 USA
(800) 638-4833 (tel)
(205) 591-4747 (tel)
(205) 591-1108 (fax)
sales@dasher.com
Visa, MasterCard
Sells demonstation model personal desktop computers, printers, plotters, and scanners.

Dataco
9001 Lenexa Dr., Overland Park, KS 66215 USA
(800) 825-1262 (tel)
Sells printers, parts, and supplies for a wide variety of computer applications.

DataCom
520 No. 30th, P.O. Box 3485, Quincy, IL 62305 USA
(217) 222-0160 (tel)
(217) 222-0912 (fax)
Manufactures radio remote controls, switches, lease line replacements, and wireless telephone lines for professional telephone and two-way radio applications.

Data Communications Service
14275 Midway Rd., #220, Dallas, TX 75244 USA
(214) 687-9011 (tel)
(214) 687-9012 (fax)
Buys and sells modems and multiplexers for a wide variety of computer applications. Offers a free brochure.

Data Display Products
445 S. Douglas St., El Segundo, CA 90245 USA
(800) 421-6815 (tel)
(310) 640-0442 (tel)
(310) 640-7639 (fax)
Manufactures optoelectronics for a wide variety of electronics applications. Offers a free catalog.

Data Exchange
3600 Via Pescador, Camarillo, CA 93012 USA
(800) 237-7911 (tel)
(805) 388-1711 (tel)
(805) 582-4856 (fax)
Clonshaugh Industrial Estate, Clonshaugh, Dublin 17, Ireland
011-353-1-848-6555 (tel)
011-353-1-848-6559 (fax)
sales@dex.com
http://www.dex.com
Sells personal computer parts and accessories, etc.

Data Express Systems
322 Paseo Sonrisa, Walnut, CA 91789 USA
(800) 878-9669 (tel)
(909) 598-0738 (tel)
(909) 598-0743 (fax)
Visa, MasterCard
Manufactures and sells personal desktop computers and network packages. Also sells computer parts and boards, such as CD-ROM drives, hard drives, network cards, motherboards, modems, etc. Offers a 30-day money-back guarantee.

Data General
4400 Computer Dr., Westboro, MA 01580 USA
(800) 67-ARRAY (tel)
(508) 366-8911 (tel)
(508) 898-7501 (fax)

Manufactures data storage systems for a wide variety of professional computer network applications.

Data General
P.O. Box 9735, Wellington, New Zealand
+64-4-495-3344 (tel)
+64-4-495-3311 (fax)
Manufactures workstations for a wide variety of professional computer network applications.

Data Hunter
5132 Bolsa Ave., Huntington Beach, CA 92649 USA
(800) 328-2468 (tel)
(714) 892-5461 (tel)
(714) 892-9768 (fax)
Sells UPC scanners, etc. for computerized business applications.

Data I/O
10525 Willows Rd. NE., P.O. Box 97046, Redmond, WA 98073 USA
(800) 332-8246 (tel)
(206) 881-6444 (tel)
Manufactures programming systems for professional electronics applications.

Datalux
155 Aviation Dr., Winchester, VA 22602 USA
(800) 328-2589 (tel)
(540) 662-1500 (tel)
(540) 662-1682 (fax)
(540) 662-1675 (fax back)
Euro House, Curtis Rd.
11 Old Water Yard, Dorking, Surrey RH4 1EJ UK
44-1-306-876718 (tel)
44-1-306-876742 (fax)
Manufactures and sells miniature personal desktop computer systems.

Dataman Programmers
22 Lake Beauty Dr., #101, Orlando, FL 32806 USA
(407) 649-3335 (tel)
(407) 649-3310 (fax)
(407) 649-3159 (BBS)
Station Rd., Maiden Newton, Dorset DT2 0AE UK
01300-320719 (tel)

Dataman Programmers

01300-321012 (fax)
01300-321095 (BBS)
sales@dataman.com
ftp://dataman.com
http://www.dataman.com
Visa, MasterCard, Discover, American Express
Manufactures and sells handheld programmer/ emulators for professional electronics applications.

Data Memory Systems (DMS)

24 Keewaydin Dr., #5, Salem, NH 03079 USA
(800) 662-7466 (tel)
(603) 898-7750 (tel)
(603) 898-6585 (fax)
Visa, MasterCard, Discover
Sells memory and math coprocessor chips for a wide variety of computer applications.

Dataram

P.O. Box 7528, Princeton, NJ 08540 USA
(800) 822-0071 (tel)
(609) 799-0071 (tel)
(609) 897-7011 (fax)
Manufactures data storage systems for a wide variety of professional computer network applications.

Data Reliance

413 W. Jefferson, Plymouth, IN 46563 USA
(219) 935-9550 (tel)
(219) 935-9555 (fax)
Sells printers and printer parts for a wide variety of computer applications.

Data Research & Applications

9041 Executive Park Dr., #200, Knoxville, TN 37923 USA
(800) 365-1356 **(tel)**
(615) 690-1345 **(tel)**
(615) 693-5488 (fax)
Manufactures data storage systems for a wide variety of professional computer network applications.

Data Rite

650 Vaqueros Ave., #E, Sunnyvale, CA 94086 USA
(800) 886-DISK (tel)
(408) 730-2194 (tel)
(408) 730-2196 (fax)
Sells hard drives for a wide variety of computer applications.

DataSouth

4216 Stuart Andrew Blvd., Charlotte, NC 28217 USA
(800) 476-2120 (tel)
(704) 523-8500 (tel)
(704) 523-9298 (fax)
Manufactures printers for professional network computer applications.

Data Station

P.O. Box 6369, Harrisburg, PA 17112 USA
(717) 545-0131 (tel)
(717) 545-7045 (retail store tel)
(717) 671-1755 (fax)
Visa, MasterCard, COD
Sells used personal desktop and laptop computers, printers, and plotters. Also sells computer parts and boards, such as floppy drives, hard drives, monitors, video cards, modems, keyboards, etc.

Data Systems Service (D.S.S.C.)

26515 Golden Valley Rd., Santa Clarita, CA 91350 USA
(800) 272-3772 (tel)
(805) 251-9064 (tel)
(805) 251-9068 (fax)
Visa, MasterCard
Sells new and refurbished printers for a wide variety of computer applications.

Data Translation

100 Locke Dr., Marlboro, MA 01752 USA
(800) 525-8528 (tel)
(508) 481-8620 (fax)
574-7126@mcimail.com
Manufactures video capture boards and video production systems for personal and professional computer video applications.

DataVision

445 Fifth Ave., New York, NY 10016 USA
(800) 482-7466 (tel)
(212) 825-1823 (customer service tel)

(212) 689-1111 (tel)
(212) 825-1970 (fax)
datavis@aol.com
Sells high-end, home, amateur, and professional audio and video equipment and computers, such as camcorders, video editors, battery packs, VCRs, cables, etc.

DataVision
30 Indian Dr., Ivyland, PA 18974 USA
(800) 517-0674 (tel)
10421 Burnham Dr., Bldg. #2, Gig Harbor, WA 98335 USA
(800) 223-5430 (tel)
2637 Enterprise Rd., Clearwater, FL 34623 USA
(800) 287-9157 (tel)
Visa, MasterCard, Discover, American Express
Sells programmable keyboards, bar code printers, UPC scanners, portable data collectors, etc. for computerized business applications.

DATEL
11 Cabot Blvd., Mansfield, MA 02048 USA
(508) 339-3000 (tel)
(508) 339-6356 (fax)
Manufactures semiconductors for a wide variety of electronics applications.

Datong Electronics
Clayton Wood Close, W. Park, Leeds LS16 6QE UK
Visa, MasterCard
Manufactures and sells antennas for shortwave listening applications.

Datum
1363 S. State College Blvd., Anaheim, CA 92806 USA
(800) 938-3286 (tel)
(714) 533-6333 (tel)
(714) 533-6345 (fax)
6541 Via del Oro, San Jose, CA 95119 USA
(800) 348-0648 (tel)
(408) 578-4161 (tel)
(408) 578-4165 (fax)
Manufactures GPS clocks for professional aviation applications and boards for a wide variety of computer applications.

DAVES (Direct Audio Video Entertainment Systems)
441 Kearny Ave., Kearny, NJ 07032 USA
(800) 308-0354 (tel)
(201) 997-7812 (tel)
Visa, MasterCard, Discover
Sells high-end home and amateur audio and video equipment, such as camcorders, VCRs, speakers, cassette decks, TVs, home theater equipment, etc.

Dave's Guitar Shop
1227 South 3rd St., La Crosse, WI 54601 USA
(608) 785-7704 (tel)
(608) 785-7703 (fax)
davesgtr@aol.com
Sells new and used electric guitars and amplifiers for professional and personal musical applications. Offers a free list.

Davilyn Corp.
13406 Saticoy St., N. Hollywood, CA 91605 USA
(800) 235-6222 (tel)
(818) 787-3334 (tel)
(818) 787-4732 (fax)
Sells a wide variety of surplus electronic gear, including test equipment, radios, radio parts, etc.

Davis
3465 Diablo Ave., Hayward, CA 94545 USA
(800) 293-2847 (tel)
(510) 670-0589 (fax)
Visa, MasterCard
Manufactures and sells instruments to monitor speed, acceleration, distance traveled, times that the vehicle moved, time of vehicle in motion, etc. Also manufactures and sells a weather monitor. Offers a free brochure and a money-back guarantee.

Davis RF
P.O. Box 230, Carlisle, MA 01741 USA
(800) 328-4773 (tel)
(508) 369-1738 (info tel)
Sells antenna tower supports, wire, cable, and DSP audio filters for amateur radio applications.

Dawn VME Products
47073 Warm Springs Blvd., Fremont, CA 94539 USA
(800) 258-DAWN (tel)
(510) 657-4444 (tel)
(510) 657-3274 (fax)
Manufactures parts and complete ruggedized computer systems for industrial computer applications.

Dayna Communications
849 W. Levoy Dr., Salt Lake City, UT 84123 USA
(800) 44-DAYNA (tel)
http://www.dayna.com
Manufactures Ethernet cards, hubs, and bridges for professional computer applications.

Daytek Electronics
2120 Hutton, Ste. 800, Carrollton, TX 75006 USA
(800) 329-8351 (tel)
(604) 270-3003 (tel)
http://www.daytek.ca
Manufactures monitors for personal desktop computer systems.

dB Communications
4007 Skippack Pike, P.O. Box 1268, Skippack, PA 19474 USA
(800) 526-2332 (tel)
(610) 584-7875 (fax)
Sells equipment for professional cable TV applications.

DB Elettronica Telecommunicazioni S.p.A.
Via Lisbona, 38, Zona Industriale Sud, 1 35020 Camin, Padova, Italy
+39-49-8700588 (tel)
+39-49-8700747 (fax)
Manufactures and sells equipment for professional TV transmitting applications, such as transmitters, modulators, converters, transposers, etc.

dB Speakerworks
1430 Dalzell, Shreveport, LA 71103 USA
(318) 425-2525 (tel)
(318) 227-8058 (fax)
Manufactures loudspeakers for car stereo applications.

DBS Satellite Television
2316 Channel Dr., Ventura, CA 93003 USA
(800) 327-4728 (tel)
(805) 652-2190 (fax)
Sells a variety of satellite TV reception equipment, including receivers, dish antennas, cabling, etc.

DB Systems
P.O. Box 460, Rindge, NH 03461 USA
(603) 899-5121 (tel)
Manufactures and sells custom active electronic crossovers for home audio applications.

dB-tronics
145 Tradd St., Spartanburg, SC 29301 USA
(800) 356-2730 (tel)
(803) 574-0155 (tel)
(803) 574-0383 (fax)
Visa, MasterCard
Sells new and used equipment for cable TV applications.

DCC Computer Parts Outlet
1412 Reynolds Rd., Maumee, OH 43537 USA
(419) 891-7516 (tel)
Visa, MasterCard, Discover, American Express
Sells personal home computer parts.

DC Electronics
P.O. Box 3203, Scottsdale, AZ 85271 USA
(800) 423-0070 (tel)
(602) 945-7736 (tel)
(602) 994-1707 (fax)
Sells a variety of new and surplus electronics parts, including ICs, semiconductors, capacitors, resistors, etc.

DC Industries
5265 Hebbardsville Rd., Athens, OH 45701 USA
(800) 537-3539 (tel)
(614) 592-4239 (tel)
(614) 592-1527 (fax)
Specializes in new and used parts, such as hard drives, motherboard upgrades, memory chips, etc., for Tandy 1000 personal computers.

DC Micros
1843 Sumner Ct., Las Cruces, NM 88001 USA
(505) 524-4029 (tel)
COD ($5)
Manufactures single-board computers for a wide variety of computer applications.

Debco Electronics
4025 Edwards Rd., Cincinatti, OH 45209 USA
(800) 423-4499 (tel)
Sells surplus electronics components, semiconductors, crystals, etc.

Decibel Instruments
3857 Breakwater Ave., Hayward, CA 94545 USA
(800) 875-8829 (tel)
(510) 264-4398 (fax)
Manufactures quiet computers for professional and industrial computer applications.

Decibel Products
P.O. Box 569610, Dallas, TX 75356-9610 USA
(800) 676-5342 (tel)
(800) 229-4706 (fax)
(214) 631-0310 (tel)
(214) 631-4706 (fax)
http://www.allentele.com
Manufactures combiners for professional cable TV applications.

Deco Industries
P.O. Box 607, Bedford Hills, NY 10507 USA
(914) 232-3878 (tel)
(914) 243-0346 (fax)
Visa, MasterCard, COD
Manufactures miniature Part 15 transmitters for UHF radio bands.

Dee Electronics
P.O. Box 1508, 2500 16th Ave. SW, Cedar Rapids, IA 52406 USA
(800) 747-3331 (tel)
(319) 365-7551 (tel)
(319) 365-8506 (fax)
Sells electronics components, such as cabinets, capacitors, connectors, crystals, semiconductors, resistors, transformers, etc., for a wide variety of electronics applications.

Dee One Systems
1550 Centre Pointe Dr., Milpitas, CA 95035 USA
(800) 831-8808 (tel)
(408) 262-8938 (tel)
(408) 956-8286 (service tel)
(408) 262-2826 (fax)
(408) 956-1813 (BBS)
Visa, MasterCard
Manufactures and sells personal desktop computers and printers; also sells computer parts and boards, such as floppy drives, hard drives, monitors, modems, video cards, tape drives, etc.

Definition Audio Video
2901 W. 182nd St., Redondo Beach, CA 90278 USA
(310) 371-0019 (tel)
Visa, MasterCard, Discover, American Express
Sells a wide variety of high-end home theater equipment.

Definitive Technology
11105 Valley Heights Dr., Baltimore, MD 21117 USA
(800) 228-7148 (tel)
(410) 363-7148 (tel)
Manufactures loudspeakers for high-end home audio and home theater applications.

De La Hunt Electronics
Hwy. 34E, Park Rapids, MN 56470 USA
Sells amateur radio equipment manufactured by ICOM.

Del Electronics
1 Commerce Park, Valhalla, NY 10595 USA
(914) 686-3600 (tel)
(914) 686-5424 (fax)
Manufactures high-voltage power supplies and transformers for industrial and laser applications.

Dell
2214 W. Braker Ln., Bldg. 3, Austin, TX 78758 USA

Dell

(800) 232-5620 (tel)
(800) 668-3021 (Canada tel)
(800) 426-5150 (sales tel)
(800) 624-9896 (service tel)
353-1-286-0500 (Northern Europe tel)
33-6706-6000 (Southern Europe tel)
81-3-5420-7386 (Japan tel)
61-2-930-3355 (Australia tel)
(512) 728-8528 (BBS)
http://www.us.dell.com/
directnews-request@dell.com (Internet price updates)
Factory outlet, 8801 Research Blvd., Austin, TX 78750 USA
(800) 336-2891 (tel)
Manufactures and sells personal desktop and laptop computers.

Delphi Components

27721A La Paz Rd., Laguna Niguel, CA 92677 USA
(714) 831-1771 (tel)
(714) 831-0862 (fax)
Manufactures CROs and DROs for professional cable TV applications.

Delta Communications

2162 S. Jupiter Rd., Garland, TX 75041 USA
(800) 880-2250 (tel)
(214) 278-5085 (fax)
Buys and sells VHF/UHF transceivers for professional two-way radio applications.

Delta Loop Antennas

P.O. Box 274, Weston, VT 05161 USA
(802) 824-8161 (tel)
Manufactures and sells antennas for amateur radio applications.

Delta Products

3225 Laurelview Ct., Fremont, CA 94538 USA
(510) 770-0660 (tel)
(510) 770-0122 (fax)
2000 Aerial Center Pkwy., #114, Morrisville, NC 27560 USA
(919) 380-8883 (tel)
(919) 380-8383 (fax)
9th Fl. Asia Enterprise Center, No. 144 Min Chuan E. Rd., Sec. 3, Taipei 10464, Taiwan
Manufactures EMI filters for a wide variety of electronics applications.

Delta Technologies

580 N. Tillman, #5, Memphis, TN 38112 USA
(800) 22-DELTA (tel)
(901) 375-9933 (fax)
Visa, MasterCard
Sells personal desktop computer systems and parts.

Delta Warranty

1775 12th Ave. NW, P.O. Box 7000, Issaquah, WA 98027 USA
(800) 472-8778 (tel)
(800) 727-1745 (fax)
Manufactures surge and power protection units for a variety of electronics applications.

Deltec Electronics

2727 Kurtz St., San Diego, CA 92110 USA
(619) 291-4211 (tel)
(619) 296-8039 (fax)
Manufactures uninterruptible power supplies for professional computer network applications.

Deltron Communications International

P.O. Box 474, NL-7900 AL Hoogeveen, Netherlands
31-0-5280-68816 (tel)
31-0-5280-72221 (fax)
Sells receivers, antennas, and accessories for shortwave radio-listening applications.

Anthony DeMaria Labs

(818) 340-0228 (tel)
(914) 256-0032 (service tel)
(818) 340-4331 (fax)
Manufactures all hollow-state compressors and direct boxes for professional studio recording and live music performances.

Demeter Amplification

2912 Colorado Ave., #204, Santa Monica, CA 90404 USA

(310) 829-4383 (tel)
(310) 829-3755 (fax)
Manufactures tube microphone preamplifiers for professional studio recording and live music performances.

Denon Electronics
222 New Rd., P.O. Box 5370, Parsippany, NJ 07054 USA
(201) 575-7810 (tel)
(201) 882-7490 (tel)
(201) 575-2532 (fax)
(201) 575-1213 (fax)
(201) 808-1608 (fax)
Manufactures CD players, amplifiers, cassette players, mobile head units, and CD changers for professional studio recording.

Densitron
3425 W. Lomita Blvd., Torrance, CA 90505 USA
(310) 530-3530 (tel)
(310) 534-8419 (fax)
Manufactures LCD modules for a wide variety of electronics applications.

Dentronics
6102 Deland Rd., Flushing, MI 48433 USA
(800) 722-5488 (tel)
(810) 659-1776 (tel)
(810) 659-1280 (fax)
Sells a wide variety of amateur radio equipment and accessories, including amateur radio transceivers, receivers, antenna tuners, antennas, linear amplifiers, etc.

Desco Industries
761 Penarth Ave., Walnut, CA 91789 USA
(909) 598-2753 (tel)
(909) 595-7028 (fax)
Manufactures overhead ionizers for a professional electronics assembly applications.

Design Acoustics
1225 Commerce Dr., Stow, OH 44224 USA
(216) 686-2600 (tel)
(216) 425-8222 (tel)
(216) 688-3752 (fax)
Manufactures loudspeakers for home audio applications.

Detection Dynamics
4700 Loyola Ln., #119, Austin, TX 78723 USA
(512) 345-8401 (tel)
(512) 926-0940 (fax)
Sells CCD cameras, time-lapse video recorders, monitors, etc. for a wide variety of security and surveillance applications. Offers a catalog for $5.

Devcom Mid America
2603 W. 22nd St., #23, Oakbrook, IL 60521 USA
(708) 574-3600 (tel)
(708) 572-0508 (fax)
Manufactures modems for a wide variety of computer telecommunications applications.

Develcon
856 51 St. E., Saskatoon, SK 57K 5C7 Canada
(800) 667-9333 (tel)
(306) 933-3300 (tel)
(306) 931-1370 (fax)
Manufactures routers for professional computer network applications.

Development Concepts
730D Louis Dr., Warminster, PA 18974 USA
(215) 443-9652 (tel)
(215) 674-0607 (fax)
Manufactures terminal servers for professional computer network applications.

Dexis
9749 Hamilton Rd., Eden Prarie, MN 55344 USA
(612) 944-7670 (tel)
(612) 942-9712 (fax)
(800) 547-4636 (fax back)
Buys and sells test equipment, scientific equipment, memory chips, data/telco equipment, etc. for a wide variety of electronics applications.

Dexon Systems
Szikla u.21, H-1025 Budapest, Hungary

Dexon Systems

+36-1-393-2617 (tel)
+36-1-393-2618 (fax)
+36-23-420-749 (fax)
Manufactures and sells Windows server boards, etc. for personal desktop computers.

DFA

1206 SE "N" St., Grants Pass, OR 97526 USA
(503) 474-4914 (tel)
(503) 474-4915 (fax)
Sells modems, multiplexers, power supplies, etc. for a wide variety of computer applications.

DFI (Diamond Flower)

135 Main St., Sacramento, CA 95838 USA
(800) 808-4334 (tel)
(916) 568-1234 (tel)
info@dfiusa.com
7 Elkins Rd., E. Brunswick, NJ 08816 USA
(908) 390-2815 (tel)
2210 MW 92nd Ave., Miami, FL 33172 USA
(305) 477-1988 (tel)
Visa, MasterCard, Discover, American Express
Manufactures and sells personal desktop and laptop computers.

DGM Electronics

13654 Metric Rd., Roscoe, IL 61073 USA
(815) 389-2040 (tel)
Manufactures and sells converters and accessories for radio fax and radio teletype applications.

DGS Pro-Audio

P.O. Box 170426, Arlington, TX 76003 USA
(800) 292-2834 (tel)
Manufactures microphone cable for professional studio recording and live performances.

DH Distributors

P.O. Box 48623, Wichita, KS 67201 USA
(316) 684-0050 (tel/fax)
Sells electrolytic capacitors for a wide variety of electronics applications.

DH Satellite

600 N. Marquette Rd., Prarie du Chien, WI 53821 USA
(800) 627-9443 (tel)
(608) 326-8406 (tel)
(608) 326-4233 (fax)
Manufactures satellite antennas and mounting systems for a variety of communications applications.

Diablo Acoustics

1944 Windward Point, Discovery Bay, CA 94514 USA
(510) 516-0864 (tel)
(510) 516-0184 (fax)
Manufactures loudspeakers for high-end home audio applications.

Dialight

1913 Atlantic Ave., Manasquan, NJ 08736 USA
(908) 223-9400 (tel)
(908) 223-5271 (fax)
Manufactures optoelectronics components for a wide variety of electronics applications.

Diamond Audio

3030 Pennsylvania Ave., Santa Monica, CA 90404 USA
(310) 582-1121 (tel)
(310) 582-1502 (fax)
Manufactures loudspeakers for car stereo applications.

Diamond Data

168 Townlight Pkwy., Woodstock, GA 30188 USA
(800) 955-0508 (tel)
(404) 928-0055 (tel)
(404) 516-0409 (fax)
Sells network equipment and printers for a wide variety of professional and personal computer applications.

Diamond Multimedia

2880 Junction Ave., San Jose, CA 95134 USA
(800) 4-MULTIMEDIA (tel)
(408) 325-7000 (tel)
(800) 380-0030 (fax)
(408) 325-7070 (fax)
tsfeedbk@diamond.mm.com (Internet e-mail tech

support)
csfeedbk@diamond.mm.com (Internet e-mail customer service)
http://www.diamondmm.com
Manufactures multimedia accelerators and software. Provides a five-year parts and labor warranty.

Dictaphone
3191 Broadbridge Ave., Stratford, CT 06497 USA
(800) 447-7749 (tel)
Sells telephone call logging systems for professional telecommunications applications.

Dielectric Communications
P.O. Box 949, Raymond, ME 04071 USA
(207) 655-4555 (tel)
(207) 655-7120 (fax)
Manufactures antennas, transmission lines and waveguides, RF systems, switches, and loads for professional broadcasting.

DigiBoard
6400 Flying Cloud Dr., Eden Prarie, MN 55344 USA
(800) 344-4273 (tel)
(612) 934-9020 (tel)
(612) 934-5398 (fax)
Manufactures routers and fax boards for professional computer network applications.

Digicomm International
14 Inverness Dr. E., #A124, Englewood, CO 80112 USA
(303) 799-3444 (tel)
(303) 799-9366 (fax)
Manufactures video titlers for professional TV broadcast and video production applications.

Digidesign
3401A Hillview Ave., Palo Alto, CA 94304 USA
(800) 332-2137 (tel)
http://www.digidesign.com
Manufactures digital audio workstations for professional audio recording applications.

Digi-key
701 Brooks Ave. S., P.O. Box 677, Thief River Falls, MN 56701 USA
(800) 344-4539 (tel)
(218) 681-3380 (fax)
Sells a variety of electronics components.

Digilog Electronics
1038 N. Ashland, Chicago, IL 60622 USA
(312) 384-8300 (tel)
(312) 384-8305 (fax)
Visa, MasterCard
Sells personal laptop computers and accessories.

Digi-Rule
#3001822, 10th SW, Calgary, AB T3C 0J8 Canada
(403) 292-0320 (tel)
http://www.digirule.com
Manufactures and sells portable digitizers for personal computer applications.

Digital Auto-Radio Sarl
88, Rue de Strasbourg, L-2560 Luxembourg
Sells head units, speakers, amplifiers, etc. for car audio applications.

Digital Communications
P.O. Box 293, 29 Hummingbird Bay, White City, SK S0G 5B0 Canada
(800) 563-5351 (tel)
(306) 781-4451 (tel)
(306) 781-2008 (fax)
Manufactures and sells a two-meter bandpass filter for amateur radio applications.

Digital Data Systems and Communications
195-28T Hillside Ave., Jamaica, NY 11423 USA
(800) 789-2343 (tel)
(718) 464-2932 (tel)
(718) 218-8249 (fax)
Buys and sells new, used, and refurbished personal and laptop computers and parts and accessories, including hard drives, printers, uninterruptible power systems, modems, monitors, etc.

Digital Electronic Systems
P.O. Box 1073, Englewood, FL 34295 USA

Digital Electronic Systems
(813) 474-9518 (tel)
(813) 474-9519 (fax)
Manufactures and sells digital demodulators, phone patches, etc.

Digital Equipment (DEC)
P.O. Box 687, Westford, MA 01886 USA
(800) DIGITAL (tel)
(508) 392-0603 (fax)
moreinfo@digital.com
http://www.digital.com
Manufactures and sells personal and professional computer systems.

Digital Integrated Systems (DIS) Computers
8142 Broadview Rd., Cleveland, OH 44147 USA
(800) 838-4130 (tel)
(216) 838-4111 (tel)
(216) 838-4169 (tech support tel)
(216) 838-4131 (fax)
Visa, MasterCard, Discover, COD
Manufactures and sells personal desktop computers; also sells computer parts and boards, such as floppy drives, hard drives, video cards, modems, motherboards, etc. Offers a 30-day money-back guarantee.

Digital Phase
2841 Hickory Valley Rd., Chattanooga, TN 37421 USA
(800) 554-7325 (tel)
Manufactures loudspeakers for high-end home audio and home theater applications.

Digital Power
41920 Christy St., Fremont, CA 94538 USA
(510) 657-2635 (tel)
(510) 657-6634 (fax)
Manufactures power supplies for a wide variety of electronics applications.

Digital Processing Systems
(606) 371-5533 (tel)
Manufactures and sells PC computer-based video recorders for home and professional video production applications.

Digital Products
134 Windstar Circle, Folsom, CA 95630 USA
(916) 985-7219 (tel)
(916) 985-8460 (fax)
digprod@aol.com
Manufactures phone-line simulator kits for a wide variety of electronics testing applications.

Digital Repair
5018 Service Center Dr., San Antonio, TX 78218 USA
(800) 776-8414 (tel)
(210) 662-0700 (tel)
(210) 662-8004 (fax)
Buys and sells refurbished terminals for professional computer applications.

Digital Semiconductor
77 Reed Rd., Hudson, MA 01749 USA
(508) 568-5102 (tel)
Manufactures semiconductors for a wide variety of electronics applications.

Digital Technology
2300 Edwin Moses Blvd., Dayton, OH 45408 USA
(513) 443-0412 (tel)
(513) 226-0511 (fax)
dti@erinet.com
Manufactures computer interfaces for a wide variety of professional VXI electronics applications.

Digitape Systems
8827 Shirley Ave., Northridge, CA 91324 USA
(800) 545-5100 (tel)
(818) 349-1865 (tel)
(818) 349-1877 (fax)
Visa, MasterCard
Sells data storage systems, including hard drives, optical drives, tape drives, etc., for a wide variety of computer applications.

DigiTech
8760 S. Sandy Pkwy., Sandy, UT 84071 USA
(800) 449-8818 (tel)
(801) 566-8919 (tel)
(800) 333-7363 (fax)

Manufactures vocal harmony processors for professional studio recording applications.

Digitrax
P.O. Box 1424, Norcross, GA 30091 USA
(770) 441-7992 (tel)
(770) 441-0759 (fax)
Manufactures and sells digital command controls for model railroads.

Dinexcom
70 Wilbur St., Unit 3, Lowell, MA 01851 USA
(508) 446-7820 (tel)
(508) 446-7830 (fax)
Manufactures dc-to-dc converters for a wide variety of electronics applications.

Ding Yih Industrial
162-5 Juen Yin St., Shulin Taipei 23804, Taiwan
886-2-6862432 (tel)
886-2-6872991 (tel)
886-2-6872579 (fax)
Manufactures electronics and video connectors for a wide variety of electronics applications.

Direct Cable Supply
Rt. 6, Plaza Mall #290, Honesdale, PA 18431 USA
(800) 808-3356 (tel)
Sells cable TV converters and decoders. Offers a 30-day money-back guarantee.

Direct CCTV
Unit 6, Carrick Ct., Forrest Grove Business Park, Middlesbrough TS2 1QE UK
Sells miniature surveillance video cameras for a wide variety of video applications. Offers a catalog for a SASE.

Direct Connections
7668 Executive Dr., Eden Prairie, MN 55344 USA
(800) 572-4305 (tel)
(612) 937-9604 (tel)
(612) 937-6285 (fax)
dconnect@skypoint.com
Visa, MasterCard, Discover
Sells data drives for a wide variety of personal and professional computer applications.

Directed Electronics
2560 Progress St., Vista, CA 92083 USA
(800) 283-1320 (tel)
(619) 598-6400 (fax)
Manufactures loudspeakers and amplifiers for car stereo applications.

Directed Energy
2301 Research Blvd., #105, Ft. Collins, CO 80526 USA
(970) 493-1901 (tel)
(970) 493-1903 (fax)
deiinfo@dirnrg.com
Manufactures high-voltage pulsers for industrial applications.

Direct Power and Water
3455 A Princeton St. NE, Albuquerque, NM 88005 USA
(800) 260-3792 (tel)
(505) 889-3585 (tel)
Sells photovoltaic (solar) electric power systems for a wide variety of electronics applications.

Direct Ware
(800) 610-9273 (tech support tel)
(310) 793-4580 (customer service tel)
(310) 793-7175 (order fax)
(310) 793-8606 (fax on request)
sales@directware.com (Internet e-mail sales)
tech_support@directware.com (Internet e-mail tech support)
customer_service@directware.com (Internet e-mail service)
http://www.directware.com
Sells personal desktop and laptop computers, printers, and scanners; also sells computer parts and boards, such as floppy drives, hard drives, monitors, processor upgrades, video cards, modems, keyboards, etc.

DirectWave
4260 E. Brickell, Ontario, CA 91761 USA
(800) 882-8108 (tel)
(800) 840-1889 (tech support tel)
(909) 390-8058 (tel)

DirectWave

(909) 390-8061 (fax)
(909) 390-8052 (BBS line)
Visa, MasterCard, Discover, American Express
Manufactures and sells personal desktop computers; also sells computer parts and boards, such as motherboards, speakers, multimedia kits, video cards, fax modems, etc. Offers a 30-day money-back guarantee.

Dirt Cheap Drives

3716 Timber Dr., Dickinson, TX 77539 USA
(800) 786-1170 (tel)
(800) 473-0976 (order inquiry tel)
(800) 786-1162 (Fortune 1000 and dealers tel)
(713) 534-4140 (tel)
(713) 534-6962 (tech support tel)
(713) 534-6452 (fax)
Visa, MasterCard, Discover, American Express
Sells hard drives, tape drives, optical drives, etc., for personal desktop computers.

Disco & Lydteknik (DLT)

Skjulhoj Allé 57, 2720 Vanlose, Denmark
38-71-69-11 (tel)
Sells high-end home and professional audio equipment.

DiscoTronix

Rugvaenget 13B, 4100 Ringsted, Denmark
57-67-18-14 (tel)
Sells lighting systems for live music performances and discos.

Discount Electronics CB Sales

P.O. Box 96, US Hwy. 1 S., Callahan, FL 32011 USA
(904) 879-9044 (tel)
(904) 879-9045 (tel)
Sells CB radios, antennas, UHF/VHF scanners, radar detectors, etc.

Discount Music Supply

41 Vreeland Ave., Totowa, NJ 07512 USA
Sells electric guitars, cables, tuners, microphones, etc. for professional and personal musical applications. Offers a free catalog.

Disk Emulation Systems

3080 Oakmead Village Dr., Santa Clara, CA 95051 USA
(408) 727-5497 (tel)
(408) 727-5496 (fax)
Manufactures disk emulators for a wide variety of professional computer network applications.

Diskette Connection

P.O. Box 1674, Bethany, OK 73008 USA
(800) 654-4058 (tel)
(405) 789-0971 (tel)
(405) 495-4598 (fax)
Visa, MasterCard, Discover
Sells a variety of computer data storage media, including floppy diskettes, CD-Rs, QIC tapes, optical disks, removable cartridges, etc.

Diskettes Unlimited

6206 Long Dr., Houston, TX 77087 USA
(800) 364-DISK (tel)
(713) 643-9939 (tel)
(713) 643-2722 (fax)
Visa, MasterCard, Discover
Manufactures diskettes and CD-ROMs for a wide variety of computer applications.

Di-Tech

48 Jefryn Blvd., Deer Park, NY 11729 USA
(516) 667-6300 (tel)
(516) 595-1012 (fax)
Manufactures routing switchers, video equalizers, and monitor amplifiers for professional broadcasting applications.

Diversified Electronics

P.O. Box 49027, Leesburg, FL 34749 USA
(800) 874-0619 (tel)
Manufactures time-delay modules for a wide variety of electronics applications.

Diversified Parts

2114 SE 9th Ave., Portland, OR 97214 USA
(800) 338-6342 (tel)
(503) 236-6140 (tel)

(800) 962-0602 (fax)
Sells replacement parts and accessories for audio equipment.

Diversified Systems Group
P.O. Box 1114, Issaquah, WA 98027 USA
(800) 255-3142 (tel)
Sells diskettes, cartridges, and optical disks for computer data storage applications.

Diversified Technology
P.O. Box 748, Ridgeland, MS 39157 USA
(800) 443-2667 (tel)
(601) 856-4121 (tel)
(601) 856-2888 (fax)
(201) 891-8719 (tel)
(201) 891-9629 (fax)
44-1753-580660 (UK tel)
Manufactures single-board computers for a wide variety of professional computer applications.

D&L Communications
3512 Cavalier Dr., Ft. Wayne, IN 46808 USA
(800) 336-6825 (tel)
Buys and sells refurbished pagers for telecommunications applications.

D-Link
5 Musick, Irvine, CA 92718 USA
(800) 326-1688 (tel)
(714) 455-1688 (tel)
(714) 455-2521 (fax)
(905) 828-0260 (Canada tel)
(905) 828-5669 (Canada fax)
44-81-203-9900 (UK tel)
44-81-203-6915 (UK fax)
49-6196-643011 (Germany tel)
49-6196-28049 (Germany fax)
33-1-45-01-55-88 (France tel)
33-1-45-01-55-66 (France fax)
46-40-16-33-30 (Scandinavia tel)
46-40-16-33-40 (Scandinavia fax)
61-2-410-9966 (Australia tel)
61-2-410-9929 (Australia fax)
Manufactures hubs, switches, and NICs for professional computer network applications.

Dobbs-Stanford
2715 Electronic Ln., Dallas, TX 75220 USA
(214) 350-4222 (tel)
Manufactures computer-to-video converters for professional, amateur, or home video playback applications.

Doc's Communications
702 Chickamauga Ave., Rossville, GA 30741 USA
(706) 866-2302 (tel)
(706) 861-5610 (tel)
(706) 866-6113 (fax)
Sells a wide variety of amateur radio equipment and accessories, including amateur radio transceivers, receivers, antenna tuners, antennas, amplifiers, etc.

Doeven Electronika
Schutstraat 58, 7921 EE Hoogeveen, Netherlands
Sells equipment for shortwave radio receiving applications.

Dolan-Jenner Industries
678 Andover St., Lawrence, MA 01843 USA
(508) 681-8000 (tel)
(508) 682-2500 (fax)
Manufactures high-temperature fiberoptics for a wide variety of professional industrial applications.

Dolby Laboratories
100 Pontrero Ave., San Francisco, CA 94103 USA
(415) 558-0200 (tel)
(415) 863-1272 (fax)
Manufactures ISDN transmission systems, digital audio coders, and spectral processors for professional broadcasting.

Dolch Computer Systems
3178 Laurelview Ct., Fremont, CA 94538 USA
(800) 995-7561 (tel)
(510) 661-2220 (tel)
(510) 490-2360 (fax)
Manufactures ruggedized portable computers for a wide variety of professional computer applications.

Dolgin Engineering
(617) 926-2608 (tel)

Dolgin Engineering

Manufactures S-VHS junction boxes for home, amateur, and professional video editing applications.

Dollar Computer
1809 E. Dyer Rd., #304, Santa Ana, CA 92705 USA
(800) 704-0550 (tel)
(714) 975-0729 (customer service tel)
(714) 975-1560 (fax)
(714) 975-1438 (fax)
Spanish: X112
French: X113
Exports: X110
Visa, MasterCard, Discover, American Express, Western Union, COD
Sells discounted personal laptop and desktop computers.

Domain Technologies
1700 Alma Dr., Plano, TX 75075 USA
(214) 985-7593 (tel)
(214) 985-8579 (fax)
Manufactures DSP boxes for a wide variety of electronics applications.

Dominator Technologies
12062 SW 117 Ct., #132, Miami, FL 33186 USA
(305) 271-2737 (tel)
American Express, COD
Manufactures auto alarm disarmers for a wide variety of professional security applications.

Doppler Systems
P.O. Box 2780, Carefree, AZ 85377 USA
(602) 488-9755 (tel)
(602) 488-1295 (fax)
Manufactures portable direction-finding systems for a variety of amateur radio and cellular microwave telephone repeater applications.

Dovetron
P.O. Box 6160, Nogales, AZ 85628 USA
(520) 281-1681 (tel)
(520) 281-1684 (fax)
Sells a noise blanker/attenuator for use with some older models of Collins amateur radio receivers and transceivers.

Down East Microwave
R.R. 1 Box 2310, Troy, ME 04987 USA
(207) 948-3741 (tel)
(207) 948-5157 (fax)
Manufactures and sells transverters, preamplifiers, antennas, etc. for amateur radio operations in the microwave frequency range.

dpi Electronic Imaging
629 Old St. Rt. 74, Cincinnati, OH 45244 USA
(800) DPI-7401 (tel)
(513) 528-8668 (tel)
(513) 528-8664 (fax)
Sells computer scanners for a variety of professional and personal desktop artwork/publishing applications.

R.L. Drake
P.O. Box 3006, Miamisburg, OH 45343 USA
(513) 866-2421 (tel sales)
(513) 746-6990 (tel service and parts)
(705) 742-3122 (tel Canada)
(800) 568-3795 (tel orders)
(513) 866-0806 (fax)
Manufactures and sells shortwave and satellite communications receivers.

Dranetz Technologies
1000 New Durham, P.O. Box 4019, Edison, NJ 08818 USA
(908) 287-3680 (tel)
http://www.dranetz.com
Manufactures power quality analyzers for a variety of professional industrial applications.

Dr. Crankenstein
1536 Winchester Ave., #215, Ashland, KY 41101 USA
(606) 325-7757 (tel)
(606) 324-3987 (fax)
Manufactures loudspeakers and power amplifiers for car stereo applications.

D&R Electronics
10 Park St., Thomaston, CT 06787 USA

(203) 283-9492 (tel)
Sells transceivers, power microphones, kits, meters, antennas, etc. for CB radio applications.

Dressler
Werther Strasse 14-16, 52224 Stolberg, Germany
2-402-71091 (tel)
2-402-71095 (fax)
Manufactures a variety of indoor and outdoor active antennas for shortwave radio receiving applications.

Drive Outlet Center
210 Terrace Dr., Mundelein, IL 60060 USA
(800) 243-9069 (tel)
(708) 970-9003 (fax)
Visa, MasterCard, Discover, COD
Sells hard disk drives, controllers, multimedia kits, CD-ROM drives, etc. for personal home and laptop computers.

DSC Communications
1000 Coit Rd., Plano, TX 75075 USA
(800) 777-6804 (tel)
Manufactures AIN systems for professional cellular microwave radio applications.

DSL Electronics
6305 N. O'Connor Blvd., Ste. 119, Irving, TX 75039 USA
(214) 869-1122 (tel)
(214) 869-1135 (fax)
Manufactures and sells MIDI converters for professional studio recording and live music performances. Offers a 30-day money-back guarantee.

DTI Factory Outlet
165 University Ave., Westwood, MA 02090 USA
(800) 366-7060 (tel)
(617) 326-4100 (tel)
(617) 326-0090 (fax)
http://www.channel1.com/business/datatrend
Sells surplus personal home and laptop computers and printers.

DTK Computer
770 Epperson Dr., City of Industry, CA 91748 USA
(818) 810-0098 (Los Angeles tel)
(713) 568-6688 (Houston tel)
(708) 593-3080 (Chicago tel)
(305) 597-8888 (Miami tel)
(908) 562-8800 (New York tel)
(404) 246-0100 (Atlanta tel)
(800) 806-1DTK (fax back)
Manufactures and sells personal desktop computers.

Dubberley's on Davie
920 Davie St., Vancouver, BC V6Z 1B8 Canada
(604) 684-5981 (tel)
Sells receivers, antennas, and accessories for shortwave listening applications.

Duplo
3050 S. Daimler St., Santa Ana, CA 92705 USA
(800) 255-1933 (tel)
(714) 752-8222 (tel)
(714) 752-7766 (fax)
Manufactures digital duplicators for a wide variety of office applications.

DuraComm
438 NW Business Park Ln., Kansas City, MO 64150 USA
(800) 467-6741 (tel)
(816) 741-7499 (fax)
Manufactures voice pagers and scanning monitors for cellular telephone applications.

Durham Radio Sales and Service
350 Wentworth St. E.
Unit #7, Oshawa, ON L1H 7R7 Canada
(905) 428-3746 (tel)
(905) 436-3231 (fax)
Sells amateur, marine, and CB radio equipment and accessories, including transceivers, receivers, antenna tuners, antennas, amplifiers, etc.

B.E. Duval
347 W. 6th St., P.O. Box 5255, San Pedro, CA 90733 USA
(310) 833-0951 (tel)
(310) 832-9970 (fax)

B.E. Duval

Sells equipment for professional cable TV applications.

D-Vision Systems
8755 W. Higgins Rd., 2nd Fl., Chicago, IL 60631 USA
(800) 838-4746 (tel)
(312) 714-1400 (tel)
Manufactures computerized digital editing systems for home and professional video applications.

DWM Enterprises
1709 N. West Ave., Jackson, MI 49202 USA
Manufactures and sells small indoor active antennas for radio hobbyists and CW transceivers for amateur radio applications.

DX Communications
10 Skyline Drive, Hawthorne, NY 10532 USA
(914) 347-4040 (tel)
Manufactures receiver/descramblers for professional cable TV applications.

Dynamic Control
P.O. Box 15232, Hamilton, OH 45014 USA
(513) 860-5094 (tel)
(513) 860-5095 (fax)
Manufactures loudspeakers for mobile audio applications.

Dynamic Electronics
P.O. Box 896, Hartselle, AL 35640 USA
(205) 773-2758 (tel)
(205) 773-7295 (fax)
Manufactures and sells Morse code computer interfaces and CW filters for amateur radio applications.

Dynamic Systems
46 Sunlight Dr., Dept. H, Leicester, NC 28748 USA
Sells sectorless Wimshurst generators and accessories for serious experimenters and classroom demonstrations.

Dynatek Automation Systems
200 Bluewater Rd., Bedford, NS B4B 1G9 Canada
(800) 267-6007 (tel)
(902) 832-3000 (tel)
(902) 832-3010 (fax)
Manufactures CD-ROM drives, CD-ROM recorders, and optical disk drives for a wide variety of professional computer network applications.

Dynaudio
P.O. Box 44283, Madison, WI 53744 USA
(608) 831-2990 (tel)
(608) 831-3771 (fax)
Manufactures loudspeakers for car stereo applications.

dynaudio acoustics
357 Liberty St., Rockland, MA 02370 USA
(617) 982-2626 (tel)
(617) 982-2610 (fax)
Manufactures reference-monitor loudspeakers for professional studio recording.

Dytec Technologies
1336 Energy Park Dr., St. Paul, MN 55108 USA
(800) 336-3732 (tel)
(612) 645-5816 (fax)
Sells printers for a wide variety of computer applications.

Eagle
P.O. Box 4010, Sedona, AZ 86340 USA
(520) 204-2597 (tel)
(520) 204-2568 (fax)
Manufactures tuneable notch filters, power combiners, diplexers, triplexers, return loss bridges, etc. for professional cellular microwave applications.

Eagle Comtronics
4562 Waterhouse Rd., Clay, NY 13041 USA
(800) 448-7474 (tel)
(315) 622-3402 (tel)
(315) 622-3800 (fax)
Manufactures telephone traps for professional cable TV applications.

Eagle Distributors
2438 Albany St., Kenner, LA 70062 USA

(800) 726-5930 (tel)
(504) 464-5930 (tel)
(504) 468-9741 (fax)
Sells replacement parts and accessories for audio and video equipment.

Eagle Electronics
1301 Railhead Blvd., #1, Naples, FL 33963 USA
(800) 259-1187 (tel)
Visa, MasterCard, Discover, American Express, COD
Sells cable TV converters and descramblers. Offers a 30-day trial and a 1-year warranty.

Eagle Picher Industries
P.O. Box 130, Seneca, MO 64865 USA
(417) 776-2256 (tel)
(417) 776-2257 (fax)
Manufactures batteries for a wide variety of electronics applications.

Eagle Resource Marketing
2500 83rd St., Bldg. #11, N. Bergen, NJ 07047 USA
(201) 662-9200 (tel)
(201) 662-0660 (fax)
Sells mainframe and personal computer systems and printers; also sells computer parts and boards, such as tape drives, CPUs, controllers, etc.

EAO Switch
198 Pepe's Farm Rd., Milford, CT 06460 USA
(203) 877-4577 (tel)
(203) 877-3694 (fax)
Manufactures switches for a wide variety of electronics applications.

EAR Professional Audio/Video
2641 E. McDowell, Phoenix, AZ 85008 USA
(602) 267-0600 (tel)
Sells complete digital audio workstations, computers, and peripherals for professional audio and video applications.

Earth Computer Technologies
27101 Aliso Creek Rd. #154, Aliso Viejo, CA 92656 USA

(714) 448-9368 (tel)
(714) 448-9316 (fax)
lcd king@ix.netcom.com
http://www.flat-panel.com
Sells LCD displays for a wide variety of electronics applications. Offers a free catalog.

Earthquake Sound
1215 O'Brien Dr., Menlo Park, CA 94025 USA
(415) 327-3003 (tel)
(415) 327-0179 (fax)
Manufactures amplifiers for mobile audio applications.

Earthworks
P.O. Box 517, Wilton, NH 03086 USA
(603) 654-6427 (tel)
(603) 654-6107 (fax)
Visa, MasterCard
Manufactures microphones for professional audio applications.

East Coast Music Mall
25 Hayestown Rd, Danbury, CT 06811 USA
(800) 901-2001 (tel)
http://www.eastcoastmusic.com
Sells a wide variety of equipment for professional audio applications.

East Coast Transistor Parts
2 Marlborough Rd., W. Hempstead, NY 11552
(800) 776-2626 (tel)
(800) 733-5904 (fax)
Sells replacement parts and accessories for audio equipment.

Eastern 21
13001 Ramona Blvd., #K, Irwindale, CA 91706 USA
(813) 813-3896 (tel)
Buys, sells, and trades memory chips for personal computer applications.

Eastern Audio
(718) 961-8256 (tel)
(718) 961-8315 (fax)
Buys and sells used equipment for home audio applications.

Eastern Avionics International
County Airport, Punta Gorda, FL 33982 USA
(813) 637-8585 (tel)
(813) 575-1819 (tel)
(813) 637-0388 (fax)
Visa, MasterCard
Sells GPS, headsets, two-way radios, etc. for professional aviation applications.

Eastern Electronic
No. 4, Shin Long Rd., Kwei-Shan Industrial Area, Tao-Yuan, Taiwan
886-3-361-1141 (tel)
886-3-362-5407 (fax)
Manufactures equipment for professional cable TV applications.

Eastman Kodak
343 State St., Rochester, NY 14650 USA
(800) 462-4307 (tel)
(716) 724-4000 (tel)
Manufactures computer digital cameras for a wide variety of computer multimedia applications and motion-analysis systems for a wide variety of industrial applications.

Easytech Computer
Flat L, 9/F, Phase 3, Camelpaint Bldg.
60 Hoi Yuen Rd., Kwun Tong, Kowloon, Hong Kong
852-2795-6728 (tel)
852-2796-1755 (fax)
Manufactures motherboards and I/O cards for a wide variety of computer applications.

Ebbett Automation
70-72 Victoria St., Petone, Wellington, New Zealand
04-568-6377 (tel)
04-568-2374 (fax)
Sells voltage regulators, uninterruptible power supplies, power inverters, etc. for a wide variety of electronics applications.

ECCO Business Systems
45 W. 45th St., #907, New York, NY 10036 USA
(212) 921-4545 (tel)
(212) 921-2198 (fax)
Manufactures paper shredders for a wide variety of office applications.

Echo Park Music
10050 Montgomery Rd., Cincinnati, OH 45242 USA
(513) 794-1660 (tel)
(513) 794-1668 (fax)
76512.227@compuserve.com
http://www.echopark.com
http://www.moog-music.com
Sells new and used synthesizer keyboards, recording equipment, etc. for professional and personal audio and musical applications. Offers a free catalog.

Eclair Engineering Services
221 Pine St., Florence, MA 01060 USA
(413) 584-6767 (tel/fax)
Manufactures hollow-state direct boxes for professional studio recording and live music performances.

Eclectic Computer Services
276 Martin Ave., Santa Clara, CA 95050 USA
(408) 653-1100 (tel)
(408) 986-1273 (fax)
http://www.eclecticusa.com
sales@electicusa.com
Sells personal desktop and laptop computer systems and printers.

Eclipse
19600 S. Vermont Ave., Torrance, CA 90502 USA
(800) 233-2216 (tel)
(800) 55-ECLIPSE (tel)
(310) 538-1272 (fax)
Manufactures amplifiers, head units, CD changers, and loudspeakers for mobile audio applications.

Eclipse Technologies
3827 Canyon Ridge Dr., San Jose, CA 95148 USA
(408) 980-7288 (tel)
(408) 270-9858 (fax)
Manufactures data storage systems for a wide variety of professional computer network applications.

Econco
1318 Commerce Ave., Woodland, CA 95695 USA
(800) 532-6626 (tel)
(800) 848-8841 (Canada tel)
(916) 662-7553 (tel)
(916) 666-7760 (tel)
Rebuilds power tubes for professional broadcast transmitters.

Ecosphere
90 Inverness Circle E., Englewood, CO 80112 USA
(800) 333-DISH (tel)
(303) 790-4445 (tel)
http://www.ecostar.com
http://www.dishnetwork.com
Manufactures complete systems, dish antennas, receivers, and accessories for home satellite TV applications.

EDCO (Electronic Distributors)
325 Mill St., Vienna, VA 22180 USA
(703) 938-8105 (tel)
(703) 938-4525 (fax)
Distributes shortwave equipment, antennas, scanners, and accessories.

Edirol
1201 Fourth Ave. S., 3rd Fl., Seattle, WA 98134 USA
(800) 380-2580 (tel)
Sells sound cards, speaker systems, keyboards, MIDI synthesizers, etc. for computer music applications. Offer a free buyer's guide and a 30-day money-back guarantee.

EDE
P.O. Box 337, Buffalo, NY 14226 USA
(716) 691-3476 (tel)
Visa, MasterCard
Sells countersurveillance electronics, including Part 15 FM transmitters, bug detectors, voice changers, covert video cameras, etc. Offers a catalog for $5.

Edmo Distributors
E. 5505 Rutter Ave., Spokane, WA 99212 USA
(800) 235-3300 (tel)
(310) 946-5534 (tel)
(310) 946-5150 (fax)
teda@edmo.com
Sells GPS antennas for professional aviation applications.

Edmund Scientific
Edscorp Blvd., Barrington, NJ 08007 USA
(609) 573-6250 (tel)
(609) 573-6879 (tel)
(609) 573-6280 (OEM pricing)
03-5800-4751 (Japan tel)
03-5800-4733 (Japan fax)
Sells a wide variety of scientific and technical products for hobby and professional applications.

Educational Insights
19560 S. Rancho Way, Dominguez Hills, CA 90220 USA
(800) 995-4436 (tel)
(800) 995-0506 (fax)
Manufactures electronics-based learning games, etc.

Scott Edwards Electronics
964 Cactus Wren Ln., Sierra Vista, AZ 85635 USA
(520) 459-4802 (tel)
(520) 459-0623 (fax)
72037.2612@compuserve.com
Visa, MasterCard, American Express, COD ($7)
Manufactures LCD display modules for a wide variety of electronics applications.

EEB (Electronic Equipment Bank)
323 Mill St. NE, Vienna, VA 22180 USA
(800) 368-3270 (tel)
(703) 938-3350 (tel)
(703) 938-6911 (fax)
(703) 938-3781 (BBS)
Sells shortwave receivers, scanners, amateur radio equipment, accessories, etc. Offers a free 96-page catalog.

EG & GIC Sensors
1701 McCarthy Blvd., Milpitas, CA 95035 USA
(800) 767-1888 (tel)
(408) 432-1800 (tel)

EG & GIC Sensors

(408) 432-7322 (fax)
Manufactures pressure sensors and accelerometers for professional industrial applications.

Eicon Technology
14755 Preston Rd., #620, Dallas, TX 75240 USA
(800) 803-4266 (tel)
(214) 239-3270 (tel)
http://www.eicon.com
Manufactures and sells network cards for professional computer applications.

Elderly Instruments
1100 N. Washington, P.O. Box 14210
Lansing, MI 48901 USA
(517) 372-7890 (tel)
Sells used equipment for professional and personal musical applications.

Eldre
1500 Jefferson Rd., Rochester, NY 14623 USA
(716) 427-7280 (tel)
(716) 272-0018 (fax)
Manufactures power bus bars for industrial electronics applications.

Electric City Electronics
521 Duanesburg Rd., Schenectady, NY 12306 USA
(800) 489-7839 (tel)
(518) 356-7839 (tel)
Sells a variety of citizens band radio equipment, including base, mobile, and handheld transceivers, antennas, power inverters, etc.

Electric Works
P.O. Box 1-36457, Ft. Worth, TX 76136 USA
(817) 625-9761 (tel)
(817) 624-9741 (fax)
Manufactures simultaneous digital recorders for professional broadcasting.

Electrified Discounters
110 Webb St., Hamden, CT 06517 USA
(800) 678-8585 (tel)
(203) 787-4246 (tel)
(203) 777-7853 (fax)
Sells discounted personal computer equipment, including laptop computers, printers, etc. Offers a 30-day money-back guarantee.

Electrim
356 Wall St., Princeton, NJ 08540 USA
(609) 683-5546 (tel)
(609) 683-5882 (fax)
Manufactures miniature video cameras for a wide variety of video and surveillance applications.

Electro
1845 57th St., Sarasota, FL 34243 USA
(800) 446-5762 (tel)
(813) 355-8411 (tel)
(813) 355-3120 (fax)
Manufactures zero-speed sensors for professional industrial applications.

Electro Automatic
599 Canal St., Lawrence, MA 01840 USA
(508) 687-6411 (tel)
(508) 687-6493 (fax)
Visa, MasterCard
Manufactures and sells a combination microphone and earphone in a single earpiece for two-way radio communications.

Electrocom Communication Systems
10400 Pioneer Blvd., Bldg E-2
Santa Fe Springs, CA 90670-3728 USA
(800) 348-1477 (tel)
(310) 946-7483 (fax)
Manufactures mobile data computers for professional law-enforcement applications.

Electro-Comm
961 E. 65th St., Tacoma, WA 98404 USA
(800) 821-9150 (tel)
(206) 473-9225 (tel/fax)
Sells a wide variety of amateur radio equipment and accessories, including amateur radio transceivers, receivers, antenna tuners, antennas, amplifiers, etc.

ElectroDynamics
9557 Crosley, Redford, MI 48239 USA
(313) 534-6514 (tel)
(313) 534-1390 (fax)
Visa, MasterCard
Manufactures NiCd testers for radio-control model airplanes.

Electrogadgets Inc.
P.O. Box 2065, Morris Township, NJ 07962 USA
Manufactures and sells digital voice recorder kits.

Electrohome Limited
809 Wellington St. N., Kitchener, ON N2G 4J6 Canada
(800) 265-2171 (tel)
Manufactures LCD projectors for computer multimedia applications.

Electrokraft
P.O. Box 598, Louisville, CO 80027 USA
Sells a kit to display Lissajous patterns on TVs. Offers a VHS video sample for $9.

Electroman
P.O. Box 24474, New Orleans, LA 70184 USA
(504) 482-3017 (tel)
Visa, MasterCard, American Express
Sells an electronic shield for cable boxes.

Electromatic Equipment
600 Oakland Ave., Cedarhurst, NY 11516 USA
(800) 645-4330 (tel)
(516) 295-4399 (fax)
Manufactures data monitor/recorders for a wide variety of professional industrial applications.

Electronic Center
2809 Ross Ave., Dallas, TX 75201 USA
(214) 969-1936 (tel)
Sells a wide variety of amateur radio equipment and accessories, including amateur radio transceivers, receivers, antenna tuners, antennas, amplifiers, etc.

Electronic Design Specialists
4647 Appalachian St., Boca Raton, FL 33428 USA
(407) 487-6103 (tel)
Visa, MasterCard, Discover
Manufactures and sells analyzers and bus line tracers for a wide variety of electronics testing applications. Offers a 60-day money-back guarantee.

The Electronic Goldmine
P.O. Box 5408, Scottsdale, AZ 85261 USA
(602) 451-7454 (tel)
(602) 451-9495 (fax)
Sells a variety of electronics kits and surplus.

Electronic Mailbox
10-12 Charles St., Glen Cove, NY 11542 USA
(800) 323-2325 (tel)
(516) 759-1943 (tech advice tel)
(516) 671-3092 (fax)
Visa, MasterCard, Discover, American Express, COD
Sells amateur and professional video editing and production equipment, including camcorders, batteries, wireless microphones, scan converters, video titlers, A/V processors, desktop video boards, etc. Offers a 10-day warranty.

Electronic Rainbow
6254 LaPas Trail, Indianapolis, IN 46268 USA
(317) 291-7262 (tel)
(317) 291-7269 (fax)
Sells and manufactures a variety of electronics kits.

Electronics Center
3921 Broaddus Ave., El Paso, TX 79904 USA
(915) 562-1000 (tel)
(915) 562-3827 (fax)
Visa, MasterCard, Discover
Buys and sells used VHF/UHF transceivers, test equipment, etc. for professional two-way radio applications.

Electronics Depot
22 Rt. 22 W., Springfield NJ 07081 USA
(800) 500-1553 (tel)
Visa, MasterCard, Discover, American Express
Sells a wide variety of audio equipment, including entire home stereo and car stereo systems. Offers free shipping and a 10-day money-back guarantee.

Electronics Design Specialists
12721 Benson, Overland Park, KS 66213 USA
(800) 793-6960 (tel)
(913) 962-4887 (tel)
Manufactures and sells wireless keyboards for a wide variety of computer applications.

Electronics Hospital
(407) 952-3838 (tel)
Manufactures and sells custom-built, high-power home audio amplifiers.

Electronics Information Systems
P.O. Box 2415, Youngstown, OH 44509 USA
(800) 456-8770 (tel)
(216) 799-7889 (tel)
Visa, MasterCard
Manufactures and sells receivers for shortwave radio-listening applications and computer speakers and world clocks. Offers information and a catalog for an SASE.

Electronics International
5289 NE Elam Yng. Pkwy., #G200, Hillsboro, OR 97124 USA
(503) 640-9797 (tel)
Manufactures a variety of panel metering systems for professional aviation applications.

Electronics Plus
10302 Southard Dr., Beltsville, MD 20705 USA
(800) 591-9009 (tel)
(301) 937-9009 (tel)
(301) 937-5092 (fax)
Sells a variety of parts for the hobbyist, including NOS components, 20,000 electron tubes, transistors, ham radio parts, etc.

Electronics Research
7777 Gardner Rd., Chandler, IN 47610 USA
(812) 925-6000 (tel)
(812) 925-4030 (fax)
Manufactures towers and antennas and antenna site monitor/reporters for professional broadcasting applications.

Electronic Switch
8491 Hospital Dr., #328, Douglasville, GA 30134 USA
(770) 920-1024 (tel)
(770) 920-0700 (fax)
Visa, MasterCard, American Express
Manufactures and sells coaxial switches and antennas for amateur radio applications.

Electron Tube Enterprises
P.O. Box 311, Essex, VT 05451 USA
(802) 879-0611 (tel)
(802) 879-7764 (fax)
Sells a variety of electron tubes. Offers a free catalog.

Electrotech Systems
115 E. Glenside Ave., Glenside, PA 19038 USA
(215) 887-2196 (tel)
(215) 887-0131 (fax)
Manufactures point-to-ground resistance meter for professional electronics testing applications.

Electro Tool
9103 Gillman, Livonia, MI 48150 USA
(800) 772-3455 (tel)
(313) 422-1221 (cust serv tel)
Visa, MasterCard
Sells tools and test equipment, including oscilloscopes, for a wide variety of electronics applications.

Electro-Voice
600 Cecil St., Buchanan, MI 49107 USA
(616) 695-6831 (US tel)
(800) 234-6831 (US fax)
(613) 382-2141 (Canada tel)
Manufactures a variety of microphones for professional studio recording.

Elenco Electronics
150 W. Carpenter Ave., Wheeling, IL 60090 USA
(708) 541-3800 (tel)
Manufactures electronics kits such as personal cassette players, telephones, motion detectors, FM wireless microphones, etc.

Eletech Electronics
16019 Kaplan Ave., Industry, CA 91744 USA
(818) 333-6394 (tel)
(818) 333-6494 (fax)
Manufactures and sells voice boards for a wide variety of computer applications.

Eleven Meter Communications
P.O. Box 3569, Poughkeepsie, NY 12603 USA
(800) 955-5960 (tel)
(914) 452-1614 (tel)
Sells a variety of citizens band radio equipment, including base, mobile, and handheld transceivers, antennas, microphones, etc.

Eli's Amateur Radio
2513 SW Ninth Ave., Ft. Lauderdale, FL 33315 USA
(800) 780-0103 (tel)
(305) 525-0103 (tel)
(305) 944-3383 (tel/fax)
Sells a wide variety of amateur radio equipment and accessories, including amateur radio transceivers, receivers, antenna tuners, antennas, amplifiers, etc.

Elite Speed Products
(614) 231-4170 (tel)
(614) 237-4126 (fax)
(614) 621-2791 (BBS)
http://205.133.101.6/esp/index.html
esp@ee.net
Visa, MasterCard, COD
Sells batteries, etc. for radio-control models.

Elite Video
321 Ouachita Ave., Hot Springs, AR 71901 USA
(800) 468-1996 (tel)
(800) 662-8125 (tel)
(501) 321-0440 (tel)
(501) 321-2327 (fax)
jonvideo@aol.com
Visa, MasterCard, COD
Manufactures video mixers, broadcast video processors, and video duplicators for home video and professional/broadcast video applications. Offers a free catalog.

Elma Electronics
44350 Grimmer Blvd., Fremont, CA 94538 USA
(510) 656-3400 (tel)
(510) 656-3783 (fax)
Manufactures ruggedized rack-mount computer equipment for professional and industrial computer applications.

El Original Electronics
1257 E. Levee, Brownsville, TX 78520 USA
(512) 546-9846 (tel)
(512) 542-8507 (tel)
Sells a wide variety of amateur radio equipment and accessories, including amateur radio transceivers, receivers, antenna tuners, antennas, amplifiers, etc.

Elpac
3711 Paoli Pike, Floyds Knobs, IN 47119 USA
(812) 923-3043 (tel)
(812) 923-3114 (fax)
Manufactures potentiometers for a wide variety of electronics applications.

El Paso Communication Systems (Epcom)
1630 Paisano Dr., El Paso, TX 79901 USA
(915) 533-5119 (tel)
(800) 524-6564 (fax)
(915) 542-4701 (fax)
Sells a wide variety of cellular and two-way radio equipment. Offers a free catalog with more than 3,000 different products.

ELSA
49-0-241-9177-211 (tel)
49-0-241-9177-112 (tel)
49-0-241-9177-600 (fax)
49-0-241-9177-981 (BBS)
Manufactures graphic cards, monitors, modems, and ISDN adapters for personal computer applications.

EMAC
P.O. Box 2042, Carbondale, IL 62902 USA
(618) 529-4525 (tel)
(618) 457-0110 (fax)
(618) 529-5708 (BBS)

EMAC

Manufactures single-board computers for a wide variety of professional computer applications.

E-Mark
4 Daniels Farm Rd., Ste. 328, Trumbull, CT 06611 USA
(800) 831-5383 (tel)
http://www.e-markinc.com
bwest@e-markinc.com
Sells components for a wide variety of electronics applications.

EMASS Storage Systems Solutions
P.O. Box 660023, 2260 Merritt Dr., Dallas, TX 75266 USA
(800) OK-EMASS (tel)
(214) 205-6855 (tel)
(214) 205-7200 (fax)
Manufactures data storage systems for a wide variety of professional computer network applications.

Emco High Voltage
11126 Ridge Rd., Sutter Creek, CA 95685 USA
(800) 546-3680 (tel)
(209) 223-2779 (fax)
emco@ix.netcom.com
Manufactures dc-to-dc converter modules for a wide variety of electronics applications.

Emerald Vision & Sound
127 W. 24th St., New York, NY 10011 USA
(800) 980-2929 (tel)
(212) 463-7663 (tel)
(212) 463-7143 (fax)
Visa, MasterCard, American Express
Sells a wide variety of video equipment, including camcorders, video editors, VCRs, battery chargers, microphones, etc.

Emergency Beacon
15 River St., New Rochelle, NY 10801 USA
(800) 382-0079 (tel)
(914) 235-9400 (tel)
(914) 576-7075 (fax)
Manufactures and sells ELTs for professional aviation applications.

Emerson Radio
9 Entin Rd., Parsippany, NJ 07054 USA
(800) 388-8333 (tech asst tel)
(800) 695-0098 (serv ctr tel)
(201) 884-5800 (tel)
(201) 428-2128 (techn asst fax)
(201) 952-0077 (consumer affairs tel)
Manufactures a wide variety of products for consumer electronics applications.

EMG
P.O. Box 4394, Santa Rosa, CA 95402 USA
(707) 525-9941 (tel)
Manufactures guitar pickups for professional and personal musical applications.

Eminent Technology
225 E. Palmer St., Tallahasse, FL 32301 USA
(904) 575-5655 (tel)
(904) 224-5999 (fax)
Manufactures loudspeakers for car stereo applications.

EMJ Data Systems
RR6 Hwy. 24 S., Guelph, ON N1H 6J3 Canada
(519) 837-2444 (tel)
(519) 837-8935 (fax)
infocan@emj.ca
http://www.emj.ca
Sells CD-ROM drives, PCs, modems, etc. for a wide variety of personal computer applications.

Emkay Enterprises
87 Spindlewick Dr., Nashua, NH 03062 USA
Sells VHF/UHF scanners. Offers information for a SASE.

Emulex
3535 Harbor Blvd., Costa Mesa, CA 92626 USA
(800) EMULEX1 (tel)
(714) 662-5600 (tel)
(714) 513-8266 (fax)
Manufactures print servers for professional computer network applications.

E-mu Systems
P.O. Box 660015, Scotts Valley, CA 95067 USA
(408) 438-1921 (tel)
#6 Adam Ferguson House, Eskmills Industrial Park, Musselburgh, EH21 7PG UK
44-131-653-6556 (tel)
Manufactures digital sampling systems and digital hard drive recorders for a wide variety of professional audio applications.

Encore Computer Systems
1216 S. 16th St., Lincoln, NE 68502 USA
(800) 279-1782 (tel)
(402) 435-8977 (tel)
(402) 435-7452 (fax)
Visa, MasterCard, Discover, American Express, COD
Sells personal desktop computers.

Encore Technology Group
311 E. Ramsey, San Antonio, TX 78216 USA
(800) 880-0202 (tel)
(210) 308-0202 (tel)
(210) 308-0255 (fax)
Manufactures data storage systems for a wide variety of professional computer network applications.

Endicott Research Group
2601 Wayne St., Endicott, NY 13760 USA
(607) 754-9187 (tel)
Manufactures dc power converters for a wide variety of electronics applications.

Energizer Power Systems
US Hwy. 441 N., P.O. Box 147114, Gainesville, FL 32614 USA
(800) 67-POWER (tel)
(904) 462-4726 (fax)
Manufactures batteries and battery chargers for a wide variety of electronics applications.

Energy-Onix
752 Warren St., Hudson, NY 12534 USA
(518) 828-1690 (tel)
(518) 828-8476 (fax)
Manufactures STLs and FM transmitters for professional broadcasting.

Engineering Design Team
1100 NW Compton, Beaverton, OR 97006 USA
(502) 690-1234 (tel)
(502) 690-1243 (fax)
Manufactures interface cards for professional network computer applications.

ENI
100 Highpower Rd., Rochester, NY 14623 USA
(716) 292-7440 (tel)
(716) 427-7839 (fax)
Manufactures power amplifiers for a wide variety of radio applications.

Ensoniq
155 Great Valley Pkwy., P.O. Box 3035, Malvern, PA 19355 USA
(800) 553-5151 (tel)
(800) 257-1439 (fax back)
(610) 647-8908 (fax)
multimedia@ensoniq.com
music_support@ensoniq.com
http://www.ensoniq.com
Manufactures synthesizers, samplers, effects processors, etc. for a wide variety of professional audio and musical applications.

Enterprise Radio Applications
P.O. Box 3144, Charlotte, NC 28210 USA
(800) 925-4735 (tel)
(704) 543-4766 (tech tel)
Manufactures and sells equipment for data communications applications.

Envisions Solutions Technology
P.O. Box 4445, Burlingame, CA 94011 USA
(800) 365-SCAN (tel)
(415) 259-3333 (customer service tel)
(415) 692-9064 (fax)
(415) 259-8145 (BBS)
http://www.ocm.com/envisions
Go Envisions (CompuServe)
Visa, MasterCard, Discover, American Express
Manufactures and sells scanners for personal desktop computers. Offers a 30-day money-back guarantee.

Envoy Data
953 E. Juniata Ave., Mesa, AZ 85204 USA
(800) 368-6971 (tel)
(602) 892-0954 (tel)
(602) 892-0029 (fax)
Manufactures memory cards and I/O cards for a wide variety of computer applications.

Epiphone
1818 Elm Hill Pike, Nashville, TN 37210 USA
(800) 444-2766 (tel)
(615) 871-4500 (tel)
Manufactures electric guitars for professional and personal musical applications. Offers a free catalog.

EPIX
381 Lexington Dr., Buffalo Grove, IL 60089 USA
(708) 465-1818 (tel)
(708) 465-1919 (fax)
Manufactures imaging boards for a wide variety of professional and industrial computer applications.

Epson
20770 Madrona Ave., Torrance, CA 90503 USA
(800) BUY-EPSON X3100 (printer samples tel)
(800) GO-EPSON (Canada tel)
(310) 787-6300 (component sales tel)
(305) 266-0092 (Latin America tel)
(310) 782-0770 (tel)
(310) 782-5220 (fax)
(310) 782-4531 (BBS)
http://www.epson.com
Manufactures and sells surface-mount crystals and oscillators for a variety of electronics and computer applications; also manufactures printers and scanners for personal desktop computer systems.

EPS Technologies
10069 Dakota Ave., Jefferson, SD 57038 USA
(800) 447-0921 (tel)
(800) 526-4258 (tech tel)
(605) 966-5586 (tel)
(605) 966-5482 (fax)
epstech@ix.netcom.com
http://www.epstech.com
Manufactures and sells personal laptop and home computers.

Equinox Systems
1 Equinox Way, Sunrise, FL 33351 USA
(800) 275-3500 (tel)
(305) 746-9000 (sales tel)
(305) 746-9101 (fax)
(305) 746-0282 (BBS)
info@equinox.com
http://www.equinox.com
ftp://equinox.com (Internet FTP site)
Manufactures and sells CPU boards and routers for a wide variety of computer applications.

Ergo Computing
1 Intercontinental Way, Peabody, MA 01960 USA
(800) 880-2494 (tel)
(800) 723-0778 (tel)
(508) 535-7510 (tel)
(508) 535-7512 (fax)
Valley Rd., W. Brigford, Nottingham NG2 6HG UK
0115-9452565 (tel)
0115-9234854 (fax)
Manufactures and sells personal desktop and laptop computers.

Erickson Communications
5456 N. Milwaukee Ave., Chicago, IL 60630 USA
(800) 621-5802 (tel)
(312) 631-5181 (tel)
Visa, MasterCard, Discover, American Express
Sells a wide variety of amateur radio equipment and accessories, including amateur radio transceivers, receivers, antenna tuners, towers, antennas, radio teletype equipment, amplifiers, etc.

Ericsson
701 N. Glenville Dr., Richardson, TX 75081 USA
(800) 431-2345 (tel)
(214) 997-6561 (tel)
(214) 680-1059 (fax)
(214) 952-8800 (tel)
(804) 528-7643 (Canada tel)
Manufactures networks, switchers, message systems, telephones, etc. for professional cellular microwave radio applications.

Erie Aviation
1607 Asbury Rd., Erie International Airport, P.O. Box

8283, Erie, PA 16505 USA
(814) 838-8934 (tel)
(814) 833-3672 (fax)
Sells communications radios, including shortwave car stereo head units.

Erin Multimedia
P.O. Box 350333, Ft. Lauderdale, FL 33335 USA
(800) 910-8187 (tel)
(305) 463-2990 (fax)
Visa, MasterCard
Sells ACER multimedia personal computers.

Eritech
740 N. Glendale Ave., Glendale, CA 91206 USA
(800) 808-6242 (tel)
Buys and sells memory chips for a wide variety of computer applications.

Erivision
+41-062-71-11-34 (Switzerland tel)
+41-062-71-48-19 (Switzerland fax)
Sells a wide variety of equipment for professional cable TV applications.

ESCR
P.O. Box 1192, Delray Beach, FL 33447 USA
(407) 735-3397 (tel)
Sells a wide variety of used and new/old stock tubes.

ESE
142 Sierra St., El Segundo, CA 90245 USA
(310) 322-2136 (tel)
(310) 322-8127 (fax)
Manufactures a variety of master clocks and digital slaves for broadcast stations. Offers a three-year warranty.

Eska & Edvis
P.O. Box 8198, S-200 41 Mamoe, Sweden
+46-40-124734 (fax)
Manufactures and sells active antennas and optional filters for shortwave radios. Sells a wide variety of electronic components. Offers a catalog on floppy disk.

Esoteric Audio USA
44 Pearl Pentecost Rd., Winder, GA 30680 USA
(770) 867-6300 (tel)
(770) 867-2613 (fax)
Manufactures wire, connectors, etc. for mobile audio applications.

ESR Electronic Components
Station Rd., Cullercoats, Tyne & Wear NE30 4PQ UK
0191-251-4363 (tel)
0191-252-2296 (fax)
Sells a wide variety of electronics components, including integrated circuits, voltage regulators, rectifiers, capacitors, etc. No mininum order.

ESS Computer
(800) 537-8981 (tel)
(701) 282-8380 (tel)
(701) 281-1129 (fax)
Visa, MasterCard
Buys and sells new and used personal desktop computers and parts and accessories, including memory chips, hard drives, etc.

Essential Data
P.O. Box 640963, San Jose, CA 95164 USA
(800) 795-4756 (tel)
(408) 955-0821 (fax)
http://www.essential-data.com
Visa, MasterCard, American Express
Sells modems, hard drives, tape drives, CD-ROM drives, and multimedia kits for a wide variety of personal computer applications. Offers a 30-day money-back guarantee.

E-Switch
26 N. 5th St., Minneapolis, MN 55403 USA
(612) 375-9639 (tel)
(612) 375-1905 (fax)
Manufactures miniature switches for a wide variety of electronics applications.

E-T-A Circuit Breakers
1551 Bishop Ct., Mt. Prospect, IL 60056 USA
(708) 827-7600 (tel)

E-T-A Circuit Breakers
(708) 827-7655 (fax)
Manufactures circuit breakers for a wide variety of electronics applications.

ETI Systems
2251 Las Palmas Dr., Carlsbad, CA 92009 USA
(619) 929-0749 (tel)
(619) 929-0748 (fax)
Manufactures potentiometers for a wide variety of electronics applications.

ETO (Ehrhorn Technological Operations)
4975 N. 30th St., Colorado Springs, CO 80919 USA
(719) 260-1191 (tel)
(719) 260-0395 (fax)
Manufactures and sells linear amplifiers for use in HF amateur radio bands. Offers a video of the equipment "in action."

ETrunk Systems
1500 Front St., Yorktown Heights, NY 10598 USA
(800) 438-7865 (tel)
(914) 245-1128 (tel)
(914) 245-2382 (fax)
(914) 245-1194 (fax back)
Manufactures and sells trunking boards for professional two-way radio applications.

Euphonix
11112 Ventura Blvd., Ste. 301, Studio City, CA 91604 USA
(818) 766-1666 (Los Angeles tel)
(818) 766-3401 (Los Angeles fax)
(212) 302-0696 (New York tel)
(212) 302-0797 (New York fax)
(615) 327-2933 (Nashville tel)
(615) 327-3306 (Nashville fax)
+44-171-602-4575 (London tel)
+44-171-603-6775 (London fax)
Manufactures digital audio workstations for professional studio recording applications.

Eurocom
148 Colonnade Rd., Units 1 & 3, Nepean, ON K2E 7R4 Canada
(613) 224-6122 (tel)
Manufactures and sells personal laptop computers.

Eurocomp
9873 Lawrence Rd., Boynton Beach, FL 33436 USA
(407) 731-0396 (tel)
(407) 738-5964 (fax)
eurocomp@flinet.com
http://www.flinet.com/_7Eeurocomp
Sells new and used personal desktop and laptop computer systems and printers.

Eurom Flashware Solution
4655 Old Ironsides Dr., Santa Clara, CA 95054 USA
(408) 748-9995 (tel)
(408) 748-8408 (fax)
Manufactures integrated circuits for a wide variety of electronics applications.

Eventide
1 Alsan Way, Little Ferry, NJ 07643 USA
(201) 641-1200 (tel)
(201) 641-1640 (fax)
postmaster@eventide.com
http://tide1.eventide.com/homepage.htm
Manufactures audio harmonizers, digital DAT audio loggers, and digital radio/video delays for professional broadcasting.

Everex
5020 Brandin Ct., Fremont, CA 94538 USA
(800) 383-7391 (tel)
(510) 683-2186 (fax)
Manufactures and sells personal laptop computers.

Exabyte
1685 38th St., Boulder, CO 80301 USA
(800) EXABYTE (tel)
(303) 442-4333 (tel)
(303) 447-7170 (fax)
31-30-548890 (Netherlands tel)
65-2716331 (Singapore tel)
Manufactures data storage systems for personal and professional desktop computers.

EXAR
2222 Qume Dr., San Jose, CA 95161 USA
(408) 434-6400 (tel)

Manufactures semiconductors for a wide variety of electronics applications.

Excalibur
981 N. Burnt Hickory Rd., Douglasville, GA 30134 USA
(770) 942-9876 (tel)
(770) 942-9632 (fax)
Manufactures loudspeakers for car stereo applications.

Excel Technology
Unit 3, 21/F, Honour Industrial Centre, 6 Sun Yip St., Chai Wan, Hong Kong
852-2515-0328 (tel)
852-2515-0340 (fax)
Manufactures floppy disks for a wide variety of computer applications.

Executive Photo and Electronics
120 W. 31st St., New York, NY 10001 USA
(800) 223-7323 (order tel)
(800) 882-2802 (service tel)
(212) 947-5290 (order tel)
(212) 947-5295 (service tel)
(212) 239-7157 (fax)
Visa, MasterCard, Discover, Federal Express, Diner's Club
Sells home, amateur, and professional photographic, audio, video, and office equipment, such as camcorders, VCRs, car audio equipment, shortwave radios, fax machines, etc.

EXFO
465 Godin Ave., Vanier, QC, G1M 3G7 Canada
(800) 663-3936 (tel)
(418) 683-0211 (tel)
(418) 683-2170 (fax)
903 North Bowser, #360, Richardson, TX 75081 USA
(214) 907-1505 (tel)
(214) 907-2297 (fax)
Centre d'Affaires--Les Metz 100,rue Albert Calmette 78353, Jouy-en-Josas, France
33-1-34-63-00-20 (tel)
33-1-34-65-90-93 (fax)
Manufactures power meters, test sets, light sources, visual fault locators, attenuators, talk sets, lab instruments, etc. for educational, military, and professional cable TV applications.

Exhibo
Vialo Vittorio Veneto 21, I-20052 Monza, Italy
Sells head units, speakers, amplifiers, receivers, cassette decks, CD players, etc. for home audio and car audio applications.

Exide Electronics
8521 Six Forks Rd., Raleigh, NC 27615 USA
(919) 872-3020 (tel)
(919) 870-3450 (fax)
http://www.exide.com/exide
Manufactures uninterruptible power supplies for a wide variety of personal computer applications.

Expen Tech Electronics
5F, No. 2, Alley 8, Syh Wei Ln., Chung Cheng Rd., Hsin Tien City, Taipei, Taiwan
886-2-218-8767 (tel)
886-2-218-0189 (fax)
Manufactures mainboards for a wide variety of computer applications.

Expert Computers
2495 Walden Ave., Buffalo, NY 14225 USA
(800) 549-9778 (tel)
Visa, MasterCard, Discover
Sells computers and computer parts and boards, such as floppy drives, hard drives, monitors, video cards, modems, motherboards, etc. Offers a 30-day money-back guarantee.

Expotech Computer
Wauconda, IL 60084 USA
(800) 215-3976 (tel)
(800) 705-6342 (govern sales tel)
(708) 438-1483 (BBS)
Manufactures and sells personal desktop computers. Offers a 30-day money-back guarantee.

Express Computer Systems
1584 N. Batavia #1, Orange, CA 92655 USA

Express Computer Systems

(800) 327-0730 (tel)
(714) 283-3398 (fax)
Sells IBM computer systems; also sells computer parts and boards, such as CPUs, hard drives, memory chips, tape drives, etc.

Express Computer & Video

726 Seventh Ave., New York, NY 10019 USA
(800) 879-5244 (tel)
(212) 397-6700 (tel)
(212) 397-6848 (fax)
Sells high-end home and amateur audio and video equipment, such as camcorders, VCRs, TVs, TV projectors, home theater equipment, etc.

Extended Systems

5777 N. Meeker Ave., Boise, ID 83704 USA
(800) 234-7576 (tel)
(208) 322-7575 (tel)
(208) 377-1906 (fax)
Manufactures print servers for professional computer network applications.

EXXUS

1645 N. Vine St., #705, Hollywood, CA 90028 USA
(800) 557-1000 (tel)
(800) 557-8000 (tech support tel)
(213) 467-7697 (tel)
(800) 557-9000 (fax)
(213) 467-7698 (fax)
Visa, MasterCard, Discover, American Express
Sells computer diskettes, data carts, optical disks, CD-ROMS, etc. Offers a 30-day money-back guarantee.

EZ Systems

5122 Bolsa Ave., #109, Huntington Beach, CA 92649 USA
(800) 392-6962 (tel)
(714) 379-8383 (tel)
(714) 379-8391 (fax)
(619) 274-9000 (fax back)
dlkezs@deltanet.com
http://register.com/drives
Visa, MasterCard
Sells data storage systems, such as optical drives and tape drives, for a wide variety of computer applications.

Fabrini Technology

87-89 Albert St., Auckland City, New Zealand
302-3476 (tel)
Unit J-16 Link Dr., N. Shore, New Zealand
444-4122 (tel)
Manufactures personal computers.

Fabtronics

5714 W. Wautoma Beach, Hilton, NY 14468 USA
(800) 294-3228 (tel)
(716) 392-1234 (fax)
Sells ring decoders for telephone/electronics applications.

Factory Direct

35 W. 35th St., New York, NY 10001 USA
(800) 695-3535 (tel)
(212) 564-4399 (tel)
(212) 564-4398 (cust serv tel)
(212) 643-9727 (fax)
Sells high-end home, amateur, and professional audio and video equipment, such as camcorders, VCRs, video editors, monitors, TVs, home theater equipment, etc.

Factory Music

962 Washington St., Hanover, MA 02339 USA
(617) 829-0004 (tel)
(617) 829-8950 (fax)
Sells instruments, equipment, and accessories for professional and personal musical applications. Offers a 30-day money-back guarantee.

Fairfax

1403 44th St., Brooklyn, NY 11219 USA
(800) 499-1998 (tel)
Sells personal laptop computers, printers, and scanners; also sells office equipment, such as fax machines and photocopiers.

Fairlight

55 Hughes Ave., 2nd Fl., Culver City, CA 90232 USA
(800) 4-FAIRLIGHT (US tel)
(310) 287-1400 (Los Angeles tel)
(310) 287-0200 (Los Angeles fax)

+44-0171-267-3323 (London tel)
+44-071-267-0919 (London fax)
+02-975-1230 (Sydney tel)
+02-975-1368 (Sydney fax)
+3-5450-8531 (Tokyo tel)
+3-5450-8530 (Tokyo fax)
http://www.fairlight.com
Manufactures digital audio workstations for professional broadcasting and professional studio recording.

Fair Radio Sales
1016 E. Eureka Box 1105, Lima, OH 45802 USA
(419) 227-6573 (tel)
(419) 227-1313 (fax)
Visa, MasterCard, Discover
Sells primarily military surplus radio equipment from as far back as World War II. The equipment includes transceivers, amplifiers, receivers, antennas, parts of all types, and test equipment. Offers a free catalog.

Falco Data Products
440 Potrero Ave., Sunnyvale, CA 94086 USA
(800) 325-2648 (tel)
(408) 745-7123 (tel)
(408) 745-7860 (fax)
Manufactures terminals for a wide variety of professional computer network applications.

Falcon Systems
1417 Market Blvd., Sacramento, CA 95834 USA
(800) 326-1002 (tel)
(916) 928-9255 (tel)
(916) 928-9355 (fax)
Manufactures file servers and data storage systems for a wide variety of professional computer network applications.

Falkner Enterprises of America Inc.
312 River St., Sabula, IA 52070 USA
(515) 683-7621 (tel)
(515) 683-7631 (tel)
Sells batteries and surplus electronic parts. Offers a free catalog.

Family Photo & Video
1961 Coney Island Ave., Brooklyn, NY 11223 USA

(800) 899-7468 (tel)
(800) 405-7468 (cust serv tel)
(718) 645-1298 (tel)
(718) 998-6650 (fax)
Visa, MasterCard, COD
Sells high-end home, amateur, and professional audio and video equipment, such as camcorders, VCRs, video editors, monitors, karaoke decks, home theater equipment, etc. Offers a free catalog.

Far Electronics
1216 Tappen Circle, Carrollton, TX 75006 USA
(214) 245-6700 (tel)
(214) 242-8856 (tel)
Buys and sells systems peripherals and parts for a wide variety of computer applications.

FarPoint Communications
104 E. Ave. K-4 #F, Lancaster, CA 93535 USA
(805) 726-4420 (tel)
(805) 726-4438 (fax)
Manufactures port cards for a wide variety of computer applications.

Fast Electronic US
One Twin Dolphin Dr., Redwood City, CA 94065 USA
(800) 684-MOVIE (tel)
(415) 802-0746 (fax)
Manufactures and sells a digital video editing package for home, amateur, and professional video editing on computer.

Federal Computer Exchange (Fedcom USA)
120 Lowlins Park, Wycoff, NJ 07841 USA
(201) 612-0800 (tel)
(201) 612-9418 (fax)
fedcom@internexus.net
Sells memory and network equipment for professional computer applications.

Feitek
2752 Walton Rd., St. Louis, MO 63114 USA
(314) 423-1770 (tel)
Visa, MasterCard
Sells used test equipment for a wide variety of electronics testing applications.

Fernandez Guitars
16123 Valerio St., Van Nuys, CA 91406 USA
(818) 988-6790 (tel)
(818) 988-3094 (fax)
Manufactures electric guitars for professional and personal musical applications. Offers a catalog for $3.

Fertik's Electronics
5400 Ella St., Philadelphia, PA 19120 USA
(215) 455-2121 (tel/fax)
Sells switches for a wide variety of electronics applications.

F & F Electronics
1 Bryant St., Woburn MA 01801 USA
(800) 462-7624 (tel)
(617) 933-5291 (tel)
(617) 933-3619 (tel)
Sells tape drives for a wide variety of computer applications.

Fibertron
P.O. Box 3130, Fullerton, CA 92634 USA
(510) 635-1970 (NW tel)
(714) 871-3344 (SW tel)
(717) 652-6565 (NE tel)
(404) 409-1700 (SE tel)
Manufactures fiberoptic cable for a wide variety of electronics applications.

Fieldpiece Instruments
231 E. Imperial Hwy., #250, Fullerton, CA 92635 USA
(714) 992-1239 (tel)
(714) 992-6541 (fax)
Manufactures multimeters and accessories that cater to field use.

FieldWorks
9961 Valley View Rd., Eden Prarie, MN 55344 USA
(612) 947-0856 (tel)
(612) 947-0137 (sales tel)
(612) 947-0859 (fax)
6320 Augusta Dr., #1400, Springfield, VA 22150 USA
(703) 913-4800 (tel)
(703) 913-4801 (fax)
Manufactures ruggedized portable computers for a wide variety of industrial and professional field applications.

Fifth Ave. Computers
3841 4th Ave., Ste. 253, San Diego, CA 92103 USA
(619) 230-1234 (tel)
(619) 230-8889 (fax)
Sells new and used personal computer equipment-- especially printers, monitors, terminals, and modems.

Figaro USA
1000 Skokie Blvd., #575, Wilmette, IL 60091 USA
(708) 256-3546 (tel)
(708) 256-3884 (fax)
Manufactures gas sensors for a wide variety of professional industrial applications.

The Filter Company (TFC)
(800) 235-8080 (tel)
Visa, MasterCard, Discover
Sells positive notch filters for cable TV applications.

Finline Technologies
180 Frobisher Dr., Unit 1C, Waterloo, ON N2V 2A2 Canada
(519) 746-1023 (tel)
Manufactures agile modulators for professional cable TV applications.

Firestik Antenna
2614 E. Adams, Phoenix, AZ 85034 USA
(602) 273-7151 (tel)
(602) 273-7152 (tech tel)
Manufactures CB antennas and accessories. Offers a free catalog.

1st CompuChoice
740 Beta Dr., Unit G, Cleveland, OH 44124 USA
(800) 345-8880 (tel)
(216) 460-1002 (tel)
(800) 421-6460 (dealer inquiry tel)
(216) 460-1065 (tech support tel)

(216) 460-1066 (fax)
Visa, MasterCard, Discover
Manufactures and sells personal desktop computers; also sells computer parts and boards, such as floppy drives, hard drives, monitors, motherboards, modems, video cards, keyboards, etc. Offers a 30-day money-back guarantee.

First Call Communications (FCC)
3 Chestnut St., Suffern, NY 10901 USA
(800) 426-8693 (tel)
(914) 357-7339 (tel)
Sells antenna towers manufactured by US Towers.

First Computer Systems
6000 Live Oak Pkwy., #107, Norcross, GA 30093 USA
(800) 325-1911 (tel)
(770) 441-1911 (tel)
(770) 447-TECH (tech support tel)
(770) 441-1856 (fax)
Visa, MasterCard, Discover, American Express, COD
Manufactures and sells personal desktop and portable computers; also sells computer parts and boards, such as hard drives, video cards, modems, motherboards, cases, etc.

First International Computer
6F, Formosa Plastics Rear Bldg.
201 Tung Hwa North Rd., Taipei, Taiwan
86-2-7174500 (tel)
86-2-7182782 (fax)
(510) 252-7777 (US tel)
(510) 252-8888 (US fax)
31-73-273300 (Europe tel)
31-73-231412 (Europe fax)
61-2-748-4566 (Australia tel)
61-2-748-4633 (Australia fax)
(813) 5461-2181 (Japan tel)
(813) 5461-2345 (Japan fax)
42-5-41122643 (Czech tel)
42-5-41213144 (Czech fax)
Manufactures laptop computers.

First Source International
7 Journey, Aliso Viejo, CA 92656 USA

(800) 942-9866 (tel)
(714) 448-7750 (tel)
(714) 448-7760 (fax)
mhs:sales@1source (CompuServe)
sales@1source.mhs.compuserve.com
Visa, MasterCard, Discover, American Express
Sells printers, scanners, hard drives, monitors, processor upgrades, computer memory, monitors, video cards, modems, keyboards, etc. Offers a free catalog.

Fisher
21350 Lassen St., Chatsworth, CA 91311 USA
(818) 998-7322 (tel)
Manufactures cassette decks, amplifiers, tuners, etc. for home audio applications.

Fittec Electronics
Block A, 13/F, Hung Cheung Industrial Centre, Phase 1, 12 Tsing Yeung Circuit, Tuen Mun, NT, Hong Kong
852-2454-9897 (tel)
852-2454-9895 (fax)
Manufactures motherboards for a wide variety of computer applications.

5D Technology
2368 W. Winton Ave., Hayward, CA 94545 USA
(800) 787-5335 (tel)
(510) 785-8909 (tel)
(510) 785-8907 (fax)
fived@hooked.net
Manufactures and sells personal laptop computers. Offers a 30-day money-back guarantee.

Flame Enterprises
10945 Osborne St., Canoga Park, CA 91304 USA
(800) 854-2255 (tel)
(818) 700-2905 (tel)
(818) 700-9168 (fax)
Sells relays, relay sockets, circuit breakers, switches, and power connectors for professional aerospace applications.

Flightcom
7340 SW Durham Rd., Portland, OR 97224 USA
(800) 432-4342 (tel)

Flightcom

(503) 620-2943 (fax)
Manufactures headsets and intercoms for professional aviation applications.

Flight Products International
140 Sherry Ln., Kalispell, MT 59901 USA
(800) 526-1231 (tel)
(406) 257-7078 (fax)
Visa, MasterCard, Discover, COD
Manufactures and sells battery-level indicators for professional aviation applications.

Fluke
P.O. Box 9090, M/S 250E, Everett, WA 98206 USA
(800) 59-FLUKE (tel)
(206) 356-5500 (tel)
(800) 44-FLUKE (tech support)
905-890-7600 (Canada tel)
31-40-644200 (Europe)
Manufactures a variety of meters, including process calibrators, display multimeters, analog/digital multimeters, etc., for a wide variety of professional and personal electronics applications. Offers a free catalog.

F&M Electronics
P.O. Box 21, Sale City, GA 31784 USA
(919) 299-3437 (tel)
Sells a wide variety of amateur and citizens band radio equipment and accessories, including transceivers, receivers, antenna tuners, antennas, amplifiers, etc.

FOB Tech-Specialties
909 Crosstimbers, Houston, TX 77022 USA
(800) 864-5391 (tel)
(713) 691-7009 (fax)
Sells computer parts and boards, such as motherboards, CPU chips, etc. Offers a free catalog.

Focal America
1531 Lookout Dr., Agoura, CA 91301 USA
(818) 707-1629 (tel)
(818) 991-3072 (fax)
Manufactures loudspeakers for high-end home audio and professional audio applications. Offers a free catalog.

Focus Computer Center
1303 46th St., Brooklyn, NY 11219 USA
(800) 223-3411 (tel)
(718) 871-7600 (tel)
(718) 438-4263 (fax)
Visa, MasterCard, Discover, American Express
Sells computer parts and boards, such as SCSI controllers, CPU upgrades, modems, etc.

Focus Enhancements
(617) 938-8088 (tel)
Manufactures and sells expansion cards for personal laptop computers.

Fordham Radio Supply
260 Motor Pkwy., Hauppauge, NY 11788 USA
(800) 695-4848 (tel)
Sells tools and electronic test equipment.

John Ford
RR #6, Lindsay, ON K9V 4R6 Canada
(705) 324-1681 (fax)
Sells a variety of transformers, PC boards, and components for electrostatic speakers for high-end audio applications. Offers a free list.

Forest Electronics
853 Fairway Dr., Bensonville, IL 60106 USA
(800) 332-1996 (tel)
(708) 860-9048 (fax)
Sells a complete line of new cable equipment. Offers a guarantee and a 30-day money-back option.

Fortex Enterprises
7712B Timberlake Rd., Lynchburg, VA 24502 USA
(804) 239-6524 (tel)
(804) 239-7255 (fax)
Manufactures mobile antennas for amateur radio applications.

Fortron/Source
2925 Bayview Dr., Fremont, CA 94538 USA
(510) 440-0188 (tel)
(510) 440-0928 (fax)
Manufactures power inverters for portable consumer electronics applications.

Forz
4400 Rt. 9, Freehold, NJ 07728 USA
(800) 531-5148 (tel)
(908) 303-7488 (tel)
(908) 303-9007 (fax)
Sells new and refurbished personal desktop and laptop computers, printers, parts, and accessories.

Foss Warehouse Distributors
285 Shenck St., N. Tonawanda, NY 14120 USA
(800) 473-0506 (tel)
(800) 488-0525 (fax)
Sells cable TV converters.

Fostex
15431 Blackburn Ave., Norwalk, CA 90650 USA
(800) 7-FOSTEX (tel)
(310) 921-1112 (tel)
(310) 802-1964 (fax)
(212) 529-2069 (tel)
Manufactures a variety of DAT recorders and a digital audio workstation for broadcast stations and a 16-channel, digital audio recorder/editor for professional studio recording.

Fotronic
P.O. Box 708, Medford, MA 02155 USA
(617) 665-1400 (tel)
(617) 665-0780 (fax)
Offers sales and service for electronic test equipment, including oscilloscopes.

Four Pi Systems
1905 Technology Pl., San Diego, CA 92127 USA
(619) 485-8551 (tel)
Manufactures x-ray inspection systems for professional industrial applications.

Fourth Dimension Industry
331G Dante Ct., Holbrook, NY 11741 USA
(800) 378-0348 (tel)
(516) 467-1220 (tel)
(516) 467-1645 (fax)
Sells telephones and accessories for cellular telephone applications.

Fox International
23600 Aurora Rd., Bedford Heights, OH 44146 USA
(800) 321-6993 (tel)
(216) 439-8500 (tel)
(800) 445-7991 (fax)
Sells replacement parts and accessories for audio equipment.

Franklin Electronic Publishers
1 Franklin Plaza, Burlington, NJ 08016 USA
(800) 266-5626 (tel)
(609) 386-2500 (tel)
Manufactures pocket computer organizers.

Free Trade Video
4718 18th Ave., #127, Brooklyn, NY 11201 USA
(800) 234-8813 (tel)
(800) 455-6677 (tel)
(718) 435-4151 (tel)
(718) 633-6890 (fax)
Visa, MasterCard, Discover, American Express, COD
Sells high-end home, amateur, and professional audio and video equipment, such as camcorders, video editors, microphones, VCRs, cables, etc. Offers a 15-day return policy.

Frequency Management Crystals
15302 Bolsa Chica St., Huntington Beach, CA 92649 USA
(800) 800-9825 (tel)
(714) 890-1832 (fax)
(800) 477-8852 (Florida tel)
(954) 755-8891 (Florida fax)
Manufactures and sells crystals for pager and two-way radio applications.

Frezzolini Electronics
5 Valley St., Hawthorne, NJ 07506 USA
(800) 345-1030 (tel)
Manufactures lighting systems for amateur and professional video production.

Friwo North America
176 Thomas Johnson Dr., #202, Fredrick, MD 21702 USA

Friwo North America
(800) 229-9430 (tel)
(301) 694-7698 (fax)
Manufactures external power supplies for a wide variety of electronics applications.

Frontgate
2800 Henkle Dr., Lebanon, OH 45036 USA
(800) 626-6488 (tel)
(800) 537-8484 (tech support tel)
(800) 436-2105 (fax)
Sells radar detectors, travel TVs, videocassette rewinders, cordless telephones, calculators, and other household products.

Frontier Technologies
10201 N. Port Washington Rd., Mequon, WI 5092 USA
(800) 929-3054 (tel)
(414) 241-4555 (tel)
(414) 241-7084 (fax)
Manufactures interface cards and controller boards for professional computer network applications.

Frydendahl Hi-Fi
Kirketorvet 4, 8900 Randers, Denmark
86-41-45-22 (tel)
Banegårdsgade 4, 8000 Århus C, Denmark
86-13-02-33 (tel)
Sells high-end home audio equipment, such as speakers, cassette decks, amplifiers, preamplifiers, etc.

FTG Data Systems
8381 Katella Ave., Stanton, CA 90680 USA
Manufactures light pens for a wide variety of computer applications.

Fujitsu Microelectronics
3545 N. First St., San Jose, CA 95134 USA
(800) 898-1320 (tel)
(800) 898-1361 (tel)
(800) 642-7616 (tel)
National Service Headquarters, 19600 S. Vermont St. Torrance, CA 90502 USA
(800) 423-8161 (tel)
Manufactures computers, communications electronics, and microelectronics, including hard disk drives.

Fujitsu Ten
(See Eclipse)

Full-Ten International
8/F, No. 32, Kanton St., Taipei, Taiwan
886-2-3020928 (tel)
886-2-3020775 (fax)
Manufactures motherboards, memory chips, CPU chips, monitors, etc. for a wide variety of computer applications.

Funai USA
100 North St., Teterboro, NJ 07698 USA
(201) 288-2666 (tel)
(201) 288-0239 (fax)
TVs, VCRs, and TV/VCR units for home video applications.

Furman Sound
30 Rich St., Greenbrae, CA 94904 USA
(415) 927-1225 (tel)
(415) 927-4548 (fax)
Manufactures several power conditioners for professional studio recording.

Futaba of America
P.O. Box 19767, Irvine, CA 92713 USA
(714) 455-9888 (tel)
(714) 455-9899 (fax)
Manufactures sport and computer radios and servos for radio-control model airplanes and boats.

Future Domain
2801 McGaw Ave., Irvine, CA 92714 USA
(714) 253-0400 (tel)
Manufactures IDE boards for a wide variety of computer applications.

Future Micro
2691 Richter Ave. #118, Irvine, CA 92714 USA
(800) 700-6507 (tel)
(714) 622-9130 (tel)
(714) 622-9136 (tech support tel)
(714) 622-9143 (fax)
Visa, MasterCard, Discover, American Express

Sells computer parts and boards, such as hard drives, motherboards, memory chips, hard disk controllers, multimedia upgrades, modems, etc.

Futuretouch
2030 E. 4th St., Ste. 152, Santa Ana, CA 92705 USA
(714) 558-6824 (tel)
Manufactures multimedia sound pads to control audio settings for computer games.

Future Video Products
28 Argonaut, #140, Aliso Viejo, CA 92656 USA
(714) 770-4416 (tel)
Manufactures edit controllers for home, amateur, and professional video editing applications via computer.

FWT
1901 E. Loop 820 S., Ft. Worth, TX 76112 USA
(800) 334-1481 (tel)
(817) 429-6010 (tel)
Manufactures monopole antennas for professional cellular microwave, PCS, and SMR radio applications.

Gabriel Video & Camera
260 W. Swedesford Rd., Berwyn, PA 19312 USA
(800) 454-1010 (tel)
(610) 889-3960 (tel)
(610) 647-5830 (fax)
Visa, MasterCard, Discover, American Express, Diner's Club
Sells high-end home, amateur, and professional audio and video equipment, such as camcorders, video editors, microphones, VCRs, cables, etc.

Gage Applied Sciences
5610 Bois Franc, Montreal, PQ H4S 1A9 Canada
(800) 567-GAGE (tel)
(514) 337-8411 (tel)
(514) 337-4317 (fax)
Manufactures data acquisition cards for a wide variety of professional computer applications.

Galaxy Computers
423 S. Lynnhaven Rd., #109, Virginia Beach, VA 23452 USA
(800) 809-1319 (tel)
(800) 771-4049 (tel)
(804) 486-8225 (tech tel)
(804) 486-8389 (tel)
(804) 498-2432 (purchasing tel)
(804) 498-1211 (fax)
(804) 486-5916 (BBS)
Visa, MasterCard, Discover, American Express
Manufactures and sells personal desktop computers; also sells computer parts and boards, such as floppy drives, hard drives, monitors, video cards, modems, keyboards, etc. Offers a 30-day money-back guarantee.

Galaxy Electronics
P.O. Box 1202, 67 Eber Ave., Akron, OH 44309 USA
(216) 376-2402 (tel)
Sells shortwave receivers, amateur and citizens band radio equipment, scanners, and accessories.

Galow Cable
(305) 593-0406 (tel)
(305) 593-7418 (tel)
(305) 593-7419 (fax)
Visa, MasterCard
Sells a wide variety of new, used, and refurbished equipment for professional cable TV applications.

Gamber-Johnson
801 Francis St., Stevens Point, WI 54481 USA
(800) 456-6868 (tel)
(800) 934-3577 (fax)
Manufactures and sells mounting hardware for professional voice/data communications and antennas in automobiles.

Francisco Ganán
Calle Fortaleza #202, San Juan, PR 00901
Sells head units, loudspeakers, amplifiers, receivers, cassette decks, CD players, etc. for home audio and car audio applications.

Gao Systems
(817) 572-1200 (tel)
(817) 483-7508 (fax)

GAO Systems
Buys and sells personal and mainframe computers and peripherals.

Gap Antenna Products
6010 N. Old Dixie Hwy., Vero Beach, FL 32967 USA
(407) 778-3728 (tel)
Visa, MasterCard, Discover, COD
Manufactures and sells vertical amateur radio antennas for all bands from shortwave through VHF frequencies.

Gardiner Communications
3605 Security St., Garland, TX 75042 USA
(214) 348-4747 (tel)
Manufactures MMDS yagi antennas for professional cable TV applications.

Garmin
9875 Widmer Rd., Lenexa, KS 66215 USA
(800) 800-1020 (tel)
(913) 599-1515 (tel)
Manufactures GPS and IFR receivers and VHF transceivers for professional aviation applications.

Gateway Electronics
8123 Page Blvd., St. Louis, MO 63130 USA
(800) 669-5810 (tel)
(314) 427-6116 (tel)
(314) 427-3147 (fax)
9222 Chesapeake Dr., San Diego, CA 92123 USA
(619) 279-6802 (tel)
2525 Federal Blvd., Denver, CO 81211 USA
(303) 458-5444 (tel)
gateway@mo.net
http://www.gatewayelex.com
Visa, MasterCard, Discover
Sells components, electronic gadgetry, surplus, etc. for a wide variety of electronics applications. Minimum order of $10.

Gateway 2000
610 Gateway Dr., P.O. Box 2000, N. Sioux City, SD 57049 USA
(800) 846-2417 (tel)
(800) 846-2080 (component add-on sales tel)
(800) 846-1778 (TDD tel)
(800) 846-4526 (fax back)
(605) 232-2000 (tel)
(605) 232-2561 (fax back)
(605) 232-2023 (fax)
Visa, MasterCard, Discover, American Express, COD
Manufactures and sells personal desktop and laptop computers; also sells printers, modems, and multimedia kits.

Gats Electronics
149 The Vale, Acton W3 7HR UK
0181-932-0144 (tel)
0181-932-0145 (fax)
Manufactures and sells kits, including timers, electronic locks, bugs, strobes, etc. Offers a kit list for a SAE.

GBC
P.O. Box 2205, Ocean City, NJ 07712 USA
(908) 774-8502 (tel)
(908) 774-8041 (fax)
Manufactures neon bulbs for a wide variety of electronics applications.

GB Labs
990 Housatonic Ave., Bridgeport, CT 06606 USA
(203) 335-1093 (tel)
(203) 331-9214 (fax)
Manufactures plug-in guitar samplers for professional and personal musical applications.

G&D Transforms
2515 E. Thomas Rd., #16-721, Phoenix, AZ 85016 USA
(602) 650-1155 (tel)
Manufactures and sells low-jitter clock kits and jitter-free transports for compact disc players.

GE
5365 A-2 Robin Hood, Norfolk, VA 23513 USA
(804) 244-3226 (tel)
Appliances/Microwave Products
Bldg. 41, Rm. 106, Louisville, KY 40225 USA
(502) 452-3568 (tel)
Manufactures a wide variety of electronics equipment, including home audio equipment (cassette

players, etc.), home video equipment (VCRs, etc.), office equipment (telephones, answering machines, etc.), and more.

Gefen Systems
6261 Variel Ave., #C, Woodlawn Hills, CA 91367 USA
(800) 545-6900 (tel)
(818) 884-6294 (tel)
(818) 884-3108 (fax)
Manufactures a touch screen extender for professional studio recording.

Gemco
1080 N. Crooks Rd., Clawson, MI 48017 USA
(800) 635-0289 (tel)
(810) 435-8120 (fax)
Manufactures linear displacement transducers for a wide variety of professional measuring applications.

Gemeni Industries
215 Entin Rd., Clifton, NJ 07014 USA
(201) 471-9050 (tel)
Manufactures accessories for a wide variety of applications.

Gems Sensors Division
1 Cowles Rd., Plainville, CT 06062 USA
(800) 321-6070 (tel)
Manufactures fluid sensors, etc. for a wide variety of professional industrial applications. Offers a 216-page catalog.

Generad
HC89 Box 210, Pocono Pines, PA 18350 USA
(800) 836-6544 (tel)
(717) 646-3475 (tel)
Manufactures a key adapter for home, amateur, and professional video editing applications.

General Communications
5157 Anton Dr., Madison, WI 53719 USA
(800) 356-3200 (tel)
(608) 271-4848 (tel)
(608) 274-2080 (fax)

Visa, MasterCard
Sells equipment for professional cellular microwave telephone applications.

General Device Instruments
(408) 241-7376 (tel)
(408) 241-6375 (fax)
(408) 983-1234 (BBS)
icdevice@best.com
Visa, MasterCard
Sells EEPROM device programmers for a wide variety of electronics applications.

Generale Electronique Services
BP 46 77542 Savigny-Le Temple, Cedex, France
33-1-64-41-7888 (tel)
33-1-60-63-2485 (fax)
Sells receivers, antennas, and accessories for shortwave radio-listening applications.

General Instrument
2200 Byberry Rd., Hatboro, PA 19040 USA
(800) 523-6678 (tel)
(215) 674-4800 (tel)
(215) 956-6497 (fax)
541-788-4567 (Argentina tel)
55-21-494-3132 (Brazil tel)
(416) 789-7831 (Canada tel)
+44-734-755555 (UK tel)
852-587-1163 (Hong Kong tel)
http://www.gi.com
Manufactures cable optics platforms for professional cable TV applications.

General Service Computers (GSCI)
4806 E. Hwy. 50, Yankton, SD 57078 USA
(800) 438-9840 (tel)
(605) 665-5123 (tel)
(605) 665-5125 (fax)
Webmaster@gsci.com
http://www.gsci.com/
Visa, MasterCard, Discover, American Express
Sells personal home computer parts and accessories, including memory chips, motherboards, hard drives, power supplies, controllers, etc.

Genesis Camera
814 W. Lancaster Ave., Bryn Mawr, PA 19010 USA
(800) 575-9977 (tel)
(610) 527-5261 (tel)
(610) 527-5148 (fax)
Sells high-end home, amateur, and professional video equipment, such as camcorders, VCRs, video editors, microphones, video switchers, mixers, etc.

Genicom
Genicom Dr., Waynesboro, VA 22980 USA
(800) 535-4364 (tel)
(703) 949-1426 (fax)
Sells personal computer printer parts.

Gentex Electro-Acoustics
5 Tinkham Ave., Derry, NH 03038 USA
(800) 258-3554 (tel)
Manufactures microphone elements for a wide variety of electronics applications.

Gentner Communications
1825 Research Way, Salt Lake City, UT 84119 USA
(801) 975-6200 (tel)
(801) 974-3676 (fax)
Manufactures multiline on-air phone systems and group teleconferencers for professional broadcasting.

Genoa Systems
(408) 362-2900 (tel)
(408) 432-8342 (tech support tel)
(408) 434-0997 (fax)
(408) 943-1256 (BBS)
Manufactures graphics boards, audio boards, tape drives, video capture boards, etc. for personal computer applications.

Gepco International
2225 Hubbard St. W., Chicago, IL 60612 USA
Manufactures coaxial cable, audio cables, snakes, patch bays, and breakout boxes for professional broadcasting.

German Physiks
Postbox 1026, D-63506 Hainburg, Germany
+49-69-4940963 (fax)
Manufactures and sells high-end loudspeakers for home audio applications.

GES
Rue de L'Industrie, 77176 Savigny le Temple, Cedex, France
Sells equipment for shortwave radio receiving applications.

Gestetner
599 W. Putnam Ave., P.O. Box 2656, Greenwich, CT 06836 USA
(203) 863-5564 (tel)
Manufactures paper shredders for a wide variety of office applications.

GETTECH
402 Riley Rd., New Windsor, NY 12553 USA
(914) 564-5347 (tel)
Manufactures and sells digital voice recorders. Offers custom designs and enclosures.

G and G Electronics of Maryland
8524 Dakota Dr., Gaithersburg, MD 20877 USA
(301) 258-7373 (tel)
Visa, MasterCard
Sells used test equipment and Commodore 64 accessories for amateur radio data applications.

G&G Technologies
350 N. St., Teterboro, NJ 07608 USA
(800) VIDEO-911 (tel)
(201) 288-8900 (tel)
(201) 288-6881 (fax)
Sells a variety of home, amateur, and professional video equipment, including VCRs, A/V mixers, camcorders, and video editors.

Gibson
641 Massman Dr., Nashville, TN 37210 USA
(800) 4-GIBSON (tel)
(800) 524-3388 (tel)
Manufactures electric guitars for professional and personal musical applications.

Gilfer Shortwave
52 Park Ave., Park Ridge, NJ 07656 USA
(800) GILFER-1 (tel)
(201) 391-7887 (tel)
(201) 391-7433 (fax)
info@gilfer.com
http://www.gilfer.com
Sells a variety of shortwave receivers, accessories, and literature. Offers a catalog for $1.

David & Ervena Gillespie
P.O. Box 788, Micaville, NC 28755 USA
(704) 675-9942 (tel)
david.gillespie@icomm.com
Manufactures and sells a Wefax/slow-scan TV converter for amateur radio applications.

Gilway Technical Lamp
800 W. Cummings Park, Woburn, MA 01801 USA
(617) 935-4442 (tel)
(617) 938-5867 (fax)
sales@gilway.com
Manufactures lamps for a wide variety of electronics applications. Offers a free catalog.

G&L
5381 Production Dr., Huntington Beach, CA 92649 USA
(714) 897-6766 (tel)
(714) 896-0736 (fax)
Manufactures electric guitars for professional and personal musical applications.

Glade Communication Equipment
2691 151st Pl. NE, Redmond, WA 98052 USA
(800) 347-0048 (tel)
(206) 869-4119 (fax)
Sells equipment for professional cable TV applications.

GL Communications
841-F Quince Orchard., Gaithersburg, MD 20878 USA
(301) 670-4784 (tel)
(301) 926-8234 (fax)
Manufactures analyzing equipment for professional cellular microwave radio applications.

Gleason Research
P.O. Box 1247, Arlington, MA 02174 USA
(617) 641-2551 (tel/fax)
gleason@tiac.net
http://www.tiac.net/users/gleason
Manufactures kits for a wide variety of robotics applications.

Glidecam Industries
676 State Rd., Plymouth, MA 02360 USA
(800) 949-2089 (tel)
http://www.glidecam.com
Manufactures Glidecam videocamera stabilizers for home, amateur, and professional video recording applications.

Global Computer Sources (GCS)
107 Northern Blvd., #408, Great Neck, NY 11021 USA
(516) 829-3774 (tel)
(516) 487-0950 (fax)
Buys, sells, and trades personal computer parts and accessories, including memory chips, system boards, hard drives, controllers, etc.

Global Computing
98-1268 Kaahumanu St., #200, Pearl City, HI 96782 USA
(800) 590-DISK (tel)
(808) 487-DISK (tel)
(808) 487-3770 (fax)
global@globalcomputing.com
http://www.planet-hawaii.com/~global
Sells more than 25,000 different products for a wide variety of computing applications.

Global Electronics International
220 E. Lake St. Ste. 300, Addison, IL 60101 USA
(800) 499-0473 (tel)
(708) 530-9123 (tel)
(708) 530-9151 (fax)
Manufactures crystals for a variety of radio applications.

Global Hobby Distributors

Global Hobby Distributors
10725 Ellis Ave., Fountain Valley, CA 92728 USA
(714) 963-0133 (tel)
(714) 962-6452 (fax)
Sells battery chargers, power panels, digital tachometers, etc. for radio-control models.

Global Video
260 W. Swedeford Rd., Berwyn, PA 19312 USA
(800) 420-5050 (tel)
(610) 889-3960 (tel)
(610) 647-5830 (fax)
Visa, MasterCard, Discover, American Express
Sells high-end and home audio and video equipment, such as camcorders, video editors, CD players, VCRs, cables, etc.

Global Village Communications
1144 E. Arques Ave., Sunnyvale, CA 94086 USA
(800) 736-4821 (tel)
(408) 523-1000 (tel)
http://www.globalvillage.com
Sells fax modem boards for a wide variety of computer applications.

Globetek
3505 E. Royalton Rd., #160, Broadview Hts., OH 44147 USA
(800) 229-4640 (tel)
(216) 526-8550 (tel)
(216) 526-8817 (fax)
Manufactures and sells I/O adapters for a wide variety of computer applications.

GlobTek
186 Veterans Dr., Northvale, NJ 07647 USA
(201) 784-1000 (tel)
(201) 784-0111 (fax)
http://grammercy.los.com/_7Eglobtek/
Manufactures plug-in wall power supplies for a wide variety of electronics applications.

GMB Sales
140 N. Belle Mead Rd., Setauket, NY 11733 USA
(800) 874-1765 (dealer tel)
(516) 689-3400 (tel)
(800) 635-0596 (fax)
Sells replacement parts and accessories for audio equipment.

GMM Research
17500 Redhill, #250, Irvine, CA 92714 USA
(714) 752-9447 (tel)
(714) 752-7335 (fax)
Manufactures communications boards for a wide variety of computer applications.

GMP Precision Products
510 E. Arrow Hwy., San Dimas, CA 91773 USA
(909) 592-5144 (tel)
(909) 599-0798 (fax)
Manufactures electric guitars for professional and personal musical applications.

GMRS Radio Sales
P.O. Box 37825, Phoenix, AZ 85069 USA
(800) 571-GMRS (tel)
(602) 878-6398 (fax)
Sells GMRS emergency radios and other accessories, including headsets, mobile antennas, speakers, microphones, etc.

GMW
Landstrasse 16, CH-5430 Wettingen, Switzerland
056-262324 (tel)
Sells equipment for shortwave radio receiving applications.

GN Nettest
109 N. Genesee St., Utica, NY 13502 USA
(800) 443-6154 (tel)
(315) 797-4449 (tel)
Manufactures fiberoptic test equipment for professional cable TV applications.

Godin
4240 Sere St., St. Laurent, PQ H4T 1A6 Canada
Manufactures electric guitars for professional and personal musical applications. Offers a catalog for $2.

Goetz/Vanderschaaf
1 Cherry Ridge Rd., Pisgah Forest, NC 28768 USA
(704) 884-9842 (tel)
Sells loudspeakers for high-end home audio applications.

Goff Professional
175 Costello Rd., Newington, CT 06111 USA
(860) 667-2358 (tel)
goffprof@aol.com
Visa, MasterCard
Sells synthesizer keyboards, MIDI modules, speakers, and parts for professional and personal musical applications.

Gold Line
P.O. Box 500, W. Redding, CT 06896 USA
(203) 938-2588 (tel)
(203) 938-8740 (fax)
Manufactures a digital rack-mount, real-time analyzer with DSP for professional studio recording and live music performances.

Gold Sound
4285 S. Broadway, Englewood, CO 80110 USA
(303) 789-5310 (tel)
(303) 762-0527 (fax)
Manufactures loudspeakers for car stereo applications.

Goldstar Electronics International
1000 Sylvan Ave., Englewood Cliffs, NJ 07632 USA
(201) 816-2000 (tel)
(201) 816-0636 (fax)
Parts Depot
201 James Record Rd., Huntsville, AL 35824 USA
(800) 222-6457 (consumer tel)
(800) 562-0244 (dealers & tech support tel)
(800) 448-4026 (fax)
(205) 772-8860 (tel)
(205) 772-8987 (fax)
Manufactures camcorders and VCRs for home, amateur, and professional video editing, recording, and viewing applications.

Gordon Products
67 Del Mar Dr., Brookfield, CT 06804 USA
(203) 775-4501 (tel)
(203) 775-1162 (fax)
Manufactures proximity sensors for a wide variety of electronics applications.

Gould Electronics
35129 Curtis Blvd., Eastlake, OH 44095 USA
(216) 953-5059 (tel)
Manufactures batteries for a wide variety of electronics applications.

Gould Fiber Optics
1121 Benfield Blvd., Millersville, MD 21108 USA
(800) 54-GOULD
(410) 987-5600 (tel)
(410) 987-1201 (fax)
Manufactures and sells custom components. Advice available from Sales Engineer.

Gould Instrument Systems
8333 Rockside Rd., Valley View, OH 44125 USA
(216) 328-7000 (tel)
(216) 328-7400 (fax)
Manufactures portable chart recorders and signal conditioners for professional industrial applications.

Gould Shawmut
374 Merrimac St., Newburyport, MA 01950 USA
(508) 462-6662 (tel)
Manufactures fuse holders for a wide variety of electronics applications.

Go Video
14455 N. Hayden Rd., #219, Scottsdale, AZ 85260 USA
(602) 998-3400 (tel)
(602) 998-8312 (fax)
Visa, MasterCard
Manufactures dual-deck VCRs for home and professional video applications.

Graffiti Electronics
3103 Bandini Blvd., Vernon, CA 90023 USA
(213) 262-4377 (tel)
(213) 262-5688 (fax)
Manufactures amplifiers for car stereo applications.

Grandma's Music & Sound
800 S-T Juan Tabo Blvd. NE, Albuquerque, NM 87123 USA
(800) 444-5252 (tel)
(505) 292-0341 (tel)
(505) 293-6184 (fax)
Sells MIDI gear, signal processors, tape decks, mixers, microphones, monitors, and blank recording tape for a wide variety of professional audio applications.

Grand Rapids Technologies
4526 Poinsettia SE, Kentwood, MI 49508 USA
(616) 531-4893 (tel)
(616) 249-0437 (fax)
Manufactures panel instruments for professional aviation applications.

Granite Digital
3101 Whipple Rd., Union City, CA 94587 USA
(510) 471-6442 (tel)
(510) 471-6267 (fax)
Manufactures SCSI terminators, diagnostic cables, internal cables, digital switches, etc. for a wide variety of computer applications.

Grayhill
561 Hillgrove Ave., P.O. Box 10373, LaGrange, IL 60525 USA
(800) 244-0559 (tel)
(708) 354-1040 (tel)
(708) 354-2820 (fax)
Manufactures switches for a wide variety of electronics applications.

Graymark
P.O. Box 2015, Tustin, CA 92861 USA
(800) 854-7393 (tel)
Visa, MasterCard, Discover, American Express
Manufactures robots for hobbyist and experimental applications. Offers a free 40-page catalog.

Grayson Electronics
306 Enterprise Dr., Forest, VA 24551 USA
(800) 800-7465 (tel)
(804) 385-7692 (fax)
Manufactures test equipment for telephone paging systems.

GRE America
425 Harbor Blvd., Belmont, CA 94002 USA
(800) 233-5973 (tel)
(415) 591-1400 (tel)
(415) 591-2001 (fax)
Manufactures and sells frequency converters, antennas, preamplifiers, etc. for scanner and VHF/UHF radio applications.

Great Southern Security
513 Bankhead Hwy., Carrolton, GA 30117 USA
(800) 732-5000 (tel)
Visa, MasterCard, Discover, COD
Manufactures and sells a "bug" detection receiver. Offers a 10-day money-back guarantee. Add $15 for CODs.

Greencorp USA
(800) 972-0707 (tel)
Sells audio cassettes.

Green Dot Audio
P.O. Box 290609, Nashville, TN 37229 USA
(888) 367-9242 (tel)
(615) 367-9242 (tel)
Sells equipment for cables, patch bays, etc. for professional studio recording applications.

Greenleaf Electronics
P.O. Box 538, Bensenville, IL 60106 USA
(800) 742-2567 (tel)
(708) 616-8050 (tel)
(708) 616-8094 (fax)
Visa, MasterCard, COD
Sells cable TV converters and descramblers. Offers a 30-day guarantee.

Green Spring Computers
1204 O'Brien Dr., Menlo Park, CA 94025 USA
(415) 327-1200 (tel)
(415) 327-3808 (fax)
support@gspring.com
Manufactures a wide variety of computer boards for professional and industrial computer applications.

Greenweld Electronic Components

27D Park Rd., Southampton SO15 3UQ UK
01703-236363 (tel)
01703-236307 (fax)
01703-231003 (accounts tel)
Sells a wide variety of surplus electronics and computer components and equipment, including capacitors, fans, LCD displays, computers, etc.

Warren Gregoire & Associates
229 El Pueblo Pl., Clayton, CA 94517 USA
(800) 634-0094 (tel)
(510) 420-5701 (tel)
Manufactures headsets for professional aviation applications.

Gre-Trading
No. 8, Long 41, Ln. 207, Sec. 2, Tsinghai Rd., Taichung, Taiwan
Sells loudspeakers, amplifiers, receivers, cassette decks, CD players, etc. for home audio applications.

Griblin Engineering
369B 3rd St., #112, San Rafael, CA 94901 USA
(800) 605-AMPS (tel)
Sells pickups, tuners, tubes, loudspeakers, components, etc. for professional and personal musical applications.

Grove Enterprises
300 S. Hwy. 64 W., Brasstown, NC 28902 USA
(800) 438-8155 (tel)
(704) 837-9200 (tel)
(704) 837-2216 (fax)
Sells a wide selection of shortwave and scanner radios, books, and accessories. Also manufactures some accessories, including a sound enhancer/speaker for shortwave listening.

Gruhn Guitars
400 Broadway, Nashville, TN 37203 USA
(615) 256-2033 (tel)
(615) 255-2021 (fax)
Buys, sells, and trades used electric guitars for professional and personal musical applications.

Grundig
Abteilung AWK, 90748 Fürth/Bay, Germany
+49-0-911-703743 (fax)
Lextronix USA, 3520 Haven Ave., Unit L, Redwood City, CA 94063 USA
(800) 872-2228 (tel)
(415) 361-1611 (tel)
(415) 361-1724 (fax)
Manufactures and sells portable shortwave receivers.

GSF Agency
122 Strand, Ste. 1, Santa Monica, CA 90405 USA
(310) 452-6216 (tel)
(310) 452-3886 (fax)
gsfa@netcom.com
Visa, MasterCard, American Express
Imports synthesizers for a wide variety of professional audio and musical applications.

GT Audio
12866 Foothill Blvd., Sylmar, CA 91342 USA
(818) 361-4500 (tel)
(818) 365-9884 (fax)
Manufactures hollow-state microphones and signal processors for professional studio recording and live music performances.

GTSI (Government Technology Services)
4100 LaFayette Center Dr., Chantilly, VA 22186 USA
(800) 999-GTSI (tel)
http://www.gtsi.com
Sells a wide variety of computer equipment and software, specifically for government computer applications.

The Guitar Broker
2455 E. Sunrise Blvd., Ft. Lauderdale, FL 33304 USA
(305) 563-5507 (tel)
(305) 563-5509 (fax)
Sells used electric guitars and amplifiers for professional and personal musical applications. Offers a free list.

Guitar Center
7425 Sunset Blvd., Hollywood, CA 90046 USA

Guitar Center

(800) 638-4280 (tel)
http://www.musician.com
Sells MIDI gear, multimedia equipment, etc. for a wide variety of professional audio and musical applications. Offers a free catalog.

Guitar Emporium

1610 Bardstown Rd., Louisville, KY 40205 USA
(502) 459-4153 (tel)
Sells used electric guitars and amplifiers for professional and personal musical applications.

Guitar Network

19 N. Market, Fredrick, MD 21701 USA
(301) 694-3231 (tel)
(301) 694-5912 (fax)
Buys and sells used equipment for professional and personal musical applications. Offers a list for a SASE.

Gulf Coast Avionics

4243 N. Westshore Blvd., Tampa Inter. Airport
Tampa, FL 33614 USA
(813) 879-9714 (tel)
(813) 875-4514 (fax)
Visa, MasterCard, Discover, American Express, COD
Sells flight panel instruments for professional aviation applications.

Gulton Statham

1644 Whittier Ave., Costa Mesa, CA 92627 USA
(714) 642-2400 (tel)
Manufactures line pressure transmitters for industrial applications.

Jo Gunn Antennas

Rt. 1 Box 32C, Hwy. 82, Ethelsville, AL 35461 USA
(205) 658-2229 (tel)
(205) 658-2259 (fax)
Visa, MasterCard, Discover
Manufactures directional CB antennas. Offers a catalog for $2.

G-Vox

400 Green St., Philadelphia, PA 19123 USA
(800) 789-GVOX (tel)
(215) 922-0880 (tel)
(215) 922-7230 (fax)
info@lyrrus.com
http://www.lyrrus.com
Manufactures guitar/computer interfaces for educational musical applications.

Haewa Communications

4357-B Park Dr., Norcross, GA 30093 USA
(800) 783-4239 (tel)
(404) 921-3272 (tel)
(404) 921-2896 (fax)
Visa, MasterCard, COD
Manufactures handheld transceivers for two-way radio communications.

Hafler Professional

628 S. River Dr., Tempe, AZ 82581 USA
(800) 366-1619 (tel)
(602) 967-8132 (fax)
Manufactures amplifiers for professional studio recording. Features a seven-year warranty.

Hagstrom Electronics

2 Green Lantern Blvd., Endicott, NY 13760 USA
(607) 786-7523 (tel)
Visa, MasterCard
Manufactures keyboard encoder boards for a wide variety of computer applications.

HAL Communications

P.O. Box 365, Urbana, IL 61801 USA
(217) 367-7373 (tel)
(217) 367-1701 (fax)
Manufactures computer DSP modem boards for digital data amateur radio communications.

Halcyon Group

P.O. Box 2264, Sarasota, FL 34230 USA
(800) 664-6999 (tel)
(813) 954-1961 (fax)
Visa, MasterCard, Discover, COD
Sells descrambler kits, converters, data blockers, etc. for home video applications.

Hales Design Group
16812 Gothard St., Huntington Beach, CA 92647 USA
(714) 841-6884 (tel)
(714) 841-2427 (fax)
Manufactures loudspeakers for high-end home audio and home theater applications.

Halladay Acoustics
73 Spring St., #8, Saratoga Springs, NY 12866 USA
(518) 581-8095 (tel)
Manufactures and sells monitor speakers. Offers a free catalog.

Hall Electronics
706 Rose Hill Dr., Charlottesville, VA 22901 USA
(804) 984-4255 (tel)
(804) 984-3299 (fax)
Sells STL transmitters and receivers, exciters, mixers, tuners, transmitters, headsets, digital effects processors, turntables, microphones, cart machines, reel-to-reel decks, etc. for professional broadcasting applications.

Doug Hall Electronics
815 E. Hudson St., Columbus, OH 43211 USA
(614) 261-8871 (tel)
(614) 261-8805 (fax)
Manufactures and sells a receiver voter and a remote base interface for repeaters for amateur radio applications.

Hallikaainen & Friends
141 Suburban E4, San Louis Obispo, CA 93401 USA
(805) 541-0200 (tel)
Manufactures digital remote controls for professional broadcasting.

Halted Specialties
3500 Ryder St., Santa Clara, CA 95051 USA
(800) 442-5833 (tel)
(408) 732-1573 (tel)
(408) 732-6428 (fax)
hsc-info@hsc.cue.com
Sells a wide variety of electronics surplus, including parts, computer assemblies, power supplies, speakers, and amateur radio kits.

Hamamatsu
360 Foothill Rd., Box 6910, Bridgewater, NJ 08807 USA
(908) 231-0960 (tel)
Manufactures power supplies for powering photomultiplier tubes.

Ham Buerger
417 Davisville Rd., Willow Grove, PA 19090 USA
(215) 659-5900 (tel)
(215) 659-5902 (fax)
Sells a wide variety of amateur radio equipment and accessories, including amateur radio transceivers, receivers, antenna tuners, antennas, amplifiers, etc.

The Ham Contact
P.O. Box 3624, Long Beach, CA 90803 USA
(800) 933-HAM4 (tel)
(310) 433-5860 (tel)
Sells rechargeable battery packs for various amateur radio transceivers.

The Ham Key Co.
8342 Olive Blvd., St. Louis, MO 63132 USA
(800) 527-0807 (tel)
Manufactures paddles, keyers, SWR meters, etc. for amateur radio use.

Ham Radio Outlet
933 N. Euclid St., Anaheim, CA 92801 USA
(800) 854-6046 (tel)
(714) 533-7373 (tel)
2492 W. Victory Blvd., Burbank, CA 91506 USA
(800) 854-6046 (tel)
(818) 842-1786 (tel)
2210 Livingston St., Oakland, CA 94606 USA
(800) 854-6046 (tel)
(510) 534-5757 (tel)
5375 Kearny Villa Rd., San Diego, CA 92123 USA
(800) 854-6046 (tel)
(619) 560-4900 (tel)
510 Lawrence Expwy., #102, Sunnyvale, CA 94086 USA

Ham Radio Outlet

(800) 854-6046 (tel)
(408) 736-9496 (tel)
kdm@hamradio.com
1509 N. Dupont Hwy., New Castle, DE 19720 USA
(800) 644-4476 (tel)
(302) 322-7092 (tel)
11705 S.W. Pacific Hwy., Portland, OR 97223 USA
(800) 854-6046 (tel)
(503) 598-0555 (tel)
8400 E. Lliff Ave., #9, Denver, CO 80231 USA
(800) 444-9476 (tel)
(303) 745-7373 (tel)
1702 W. Camelback Rd., Phoenix, AZ 85015 USA
(800) 444-9476 (tel)
(602) 242-3515 (tel)
6071 Buford Hwy., Atlanta, GA 30340 USA
(800) 444-7927 (tel)
(404) 263-0700 (tel)
14803 Build America Dr., Woodbridge, VA 22191 USA
(800) 444-4799 (tel)
(703) 643-1063 (tel)
224 N. Broadway, Salem, NH 03079 USA
(800) 444-0047 (tel)
(603) 898-3750 (tel)
clw@hamradio.com
http://www.hamradio.com
Sells a wide variety of amateur radio equipment and accessories, including amateur radio transceivers, receivers, antenna tuners, towers, antennas, radio teletype equipment, amplifiers, etc. Offers a free catalog.

The Ham Station

P.O. Box 6522, 220 N. Fulton Ave.
Evansville, IN 47719 USA
(800) 729-4373 (tel)
(812) 422-0231 (tel)
(812) 422-0252 (service tel)
(812) 422-4253 (fax)
(812) 424-3614 (BBS)
Visa, MasterCard, Discover, COD
Sells a wide variety of amateur radio equipment and accessories, including amateur radio transceivers, receivers, antenna tuners, antennas, amplifiers, etc.

Hamtronics

65 Moul Rd., Hilton, NY 14468 USA
(716) 392-9430 (tel)
(716) 392-9420 (fax)
Manufactures receiver kits, receivers, exciters, and converters for amateur radio applications.

Hamtronics Trevose

4033 Brownsville Rd., Trevose, PA 19053 USA
(215) 357-1400 (tel)
(215) 355-8959 (fax)
Sells a wide variety of amateur radio equipment and accessories, including amateur radio transceivers, receivers, antenna tuners, antennas, amplifiers, etc.

Hand Held Products

7510 E. Independence Blvd., Charlotte, NC 28227 USA
(800) 582-4263 (tel)
Manufactures bar code scanning systems for professional computerized business applications.

Njal Hansson

Tvetenveien 156, P.O. Box 6031 Ettterstad,
N-0601-Oslo 6 Norway
Sells head units, loudspeakers, amplifiers, receivers, cassette decks, CD players, etc. for home audio and car audio applications.

Hantronix

10080 Bubb Rd., Cupertino, CA 95014 USA
(408) 252-1100 (tel)
(408) 252-1123 (fax)
Manufactures LCD modules for a variety of electronics and computer display applications.

The Happy Medium

430 State St., Madison, WI 53703 USA
(608) 255-2887 (tel)
Visa, MasterCard, Discover, American Express
Sells a wide variety of high-end home audio equipment.

Harborsound & Video

180A Central St., Saugas, MA 01906 USA
(617) 231-0095 (tel)
(617) 231-0295 (fax)

Buys and sells new and used mixing consoles, multi-track recorders, audio processors, compressors, microphones, etc. for professional studio recording and live music performances. Offers a free list of equipment for sale.

Hard Drives & Accessories (HDA)
18601 Hatteras St., #139, Tarzana, CA 91356 USA
(818) 758-9544 (tel)
(818) 343-3985 (fax)
Sells data storage systems, such as hard drives, floppy drives, and tape drives, for a wide variety of computer applications.

Hardin Electronics
5635 E. Rosedale St., Ft. Worth, TX 76112 USA
(800) 433-3203 (tel)
(817) 429-9761 (tel)
(817) 457-2200 (tel)
(817) 457-2429 (fax)
Sells a wide variety of amateur radio equipment and accessories, including amateur radio transceivers, receivers, antenna tuners, antennas, amplifiers, etc.

John Hardy
P.O. Box 631, Evanston, IL 60204 USA
(847) 864-8060 (tel)
(847) 864-8076 (fax)
Manufactures microphone preamplifiers for professional studio recording applications.

Harmonic Lightwaves
3005 Bunker Hill Ln., Santa Clara, CA 95054 USA
Manufactures optical return path systems for professional cable TV applications.

Harmon/Kardon
240 Crossways Pk., Woodbury, NY 11797 USA
(516) 496-3400 (tel)
8380 Balboa Blvd., Northridge, CA 91325 USA
(818) 895-8100 (tel)
Manufactures audio amplifiers, receivers, cassette decks, CD players, etc. for high-end home audio and home theater applications.

Harmony Computers
1801 Flatbush Ave., Brooklyn, NY 11210 USA
(800) 441-1144 (tel)
(718) 692-3232 (tel)
(718) 692-2828 (cust service)
Visa, MasterCard
Manufactures and sells personal home and laptop computers; also sells parts and accessories, including modems, monitors, printers, scanners, disk drives, etc.

Harper's Guitars
P.O. Box 2877, Apple Valley, CA 92307 USA
(619) 240-1792 (tel/fax)
Visa, MasterCard
Manufactures and sells electric guitars for professional and personal musical applications.

Harris Allied
P.O. Box 883, Melbourne, FL 32901 USA
(800) 442-7747 (tel)
(217) 222-8200 (tel)
(217) 224-1439 (fax)
(217) 222-8290 (interntl)
(217) 224-2764 (interntl fax)
Manufactures a variety of integrated circuits; FM, ATV, TV, and satellite exciters and transmitters; mixing consoles; sound enhancement systems; sound editors; DDSes; and microphones for broadcast stations.

Harris Computer Systems
2101 W. Cypress Creek Rd., Ft. Lauderdale, FL 33309 USA
(800) 666-4544 (tel)
(305) 977-5513 (tel)
(305) 977-5580 (fax)
cyberguard@csd.harris.com
http://www.csd.harris.com
Manufactures protection systems for professional computer applications.

Harrison
7104 Crossroads Blvd., Ste 118, Brentwood, TN 37027 USA
(615) 370-9001 (tel)

Harrison

(615) 370-4906 (fax)
Manufactures audio mixing consoles for professional broadcast, film, and post-production applications.

Harrison Laboratory
6466 E. Ponderosa Dr., P.O. Box 1349, Parker, CO 80134 USA
(303) 841-5360 (tel)
Manufactures loudspeakers and amplifiers for home and mobile audio applications.

Hart Electronic Kits
6 Penylan Hill, Oswestry, Shropshire, SY10 9AF UK
01691-652894 (tel)
Manufactures and sells kits, including preamplifiers and headphone amplifiers, for high-end audio applications. Also sells components for audio systems. Offers a free kit list.

Hartford Computer Group
(800) 617-4424 (tel)
(708) 934-3380 (tel)
(708) 827-5900 (tech support tel)
(708) 934-0157 (fax)
http://www.awa.com/hartford
Visa, MasterCard, Discover, American Express
Sells new and used personal home and laptop computers and parts and accessories, including memory chips, video cards, hard disk drives, modems, monitors, etc.

Harting Elektronik
2155 Stonington Ave., #212, Hoffman Estates, IL 60195 USA
(708) 519-7700 (tel)
(708) 519-9771 (fax)
Manufactures connectors for a wide variety of electronics applications.

Hart Scientific
220 N. 1300 W., Pleasant Grove, UT 84062 USA
(801) 785-1600 (tel)
(801) 785-7118 (fax)
Manufactures thermometers and sensor calibrators for professional electronics testing applications.

Harvard Radio
2 Mount Royal Ave., Marlborough, MA 01752 USA
(508) 460-4000 (tel)
(508) 460-4099 (fax)
Manufactures mobile data systems for law enforcement agencies.

Harz
22 Parkway Circle, #6, New Castle, DE 19720 USA
(302) 323-6100 (tel)
(302) 323-6101 (fax)
Manufactures video monitors for a wide variety of applications.

Hatry Electronics
500 Ledyard St., Hartford, CT 06114 USA
(203) 296-1881 (tel)
(203) 296-7110 (fax)
Sells a wide variety of amateur radio equipment and accessories, including amateur radio transceivers, receivers, antenna tuners, antennas, amplifiers, etc.

HAVE
309 Power Ave., Hudson, NY 12534 USA
(800) 999-4283 (tel)
Sells audio and video tape, cable, accessories, equipment, and supplies for professional recording studios and live music performances. Offers a free catalog.

Haven Industries
2950 Lake Emma Rd., Lake Mary, FL 32746 USA
46-23 Crane St., LIC, NY 11101 USA
(800) 231-0031 (tel)
Sells a wide variety of electronics products, including computers, cellular phones, camcorders, home and car audio components, adapters, connectors, etc.

Hawk
2124 Zanker Rd., San Jose, CA 94131 USA
(800) 875-0333 (tel)
(408) 437-1583 (fax)
Manufactures systems for car security applications.

Hayes Microcomputer Products
P.O. Box 105203, Atlanta, GA 30348 USA

(800) 374-8388 (tel)
(800) HAYES-FX (fax back)
(770) 446-6336 (BBS)
support@os.hayes.com
http://www.hayes.com
Manufactures modems for personal and professional computer applications.

HCM Audio
(800) 222-3465 (tel)
(916) 345-1341 (tel)
(916) 345-7269 (fax)
Sells speakers, cassette decks, amplifiers, audio cable, etc. for home audio applications.

H. Co. Computer Products
16812 Hale Ave., Irvine, CA 92714 USA
(800) 347-1273 (tel)
(714) 833-3222 (tel)
(714) 833-3389 (fax)
Visa, MasterCard, Discover, American Express
Sells computer parts and boards, such as hard drives, memory, processor, etc.

HC Power
17032 Armstrong, Irvine, CA 92714 USA
(800) 486-4427 (tel)
(714) 261-2200 (tel)
(714) 261-6584 (fax)
Manufactures power supplies for a wide variety of electronics applications.

HD Computer Products
2031 E. Cerritos Ave., #7K, Anaheim, CA 92806 USA
(714) 490-0207 (tel)
(714) 490-0294 (fax)
Sells laptop computer parts, including displays, system boards, hard drives, etc., for a wide variety of computer applications.

HDSS Computer Products
6625 Jarvis Ave., Newark, CA 94560 USA
Visa, MasterCard, Discover, American Express, COD
(800) 252-9777 (tel)
(510) 494-8100 (corp accounts)
(510) 494-8045 (tech support)
(510) 494-8501 (fax)
Specializes in the sales of storage devices and multimedia, including hard drives, tape drives, CD-ROM drives, disc drives, monitors, video cards, sound cards, and multimedia packages. Offers a 30-day money-back guarantee.

Heartwood Computer
2103 Stafford St. Extension, Monroe, NC 28110 USA
(704) 283-6447 (tel)
(704) 289-9229 (fax)
Sells refurbished Memorex/Telex and IBM computer systems and parts for a wide variety of computer applications.

Harold Heaster
84 N. Timber Creek Rd., Ormond Beach, FL 32174 USA
(904) 672-2878 (tel)
Sells a wide variety of amateur radio equipment and accessories, including amateur radio transceivers, receivers, antenna tuners, antennas, amplifiers, etc.

Heath
P.O. Box 1288, 455 Riverview Dr., Benton Harbor, MI 49022 USA
(616) 925-6000 (tel)
(616) 925-2898 (fax)
Manufactures and sells electronics educational courses.

Heil Sound
2 Heil Dr., Marissa, IL 62257 USA
(618) 295-3000 (tel)
(618) 295-3030 (fax)
Manufactures and sells microphones, speakers, and 10-meter FM transceiver kits for amateur radio applications.

Helios Systems
1996 Lundy Ave., San Jose, CA 95131 USA
(800) 366-2983 (tel)
(408) 432-0292 (tel)
(408) 452-4549 (fax)

Helios Systems
Sells memory upgrades for professional network computer applications.

Hellfire
51 Gravel St., Wilkes-Barre, PA 18705 USA
(800) 422-5050 (tel)
(717) 825-0540 (tel)
(717) 823-6343 (fax)
Manufactures systems for car security applications.

Henry Engineering
(818) 355-3656 (tel)
(818) 355-0077 (fax)
(818) 355-4210 (fax back)
Manufactures sound card-to-line level interface amplifiers for professional broadcasting applications.

Henry Radio
2050 S. Bundy Dr., Los Angeles, CA 90025 USA
(800) 877-7979 (tel)
(310) 820-1234 (tel)
(310) 826-7790 (fax)
Sells a wide variety of amateur radio equipment and accessories, including amateur radio transceivers, receivers, antenna tuners, antennas, amplifiers, etc. Also manufactures amateur radio amplifiers.

Herman Electronics
1365 NW 23rd St., Miami, FL 33142 USA
(800) 938-4376 (tel)
(305) 634-6591 (tel)
(305) 634-6247 (fax)
Sells replacement parts and accessories for audio equipment.

Herman Elig Computers
P.O. Box 180, Amelia, OH 45102 USA
(800) 878-8806 (tel)
Sells McDonnell Douglas tape drives for computer data storage applications.

Hero Communications
7291 NW 74th St., Medley, FL 33166 USA
(305) 887-3203 (tel)
(305) 885-8532 (fax)
Manufactures and sells equipment for professional cable TV applications.

Hewlett Packard
5301 Stevens Creek Blvd., P.O. Box 58059, Santa Clara, CA 95051 USA
(800) 387-3867 (tel)
(800) 452-4844 (tel)
(800) 322-HPPC (tel)
(800) 810-0134 (tel)
(970) 635-1500 (tel)
(916) 785-3090 (tech support fax)
(916) 785-2689 (BBS)
http://www.hp.com
Manufactures RF and measurement equipment for professional video and cable TV applications, other test equipment (such as spectrum analyzers and digital multimeters), CD-ROM drives and recorders, personal and professional computer systems, computer printers, and image scanners.

HHB Communications
73-75 Scrubs Lane, London NW10 6QU UK
+44-0181-962-5000 (tel)
+44-0181-962-5050 (fax)
Manufactures DAT decks, DATs, and recordable CDs for consumer and professional studio recordings.

Hialeah Communications
801 Hialeah Dr., Hialeah, FL 33010 USA
(305) 885-9929 (tel)
(305) 888-5040 (tel)
(305) 888-8768 (fax)
Sells a wide variety of amateur radio equipment and accessories, including amateur radio transceivers, receivers, antenna tuners, antennas, amplifiers, etc.

Hi-Fi Entusiasten
Tagensvej 162, 2200 Kobenhavn N, Denmark
31-83-22-00 (tel)
Sells high-end home audio equipment, such as speakers, cassette decks, amplifiers, preamplifiers, etc.

Hi Fi Exchange
Foreside Mall, Rt. 1, Falmouth, ME 04105 USA

(207) 781-2326 (tel)
Visa, MasterCard, Discover, American Express
Sells a wide variety of new and used high-end home audio equipment.

Hi-Fi Farm
2039 Electric Rd., Roanoke, VA 24018 USA
(800) 752-4018 (tel)
(703) 772-4434 (tel)
Visa, MasterCard, Discover, American Express
Sells high-end home audio equipment, such as speakers, cassette decks, TVs, home theater equipment, etc.

HiFonics Corp.
501 Broad Ave. S., Ridgefield, NJ 07657 USA
(201) 945-8880 (tel)
(201) 945-1218 (fax)
Manufactures amplifiers and loudspeakers for mobile audio applications.

High Sierra Antennas
P.O. Box 2389, Nevada City, CA 95959 USA
(916) 273-3415 (tel)
(916) 273-7561 (fax)
high_sierra@psyber.com
http://www.psyber.com/biz/high_sierra
Manufactures and sells an all-band mobile HF transceiving antenna for amateur radio applications.

High Tech Cable
1055 Bobby Brown State Park Rd., Elberton, GA 30635 USA
(407) 375-9458 (tel)
(706) 213-0385 (tel)
(407) 375-8631 (fax)
Sells new and used equipment for cable TV applications.

High Tech Mart
1590 Oakland Rd., #B115, San Jose, CA 95131 USA
(800) 998-4359 (tel)
(408) 453-8867 (fax)
Sells cards and hubs for professional computer network applications.

Highway 155 CB Shop
18888 State Hwy. 155 S., Flint, TX 75762 USA
(800) 582-0426 (tel)
(903) 825-2070 (tel)
Sells a variety of citizens band radio equipment, including base, mobile, and handheld transceivers, antennas, power inverters, etc.

Hi-MU Amplifiers
52 Wheeler Ave., Cranston, RI 02905 USA
(401) 781-7314 (tel)
Manufactures amplifiers for personal and professional musical applications.

Hi-Q Products
506 N. Garfield Ave., #200, Alhambra, CA 91801 USA
(800) 996-6873 (tel)
(818) 308-4400 (tel)
(818) 308-4406 (fax)
Visa, MasterCard
Manufactures lasers for a wide variety of electronics applications.

Hirose Electric
2688 Westhills Ct., Simi Valley, CA 93065 USA
(800) 879-8071 (tel)
(805) 522-7958 (tel)
(805) 522-3217 (fax)
Manufactures connectors for a wide variety of electronics and computer applications. Offers a free catalog on CD-ROM.

Richard Hirschman
Postbox 1649, D-72606, Nurtingen, Germany
+49-7127-141499 (tel)
+49-7127-141502 (fax)
Manufactures multimeter sets for professional electronics testing applications.

Hirsch Sales
219 California Dr., Williamsville, NY 14221 USA
(716) 632-1189 (tel)
(716) 634-0634 (tel)
(716) 632-6304 (fax)

Hirsch Sales

Sells a wide variety of amateur radio equipment and accessories, including amateur radio transceivers, receivers, antenna tuners, antennas, amplifiers, etc.

Hitachi
2000 Sierra Point Pkwy., Brisbane, CA 94005 USA
(800) 285-1601 (tel)
3890 Steve Reynolds, Norcross, GA 30093 USA
(800) 241-6558 (tel)
(404) 279-5600 (tel)
110 Summit Ave., Montvale, NJ 07645 USA
(201) 573-0774 (tel)
(401) W. Artesia Blvd., Compton, CA 90220 USA
(800) 448-2244 (tel)
371 Van Ness Way, #120, Torrance, CA 90501 USA
(310) 328-6116 (tel)
(310) 328-6252 (fax)
(516) 921-7200 (tel)
(516) 921-0993 (fax)
http://www.hitachi.co.jp
Manufactures video monitors and data storage systems for computer applications; also manufactures oscilloscopes for professional electronics testing applications.

Hitachi Denchi America
371 Van Ness Way, #120, Torrance, CA 90501 USA
(310) 328-6116 (tel)
Manufactures DSOs for professional electronics testing applications.

Hi-Tech Component Distributors
59 S. La Patera Ln., Goleta, CA 93117 USA
(800) 708-2867 (tel)
(805) 681-9961 (tel)
(805) 681-9971 (fax)
Visa, MasterCard, American Express
Sells computer parts and boards, such as hard disk drives, monitors, surplus computers, etc.

Hitec RCD
10729 Wheatlands Ave., #C, Santee, CA 92071 USA
(619) 258-4940 (tel)
(619) 449-1002 (fax)
Sells sport and computer radios, chargers, and servos for radio-control model airplanes and boats.

Hi-Tech USA
1562 Centre Pointe Dr., Milpitas, CA 95035 USA
(800) 831-2888 (tel)
(408) 262-8688 (tel)
(408) 262-8772 (fax)
(800) 869-8868 (tech support tel)
(408) 956-8285 (tech support tel)
(408) 956-8243 (BBS)
Manufactures and sells customized complete systems and a variety of parts and components, including motherboards, cases, controllers, modems, multimedia packages, monitors, math coprocessors, mouses, printers, tape drives, SVGA cards, etc. Offers a 30-day money-back guarantee.

Hitron
933 E. 11th St., Los Angeles, CA 90021 USA
(213) 623-1071 (tel)
(213) 623-2340 (fax)
Manufactures loudspeakers and amplifiers for car stereo applications.

H & J Electronics
2700 W. Cypress Creek Rd., Ft. Lauderdale, FL 33309 USA
(800) 275-2447 (tel)
(305) 971-7750 (tel)
(305) 979-9028 (fax)
Visa, MasterCard, Discover, American Express
Sells personal desktop and laptop computers; also sells computer parts and boards, such as memory chips, hard drives, motherboards, video cards, modems, etc.

Hobbs
P.O. Box 19424, Springfield, IL 62794 USA
(217) 753-7773 (tel)
(217) 753-7789 (fax)
Manufactures pressure and vacuum switches for a wide variety of professional applications.

Hobby Club
23141 Arroyo Vista #210, R.S. Margarita, CA 92688 USA
(714) 459-1750 (tel)
(714) 459-1760 (fax)

Sells radio-control systems, etc. for radio-control model airplanes.

Hobby Mart
6889 Peachtree Industrial Blvd., Norcross, GA 30092 USA
(800) 959-6278 (tel)
(707) 840-8495 (tel)
Visa, MasterCard, COD ($5.95)
Sells radios, batteries, battery chargers, etc. for radio-control models.

Hobby Radio Stop
P.O. Box 291849, Kettering, OH 45429 USA
(800) 423-1331 (tel)
(513) 299-6440 (tel)
Sells scanners, CB transceivers, and antennas for a wide variety of electronics applications.

Hobby Shack
18480 Bandilier Circle, Fountain Valley, CA 92728 USA
(800) 854-8471 (tel)
(714) 963-9634 (tel)
(714) 962-6452 (fax)
Visa, MasterCard, Discover, COD ($5.50)
Manufactures and sells sport and computer radios, batteries, power panels, tachometers, and servos for radio-control model airplanes.

Hobbytech
34 Joslyn Dr., Elgin, IL 60120 USA
(708) 695-5903 (tel)
(708) 837-6235 (fax)
Manufactures maximum airspeed indicators for radio-control model airplanes.

Hobby Warehouse of Sacramento
8950 Osage Ave., Sacramento, CA 95828 USA
(800) 333-3640 (tel)
(916) 381-7588 (customer service tel)
(916) 381-7589 (fax)
Visa, MasterCard, Discover, COD ($5.99)
Sells radios, batteries, battery chargers, speed controls, etc. for radio-control models.

C. Hoelzle Associates
17321 Eastman St., Irvine, CA 92714 USA
(800) 959-9601 (tel)
(714) 251-9000 (tel)
(714) 251-9291 (fax)
Sells printer parts (more than 60,000 in stock) for a wide variety of computer applications.

Holaday Industries
14825 Martin Dr., Eden Prarie, MN 55344
(612) 934-4920 (tel)
(612) 934-3604 (fax)
Manufactures RF hazard meters, RF-induced current meters, EMF meters, and dc-to-microwave meters, and test equipment.

Holland Electronics
(800) 628-4511 (tel)
(805) 339-9060 (tel)
(805) 339-9064 (fax)
Manufactures modulators for professional cable TV applications.

Home Theatre Systems
44 Rt. 23N, Little Falls, NJ 07424 USA
(800) 978-7768 (tel)
(201) 890-5959 (tech support)
(201) 890-9142 (fax)
Sells a wide variety of mid- to- high-end audio and home theater equipment, including audio and video speakers, receivers, video projectors, TVs, VCRs, laser discs, CD players, preamps, power amplifiers, cassette decks, satellite dishes, and accessories

Honolulu Electronics
870 Kawaiahao St., Honolulu, HI 96813 USA
(808) 949-5564 (tel)
(808) 949-5565 (tel)
(808) 949-1209 (fax)
Sells a wide variety of amateur radio equipment and accessories, including amateur radio transceivers, receivers, antenna tuners, antennas, amplifiers, etc.

Hooper Electronics
1702 Pass Rd., Biloxi, MS 39531 USA
(601) 432-1100 (tel)

Hooper Electronics
(601) 432-0584 (tel)
(601) 432-7651 (fax)
1700 Terry Rd., Jackson, MS 38204 USA
(601) 353-0922 (tel)
(601) 354-4531 (tel)
(601) 948-3807 (fax)
Sells a wide variety of amateur radio equipment and accessories, including amateur radio transceivers, receivers, antenna tuners, antennas, amplifiers, etc.

Horizon Hobby Distributors
4105 Fieldstone Rd., Champaign, IL 61821 USA
(217) 355-9511 (tel)
(217) 355-8734 (fax)
Sells sport and computer radios and servos for radio-control model airplanes and boats.

Hostlink
Rm. 14-15, 1/F, Block B, Proficient Industrial Centre, 6 Wang Kwun Rd., Kowloon Bay, Kowloon, Hong Kong
852-2798-6699 (tel)
852-2753-6226 (fax)
5F-1, No. 65, Seng Teh Rd., Taipei, Taiwan
8862-643-8502 (tel)
8862-643-8507 (fax)
Manufactures motherboards, sound cards, CD-ROM drives, print servers, routers, etc. for a wide variety of computer applications.

Hotronic
1875 S. Winchester Blvd., Campbell, CA 95008 USA
(408) 378-3883 (tel)
(408) 378-3888 (fax)
Manufactures a video frame synchronizer for amateur and professional video editing applications.

Howell Instruments
3479 W. Vickery Blvd., Ft. Worth, TX 76107 USA
Manufactures an engine monitor for professional aviation applications.

Hoyt Electrical Instrument Works
19 Linden St., P.O. Box 8798, Penacook, NH 03303 USA
(603) 753-6321 (tel)
(603) 753-9592 (fax)
Manufactures analog panel meters for a wide variety of electronics applications.

HPR
P.O. Box 621136C, Littleton, CO 80162 USA
Sells base, mobile, and handheld scanners, CB transceivers, antennas, and accessories.

HSB Computer Laboratories
34208 Aurora Rd. #297, Cleveland, OH 44139 USA
(800) 497-0401 (tel)
(216) 498-0382 (tel)
(216) 498-1374 (fax)
Visa, MasterCard, Discover, American Express, COD
Manufactures motherboards, video cards, main boards, etc. for a wide variety of computer applications. Offers a 60-day money-back guarantee.

HSU Research
20013 Rainbow Way, Cerritos, CA 90701 USA
(800) 554-0150 (tel)
(310) 924-7550 (tel/fax)
Manufactures and sells subwoofer loudspeakers for high-end home audio applications. Offers a 30-day money-back guarantee and a 5-year defect warranty.

HT Communications
4480 Shopping Ln., Simi Valley, CA 93063 USA
(805) 579-1700 (tel)
(805) 522-5295 (fax)
Manufactures multiplexers for professional computer network applications.

HTP
1620 S. Louis St., Anaheim, CA 92805 USA
(800) 777-9948 (tel)
Manufactures high-end speakers for home theater applications.

Human Designed Systems
400 Feheley Dr., King of Prussia, PA 19406 USA
(800) 437-1551 (tel)
(610) 277-8300 (tel)
(610) 275-5739 (fax)

Manufactures terminals for a wide variety of professional computer network applications.

Huntsville Microsystems
P.O. Box 12415, Huntsville, AL 35815 USA
(205) 881-6005 (tel)
(205) 882-6701 (fax)
sales@hmi.com
http://www.hmi.com
Manufactures emulators for professional and industrial computer applications.

Hustler Antennas
One Newtronics Pl., Mineral Wells, TX 76067 USA
(800) 949-9490 (tel)
(817) 325-1386 (tel)
(817) 328-1409 (fax)
Manufactures vertical base and mobile antennas, antenna mounts, and accessories for amateur radio and citizens band radio applications.

Hutton Communications
5470 Oakbrook Pkwy., Ste. G., Norcross, GA 30093 USA
(800) 741-3811 (tel)
(770) 729-9413 (tel)
(770) 729-9567 (tel)
(800) 442-3811 (Dallas tel)
(214) 239-0580 (Dallas tel)
(214) 239-5264 (Dallas fax)
(800) 726-6245 (Denver tel)
(303) 820-2929 (Denver tel)
(303) 820-2809 (Denver fax)
(800) 426-2964 (Seattle tel)
(206) 453-2132 (Seattle tel)
(206) 453-1558 (Seattle fax)
(800) 435-9313 (Chicago tel)
(815) 744-6444 (Chicago tel)
(815) 744-8996 (Chicago fax)
(800) 265-8685 (Canada tel)
(416) 255-6063 (Canada tel)
(416) 255-9179 (Canada fax)
Sells photovoltaic (solar) electric systems for a wide variety of professional electronics applications.

Hybrids International
707 N. Lindenwood Dr., Olathe, KS 66062 USA
(913) 764-6400 (tel)
(913) 764-6409 (fax)
Manufactures crystal oscillator modules for a wide variety of electronics applications. Offers a free 60-page catalog.

Hyde Park Electronics
4547 Gateway Circle, Dayton, OH 45440 USA
(513) 435-2121 (tel)
(513) 435-6375 (fax)
Manufactures ultrasonic sensors for a wide variety of professional industrial applications. Offers an 8-page brochure.

Hydra Audio
(915) 857-1191 (tel)
Manufactures custom-wired patch bays and wiring harnesses for professional studio recording and live music performances.

HyperData
801 S. Lemon Ave., Walnut, CA 91789 USA
(800) 786-3343 (tel)
(800) 786-3343 (tech support tel)
(909) 468-2955 (tel)
(909) 468-2961 (fax)
(909) 594-3645 (BBS)
Visa, MasterCard, American Express
Manufactures and sells personal laptop computers. Offers a 30-day money-back guarantee.

Hy-Q International
1438 Cox Ave., Erlanger, KY 41018 USA
(606) 283-5000 (tel)
(606) 283-0883 (fax)
Manufactures pager and communications crystals for cellular microwave radio applications.

Hyundai Electronics
1955 Hyundai Ave., San Jose, CA 95131 USA
(800) 568-0060 (tel)
(408) 232-8342 (tel)
(800) 501-4986 (fax back)
(408) 232-8125 (fax)
http://www.hea.com

Hyundai Electronics

Manufactures monitors for personal desktop computers.

Ibanez
P.O. Box 886, Bensalem, PA 19020 USA
P.O. Box 2009, Idaho Falls, ID 83403 USA
2165 46th Ave., Lachine, Quebec H8T 2P1 Canada
Manufactures electric guitars for a wide variety of musical applications. Offers a catalog for $5.

IBIP
2601 Westball Rd., #105, Anaheim, CA 92804 USA
(714) 229-9949 (tel)
(714) 229-8451 (fax)
Buys and sells new, refurbished, and used personal laptop computers and parts.

IBI Systems
6842 NW 20 Ave., Ft. Lauderdale, FL 33309 USA
(305) 978-9225 (tel)
(305) 978-9226 (fax)
Manufactures ruggedized and rack-mount equipment for professional and industrial computer applications.

IBM
3039 Corn Wallis Rd., Research Triangle, NC 27709 USA
(800) 426-7255 X4821 (computer order tel)
(800) 388-7080 (parts sales tel)
(800) IBM-3333 XG111 (virus protection tel)
http://www.pc.ibm.com
Visa, MasterCard, Discover, American Express
Manufactures and sells a wide variety of personal and network computer equipment.

IBM Credit
(800) IBM-5440 (tel)
(203) 973-5649 (fax)
Visa, MasterCard, Discover, American Express
Sells refurbished IBM personal and network computer equipment.

IC Engineering
16350 Ventura Blvd., #125, Encino, CA 91436 USA
(800) 343-5358 (tel)
(818) 345-1692 (tel)
Visa, MasterCard, American Express
Manufactures and sells field-strength meters for RF and/or microwave radiation measuring applications.

ICL
9801 Muirlands Blvd., Irvine, CA 92718 USA
(714) 855-5500 (tel)
(714) 458-6257 (fax)
Manufactures workstations for a wide variety of professional computer network applications.

ICOM
2380 116th Ave. NE, Bellevue, WA 98004 USA
(206) 454-8155 (tel)
(206) 450-6088 (brochure tel)
(206) 454-7619 (tech support tel)
177 Phoenix Pkwy., #201, Atlanta, GA 30349 USA
(404) 991-6166 (tel)
18102 Skypark S. #52B, Irvine, CA 92714 USA
(714) 852-8026 (tel)
3071 #S Rd. Unit 9, Richmond, BC Canada
(604) 273-7400 (tel)
77540.525@compuserve.com (Internet tech support)
Manufactures a variety of amateur radio and shortwave equipment and accessories. Offers free brochures.

ICS Electronics
473 Los Coches St., Milpitas, CA 95035 USA
(800) 952-4499 (tel)
(408) 263-5896 (fax)
Manufactures VXI interface modules for professional industrial applications.

IDA
1345 W. Main Ave., Fargo, ND 58103 USA
(800) 627-4432 (tel)
(701) 280-1122 (tel)
(218) 233-1886 (fax)
Manufactures trunking logic and interconnects, repeater logic controllers, etc. for professional cellular microwave radio applications.

Ideal Acoustics
(601) 323-4001 (tel)
Sells new and used high-end home audio equipment, such as speakers, cassette decks, amplifiers, preamplifiers, etc.

Idea/onics
879 W. Main St., P.O. Box 369, Mayville, ND 58257 USA
(701) 786-3904 (tel)
(701) 786-4294 (fax)
Manufactures emergency alert systems for cable TV applications.

IDER (International Digital Equipment Research)
9080 Telstar Ave., #303, El Monte, CA 91731 USA
(818) 288-0709 (tel)
(818) 288-4008 (tech support tel)
(818) 288-0797 (fax)
Visa, MasterCard, Discover
Sells computer parts and boards, such as computer cases, floppy drives, motherboards, power supplies, video cards, modems, keyboards, etc.

IEM
P.O. Box 1889, Ft. Collins, CO 80522 USA
1629 Blue Spruce Dr., Ft. Collins, CO 80524 USA
(800) 321-4671 (tel)
(970) 221-3005 (tel)
(970) 221-1909 (fax)
info@iem.com
Colorado House, Cromwell Park, Banbury Rd. Chipping Norton, Oxfordshire, OX7 5SR UK
44-01608-645000 (tel)
44-01608-645155 (fax)
Sells network systems and data storage systems for professional computer applications.

IFR Systems
10200 W. York St., Wichita, KS 67215 USA
(800) 835-2352 (tel)
(316) 522-4981 (tel)
(316) 524-2623 (fax)
(316) 522-1360 (fax)
Manufactures portable spectrum analyzers; also manufactures test sets for professional cellular microwave radio applications.

IGEL
31 Stonecroft Dr., #105, Easton, PA 18045 USA
(610) 258-4290 (tel/fax)
Manufactures terminals for a wide variety of professional computer network applications.

I & G Hobbies
91-08 87th St., Woodhaven, NY 11421 USA
(800) 337-3228 (tel)
(718) 850-0765 (tel)
(718) 850-0684 (fax)
Visa, MasterCard, COD ($5.95)
Sells batteries, battery chargers, radios, etc. for radio-control models.

IGS
Pfeifferstreet 7, A-4041 Linz, Austria
0732-233128 (tel)
Sells equipment for shortwave radio receiving applications.

II Morrow
P.O. Box 13549, Salem, OR 97309 USA
(503) 581-8101 (tel)
Manufactures GPS and Loran systems for professional aviation applications.

IIX Equipment
P.O. Box 9, Oak Lawn, IL 60454 USA
(708) 423-0605 (tel)
(708) 423-1691 (fax)
Visa, MasterCard
Manufactures and sells a variety of amateur radio antenna towers, mounts, and accessories, including pulleys, standoff brackets, gin poles, quad pods, etc. Offers a free catalog.

Ikelight
P.O. Box 88100, Indianapolis, IN 46208 USA
(317) 923-4523 (tel)
Manufactures lighting systems for amateur and professional video production.

IKON
2617 Western Ave., Seattle, WA 98121 USA
(206) 728-6465 (tel)
(206) 728-1633 (fax)
Manufactures printer interfaces for professional network computer applications.

ILC Data Device
105 Wilbur Pl., Bohemia, NY 11716 USA
(516) 567-5600 (tel)
Manufactures synchro/resolver instrumentation for industrial electronics applications.

ILC Technology
399 W. Java Dr., Sunnyvale, CA 94089 USA
(408) 745-7900 (tel)
(408) 744-0829 (fax)
Manufactures backlighting for a wide variety of electronics applications.

ILIADIS
1760 Dibble Circle W., Jacksonville, FL 32246 USA
(904) 641-9942 (tel)
Manufactures contact microphones for a wide variety of electronics applications.

Illinois Capacitor
3757 W. Touhy Ave., Lincolnwood, IL 60645 USA
(708) 675-1760 (tel)
(708) 673-2850 (fax)
Manufactures capacitors for a wide variety of electronics applications. Offers a free capacitor engineering guide.

Image Communications
4301 W. 69th St., Chicago, IL 60629 USA
(800) 552-1639 (tel)
(312) 585-1212 (tel)
(312) 585-7847 (fax)
Sells a wide variety of components, such as loudspeakers, repair kits, crossovers, etc., for home audio and professional audio applications.

Image Computers
46525 Continental Dr., Chesterfield Twp., MI 48047 USA
(800) 858-9578 (tel)
(810) 795-4044 (fax)
Sells new, used, and surplus personal desktop computer parts and accessories.

ImageNation
P.O. Box 276, Beaverton, OR 97075 USA
(503) 641-7408 (tel)
(503) 643-2458 (fax)
(503) 626-7763 (BBS)
Manufactures video frame capture computer boards for a variety of professional industrial applications.

Image Systems (ISI)
(800) 486-1996 (tel)
(770) 623-0111 (tel)
(770) 623-5741 (fax)
Buys and sells mainframes, peripherals, workstations, banking equipment, parts, etc. for a wide variety of computer applications.

Imaging Technology
55 Middlesex Twpk., Bedford MA 01730 USA
(800) 333-3035 (tel)
(617) 275-9590 (fax)
kellyr@imaging.com
Manufactures video capture boards for a wide variety of computer applications.

Imagraph
11 Elizabeth Dr., Chelmsford, MA 01824 USA
(508) 256-4624 (tel)
(508) 250-9155 (fax)
Manufactures terminals for a wide variety of professional computer network applications.

IME
P.O. Box 170415, Arlington, TX 76003 USA
(817) 473-1730 (tel)
Sells new, overstocked electronics parts. Offers a free catalog.

IMO TransInstruments
1 Cowles Rd., Plainville, CT 06062 USA

(203) 793-4516 (tel)
(203) 793-4514 (fax)
Manufactures pressure transducers for professional industrial applications.

InBus Engineering
4569 Las Positas Rd., Unit C, Livermore, CA 94550 USA
(510) 447-0101 (tel)
(510) 447-0448 (fax)
Sells multibus boards and systems, in-circuit emulators, etc. for a wide variety of computer applications.

Independent Audio
295 Forest Ave., #121, Portland, ME 04101 USA
(207) 773-2424 (tel)
(207) 773-2422 (fax)
75671.3316@compuserve.com
Sells HHB DAT decks and tapes in North America.

Index Laboratories
9318 Randall Dr. N.W., Gig Harbor, WA 98332 USA
(206) 851-5725 (tel)
Manufactures a low-power HF SSB/CW transceiver for amateur radio applications.

Indus Group
8 Hammond Dr., #110, Irvine, CA 92718 USA
(800) 644-6387 (tel)
(714) 597-0430 (tel)
(714) 597-0433 (fax)
Visa, MasterCard, COD
Sells computer parts and boards, such as floppy drives, hard drives, tape drives, monitors, memory chips, motherboards, etc.

Industrial Communications Engineers
P.O. Box 18495, Indianapolis, IN 46218 USA
(800) 423-2666 (tel)
(317) 547-1398 (tel)
(317) 545-9645 (fax)
Manufactures and sells receiving preamplifiers, lightning arrestors, filters, RF switches, etc. for shortwave and amateur radio applications.

Information Data Products

Industrial Electronic Engineers
7740 Lemona Ave., Van Nuys, CA 91409 USA
(818) 787-0311 (tel)
(818) 901-9046 (fax)
Manufactures LCD display panels for a wide variety of electronics applications.

Infinity
9409 Owensmouth, Chatsworth, CA 91311 USA
(800) 553-3332 (tel)
(818) 407-0228 (tel)
Manufactures amplifiers and loudspeakers for mobile audio applications.

InFocus Systems
2700B SW Parkway Ave., Wilsonville, OR 97070 USA
(800) 294-6400 (tel)
(503) 685-5555 (tel)
(503) 685-8631 (fax)
31-0-2503-23200 (Europe tel)
31-0-2503-24388 (Europe fax)
Manufactures LCD projection panels for computer multimedia applications.

Infogrip
1141 E. Main St., Ventura, CA 93001 USA
(800) 397-0921 (tel)
Manufactures and sells one-handed keyboards for a wide variety of personal computer applications.

Infomatic
9945 S. Pioneer Blvd., S. San Francisco, CA 90670 USA
(800) 948-2217 (tel)
(310) 948-5264 (fax)
Visa, MasterCard
Sells laptops computers, hard drives, modems, etc. for portable personal computer applications.

Information Data Products
28 Bowling Green Pkwy., Lake Hopatcong, NJ 07849 USA
(800) 362-3770 (tel)
(201) 663-3700 (tel)
(201) 663-3710 (fax)

Information Data Products

wan-info@planet.net
http://www.planet.net/idpc
Sells data switches, modems, multiplexers, line drivers, etc. for a wide variety of computer applications.

Information Unlimited
P.O. Box 719, Amherst, NH 03031 USA
(800) 221-1705 (tel)
(603) 673-4730 (tel)
(603) 672-5406 (fax)
Sells Part 15 FM voice and telephone bug kits and an extended-play telephone recording system. Offers a catalog for $1.

Infotec International
P.O. Box 65 332, Mairangi Bay, Auckland 10, New Zealand
64-09-478-7589 (tel)
64-09-478-5689 (fax)
Manufactures video capture boards for a wide variety of personal computer applications.

Infotronic America
8834 N. Capital of Texas Hwy., #200, Austin, TX 78759 USA
(800) 944-4545 (tel)
(512) 345-9646 (tel)
(512) 345-9895 (fax)
Manufactures video accelerator boards for a wide variety of computer applications.

Inframetrics
16 Esquire Rd., N. Billerica, MA 01862 USA
(508) 670-5555 (tel)
(508) 667-2702 (fax)
Mechelse Steenweg 277, B-1800 Belgium
32-2-252-5712 (tel)
32-2-252-5388 (fax)
358-200-740-760 (fax)
Manufactures infrared cameras for a wide variety of professional industrial applications.

Inmark Industrial
1A Man Foong Ind., Bldg. 7 Cheung Lee St.
Chai Wan, Hong Kong
8522-558-2203 (tel)
8522-897-3700 (fax)
Manufactures floppy diskettes for a wide variety of computer applications.

Innovation Specialties
2625 Alcatraz #243, P.O. Box 5899-243
Berkeley, CA 94705 USA
(800) 222-8228 (tel)
Visa, MasterCard, COD
Sells Nady audio/video equipment.

Innovation West
2275 Huntington Dr., #265A, San Marino, CA 91108 USA
(818) 309-6085 (tel)
(818) 309-9972 (fax)
Manufactures data acquisitions boards for a wide variety of computer applications.

Inovonics
1305 Fair Ave., Santa Cruz, CA 95960 USA
(408) 458-0552 (tel)
(408) 458-0554 (fax)
inovonics@aol.com
Manufactures stereo generators, FM receivers, and processor/generators for broadcast stations.

Inphone Electronics Enterprise
3F, No. 42, Yin-Shan St., Kaohsiung, Taiwan
886-7-385-9883 (tel)
886-7-389-5322 (tel)
886-7-385-9890 (fax)
886-7-389-5341 (fax)
Manufactures amplifiers, converters, modulators, etc. for professional cable TV applications.

Insight
1912 W. 4th St., Tempe, AZ 85281 USA
(800) 927-3010 (tel)
(800) 377-3000 (tel)
(800) 927-3004 (tel)
(602) 902-1000 (tel)
(602) 902-5929 (BBS)
info@insight.com
Sells personal desktop and laptop computers, print-

ers, and scanners; also sells computer parts and boards, such as network hardware, hard drives, monitors, surge protectors, video cards, modems, etc. Offers a free catalog.

Insight
P.O. Box 194, Ellicot Stn., Buffalo, NY 14205 USA
(716) 852-3217 (tel)
Manufactures graphic engine monitors for professional aviation applications.

Insight Electronics
9980 Huennekenna St., San Diego, CA 92121 USA
(800) 677-6011 (tel)
http://ikn.com
Sells semiconductors for a wide variety of electronics applications.

Insite Technology
77 Montreal St., P.O. Box 19678, Christchurch, New Zealand
03-365-6190 (tel)
03-365-8870 (fax)
04-385-9240 (Wellington tel)
04-385-9830 (Wellington fax)
user id@insite.co.nz
Distributes US Robotics computer products in New Zealand.

Instek
1205 John Reed Ct., City of Industry, CA 91745 USA
(818) 336-6537 (tel)
(818) 369-1748 (fax)
Manufactures test and measuring instruments for a variety of professional industrial applications.

Intacta
P.O. Box 35-590, Browns Bay, Auckland 10, New Zealand
09-479-1100 (tel)
09-479-8009 (fax)
Sells external tape drives for computer data storage applications.

Intec Inoventures
2751 Arbutus Rd., Victoria, BC V8N 5X7 Canada
(604) 721-5150 (tel)
(604) 721-4191 (fax)
Manufactures data acquisitions boards for a wide variety of computer applications.

Integrand Research
8620 Roosevelt Ave., Visalia, CA 93291 USA
(209) 651-1203 (tel)
(209) 651-1353 (fax)
Manufactures ruggedized rack-mount computer equipment for professional and industrial computer applications.

Integrated Circuit Systems (ICS)
2345 Boulevard of the Generals, P.O. Box 968, Valley Forge, PA 19482 USA
(800) 220-3366 (tel)
(610) 630-5300 (tel)
1271 Parkmoor Ave., San Jose, CA 95126 USA
(408) 297-1201 (tel)
http://www.icsinc.com
Manufactures semiconductors for a wide variety of electronics applications.

Integrated Device Technology (IDT)
2975 Stender Way., Santa Clara, CA 95054 USA
(800) 345-7015 (tel)
(408) 492-8674 (fax)
http://www.idt.com
Manufactures semiconductors for a wide variety of electronics applications.

Integrated Solutions
836 North St., Bldg. 5, Tewksbury, MA 01876 USA
(508) 640-1400 (tel)
Manufactures probe testers for professional industrial applications.

Integrix
1200 Lawrence Dr., #150, Newbury Park, CA 91320 USA
(800) 300-8277 (tel)
(805) 375-1055 (tel)
(805) 375-2799 (fax)
8610-253-5305 (China tel)
8610-253-5306 (China fax)

Integrix

822-515-5303 (South Korea tel)
822-515-5302 (South Korea tel)
sales@integrix.com
http://www.integrix.com
Manufactures and sells SPARC desktop and server computer systems.

Intel

P.O. Box 58119, Santa Clara, CA 95052 USA
(800) 538-3373 (tel)
(800) 525-3019 (fax back)
http://www.intel.com/
Manufactures integrated processor chips for personal computer applications.

Intellicomp Technologies

9698 Telstar Ave. #310, El Monte, CA 91731 USA
(800) 468-3696 (tel)
(818) 582-8096 (tel)
(818) 582-8213 (fax)
Manufactures and sells personal desktop computers and memory modules.

Intelligent Barcode Systems (IBS)

16031 Kaplan Ave., Industry, CA 91744 USA
(800) 765-2271 (tel)
(818) 968-6265 (tel)
(818) 968-5527 (fax)
Sells UPC scanners, etc. for business applications.

Intelligent Instrumentation

6550 S. Bay Colony Dr., MS130, Tucson, AZ 85706 USA
(800) 685-9911 (tel)
Manufactures data acquisitions equipment for a variety of professional applications.

IntelliMedia

455 Riverview Dr., Benton Harbor, MI 49022 USA
(616) 925-3675 (tel)
Manufactures LCD projection panels for computer multimedia applications.

Intellipower

10A Thomas St., Irvine, CA 92718 USA
(714) 587-0155 (tel)
(714) 587-0230 (fax)
Manufactures uninterruptible power supplies for a wide variety of personal computer applications.

Intensitronics

P.O. Box 562, Hales Corners, WI 53130 USA
(800) 382-4155 (tel)
(317) 883-7555 (fax)
Manufactures adjustable radio antennas for AM radio listening applications.

Interactive Light

1202 Olympic Blvd., Santa Monica, CA 90404 USA
(800) 213-3752 (tel)
(310) 452-7443 (fax)
Manufactures infrared controllers for professional and personal musical applications.

Interconnections

Unit 51, InShops, Wellington Centre, Aldershot, Hants GU11 5DB UK
01252-341900 (tel)
01293-822786 (fax)
Sells bar-code readers for computerized business applications.

Interface Data

600 W. Cummings Park, #3100, Woburn, MA 01801 USA
(800) 370-3282 (tel)
(617) 938-6333 (tel)
(617) 938-0626 (fax)
Manufactures data storage systems for a wide variety of professional computer network applications.

Interface Systems

3601 Old Baltimore Dr., Olney, MD 20832 USA
(800) 362-6250 (tel)
(301) 424-4419 (fax)
isi6250@erols.com
Visa, MasterCard
Sells a wide variety of computer data recording media, including cartridges, etc.

Interface Technology
300 S. Lemon Creek Dr., #A, Walnut, CA 91789 USA
(909) 595-6030 (tel)
(909) 595-7177 (fax)
Manufactures digital test subsystems for a variety of professional industrial applications.

Intergraph Computer Systems
1 Madison Industrial Park, Huntsville, AL 35807 USA
(800) 763-0242 (tel)
(205) 730-5441 (tel)
http://www.intergraph.com
Manufactures and sells personal desktop computers.

Interlink Computer Sciences
47370 Fremont Blvd., Fremont, CA 94538 USA
(800) 422-3711 (tel)
(510) 657-9800 (tel)
(510) 659-6381 (fax)
Manufactures controller boards for professional computer network applications.

Interlynx Computer Systems (ICS)
300 E. New Hope Rd. 204, Cedar Park, TX 78613 USA
(512) 259-9090 (tel)
(512) 259-9153 (fax)
Visa, MasterCard, American Express
Sells terminals, monitors, printers, modems, etc. for a wide variety of computer applications.

International Business Systems
2905-E Amwiler Rd., Atlanta, GA 30360 USA
(800) 684-8006 (tel)
(404) 416-9495 (tel)
(404) 416-9474 (fax)
Sells personal desktop computer systems and printers.

International Cellular Products
2082 Foothill Blvd., Laverne, CA 91750 USA
(800) 573-4ICP (tel)
(909) 593-3386 (tel)
(909) 593-7455 (fax)
Sells telephones and accessories for cellular telephone applications. Offers a free catalog.

International Components
105 Maxess Rd., Melville, NY 11747 USA
(800) 645-9154 (tel)
(516) 293-1500 (tel)
(516) 293-4983 (fax)
221576 (telex)
Manufactures a variety of electron tubes for various audio, receiving, and transmitting applications. Offers a free price list and accessories catalog.

International Components Marketing (ICM)
1226 3rd St. Promenade, Ste. #206, Santa Monica, CA 90401 USA
(800) 748-6232 (tel)
(310) 260-1444 (tel)
(310) 451-8727 (fax)
Visa, MasterCard, Discover, American Express
Sells computer monitor testers for a wide variety of computer applications.

International Computer Purchasing (ICP)
Beverly Technology Center, 4 Federal St., Beverly, MA 01915 USA
(508) 921-0603 (tel)
(508) 921-0650 (fax)
Sells memory chips for a wide variety of computer applications.

International Computers
12021 W. Bluemound Rd., Wauwatosa, WI 53226 USA
(800) 992-9000 (tel)
(414) 764-9000 (tel)
(414) 281-3522 (fax)
Manufactures ruggedized computers for a wide variety of industrial computer applications.

International Crystal
10 N. Lee St., Oklahoma City, OK 73102 USA
(800) 725-1426 (tel)
(405) 236-3741 (tel)
(800) 322-9426 (fax)
(405) 235-1904 (fax)
Manufactures a variety of electronics crystals.

International Datacasting
3850 Holcomb Bridge Rd., #420, Norcross, GA 30092 USA
(770) 446-9684 (tel)
(770) 448-6396 (fax)
Manufactures satellite equipment and data broadcasting systems for professional broadcasting.

International Electronic
2260A Lambert St., Unit 703, Lake Forest, CA 92630 USA
(800) 274-3774 (tel)
(714) 380-3781 (fax)
Sells printers, fax machines, and parts for a wide variety of computer applications.

International Electronic Wire & Cable
89½ O'Leary Dr., Bensenville, IL 60106 USA
(800) 323-0210 (tel)
(708) 299-0021 (tel)
(708) 298-8433 (fax)
Manufactures wire, coaxial cable, rotor cable, antenna wire, etc. for a variety of electronics applications.

International Instruments
P.O. Box 185, N. Branford, CT 06471 USA
(203) 481-5721 (tel)
(203) 481-8937 (fax)
Manufactures bar-graph displays for a wide variety of electronics applications.

International Microelectronics
P.O. Box 170415, Arlington, TX 76003 USA
(800) 999-0463 (tel)
(817) 473-0525 (fax)
Sells surplus electronics components for a wide variety of electronics applications. Offers a free catalog.

International Parts & Systems (IPSI)
1445 W. 12th Pl., Tempe, AZ 85281 USA
(800) 451-4774 (tel)
(602) 894-0450 (fax)
Sells personal desktop and laptop computer parts.

International Power Sources
200 Butterfield Dr., Ashland, MA 01721 USA
(508) 881-7434 (tel)
(800) 226-2100 (fax)
Manufactures dc-to-dc power converter modules for a wide variety of electronics applications. Offers a free catalog.

International Radio & Computer
3804 S. US #1, Ft. Pierce, FL 34982 USA
(407) 489-0956 (tel)
(407) 464-6386 (fax)
Sells a variety of amateur radio equipment, including crystal filters, transceiver upgrade kits, etc.

Internet Memory Exchange
3925 N. Clarey St., Eugene, OR 97402 USA
(503) 688-9263 (tel)
(503) 688-9775 (fax)
dany@efn.org
http://www.teleport.com/_7Edany
Visa, MasterCard
Buys, sells, and trades memory chips for computer applications.

Interphase
13800 Senlac, Dallas, TX 75234 USA
(214) 919-9000 (tel)
(214) 919-9200 (fax)
Manufactures disk controller boards and concentrators for a wide variety of professional computer network applications.

Interpoint
P.O. Box 97005, Redmond, WA 98073 USA
(800) 882-8782 (tel)
(206) 882-3100 (tel)
(206) 882-1990 (fax)
power@intp.com
44-1252-815511 (UK tel)
44-1252-815577 (UK fax)
poweruk@intp.com
33-1-34285455 (France tel)
33-1-34282387 (France tel)
powerfr@intp.com
Visa, MasterCard

Manufactures dc-to-dc converters for a wide variety of electronics applications.

InterPro Microsystems
46560 Fremont Blvd., #417, Fremont, CA 94538 USA
(800) 226-7216 (tel)
(510) 226-7226 (tel)
(510) 226-7219 (fax)
Visa, MasterCard
Manufactures and sells personal desktop computers and scanners; also sells computer parts and boards, such as hard drives, monitors, video cards, multimedia accessories, memory chips, etc. Offers a 30-day money-back guarantee.

Interstate Electronics
P.O. Box 3117, Anaheim, CA 92803 USA
(800) 854-6979 (tel)
(714) 758-4115 (tel)
(714) 758-4148 (fax)
Manufactures GPS flight management and navigation systems for professional aviation applications.

Interstate Musician Supply
P.O. Box 315, New Berlin, WI 53151 USA
(414) 786-6210 (tel)
Sells electric guitars, synthesizer keyboards, etc. for professional and personal musical applications. Offers free catalogs.

Intronics
P.O. Box 13723, Edwardsville, KS 66113 USA
(913) 422-2094 (tel)
(913) 441-1623 (fax)
Visa, MasterCard, COD ($4.50)
Manufactures and sells several EPROM programmers for both IBM compatible and Macintosh computers.

I/O Concepts
2125 112th Ave. NE, #303, Bellevue, WA 98004 USA
(206) 450-0650 (tel)
(206) 622-0058 (fax)
Manufactures interface cards for professional computer network applications.

I/O Controls
1359 W. Foothill Blvd., Azusa, CA 91702 USA
(818) 812-5333 (tel)
(818) 812-5332 (fax)
Visa, MasterCard
Manufactures and sells DINEX adapters, analog-to-digital converters, network controllers, etc. Offers a 15-day money-back guarantee.

Iomega
1821 W. Iomega Way, Roy, UT 84067 USA
(800) 697-8833 (tel)
(800) 456-5522 (tech support tel)
(801) 778-1000 (tel)
(801) 778-3450 (fax)
(801) 778-5763 (fax back)
(801) 394-9819 (BBS)
info@iomega.com
Manufactures data storage systems for a wide variety of professional computer network applications.

IOtech
25971 Cannon Rd., Cleveland, OH 44146 USA
(216) 439-4091 (tel)
(216) 439-4093 (fax)
Manufactures data acquisition cards for a wide variety of professional computer applications.

IPC Business Centre
29 East St., P.O. Box 68 227, Newton, Auckland, New Zealand
09-377-8998 (tel)
09-377-8996 (fax)
Manufactures personal computers.

IPC Technologies
10300 Metric Blvd., Austin, TX 78758 USA
(800) 240-9497 (tel)
(512) 339-3500 (tel)
(512) 454-1357 (fax)
Visa, MasterCard, Discover, American Express
Manufactures and sells personal computers.

Ipitek
2330 Faraday Ave., Carlsbad, CA 92008 USA

Ipitek

(619) 438-8362 (tel)
(619) 438-2412 (fax)
Manufactures repeaters, optical receivers, headend transceivers, etc. for professional cable TV applications.

IQD
Station Rd., Crewkerne, Somerset TA18 7AR UK
01460-77155 (tel)
Manufactures oscillator chips for a wide variety of electronics applications.

IRC
4222 S. Staples St., Corpus Christi, TX 78411 USA
(512) 992-7900 (tel)
Manufactures surface-mount resistors for a wide variety of electronics applications.

IRI Amateur Electronics
918 Plantation Farms Rd., Greensboro, NC 27409 USA
(910) 299-1725 (tel)
Visa, MasterCard, COD
Sells amateur radio antennas and accessories.

Ironwood Electronics
P.O. Box 21151, St. Paul, MN 55121 USA
(612) 431-7025 (tel)
(612) 432-8616 (fax)
Manufactures XILINX test adapters for professional computer applications.

Irwin Performance Products
7809 S. Sun Mor Dr., Muncie, IN 47302 USA
(317) 747-4190 (tel)
Manufactures battery chargers for radio-control model airplanes and boats.

ISA
P.O. Box 6955, Laguna Niguel, CA 92607 USA
(714) 488-0010 (tel)
(714) 488-0299 (fax)
Sells surplus electron tubes and test equipment for a wide variety of electronics applications.

Ishizuka Electronics
7-7 Kinshi 1-chome, Sumida-ku, Tokyo 130, Japan
3-3621-2704 (tel)
3-3623-7776 (fax)
02-509-9855 (Taiwan tel)
02-509-9599 (Taiwan fax)
(516) 587-4086 (US tel)
(516) 321-9604 (US fax)
02-851-3127 (Korea tel)
02-851-3129 (Korea fax)
Manufactures thermistors, etc. for a wide variety of electronics applications.

i Sight
14 Ridgedale Ave., #125, Cedar Knolls, NJ 07927 USA
(201) 326-6720 (tel)
(201) 326-8943 (fax)
Manufactures computer digital cameras for a wide variety of computer multimedia applications.

Island Micro Systems
4828 Old National Hwy., College Park, GA 30337 USA
(404) 305-0060 (tel)
(404) 305-0960 (fax)
Visa, MasterCard
Sells personal desktop and laptop computer systems and printers; also sells computer parts and boards, such as system boards, power supplies, monitors, keyboards, etc.

ITAC Systems
3113 Benton St., Garland, TX 75042 USA
(800) 533-4822 (tel)
sales@mousetrak.com
Manufactures and sells trackballs for computer applications.

Item
5509 Vine St., Alexandria, VA 22310 USA
(800) 367-4836 (tel)
(703) 971-5700 (tel)
(703) 971-0070 (fax)
Sells printers for a wide variety of computer applications.

ITT Barton
P.O. Box 1882, City of Industry, CA 91749 USA
(800) 522-7866 (tel)
(818) 961-2547 (tel)
(818) 961-4452 (fax)
Manufactures electronic flow measurement equipment for industrial applications.

ITT Pomona
1500 E. Ninth St., Pomona, CA 91766 USA
(909) 623-3463 (tel)
(909) 629-3317 (fax)
Manufactures probe kits for a wide variety of electronics applications.

ITT Schadow
8081 Wallace Rd., Eden Prarie, MN 55344 USA
(612) 934-4400 (tel)
(612) 934-9121 (fax)
Manufactures surface-mount switches for a wide variety of electronics applications. Offers a free catalog.

ITT Pomona Electronics
1500 E. Ninth St., Pomona, CA 91766 USA
(909) 469-2900 (tel)
(909) 629-3317 (fax)
Sells a kit to convert analog oscilliscopes into an eight-channel, real-time logic analyzer, trigger probe, or MUX-mode tool.

ITU Technologies
3477 Westport Ct., Cincinnati, OH 45248 USA
(513) 574-7523 (tel)
Manufactures and sells PIC programmers and EEPROM-based microcontrollers.

ITW Paktron
P.O. Box 4539, 1205 McConville Rd., Lynchburg, VA 24502 USA
(804) 239-6941 (tel)
(804) 239-4730 (fax)
Manufactures capacitors for a wide variety of electronics applications.

Izzy's PC Systems
4734 Larkspur Sq. Shopping Ctr., Virginia Beach, VA 23462 USA
(804) 495-6945 (tel)
(804) 495-6946 (fax)
Visa, MasterCard
Sells personal desktop computers and parts, such as hard drives, motherboards, memory chips, modems, etc.

J.A. Air Center
(800) 323-5966 (tel)
(708) 584-3200 (tel)
(708) 584-7883 (fax)
Sells instruments and radios for professional aviation applications.

Jabbour Electronic Suppliers
345 Fountain St., Pawtucket, RI 02860 USA
(800) 321-5018 (tel)
(401) 728-4600 (tel)
(401) 727-3374 (fax)
Sells replacement parts and accessories for audio equipment.

JAC Industrial
4564 Enterprise St., Fremont, CA 94538 USA
(510) 438-6668 (tel)
(510) 438-6669 (fax)
Sells floppy diskettes for a wide variety of computer applications.

Jackson Guitar
P.O. Box 2344, Ft. Worth, TX 76113 USA
(817) 336-5114 (tel)
(817) 870-1271 (fax)
Manufactures electric guitars for professional and personal musical applications.

Jade Computer
18503 Hawthorne Blvd., Box 2874, Torrance, CA 90504 USA
(800) 421-5500 (tel)
(310) 370-7474 (tel)
(310) 371-4288 (fax)
Visa, MasterCard, Discover, American Express
Manufactures and sells personal desktop computers; also sells computer parts and boards, such as floppy

Jade Computer

drives, hard drives, monitors, motherboards, video cards, modems, keyboards, etc. Offers a 30-day money-back guarantee.

Jade Products
E. Hampstead, NH 03826-0368 USA
(800) JADE-PRO (tel)
(603) 329-6995 (tel)
(603) 329-4499 (fax)
Visa, MasterCard
Sells battery controllers to prevent battery overcharging (for gel cells and lead-acid batteries) and amateur radio keyers and antennas.

Jameco Electronics
1355 Shoreway Rd., Belmont, CA 94002 USA
(800) 831-4242 (tel)
(415) 592-8097 (tel)
(800) 237-6948 (fax)
(415) 592-2503 (fax)
Visa, MasterCard, Discover, American Express
Sells a variety of electronic components, kits (including amplifiers, power supplies, etc.), test equipment, and computer assemblies. Mininum order of $30. Offers a free catalog.

Jamo Hi-Fi
425 Huehl Rd., Bldg. 8, Northbrook, IL 60062 USA
(847) 498-4648 (tel)
(847) 498-1948 (fax)
jamoinfo@jamospeakers.com
435 SW 6th Ave., Miami, FL 33144 USA
(305) 266-7230 (tel)
Elmevej 8, P.O. Box 232, DK-7870 Glyngoere, Denmark
45-99-767676 (tel)
45-99-767600 (tel)
speakers@jamo.dk
Koronakatu 1, FIN-02210 Espoo, Finland
+358-08037422 (tel)
+358-08037441 (fax)
P.A. des Bellevues, BP 235 Eragny
F-95614 Cergy-Pontoise Cedex, France
+33-1-34214636 (tel)
+33-1-34214630 (fax)
Heideweg 2, D-31199 Diekholzen, Germany
49-5121264518 (tel)
49-5121263582 (fax)
P.O. Box 258, N-1360 Nesbru, Norway
+47-66980941 (tel)
+47-66982425 (fax)
Ronda General Mitre 126, Atico 2a, E-08021 Barcelona, Spain
+34-3-4189212 (tel)
+34-3-4187718 (fax)
Box 22078, S-250 22 Helsinborg, Sweden
+46-42201110 (tel)
+46-42201964 (fax)
Jamo House, 5 Faraday Close, Drayton Fields, Daventry, Northamptonshire NN11 5RD UK
+44-1327-301300 (tel)
+44-1327-300511 (fax)
http://www.jamospeakers.com
Manufactures loudspeakers and amplified loudspeakers for home audio and home theater applications.

Jampro Antennas & RF Systems
P.O. Box 292880, Sacramento, CA 95829, USA
(916) 383-1177 (tel)
(916) 383-1182 (fax)
Manufactures antennas, filters, transmission panels, and patch panels for broadcast stations.

JAN Crystals
2341 Crystal Dr., P.O. Box 06017, Ft. Myers, FL 33906
(800) 529-9825 (tel)
(813) 936-2397 (tel)
(813) 936-3750 (fax)
Manufactures a variety of electronics crystals.

Japan Radio (JRC)
430 Park Ave., 2nd Fl., New York, NY 10022 USA
(212) 355-1180 (tel)
(212) 319-5227 (fax)
Akasaka Twin Tower (Main), 17-22, Akasaka 2-chome, Minato-ku, Tokyo 107, Japan
03-3584-8836 (tel)
03-3584-8878 (fax)
(800) 231-5100 (US cellular group tel)
(800) 881-5722 (Canada cellular group tel)
Manufactures receivers, transceivers, RF amplifiers, etc. for amateur radio, shortwave listening, and

commercial monitoring applications; also manufactures cellular telephones.

Jasco
311 NW 122nd St., Oklahoma City, OK 73114 USA
P.O. Box 466, Oklahoma City, OK 73101 USA
(800) 654-8483 (tel)
(405) 752-0710 (tel)
Manufactures lighting systems for amateur and professional video production.

Javanco
501, 12th Ave. S., Nashville, TN 37203 USA
(800) 528-2626 (tel)
(615) 244-4444 (tel)
(615) 244-4446 (fax)
(615) 244-4445 (BBS)
Visa, MasterCard, Discover
Sells surplus electronics components, fans, meters, computer parts, terminals, speakers, circuit boards, etc. for a wide variety of electronics applications. Offers a free catalog.

Javiation
Carlton Works, Carlton St., Bradford BD7 1DA UK
1274-732146 (tel)
1274-722627 (fax)
100117.535@compuserve.com
info@javiation.demon.co.uk
http://www.demon.co.uk/javiation (Internet online catalog)
Sells many different shortwave, VHF, and UHF receivers and accessories.

J.B. Electronics
3446 Dempster St., Skokie, IL 60076 USA
(708) 982-1973 (tel)
Sells test equipment, including signal generators and oscilloscopes.

JBL
240 Crossways Pk., Woodbury, NY 11797 USA
(516) 496-3400 (tel)
Manufactures loudspeakers for home and car audio applications.

JBRO Batteries
1938-A University Lane, Lisle, IL 60532-2150 USA
(800) 323-3779 (tel)
(800) 237-6435 (fax)
(708) 964-9358 (tel)
(708) 964-9081 (fax)
25700 I-45 North #111, Spring, TX 77386 USA
(800) 245-1138 (tel)
(713) 367-9393 (tel)
(713) 292-7139 (fax)
Manufactures batteries and conditioner/analyzers for mobile communications.

JCC
1 Bridge Plaza, #400, Ft. Lee, NJ 07024 USA
(800) 522-8376 (tel)
(201) 592-5023 (tel)
(201) 592-1549 (fax)
Manufactures workstations for a wide variety of professional computer network applications.

JCG Electronic Projects
P.O. Box HP79, Woodhouse St., Headingley, Leeds LS6 3XN UK
Manufactures and sells a wide variety of electronics kits, including transmitters, remote controls, radios, preamplifiers, amplifiers, etc. Offers a free list.

J & C Hobbies
100 A St., Penn Hills, PA 15235 USA
(800) 309-8314 (tel)
(412) 795-9344 (tel)
(412) 798-9857 (fax)
Visa, Mastercard
Sells sport and computer radios, batteries, and servos for radio-control model airplanes.

JC Marketing
#103, 2300 N. Clybourn, Chicago, IL 60614 USA
(312) 281-4903 (tel)
Sells caller ID and call routing systems for a wide variety of telephone applications.

J-Com
P.O. Box 194, Ben Lomond, CA 95005 USA
(408) 335-9120 (tel)

J-Com

(408) 335-9121 (fax)
Sells amateur radio antennas and accessories.

J Comm
793 Canning Pkwy., Victor, NY 14564 USA
(716) 924-0422 (tel)
(716) 924-4555 (fax)
Manufactures and sells low-power HT kits, computer interfaces, etc. for amateur radio applications.

JCPenney
6840 Barton Rd., Morrow, GA 30260 USA
(800) 933-7115 (tel)
(404) 961-8408 (fax)
Visa, MasterCard, Discover, American Express
Sells replacement parts and accessories for audio and video equipment.

JDR Micro Devices
2233 Branham Ln., San Jose, CA 95124 USA
(800) 538-5000 (tel)
(800) 538-5001 (tel)
(800) 538-5005 (fax)
(408) 494-1430 (BBS)
Visa, MasterCard, Discover, American Express
Sells personal desktop and laptop computers, printers, and scanners; also sells computer parts and boards, such as floppy drives, hard drives, monitors, motherboards, video cards, modems, keyboards, etc.

JDS Microprocessing
22661 Lambert St., #206, Lake Forest, CA 92630 USA
(800) 554-9372 (tel)
(714) 770-2263 (tel)
(714) 770-4826 (fax)
Manufactures computer network hubs for professional computer network applications.

Jebsee Electronics
P.O. Box 57, Tainan, Taiwan
886-6-264-7622 (tel)
886-6-263-8446 (fax)
Manufactures power supplies, distribution amplifiers, converters, splitters, taps, etc. for professional cable TV applications.

JEM R/C Electronics Research
P.O. Box 26135, Nepean, ON K2H 9R0 Canada
(613) 253-0204 (tel)
Manufactures channel expanders for radio-control model airplanes and boats.

Jensen
25 Tri State Int'l Ctr., #400, Lincolnshire, IL 60069 USA
(800) 323-0221 (tel)
(800) 323-4815 (tel)
(708) 317-3700 (tel)
(708) 317-3826 (fax)
Manufactures head units, crossovers, equalizers, loudspeakers, and power amplifiers for car stereo applications.

Jensen Tools
7815 S. 46th St., Phoenix, AZ 85044 USA
(800) 426-1104 (tel)
Manufactures a power consumption monitor for ac voltages.

Jeremy Communications
169 Jeremy Hill Rd., Pelham, NH 03076 USA
(603) 635-3048 (tel)
Sells amateur radio equipment, including antennas, towers, transmitting electron tubes, components, and transceiver kits.

Jescom International
1 Waters Park Dr. #117, San Mateo, CA 94403 USA
(415) 574-1421 (tel)
(415) 574-5297 (fax)
Sells a wide variety of equipment for professional tone-signalling applications.

JetFax
1376 Willow Rd., Menlo Park, CA 94025 USA
(800) 7JETFAX (tel)
(415) 326-6003 (fax)
Manufactures fax machines for a wide variety of office applications.

Jinco Computers
5122 Walnut Grive Ave., San Gabriel, CA 91776 USA
(800) 253-2531 (tel)
(818) 309-1108 (tel)
(818) 309-1107 (fax)
Visa, MasterCard, Discover, American Express
Sells personal home and laptop computers, and parts and accessories, including cases, scanners, power supplies, speakers, etc.

J&J International
211 Parsippany Rd., Parsippany, NJ 07054 USA
(800) 627-4368 (tel)
(800) 627-4366 (fax)
Sells replacement parts and accessories for audio equipment.

JK Microsystems
1275 Yuba Ave., San Pablo, CA 94806 USA
(510) 236-1151
jkmicro@dsp.com
Manufactures single-board, embedded computers for professional computer applications.

JL Audio
2321 SW 60th Way, Miramar, FL 33023 USA
(305) 981-9497 (tel)
(305) 987-7855 (fax)
(305) 981-2663 (tech fax)
http://www.jlaudio.com
Manufactures loudspeakers for mobile audio applications.

J&L Information Systems
9600 Topanga Canyon Blvd., Chatsworth, CA 91311 USA
(800) 456-1333 (tel)
(818) 882-9134 (fax)
Manufactures applications servers for professional computer network applications.

J&M Microtek
83 Seaman Rd., W. Orange, NJ 07052 USA
(201) 325-1892 (tel)
(201) 736-4567 (fax)
Manufactures PIC programmers for professional computer applications.

JMS
35 Hilltop Ave., Stamford, CT 06907 USA
Manufactures and sells flexible ground straps for electronic equipment.

J & N Factors
Pilgrim Works, Stairbridge Ln., Bolney, Sussex RH17 5PA UK
01444-0881965 (tel)
Sells a wide variety of surplus electronic packs, including capacitors, transformers, meters, solar panels, switches, etc., for electronic applications.

Jo Gunn Enterprises
Rt. 1 Box 32C, Hwy. 82, Ethelsville, AL 35461 USA
(205) 658-2229 (tel)
(205) 658-2259 (fax)
Manufactures and sells mobile and base antennas for citizens band and amateur radio applications.

Johnson Shop Products
P.O. Box 160113, Cupertino, CA 95016 USA
(408) 257-8614 (tel)
(408) 253-6288 (fax)
Sells new and used electronic parts. Offers a catalog for $1 or 2 stamps.

Johnson Yokogawa
4 Dart Rd., Newnan, GA 30265 USA
(800) 393-9134 (tel)
(404) 251-6427 (fax)
Manufactures chart recorders, interface systems, etc. for industrial applications.

Johns Radio
Whitehall Works, 84 Whitehall Rd. E., Birkenshaw, Bradford BD11 2ER UK
+44 274 684007 (tel)
+44 274 651160 (fax)
Sells surplus radios, components, repairs, converters, etc.

Jolly R. Electronics
(800) 279-0614 (tel)
Visa, MasterCard, COD
Sells cable TV descramblers.

Jolly Sound
13A Fl., Unit A, Gemstar Tower, 23 Man Lok St. Hunghom, Kowloon, Hong Kong
Sells head units, loudspeakers, amplifiers, receivers, cassette decks, CD players, etc. for home audio and car audio applications.

G.E. Jones Electronics
P.O. Box 467, Buckeye Lake, OH 43008 USA
(614) 928-8961 (tel)
Sells crystals and antennas for scanner radio applications.

Jones Surplus
302 Oceanside Blvd., Oceanside, CA 92054 USA
(619) 757-3008 (tel)
(619) 757-4048 (fax)
Buys and sells a variety of cable TV equipment, including line amplifiers, connectors, converters, headend equipment, etc.

JPG Electronics
276-278 Chatsworth Rd., Chesterfield S40 2BH UK
01246-211202 (tel)
01246-550959 (fax)
Sells a wide variety of surplus electronics components and assemblies, including rechargeable batteries, switches, capacitors, remote controls, power converters, multimeters, etc.

JP Instruments
P.O. Box 7033, Huntington Beach, CA 92615 USA
(800) 345-4574 (tel)
(714) 557-5434 (tel)
(714) 557-9850 (fax)
Manufactures EGT/CHT systems for professional aviation applications.

JPS Communications
P.O. Box 97757, Raleigh, NC 27624 USA
(800) 533-3819 (tel)
(919) 790-1048 (tel)
(919) 790-1456 (fax)
71673.1254 (CompuServe)
jps@nando.net
Visa, MasterCard
Manufactures and sells DSP (digital sound processing) noise-reduction devices for two-way communications and shortwave listening.

J & R Music World
59-50 Queens Midtown Expwy., Maspeth, Queens, NY 11378 USA
(800) 221-8180 (tel)
Visa, MasterCard, Discover, American Express
Sells most everything electronic, including audio and video equipment, computers, office equipment, and much more. Offers a free catalog.

J.R.S. Distributors
646 W. Market St., York, PA 17404 USA
(800) 577-4261 (tel)
(717) 854-8624 (tel)
Sells a wide variety of amateur radio equipment and accessories, including amateur radio transceivers, receivers, antenna tuners, antennas, amplifiers, etc.

J.S. Computer
1514 7th Ave. W., Bradenton, FL 34205 USA
(800) 659-9763 (tel)
(813) 749-1684 (fax)
Visa, MasterCard, Discover, American Express
Buys, sells, leases, and trades computer terminals, printers, and drives.

JSC Wire & Cable
P.O. Box 248, Wayne, NJ 07474 USA
(800) 572-9473 (tel)
(201) 694-6200 (tel)
(201) 694-8297 (fax)
Manufactures wire, coaxial cable, open-line/ladder-line transmission line, multiconductor cable, etc. for a variety of electronic applications.

JT Communications
579 NE 44th Ave., Ocala, FL 34470 USA
(904) 236-0744 (tel)

(904) 236-5130 (fax)
Manufactures replacement transmitter modules for professional broadcasting applications.

J.T. Computer Marketing
3731 Conflans Rd., Irving, TX 75061 USA
(800) 536-2662 (tel)
(214) 313-0105 (tel)
(214) 313-0730 (fax)
jtcomp@aol.com
Sells new and refurbished personal and network computer systems.

Jun's Electronics
5563 Sepulveda Blvd., Culver City, CA 90230 USA
(800) 882-1343 (tel)
(310) 390-8003 (tel)
(310) 390-4393 (fax)
Sells a wide variety of amateur radio equipment and accessories, including amateur radio transceivers, receivers, antenna tuners, antennas, amplifiers, etc.

Just Hobbies
202 Sawdust Rd. #110, The Woodlands, TX 77380 USA
(800) 799-JUST (tel)
(713) 292-5878 (tel)
(713) 292-5883 (fax)
Visa, MasterCard, COD
Sells radio-control systems, etc. for radio-control model airplanes.

JVC Company of America
41 Slater Dr., Elmwood Park, NJ 07407 USA
(800) JVC-5825 (tel)
(800) 247-3608 (tel)
(201) 794-3900 (tel)
107 Little Falls Rd., Fairfield, NJ 07004 USA
(800) 882-2345 (tel)
(201) 808-2100 (tel)
(800) 446-0085 (fax)
Manufactures CD-ROM drives for a wide variety of professional computer network applications.

J.V. Electronics
(800) 604-8777 (tel)

(716) 342-7629 (tel)
Sells cable TV testing and descrambling equipment.

The Kaba Group
24 Commercial Blvd., Novato, CA 94949 USA
(800) 231-TAPE (tel)
(415) 883-5041 (tel)
(415) 883-5222 (fax)
Manufactures cassette duplicators for professional audio applications.

Kahlon
22699 Old Canal Rd., Yorba Linda, CA 92687 USA
(800) 938-5060 (tel)
(714) 637-5060 (tel)
(714) 637-5597 (fax)
Visa, MasterCard, Discover, American Express
Specializes in the sales of IBM, Apple, and Compaq computer parts.

Kaitek Engineering
9051 Pelican Ave., Fountain Valley, CA 92708 USA
(714) 964-6405 (tel)
(714) 965-9935 (fax)
Manufactures PCI extender cards for a wide variety of computer applications.

Kalimar
622 Goddard Ave., Chesterfield, MO 63005 USA
(314) 532-4511 (tel)
Manufactures lighting systems for amateur and professional video production.

Kanga US
3521 Spring Lake Dr., Findlay, OH 45840 USA
(419) 423-5643 (tel)
Sells low-power (QRP) amateur radio kits, projects, etc.

Kantronics
1202 E. 23rd St., Lawrence, KS 66046 USA
(913) 842-7745 (tel)
(913) 842-4476 (tech tel)
(913) 842-2021 (fax)
Manufactures digital data interfaces for a variety of

Kantronics

transmission modes on the shortwave and VHF radio bands.

Kathrein-Werke
P.O. Box 100 444, D-83004, Germany
+49-8031-184439 (tel)
Manufactures receivers for satellite TV applications.

Kaul-Tronics (KTI)
1140 Sextonville Rd., Richland Center, WI 53581 USA
(608) 647-8902 (tel)
(608) 647-4953 (fax)
Manufactures antennas and accessories for professional cable TV applications.

K-COM
P.O. Box 82, Randolph, OH 44265 USA
(216) 325-2110 (tel)
(216) 325-2525 (fax)
Visa, MasterCard
Manufactures and sells modular filters for amateur radio applications.

KComm
5730 Mobud, San Antonio, TX 78238 USA
(800) 344-3144 (tel)
(512) 680-6110 (tel)
Sells a wide variety of amateur radio equipment and accessories, including amateur radio transceivers, receivers, antenna tuners, antennas, amplifiers, etc.

KDC Sound
17294 3083, Conroe, TX 77302 USA
(800) 256-9895 (tel)
Visa, MasterCard, Discover, American Express, COD
Sells transceivers for amateur and citizens band radio applications.

KDE Electronics
P.O. Box 1494, Addison, IL 60101 USA
(800) 361-4586 (tel)
(708) 889-0281 (tel)
(708) 889-0283 (fax)

Sells cable TV converters, filters, and decoders. Offers a 30-day money-back guarantee.

K&D Electric
P.O. Box 659, Newcastle, OK 73065 USA
Sells high-voltage transformers, insulators, Tesla coil parts, wire, etc. Offers a catalog for $1.

K.D.S.
8824 Shirley Ave., #10, Northridge, CA 91324 USA
(818) 700-9945 (tel)
(818) 700-8895 (fax)
Visa, MasterCard
Sells data storage systems, such as hard drives and tape drives, for a wide variety of computer applications.

KD4YBC
197 Chicksaw Ln., Myrtle Beach, SC 29577 USA
Manufactures shortwave transmitter/transceiver amplifier kits for amateur radio applications.

KEF Electronics of America
89 Doug Brown Way, Holliston, MA 01746 USA
(508) 429-3600 (tel)
Manufactures loudspeakers for home and car audio applications.

Keithley Instruments
28775 Aurora Rd., Solon, OH 44139 USA
(800) 552-1115 (tel)
(800) 936-3300 (specifications tel)
(216) 248-6168 (fax)
product_info@keithley.com
http://www.keithley.com
Manufactures test equipment, including digital source meters, for a wide variety of electronics testing applications.

Keltronix
(408) 378-5551 (tel)
(408) 378-4748 (fax)
Sells terminals for professional computer applications.

Kelvin Electronics
10 Hub Dr., Melville, NY 11747 USA
(800) 645-9212 (tel)
(516) 756-1750 (tel)
(516) 756-1763 (fax)
Visa, MasterCard
Manufactures and sells a variety of test meters. Offers a catalog for $3. Mininum order of $20.

Kendrick Amplifiers
P.O. Box 160, Pflugerville, TX 78660 USA
(512) 990-5486 (tel)
(512) 990-0548 (fax)
Manufactures and sells amplifiers for professional and personal musical applications.

Kenford
16001 Gothard St., Huntington Beach, CA 92647 USA
(714) 841-3200 (tel)
(714) 842-3082 (fax)
Manufactures head units, equalizers, crossovers, amplifiers, and loudspeakers for mobile audio applications.

Kenosha Computer Center
2133 91st St., Kenosha, WI 53143 USA
(800) 255-2989 (tel)
(414) 697-9595 (tel)
(414) 697-0620 (fax)
Visa, MasterCard, Discover, American Express
Sells personal desktop and laptop computers, scanners, and printers; also sells computer parts, such as floppy drives, hard drives, monitors, motherboards, video cards, modems, keyboards, etc.

Kenwood
P.O. Box 22745, 2201 E. Dominguez St., Long Beach, CA 90801 USA
(800) 950-5005 (tel)
(213) 639-9000 (tel)
(310) 639-4200 (tel)
(310) 761-8246 (fax)
(310) 609-2127 (fax)
6070 Kestrel Rd., Mississauga, ON L5T 1S8 Canada
(310) 639-5300 (customer support/brochure tel)
(800) KENWOOD (service location/parts tel)
(310) 761-8284 (BBS)
Manufactures amateur radio equipment (including transceivers, receivers, amplifiers, two-ways, and accessories) and stereo equipment (including receivers, speakers, cassette decks, CD changers, car head units, etc.).

Kepco
131-38 Sanford Ave., Flushing, NY 11352 USA
(718) 461-7000 (tel)
(718) 767-1102 (fax)
kepcopower@aol.com
Manufactures programmable power supplies for a wide variety of electronics applications.

Kern Pager Repair
834 Foley St., Jackson, MS 39202 USA
(800) 844-8287 (tel)
(601) 357-4138 (tel)
(601) 948-8257 (fax)
Sells new and refurbished pagers for telecommunications applications.

Kerotec
301 Leominster Rd., Sterling, MA 01564 USA
(508) 422-9797 (tel)
(508) 422-0212 (fax)
Buys and sells IBM memory and networking products for a wide variety of computer applications.

Kessler-Ellis Products
120 First Ave., Atlantic Highlands, NJ 07716 USA
(800) 631-2165 (tel)
(908) 291-0500 (tel)
(908) 291-8097 (fax)
(908) 291-1615 (fax)
Manufactures mass flow computers, "smart" transmitters, rate and total displays, DPMs, batchers, etc. for industrial flow instrumentation applications.

Key Computers
1401 Johnson Ferry Rd., Ste. 328-A2, Marietta, GA 30062 USA
(770) 565-0089 (tel)
Buys and sells IBM and Compaq computer parts.

Key Solutions
105 Santa Rosa Pl., Santa Barbara, CA 93109 USA
(805) 882-1967 (tel)
(805) 882-1977 (fax)
Sells data storage systems, such as hard drives and tape drives, for a wide variety of computer applications.

Keystone Electronics
31-07 20th Rd., Astoria, NY 11105 USA
(718) 956-8900 (tel)
(718) 956-9040 (fax)
Manufactures battery connectors and holders for a wide variety of electronics applications.

K40 Electronics
1500 Executive Dr., Elgin, IL 60123 USA
(800) 323-6768 (tel)
(800) 323-4458 (tech tel)
Manufactures transceivers and antennas for citizens band applications; also sells radar and laser detectors.

KIC
15 Trade Zone Ct., Ronkonkoma, NY 11779 USA
(800) 468-7264 (tel)
(516) 981-8788 (tel)
(516) 981-8826 (fax)
Manufactures LEDs for a wide variety of electronics applications.

KIC Thermal Profiling
15950 Bernardo Center Dr., San Diego, CA 92127 USA
(619) 673-6050 (tel)
Manufactures thermal monitoring systems for industrial electronics applications.

Kilo-Tec
P.O. Box 10, Oakville, CA 93022 USA
Manufactures and sells a 2-kW antenna tuner kit and a J-pole antenna kit for amateur radio applications.

Kilovac
P.O. Box 4422, Santa Barbara, CA 93140 USA
(805) 684-4560 (tel)
(805) 684-9679 (fax)
Manufactures high-voltage relays for a wide variety of professional, medical, and industrial electronics applications.

Kimber Kable
2752 S. 1900 W., Ogden, UT 84401 USA
(801) 621-5530 (tel)
Manufactures power and audio wire and cable for electronics applications.

Kinergetics Research
4260 Charter St., Vernon, CA 90058 USA
(213) 582-9349 (tel)
(213) 582-9434 (fax)
Manufactures audio preamplifiers and amplifiers for high-end home audio and home theater applications.

Kingston Technology
17600 Newhope St., Fountain Valley, CA 92708 USA
(800) 435-2620 (tel)
(714) 435-2667 (tel)
(714) 435-2600 (tel)
(714) 435-2636 (BBS)
(714) 435-2699 (fax)
sales@kingston.com
techsupport@kingston.com
http://www.kingston.com
Manufactures data storage systems, cards, and upgrades for a wide variety of personal and professional computer applications.

Kinovox
Industrivej 9, 3540 Lynge, Denmark
4218-7617 (tel)
4218-9024 (fax)
Sells high-end home audio equipment, such as speakers, cassette decks, amplifiers, preamplifiers, etc.

KinTronic Labs
P.O. Box 845, Bristol, TN 37621 USA
(615) 878-3141 (tel)
(615) 878-4224 (fax)

kintronics@delphi.com
Visa, Master Card
Manufactures and sells a variety of antenna systems, transmitter combiners, inductors, RF switchs, dummy loads, etc. for professional broadcasting.

Kirby
298 W. Carmel Dr., Carmel, IN 46032 USA
Sells a variety of electron tubes. Offers a tube list for a SASE.

Kiwa Electronics
612 S. 14th Ave., Yakima, WA 98902 USA
(800) 398-1146 (tel)
(509) 453-KIWA (tel)
(509) 966-6388 (fax)
kiwa@wolfe.net (Internet/full catalog)
http://www.wolfe.net/_7Ekiwa
Manufactures and sells IF filters, loop antennas, and receiver accessories for shortwave and MW broadcast band listening applications.

KLH
P.O. Box 1085, Sun Valley, CA 91352 USA
(818) 767-2843 (tel)
(213) 875-0480 (tel)
Manufactures loudspeakers for home audio applications.

Klipsch
8900 Keystone Crossing, #1220, Indianapolis, IN 46240 USA
(800) KLIPSCH (tel)
(317) 581-3199 (fax)
Manufactures loudspeakers for high-end home audio and home theater applications.

Kloot Computer
1867 NW 97th Ave., Ste. 103, Miami, FL 33172 USA
(800) 743-8888 (tel)
Visa, MasterCard
Sells computer parts and boards, such as hard drives, CPUs, and memory chips, etc.

Klungness Electronic Supply
101 Merritt Ave., P.O. Box 885, Iron Mountain, MI 49801 USA
(800) 338-9292 (tel)
(906) 774-1755 (tel)
(906) 774-6117 (fax)
Sells a wide variety of equipment for professional cable TV applications.

Knaus Systems
14410 Carlson Circle, Tampa, FL 33626 USA
(800) 970-0444 (tel)
(813) 855-0444 (tel)
(813) 855-0449 (fax)
Sells Honeywell equipment for professional computer applications.

Knight Industries
P.O. Box 1525, Woodinville, WA 98072 USA
(206) 481-9491 (tel)
Manufactures stun batons for security applications.

Korbon Trading
6800 Kitimat Rd., Unit 19 & 20, Mississauga, ON L5N 5M1 Canada
Sells head units, loudspeakers, amplifiers, receivers, cassette decks, CD players, etc. for home audio and car audio applications.

Korg
89 Frost St., Westbury, NY 11590 USA
Manufactures keyboard synthesizers for a wide variety of musical applications.

Kortesis Marketing
P.O. Box 460543, Aurora, CO 80046 USA
(303) 548-3995 (tel)
(303) 699-6660 (fax)
Visa, MasterCard, COD
Manufactures radar jammers for a wide variety of vehicular applications.

Koss
4129 N. Pt. Washington, Milwaukee, WI 53212 USA
(414) 964-5000 (tel)

Koss

Manufactures loudspeakers for home audio and home theater applications.

Kramer
3320 Rt. 66, Neptune, NJ 07553 USA
(908) 922-8600 (tel)
Manufactures electric guitars for professional and personal musical applications.

Kress Jets
4308 Ulster Landing Rd., Saugerties, NY 12477 USA
(914) 336-8149 (tel)
Sells fan kits and electronics for radio-control model airplanes.

Jeffrey A. Krinsky
(206) 228-6003 (tel)
(206) 228-3318 (fax)
Sells printers, monitors, and plotters for a wide variety of computer applications.

KRK Monitoring Systems
16462 Gothard St., Unit D, Huntington Beach, CA 92647 USA
(714) 841-1600 (tel)
(714) 375-6496 (fax)
Manufactures reference monitor loudspeakers for professional studio recording.

Kroy
14555 N. Hayden Rd., Scottsdale, AZ 85260 USA
(800) 733-5769 (info request tel)
(800) 776-5769 (tel)
(602) 948-2222 (tel)
info@kroy.com
Manufactures and sells label printers for use with personal computer systems.

K2AW's "Silicon Alley"
175 Friends Ln., Westbury, NY 11590 USA
(516) 334-7024 (tel)
Manufactures high-voltage modules for amateur radio applications.

K-Tec
9831 S. 51 St. Bldg. C-112, Phoenix, AZ 85044 USA
(800) 468-5832 (tel)
(602) 598-0620 (fax)
Manufactures body voltage measurers, static-electricity measuring equipment, soldering irons, etc. for a variety of professional industrial applications.

KT Radio
Vesterbogade 179-181, Kobenhavn, Denmark
31-31-14-40 (tel)
Sells high-end home audio equipment, such as speakers, cassette decks, amplifiers, preamplifiers, etc.

Kurluff Enterprises
4331 Maxson Rd., El Monte, CA 91732 USA
(818) 444-7079 (tel)
(818) 444-6863 (fax)
Sells new old stock (NOS) electronics, such as paper capacitors, electron tubes, etc. for a variety of electronics applications.

KurlyTie
(800) 587-5984 (tel)
Manufactures spring coil cable wraps for audio and radio cables.

Kurzweil
13336 Alondra Blvd., Cerritos, CA 90703 USA
(310) 926-3200 (tel)
(310) 404-0748 (fax)
kurzweil@aol.com
http://www.musicpro.com/kurzweil
Manufactures sampling synthesizers for professional studio recording.

Kyocera Electronics
Industrial Ceramics Department
5713 E. Fourth Plain Blvd., Vancouver, WA 98661 USA
(800) 232-6797 (tel)
100 Randolph Rd., Somerset, NJ 08875 USA
(908) 560-0060 (tel)
(908) 560-3400 (tel)
(908) 560-0350 (fax)

Manufactures laser printers, LCD displays, and printer thermal heads for personal desktop computer applications.

Jim Laabs Music Superstore
1055 Main St., Stevens Point, WI 54481 USA
(800) 657-5125 (tel)
Sells synthesizer keyboards, digital pianos, sound modules, etc. for professional and personal musical applications.

Labtech
3801 NE 109th Ave, #J, Vancouver, WA 98692 USA
Manufactures speakers for computer multimedia configurations.

Lago Systems
151 Albright Way, Los Gatos, CA 95030 USA
(800) 866-LAGO (tel)
(408) 376-2750 (tel)
(408) 374-2330 (fax)
Manufactures data storage systems for a wide variety of professional computer network applications.

Laguna Data
201 E. Colorado Ave., Las Vega, NV 89104 USA
(800) 938-8273 (tel)
(702) 382-3964 (fax)
Sells recordable CD-ROM drives, optical hard drives, etc.

Lake Erie Systems and Services
5321 Buffalo Rd., Erie, PA 16510 USA
(814) 899-1384 (tel)
(814) 898-0704 (fax)
2694 Sawbury Blvd., Columbus, OH 43235 USA
(614) 766-5077 (tel)
(614) 766-0549 (fax)
Sells new and refurbished personal desktop computers, printers, and terminals.

Lake Monitors
3828 W. Mitchell St., Milwaukee, WI 53215 USA
Manufactures flow-rate measuring equipment for professional industrial applications. Offers a free, 16-page new product data folder.

Lakeview
3620 Whitehall Rd., Anderson, SC 29624 USA
(803) 226-6990 (tel)
(803) 225-4565 (fax)
Visa, MasterCard
Manufactures mobile antennas for shortwave and VHF/UHF amateur radio applications.

Lambda Electronics
515 Broad Hollow Rd., Melville, NY 11747 USA
(800) LAMBDA-4 (tel)
(516) 694-4200 (tel)
(516) 293-0519 (fax)
Visa, MasterCard
Manufactures power supplies for a wide variety of computer and electronics applications.

Lamp Technology
1645 Sycamore Ave., Bohemia, NY 11716 USA
(516) 567-1800 (tel)
Manufactures LEDs for a wide variety of electronics applications.

Lance Technology
5th Fl., No. 164-2, Lien Cheng Rd.
Chung-Ho, Taipei Hsien, Taiwan
886-2-2453124-5 (tel)
886-2-2489697 (tel)
886-2-2489547 (fax)
Manufactures loudspeakers and game controllers for computer multimedia applications.

Länggasse
Länggass Str. 16, CH-3012 Bern, Switzerland
031-266464 (tel)
Sells equipment for shortwave radio receiving applications.

Lanier Information Systems
2300 Park Lake Dr. NE, Atlanta, GA 30345 USA
(800) 708-7088 (tel)
(800) 252-9703 (fax)

Lanier Information Systems

http://www.lanier.com
Manufactures fax machines and photocopiers for a wide variety of office applications.

Lansing Instrument
P.O. Box 730, Ithaca, NY 14851 USA
(800) 847-3535 (tel)
(607) 277-1953 (tel)
(607) 272-3265 (fax)
Manufactures enclosures for a wide variety of electronics applications.

Lantronix
15353 Barranca Pkwy., Irvine, CA 92718 USA
(800) 422-7015 (resellers tel)
(800) 422-7055 (tel)
(714) 453-3990 (tel)
(714) 453-3995 (fax)
sales@lantronix.com
Manufactures bridges, switches, and routers for professional computer network applications. Offers a free tutorial catalog.

Lanzar
21025 Osborne St., Canoga Park, CA 91304 USA
(818) 882-1634 (tel)
(818) 882-4094 (fax)
Manufactures loudspeakers and amplifiers for car stereo applications.

La Paz Electronics International
P.O. Box 261095, San Diego, CA 92129 USA
(800) 586-4199 (tel)
(619) 586-1482 (fax)
Sells RAM chips, CPU chips, EPROM programmers, etc. for a wide variety of computer applications.

Larsen Electronics
3611 NE 112th Ave., Vancouver, WA 98668 USA
(800) 426-1656 (tel)
(206) 944-7556 (fax)
(360) 944-7551 (tel)
5049 Still Creek Ave., Burnaby, BC V5C 5V1 Canada
(800) 663-6734 (tel)
(604) 299-8517 (tel)
Manufactures and sells antennas for amateur radio and cellular microwave telephone applications.

LaRue Electronics
1112 Grandview St., Scranton, PA 18509 USA
(717) 343-2124 (tel)
Sells a wide variety of amateur radio equipment and accessories, including amateur radio transceivers, receivers, antenna tuners, antennas, amplifiers, etc.

Laserdisken
Prinsensgade 38, 9000 Ålborg, Denmark
98-13-22-22 (tel)
98-11-49-59 (fax)
Skt. Pedersstraede 49, 1453 Kobenhavn K, Denmark
33-93-10-80 (tel)
Manufactures closed-caption decoders for LaserDisc systems.

Laser Graphics
20 Ada, Irvine, CA 92718 USA
(714) 753-8282 (tel)
Manufactures a computer digital film recorder for professional photographic applications.

Laser Impact
10435 Burnet Rd., #114, Austin, TX 78758 USA
(800) 777-4323 (tel)
(512) 832-9151 (tel)
(512) 832-9321 (fax)
3400 14th Ave., Unit 34, Markham, ON L3R OH7 Canada
(800) 263-0771 (tel)
(905) 470-0771 (tel)
(905) 470-9985 (fax)
Unit 12 First Quarter, Blenheim Rd., Epsom, Surrey KT19 9QN UK
0500-747-444 (tel)
01372-747-555 (tel)
01372-747-666 (fax)
Sells parts for computer laser printers.

Laser Master
6900 Shady Oak Rd., Eden Prarie, MN 55344 USA
(800) 428-5936 (tel)
(800) 688-8342 (tel)

(612) 944-9330 (tel)
(612) 944-1244 (fax)
(612) 943-3737 (fax tech support)
31-2503-22000 (Netherlands tel)
Manufactures professional desktop publishing printers for paper and chemical-free film.

Laser Products
1010 E. 18th Ave., Kansas City, MO 64116 USA
(800) 786-8897 (tel)
(816) 421-7830 (tel)
(816) 471-1677 (fax)
Visa, MasterCard
Sells laser printers and printer parts for a wide variety of computer applications.

Laser Sensor Technology
3873 Easy Circle, Marietta, GA 30066 USA
(770) 928-2867 (tel/fax)
Manufactures laser range finders for a wide variety of outdoor measurement applications.

Laserstar
(800) 432-9989 (tel)
(617) 937-0564 (fax)
Visa, MasterCard
Buys, sells, and trades used and refurbished printer cartridges for a wide variety of computer applications.

Lasiris
3549 Ashby, St. Laurent, PQ H4R 2K3 Canada
(800) 814-9552 (tel)
(514) 335-4576 (fax)
Manufactures a laser crosshair projector for a wide variety of applications.

LAT International
317 Provincetown Rd., Cherry Hill, NJ 08034 USA
(800) 321-2108 (tel)
Sells high-end home audio equipment, such as speakers, cassette decks, amplifiers, preamplifiers, etc. Offers a free catalog.

L.A. Trade
22825 Lockness Ave., Torrance, CA 90501 USA
(800) 433-3726 (tel)
(310) 539-0019 (tel)
(310) 539-5844 (fax)
Visa, MasterCard, Discover, American Express
Sells computer parts and boards, such as math processors and memory chips. Offers a 30-day money-back guarantee.

Lattice Semiconductor
5555 NE Moore Ct., Hillsboro, OR 97124 USA
(800) FASTGAL (tel)
http://www.latticesemi.com
Manufactures semiconductors for a wide variety of electronics applications.

Laurel Computer Systems
P.O. Box 2111, Uniontown, PA 15401 USA
(800) 589-6601 (tel)
(412) 438-3358 (fax)
http://www.lcsys.com
Visa, MasterCard, Discover
Manufactures and sells personal desktop computers. Offers a 30-day money-back guarantee.

Lazer Industries
111 Market St., Warner Robbins, GA 31093 USA
(912) 329-0190 (tel)
(912) 329-0290 (fax)
Manufactures amplifiers for car stereo applications.

LDG Electronics
1445 Parran Rd., St. Leonard, MD 20685 USA
(410) 586-2177 (tel/fax)
Visa, MasterCard
Manufactures microcontrollers for a wide variety of electronics applications. Offers a free catalog.

Leader Instruments
380 Oser Ave., Hauppauge, NY 11788 USA
(800) 645-5104 (tel)
(516) 231-6900 (NY tel)
Omnitronix, Canada
(416) 828-6221 (tel)
Manufactures spectrum meters for professional cable TV applications.

Leader Tech
14100 McCormick Dr., Tampa, FL 33626 USA
(813) 855-6921 (tel)
(813) 855-3291 (fax)
Manufactures RFI enclosures for a variety of applications.

LeCroy
700 Chestnut Ridge Rd., Chestnut Ridge, NY 10977 USA
(800) 4-LECROY (tel)
Manufactures test equipment, including oscilloscopes, for a wide variety of electronics testing applications.

Lectro Products
420 Athena Dr., Athens, GA 30603 USA
(800) 551-3790 (tel)
Manufactures power supplies for a wide variety of electronics applications.

Ledtronics
4009 Pacific Coast Hwy., Torrance, CA 90505 USA
(310) 534-1505 (tel)
(310) 534-1424 (fax)
Manufactures LEDs for a wide variety of electronics applications.

Legacy Audio
3021 Sangamon Ave., Springfield, IL 62702 USA
(800) 283-4644 (tel)
(217) 744-7269 (fax)
Visa, MasterCard, Discover, American Express
Manufactures amplifiers, preamplifiers, and speakers; also sells a wide variety of high-end audio equipment. Offers a free brochure, free shipping, and a 10-day money-back guarantee.

Legato Systems
3145 Porter Dr., Palo Alto, CA 94304 USA
(415) 812-6000 (tel)
(415) 812-6032 (fax)
Manufactures disk controller boards for a wide variety of professional computer network applications.

Legend Micro
5590 Lauby Rd., Canton, OH 44720 USA
(800) 366-6333 (tel)
(216) 497-8620 (tel)
(216) 497-3156 (fax)
http://www.industry.net/legend
Visa, MasterCard, Discover, COD
Manufactures and sells personal desktop computers; also sells computer parts and boards, such as floppy drives, hard drives, monitors, video cards, modems, keyboards, etc. Offers a 30-day money-back guarantee.

Lehman Scientific
R.R. 1 Box 580, Wrightsville, PA 17368 USA
(717) 244-7540 (tel)
(717) 252-4266 (fax)
lscience@yrkpa.kias.com
Visa, MasterCard
Sells test equipment, including oscilloscopes, voltmeters, pulse generators, etc., for a wide variety of electronics testing applications.

Leica
3155 Medlock Bridge Rd., Norcross, GA 30071 USA
(800) 367-9453 (tel)
(404) 447-6361 (tel)
(404) 447-0710 (fax)
Manufactures electronic distance-measuring equipment for professional industrial applications.

Leigh's Computers
129 W. Eagle Rd., Havertown, PA 19083 USA
(800) 321-6434 (tel)
(610) 658-2323 (tel)
(610) 896-4414 (fax)
leighs@ix.ix.netcom.com
75213.3615.compuserve.com
http://soho.ios.com/~leighs
Sells MIDI equipment, CD-ROM recorders, sound cards, etc. for a wide variety of professional audio and musical applications. Offers a free MIDI/multimedia catalog.

Leisure Electronics
2950 Airway Bldg. A-4, Costa Mesa, CA 92626 USA

(714) 435-9218 (tel)
Visa, MasterCard
Manufactures and sells electronic motors, etc. for radio-control model airplanes.

Lemo USA
P.O. Box 11488, Santa Rosa, CA 95406 USA
(800) 444-5366 (tel)
(707) 578-0869 (fax)
Manufactures connectors for a wide variety of electronics applications. Offers a free catalog.

Lenmar Enterprises
31328 Via Colinas, #102, Westlake Village, CA 91362 USA
(818) 879-2700 (tel)
Manufactures lighting systems for amateur and professional video production.

Lentini Communications
21 Garfield St., Newington, CT 06111 USA
(800) 666-0908 (tel)
(203) 667-9479 (tel)
(203) 666-6227 (tel)
(203) 667-3561 (fax)
Visa, MasterCard, Discover, COD
Sells a wide variety of amateur radio equipment and accessories, including amateur radio transceivers, receivers, antenna tuners, antennas, amplifiers, etc.

Less Buster's Electronics
2039 Civic Center Dr., #176V, N. Las Vegas, NV 89030 USA
(702) 642-0325 (tel)
Sells power inverters and other electronics. Offers a free catalog.

Leunig Communications
1630 Oakland Rd., #A-211, San Jose, CA 95131 USA
(800) 441-6562 (tel)
(408) 441-6580 (tel)
Manufactures network hubs for professional computer network applications.

David Lewis Audio
8010 Bustleton Ave., Philadelphia, PA 19152 USA
(215) 725-4080 (tel)
Visa, MasterCard, Discover, American Express
Sells home and amateur audio equipment, such as speakers, cassette decks, TVs, home theater equipment, etc.

Lexicon
100 Beaver St., Waltham, MA 02154 USA
(617) 736-0300 (tel)
(617) 891-0340 (fax)
71333.434@compuserve.com
Manufactures digital audio processors for professional audio applications.

Lexmark International
740 New Circle Rd., Lexington, KY 40511 USA
(800) 891-0411 (tel)
(800) 891-0331 (tel)
(800) 891-0399 (tel)
(606) 232-2000 (tel)
http://www.lexmark.com
Manufactures and sells printers for personal computers.

LF Engineering
17 Jeffry Rd., E. Haven, CT 06513 USA
(203) 248-8851 (tel)
76715.2361@compuserve.com
Manufactures and sells active antennas for LF/MF/HF radio receiving applications.

LG Precision
13013 E. 166th St., Cerritos, CA 90701 USA
(310) 404-0101 (tel)
(310) 921-6227 (fax)
Sells analog and digital storage oscilloscopes, sweep function generators, frequency counters, bench power supplies, digital multimeters, etc.

LG Semiconductor
3003 N. 1st St., San Jose, CA 95134 USA
(408) 432-5000 (tel)
(408) 432-6067 (fax)

LG Semiconductor

81-3-3224-0123 (Japan tel)
81-3-3224-0692 (Japan fax)
886-2-757-7022 (Taiwan tel)
886-2-757-7013 (Taiwan fax)
65-226-1191 (Singapore tel)
65-221-8575 (Singapore fax)
49-2154-492-172 (Germany tel)
49-2154-429-424 (Germany fax)
Manufactures semiconductors for a wide variety of electronics applications.

Liberty Electronic

48089 Fremont Blvd., Fremont, CA 94538 USA
(800) 745-7011 (tel)
(510) 623-6000 (tel)
(510) 623-7021 (fax)
Manufactures terminals for a wide variety of professional computer network applications.

Liberty Systems

375 Saratoga Ave., #A, San Jose, CA 95129 USA
(408) 983-1127 (tel)
(408) 243-2885 (fax)
Manufactures data storage systems for a wide variety of professional computer network applications.

Liebert

1050 Dearborn Dr., P.O. Box 29186, Columbus, OH 43229 USA
(800) 877-9222 (tel)
(614) 888-0246 (tel)
(614) 841-6022 (fax)
Manufactures uninterruptible power supplies for a wide variety of personal computer applications.

Lightline Engineering

P.O. Box 24, Mullica Hill, NJ 08062 USA
Sells fiberoptics supplies, including fiber, cable, splices, detectors, lasers, kits, etc. Offers a catalog for $2.

Lightning Audio

1835 E. 6th St., Ste. 6., Tempe, AZ 85281 USA
(602) 966-8278 (tel)
(602) 966-0393 (fax)
Manufactures batteries, capacitors, and connectors for mobile audio applications.

Ligon Guitars

P.O. Box 27, Orient, IL 62874 USA
(618) 937-2532 (tel)
Sells electric guitars for professional and personal musical applications.

The Likes Line

728 Wikie St., Dunedin, FL 34698 USA
(813) 733-0537 (tel)
Sells electrical retracts for radio-control model airplanes.

Linco Computer

13210 Hempstead Hwy., Ste. 2A, Houston, TX 77040 USA
(713) 895-8880 (tel)
(713) 895-8885 (fax)
Visa, MasterCard, American Express
Sells terminals for professional computer applications.

Lindsay Electronics

50 Mary St. W., Lindsay, ON K9V 4S7 Canada
(800) 465-7046 (tel)
(705) 324-2196 (tel)
(705) 324-5474 (fax)
Manufactures amplifiers, hardline passives, multitaps, etc. for professional cable TV applications.

Linear Laboratories

49050 Milmont Dr., Fremont, CA 94538 USA
(510) 226-0488 (tel)
(510) 226-1112 (fax)
Manufactures temperature sensors for a wide variety of professional industrial applications.

Linear Systems

310 S. Milpitas Blvd., Milpitas, CA 95035 USA
(408) 263-8401 (tel)
(408) 263-7280 (fax)
Manufactures and sells a variety of linear semiconductor products, including JFET.

Linear Technology
1630 McCarthy Blvd., Milpitas, CA 95035 USA
(800) 4-LINEAR (literature tel)
(408) 432-1900 (tel)
(408) 434-6441 (fax)
Manufactures semiconductors for a wide variety of electronics applications.

Link Communications
115 2nd Ave. NE, Sidney, MT 59270 USA
(800) 610-4085 (tel)
(406) 482-7515 (tel)
(406) 482-7547 (fax)
Manufactures and sells repeater controllers for amateur radio applications.

Link Technologies
46595 Landing Pkwy., Fremont, CA 94538 USA
(800) 448-LINK (tel)
(510) 651-8000 (tel)
(510) 651-8808 (fax)
Manufactures terminals for a wide variety of professional computer network applications.

Linn Hi-Fi
8709 Castle Park Dr., Indianapolis, IN 46256 USA
(800) 546-6443 (tel)
(317) 841-4107 (fax)
Manufactures high-end amplifiers for audio and home theater applications.

Lintel Security
Avenue de Jette 32, B-1080, Brussels, Belgium
32-2-425-77-67 (tel)
32-2-425-37-22 (fax)
Manufactures handheld personal authenticators for a wide variety of professional security applications.

Littelfuse
800 E. Northwest Hwy., Des Plaines, IL 60016 USA
(708) 824-0400 (tel)
Manufactures surface-mount fuses for a wide variety of electronics applications.

The Live Wire Companies
P.O. Box 20242, New York, NY 10021 USA
livewire@inch.com
http://www.inch.com/_7Elivewire
Manufactures Y audio cables for a wide variety of audio applications.

L&L Systems
11437 E. 20th St., Tulsa, OK 74128 USA
(918) 234-1200 (tel)
(918) 234-1276 (fax)
Sells Memorex telex equipment and parts for professional computer applications.

L&M Music
6228 Airpark Dr., Chatanooga, TN 37421 USA
(800) 8-SOUND-8 (tel)
http://www.lmmusic.com
Sells recording equipment, amplifiers, synthesizers, loudspeakers, electric guitars, etc. for a wide variety of professional audio and musical applications.

LNJ Enterprises
2080-C Tigertail Blvd., Dania, FL 33004 USA
(954) 927-7339 (tel)
(954) 927-7349 (fax)
Visa, MasterCard, Discover, American Express
Buys, sells, and trades personal desktop computers and printers and parts and accessories, including monitors, terminals, CD-ROM drives, etc.

Logan Enterprises
P.O. Box 839, 8844 US 68 N., West Liberty, OH 43357 USA
(800) 473-9880 (tel)
(513) 465-8170 (tel)
(513) 465-9140 (fax)
Manufactures temperature sensors for a wide variety of professional industrial applications.

Logical Design Group
6301 Chapel Hill Rd., Raleigh, NC 27607 USA
(919) 851-1101 (tel)
(919) 851-2844 (fax)
Manufactures single-board computers for professional and industrial computer applications.

Logical Design Tools
5403 Everhart, Box 55, Corpus Christi, TX 78411 USA
(512) 888-4535 (tel)
Sells ring decoders for telephone/electronics applications.

Logical Systems
P.O. Box 6184, Syracuse, NY 13217 USA
(315) 478-0722 (tel)
Manufactures EPROM adapters for professional electronics applications.

Logitech
6505 Kaiser Dr., Fremont, CA 94555 USA
(800) 732-2945 (tel)
(800) 732-3188 (order tel)
(800) 245-0000 (fax)
http://www.logitech.com
Manufactures and sells trackballs, digital cameras, etc. for personal and professional computer applications.

Longs Electronics
2630 S. 5th Ave., Irondale, AL 35210 USA
(800) 633-3410 (tel)
(800) 292-8668 (tel)
Sells a wide variety of amateur radio equipment and accessories, including amateur radio transceivers, receivers, antenna tuners, antennas, amplifiers, etc.

Longshine Microsystems
10400-9 Pioneer Blvd., Santa Fe Springs, CA 90670 USA
(310) 903-0899 (tel)
Manufactures CD-ROM drives and Ethernet hubs for a wide variety of professional computer network applications.

Loral Microwave-Narda
435 Moreland Rd., Hauppauge, NY 11788 USA
(516) 231-1700 X230 (tel)
(516) 231-1390 (inter tel)
(516) 231-1711 (fax)
Manufactures SWR/power meters for professional cellular microwave radio applications.

Loral Test & Information Systems
15378 Avenue of Science, San Diego, CA 92128 USA
(800) 644-3334 (tel)
(619) 674-5100 (tel)
(619) 674-5145 (fax)
Manufactures routers for professional computer network applications.

Los' Music Center
1045 Airline Rd., Corpus Christi, TX 78412 USA
(512) 993-7302 (tel)
Sells equipment for professional and personal musical applications.

Loveland CB
329 E. 5th St., #2, Loveland, CO 80537 USA
(800) 619-9527 (tel)
(970) 667-9527 (tel)
Sells a wide variety of transceivers, antennas, and accessories for CB radio applications.

Lowe Electronics
Chesterfield Rd., Matlock, Derbyshire DE4 5LE UK
629-580800 (tel)
629-580020 (fax)
Manufactures radios, preselectors, and accessories for shortwave radio applications.

Lowel-Light Manufacturing
140 58th St., Brooklyn, NY 11220 USA
(718) 921-0600 (tel)
(718) 921-0303 (fax)
Manufactures lighting kits for amateur and professional video production applications.

LSC
4211 Lexington Ave. N., St. Paul, MN 55126 USA
(800) 831-9482 (tel)
(612) 482-4535 (tel)
(612) 482-4595 (fax)
Manufactures file servers for a wide variety of professional computer network applications.

LT Sound
7988 LT Pkwy, Lithonia, GA 30058 USA

(404) 482-2485 (tel)
Manufactures equipment to eliminate vocals from standard tapes, records, and CDs.

Lucas Transformer
7113 N. 9 Mile, Lake City, MI 49651 USA
(616) 229-4318 (tel)
Visa, MasterCard
Sells transformers, chokes, and capacitors--especially for high-power amateur radio transmitters and amplifiers.

Lumberg
11352 Business Center Dr., Richmond, VA 23236 USA
(804) 379-2010 (tel)
(804) 379-3232 (fax)
Manufactures central power interfaces for a variety of professional electronics applications.

Lumex Opto/Components
292 E. Hellen Rd., Palatine, IL 60067 USA
(708) 359-2790 (tel)
(800) 944-2790 (fax)
Manufactures LEDs for a wide variety of electronics applications.

Lumitex
8443 Dow Circle, Strongsville, OH 44136 USA
(800) 969-5483 (tel)
(216) 243-8402 (fax)
Manufactures backlighting for electronics displays.

Luxman Electronics
915 Washington Ave. S., Minneapolis, MN 55415 USA
(612) 333-1150 (tel)
(612) 338-8129 (fax)
10-27, 4-Chome, Higashi-Shinagawa, Shinagawa-Ku, Tokyo 140 Japan
03-5462-0405 (tel)
03-5462-0404 (fax)
Manufactures receivers, amplifiers, CD players, cassette decks, etc. for home audio applications.

Lyben Computer Systems
5545 Bridgewood, Sterling Heights, MI 48310 USA
(800) 493-5777 (tel)
(810) 268-8899 (fax)
Sells a wide variety of computer supplies and accessories. Offers a free 268-page catalog.

Lyle Cartridges
115 S. Cornona Ave., Valley Stream, NY 11582 USA
(800) 221-0906 (tel)
(516) 599-1112 (tel)
Visa, MasterCard, Discover, American Express
Sells a wide variety of high-end home audio equipment.

Lynnhaven Custom Computers
1060 Lynnhaven Pkwy., #111, Virginia Beach, VA 23452 USA
(804) 430-2801 (tel)
(804) 430-3282 (fax)
Sells personal desktop computers and parts, such as hard drives, motherboards, video boards, cases, etc.

Lyra
Fylis St. 16, GR-Athens 104 33 Greece
Sells head units, loudspeakers, amplifiers, receivers, cassette decks, CD players, etc. for home audio and car audio applications.

LZR Electronics
8051 Cessna Ave., Gaithersburg, MD 20879 USA
(301) 921-4600 (tel)
(301) 670-0436 (fax)
Manufactures external power supplies for a wide variety of electronics applications.

TA Macalister
Private Bag 92-146, Auckland, New Zealand
0-9-303-4334 (tel)
0-9-309-6502 (fax)
Sells computer equipment, including image scanners. Offers a free brochure.

Mackie
16220 Wood-Red Rd., Woodinville, WA 98072 USA

Mackie

(800) 898-3211 (tel)
(800) 363-8855 (tel)
(206) 487-4333 (tel)
(206) 487-4337 (fax)
Manufactures mixing consoles for professional recording studios and live music performances.

MaCo Manufacturing

4091 V St., Memphis, TN 38118 USA
(901) 794-9494 (tel)
(901) 794-9497 (tech tel)
Manufactures and sells antennas, power supplies, and accessories for citizens band radio applications.

Macrolink

1500 N. Kellogg Dr., Anaheim, CA 92807 USA
(714) 777-8800 (tel)
(714) 777-8807 (fax)
Manufactures ruggedized enclosures for industrial computer applications.

Macrom

120 Blue Ravine Rd., Folsom, CA 95630 USA
(916) 985-7645 (tel)
(916) 985-7647 (fax)
Manufactures loudspeakers for car stereo applications.

Madge Networks

(800) 876-2343 (tel)
(408) 955-0700 (tel)
(408) 955-0970 (tech support fax)
(408) 383-1002 (fax back)
(408) 955-0262 (BBS)
852-593-9888 (Asia tel)
61-2-256-2738 (Oceania tel)
61-2-256-2739 (Oceania tech support fax)
44-1345-125539 (UK tel)
44-1628-810-607 (UK tech support fax)
44-1628-858-008 (UK BBS)
71333.2103@compuserve.com
support@madge.com (e-mail tech support)
Manufactures token ring equipment for professional computer network applications.

Madison Electronics

12310 Zavella St., Houston, TX 77085 USA

(800) 231-3057 (tel)
(713) 729-7300 (tel)
(713) 358-0051 (fax)
Sells amateur radio equipment and accessories, including amateur radio transceivers, receivers, antenna tuners, antennas, amplifiers, industrial surplus electronics, etc.

Madisound Speaker Components

8608 University Green, P.O. Box 44283, Madison, WI 53744 USA
(608) 831-3433 (tel)
(608) 831-3771 (fax)
Sells a variety of hi-fi output components, including speakers, custom speaker components and terminals, etc.

Madrigal Audio Laboratories

P.O. Box 781, Middletown, CT 06457 USA
(203) 346-1540 (fax)
Manufactures high-end preamplifiers for home theater applications.

Magellan

960 Overland Ct., San Dimas, CA 91773 USA
(909) 394-5000 (tel)
Manufactures GPS systems for professional aviation applications.

Magellan's

P.O. Box 5485 Santa Barbara, CA 93150 USA
(800) 962-4943 (tel)
(805) 568-5400 (tel)
(805) 568-5406 (fax)
75713.2463@compuserve.com
Sells a wide variety of products for travelers, including modem adapters, plug adapters, power adapters and converters, power transformers, digital clocks, electronic translators, etc.

Magenta Electronics

135 Hunter St., Burton-on-Trent, Staffordshire DE14 2ST UK
01283-565435 (tel)
01283-566932 (fax)

Visa, Access, Switch
Manufactures and sells a wide variety of electronics kits, including power supplies, metal detectors, thermostats, EPROM erasers, etc.

MAG InnoVision
2801 S. Yale St., Santa Ana, CA 92704 USA
(800) 827-3998 (tel)
(714) 751-2008 (tel)
(714) 751-5522 (fax)
(714) 751-0166 (fax)
http://www.maginnovision.com
Manufactures monitors for personal computer applications.

Magitronic
1630 Pleasant Hill Rd., #180-180, Duluth, GA 30136 USA
(800) 977-6267 (tel)
(770) 416-7530 (fax)
Visa, MasterCard, Discover, American Express
Manufactures and sells personal desktop and laptop computers.

Magnatek
2 McLaren, #E, Irvine, CA 92718 USA
(714) 457-0706 (tel)
(714) 457-0605 (fax)
Sells data storage systems, such as hard drives and CD-ROM drives, for a wide variety of computer applications.

Magnavox
1 Philips Dr., Knoxville, TN 37914 USA
(800) 835-3506 (tel)
(615) 521-4391 (tel)
(432) 475-0317 (tel)
http://www.magnavox.com
Manufactures monitors for personal desktop computer systems; CD players, TVs, cassette decks, loudspeakers, VCRs, etc., for home video and home audio applications; etc.

Magnum Computer Products
403-5 Bloomfield Dr., W. Berlin, NJ 08091 USA
(800) 562-1555 (tel)

(800) 220-1125 (fax)
Buys and sells used equipment and parts for a wide variety of computer applications.

Magtrol
70 Gardenville Pkwy. W., Buffalo, NY 14224 USA
(800) 828-7844 (tel)
http://www.magtrol.com
Manufactures power analyzers for a variety of professional industrial applications.

Maha Communications
15356A Valley Blvd., City of Industry, CA 91746 USA
(818) 333-4497 (tel)
(818) 333-4987 (fax)
Sells batteries, antennas, and accessories for amateur radio applications.

Mahogany Sound
2610 Schillingers Rd., #488, Mobile, AL 36695 USA
(205) 633-2054 (tel)
Specializes in transmission line for professional audio applications.

Majestic
14614 Lanark St., Panorama City, CA 91402 USA
(818) 781-8200 (tel)
(818) 989-3154 (fax)
Manufactures terminals, cables, distribution blocks, etc. for car stereo applications.

Major Hobby
1520 B Corona Dr., Lake Havasu City, AZ 86403 USA
(800) 625-6772 (tel)
(520) 855-7901 (tech tel)
(520) 855-5930 (fax)
Visa, MasterCard, Discover, American Express, COD ($4.95)
Sells sport and computer radios, batteries, and servos for radio-control model airplanes. Offers a free catalog.

Maldol Antennas
Hokushin Industry Co., 1111-1 Nagasaku Cho, Hanamigawa-ku, Chiba City, Chiba, Japan

Maldol Antennas

043-257-1581 (tel)
043-259-6000 (fax)
Manufactures mobile and base antennas, SWR meters, speaker microphones, speakers, and accessories for amateur radio applications.

Management Products

2150 W. 6th Ave., Unit N, Broomfield, CO 80020 USA
(800) 245-9933 (tel)
(303) 465-0651 (tel)
Manufactures and sells telephone/fax/modem switching systems for personal telecommunications applications.

Mandeno Electronic Equipment

463 Mt. Eden Rd., Mt. Eden, Auckland 3, New Zealand
0-9-630-7871 (tel)
0-9-630-1720 (fax)
14 Freyberg Cres., Walkanae, New Zealand
0-4-293-2453 (tel)
Sells plotters for professional computer graphics applications.

Manley Laboratories

13880 Magnolia Ave., Chino, CA 91710 USA
(909) 627-4256 (tel)
(909) 628-2482 (fax)
Manufactures reference microphones and all-tube amplifiers for professional studio recording.

Mann Endless Cassette Industries

3700 Sacramento, San Francisco, CA 94118 USA
(415) 221-2000 (tel)
(800) 683-7569 (fax)
(415) 387-2425 (fax)
Manufactures endless loop cassette tapes.

Manny's Mailbox Music

156 W. 48th St., New York, NY 10036 USA
(212) 869-5172 (tel)
Sells a variety of audio equipment for musicians. Offers a free catalog.

Man Technologie

Dieselstrasse 8, D-85757 Karlsfeld Germany
+49-8131-89-1686 (tel)
+49-8131-89-1900 (fax)
Manufactures GPS/GLONASS receiving systems for professional aviation applications.

Maplin Electronics

P.O. Box 3, Rayleigh, Essex SS6 8LR UK
01702-554161 (tel)
Sells security products, GPS systems, clocks, test equipment, more than 300 electronic kits, radios, and computer components. Offers a 1,100-page catalog for £2.95.

Marantz

440 Medinah Rd., Roselle, IL 60172 USA
1150 Feehanville Dr., Mt. Prospect, IL 60056 USA
(708) 307-3100 (tel)
(708) 299-4000 (tel)
(708) 307-2687 (fax)
(708) 299-4004 (fax)
Manufactures a variety of equipment for audio and home theater applications, including CD players, receivers, power amplifiers, and cassette decks.

Marathon Computer

1625 Washington St., Holliston, MA 01746 USA
(508) 429-3330 (tel)
(508) 429-2660 (fax)
Sells personal desktop computer parts and accessories, including system boards, hard drives, controllers, power supplies, etc.

Marchand Electronics

P.O. Box 473, Webster, NY 14580 USA
(716) 872-0980 (tel)
(716) 872-1960 (fax)
Manufactures high-end audio kits and fully assembled equipment, including power supplies, crossovers, power amplifiers, equalizers, etc. Offers free literature.

Marconi Instruments

3 Pearl Ct., Allendale, NJ 07401 USA

(800) 888-4114 (tel)
(201) 934-9050 (tel)
(201) 934-9229 (fax)
(514) 341-7630 (Canada tel)
Manufactures test equipment for a wide variety of electronics applications.

Marine Electronics
Rt. 33, Box 160, Hartfield, VA 23071 USA
(800) 654-9251 (tel)
(804) 776-9500 (tech tel)
5760 Northampton Blvd., #110, Virginia Beach, VA 23455 USA
(800) 344-6388 (tel)
(804) 363-2002 (tel)
303B Second St., Annapolis, MD 20403 USA
Sells a variety of marine electronics, including marine radio transceivers, antennas, stereo head units, speakers, etc.

Marine Park Camera & Video
3126 Ave. U., Brooklyn, NY 11229 USA
(800) 448-8811 (tel)
(800) 360-1722 (order inquiry)
(718) 891-1878 (tel)
Visa, MasterCard, Discover, American Express, Diner's Club, Canon, COD
Sells high-end home, amateur, and professional audio and video equipment, such as camcorders, VCRs, video editors, speakers, TVs, home theater equipment, etc.

Markertek Video Supply
4 High St., Saugerties, NY 12477 USA
(800) 522-2025 (tel)
(914) 246-3036 (tel)
(914) 246-1757 (fax)
Sells a wide variety of equipment (including standards converters, camcorders, video editors, digital standards converters, etc.) for professional audio and video applications. Offers a free 174-page catalog.

Marketing by Design
P.O. Box 84, Port Murray, NJ 07865 USA
Sells a digital video stabilizer to eliminate video copyguards.

Market Point
916 Springdale Dr., Exton, PA 19341 USA
(610) 594-1880 (tel)
(610) 594-1881 (fax)
Buys, sells, and trades printers for a wide variety of computer applications.

Market Value Programs
19 Noxon Rd., Poughkeepsie, NY 12603 USA
(800) 434-3880 (tel)
Visa, MasterCard
Sells used IBM computer systems and parts.

Marktech International
5 Hemlock St., Latham, NY 12110 USA
(518) 786-6591 (tel)
(518) 786-6599 (fax)
Manufactures LED displays for a wide variety of electronics applications.

Mark V Electronics
8019 E. Slauson Ave., Montebello, CA 90640 USA
(800) 521-MARK (tel)
(800) 423-FIVE (tel)
(213) 888-8988 (tel)
(213) 888-6868 (fax)
Visa, MasterCard
Manufactures and sells a number of hi-fi amplifier kits. Offers a free catalog. Mininum order of $20.

Marrick
P.O. Box 950940, Lake Mary, FL 32795 USA
(407) 323-4467 (tel)
(407) 324-1291 (fax)
(407) 322-1429 (BBS)
Sells co-processor interface boards and other kits and boards.

Marshall
89 Frost St., Westbury, NY 11590 USA
2165 46th Ave., Lachine, PQ H8T 2P1 Canada
Manufactures amplifiers for professional and personal musical applications.

Marshall Industries
9320 Telstar Ave., El Monte, CA 91731 USA

Marshall Industries
(800) 877-9839 (tel)
(800) 833-9910 (tel)
http://www.marshall.com
Distributes industrial electronic components and production supplies, including semiconductors, connectors, computer systems and peripherals, etc.

Glen Martin Engineering
13620 Old Hwy. 40., Booneville, MO 65233 USA
(816) 882-2734 (tel)
Visa, MasterCard
Manufactures and sells aluminum and steel antenna towers for amateur radio applications.

Martin Logan
P.O. Box 707, Lawrence, KS 66044 USA
(913) 749-0133 (tel)
(913) 749-5320 (fax)
Manufactures loudspeakers for high-end home audio and home theater applications.

Marty's Emergency Products
P.O. Box 93, Baldwin, NY 11510 USA
(516) 378-4814 (tel)
Sells transceivers for a wide variety of two-way radio applications.

Marvel Communications
6000-D Old Hemphill Rd., Ft. Worth, TX 76134 USA
(817) 568-0177 (tel)
(817) 293-4441 (fax)
Sells mobile antennas for CB, marine, scanner, and cellular telephone applications.

Maryland Radio
8576 Laureldale Dr., Laurel, MD 20707 USA
(800) 447-7489 (tel)
(301) 725-1212 (Laurel, MD tel)
(301) 470-4266 (Washington tel)
(410) 880-4266 (Baltimore tel)
(301) 725-1198 (fax)
Sells a wide variety of amateur radio equipment and accessories, including amateur radio transceivers, receivers, antenna tuners, towers, antennas, radio teletype equipment, amplifiers, etc.

Marymac
22511 Katy Fwy., Katy, TX 77450 USA
(800) 231-3680 (tel)
(713) 392-0747 (tel)
(713) 574-4567 (fax)
Sells Radio Shack components and equipment at discount prices.

MasPar Computer
749 N. Mary Ave., Sunnyvale, CA 94086 USA
(800) 624-6427 (tel)
(408) 736-3300 (tel)
(408) 736-9560 (fax)
Manufactures computer systems for a wide variety of professional computer network applications.

Master Lease
1055 Westlake Dr., Berwyn, PA 19312 USA
(800) 767-5004 (tel)
Sells reposessed personal desktop computer equipment.

Alex Mastin
P.O. Box 8802, Canton, OH 44711 USA
(216) 456-8177 (tel)
Sells parts and accessories for a wide variety of telephone applications.

Matejka Cable Recovery
818 W. 3rd St., P.O. Box 1224, Winona, MN 55987 USA
(800) 831-2925 (tel)
Buys and sells scrap cable TV cables.

Matrix Portable Computer Services
13581 Pond Springs Rd., #315, Austin, TX 78729 USA
(800) 726-1503 (tel)
Sells parts for personal laptop computers.

Matsushita (Panasonic)
50 Meadowland Pkwy., Seacaucus, NJ 07094 USA
(201) 348-7000 (tel)
(201) 348-7527 (fax)
Manufactures head units, loudspeakers, power am-

plifiers, etc. for car stereo applications; camcorders, televisions, VCRs, etc. for home video applications; cassette decks, CD players, amplifiers, etc. for home audio applications, etc.

Maxcom
1309 SW 5th Ct., P.O. Box 502, Ft. Lauderdale, FL 33302 USA
(305) 527-5172 (tel)
(305) 523-6369 (tel)
(305) 522-8159 (fax)
Manufactures and sells antennas matchers and antennas for shortwave and amateur radio applications.

Maxell
22-08 Rt. 208, Fair Lawn, NJ 07410 USA
(800) 533-2836 (tel)
Manufactures a variety of analog and digital tape formats and recordable CDs for consumer and professional studio recordings.

Maxim Integrated Products
120 San Gabriel Dr., Sunnyvale, CA 94086 USA
(408) 737-7600 (tel)
Manufactures integrated circuits for a wide variety of electronics applications.

Maxim
120 San Gabriel Dr., Sunnyvale, CA 94086 USA
(800) 998-8800 (tel)
(408) 737-7600 (tel)
(408) 737-7194 (fax)
Visa, MasterCard
Manufactures semiconductors for a wide variety of electronics applications. Offers free interface design guides.

Maximum Strategy
801 Buckeye Ct., Milpitas, CA 95035 USA
(800) 352-1600 (tel)
(408) 383-1600 (tel)
(408) 383-1616 (fax)
Manufactures file servers for a wide variety of professional computer network applications.

Maxim Technology
3930 W. 29th St. S., #35, Wicthita, KS 67217 USA
(800) 755-1008 (tel)
(316) 941-9799 (cust serv tel)
(316) 941-9883 (fax)
Visa, MasterCard, Discover
Sells refurbished personal desktop and notebook computers, monitors, printers, etc.

Maximus
710 E. Cypress Ave. Unit A, Monrovia, CA 91016 USA
(800) 888-MAXI (tel)
(818) 305-5925 (tel)
(818) 357-9140 (fax)
Visa, MasterCard, Discover, American Express, COD
Manufactures and sells personal desktop and laptop computers. Offers a 30-day money-back guarantee.

Max Machinery
Healdsburg, CA 95448 USA
(707) 433-7281 (tel)
(707) 433-0571 (fax)
Manufactures flow meters for industrial applications.

Maxon Systems
10828 NW Air World Dr., Kansas City, MO 64153 USA
Manufactures and sells weather receivers, portable car alarms, and transceivers for GMRS, 49-MHz, and citizens band radio applications.

Maxpeed
1120 Chess Dr., Foster City, CA 94404 USA
(800) 877-7998 (tel)
(415) 345-5447 (tel)
(415) 345-6398 (fax)
Manufactures terminals for a wide variety of professional computer network applications.

Maxpoint Computers
20653 Lycoming St., Unit A1 & A2, Walnut, CA 91789 USA
(800) 484-5983 (tel)
(909) 468-2788 (tel)

Maxpoint Computers

(909) 468-3811 (fax)
Sells computer parts and boards, such as cases, power supplies, and multimedia speakers.

MAX System Antennas
4 Gerring Rd., Gloucester, MA 01930 USA
Manufactures and sells ground plane antennas for scanners.

Maxtech
P.O. Box 8086, New York, NY 10150 USA
(718) 547-8244 (tel)
American Express, COD
Sells transceivers, microphones, scanners, antennas, etc. for citizens band radio applications. Offers a catalog for $3.

Maxtek Components
P.O. Box 1480, Beaverton, OR 97075 USA
(503) 627-4133 (tel)
(503) 627-4651 (fax)
Manufactures CRT driver cards for a wide variety of personal computer applications.

Maxxima Marine
125 Cabot Ct., Hauppauge, NY 11788 USA
(516) 434-1200 (tel)
(516) 434-1457 (fax)
Manufactures amplifiers for car stereo applications.

MBI
507 Highview St., Newbury Park, CA 91320 USA
(805) 499-4993 (tel)
(805) 499-9142 (fax)
74667.1312@compuserve.com
Sells data storage systems, such as hard drives, CD-ROM drives, and tape drives, for a wide variety of computer applications.

MBM
3134 Industry Dr., N. Charleston, SC 29418 USA
(800) 223-2508 (tel)
Manufactures paper shredders for a wide variety of office applications.

MB Quart
25 Walpole Pk. S., Walpole, MA 02081 USA
(800) 962-7757 (tel)
(508) 668-8973 (tel)
Manufactures loudspeakers for high-end home and auto and audio and home theater applications.

MBS
7466 Early Dr., Mechanicsville, VA 23111 USA
(800) 944-3808 (tel)
Visa, MasterCard, Discover, American Express
Sells personal home and laptop computers; also sells parts and boards, such as monitors, scanners, terminals, video cards, modems, power supplies, CD-ROM drives, etc.

McCallie
P.O. Box 77, Brownsboro, AL 35741 USA
Sells electronic lightning sensors for home protection.

McClaran Sales
P.O. Box 2513, Vero Beach, FL 32961 USA
(407) 778-7584 (tel)
Sells towers and antennas for amateur radio applications.

McCormick Computer Resale
6950 W. 146th St., #108, Apple Valley, MN 55124 USA
(612) 891-2322 (tel)
(612) 891-2311 (fax)
Visa, MasterCard
Sells new and used personal desktop computers and parts and accessories, including memory chips, hard drives, monitors, etc.

MCI
P.O. Box 467279, Atlanta, GA 30346 USA
(800) 555-2772 (tel)
http://www.mci.com
Manufactures pagers for personal paging applications.

McIntosh
2 Chambers St., Binghampton, NY 13903 USA

(607) 723-3512 (tel)
(607) 724-0549 (fax)
20640 Bahama St., Chatsworth, CA 91311 USA
(818) 718-5787 (tel)
(818) 772-8387 (fax)
Manufactures loudspeakers, amplifiers, and tuners for high-end home audio, mobile audio, and home theater applications.

MCM Electronics
650 Congress Park Dr., Centervile, OH 45459 USA
(800) 543-4330 (tel)
(800) 824-8324 (tech tel)
Sells a wide variety of test equipment, communications products (including amateur radio equipment and accessories), consumer electronics repair components, and computer components and products. Offers a free 324-page catalog.

McMillan
P.O. Box 1340, Georgetown, TX 78627 USA
(800) 861-0231 (tel)
(512) 863-0231 (tel)
(512) 863-0671 (fax)
Manufactures flow sensors for professional industrial applications.

MC2
P.O. Box 1211, Mt. Laurel, NJ 08054 USA
(609) 231-1863 (tel)
(609) 722-0001 (fax)
Sells batteries for radio-control models.

MD Electronics
875 S. 72 St., Omaha, NE 68114 USA
(800) 624-1150 (tel)
Visa, MasterCard, Discover
Sells video converters and descramblers. Offers a free catalog.

MD & I Computers
9682 Telstar Ave., #102, El Monte, CA 91731 USA
Visa, MasterCard
Manufactures and sells personal desktop computers.

MDL
14940 NE 95th St., Redmond, WA 98052 USA
(800) 800-3766 (tel)
(206) 861-6767 (fax)
sales@mdlcorp.com
http://www.mdlcorp.com/mdlcorp/welcome.html
Sells SCSI controllers, EISA bus expansion boxes, etc. for professional computer applications.

MDM Radio
7112 W. Roosevelt Rd., Oak Park, IL 60304 USA
(708) 848-4210 (tel)
(708) 848-0230 (fax)
Visa, MasterCard
Sells pagers for personal cellular communications applications and VHF/UHF transceivers for commercial two-way radio applications.

M/D Totco Instrumentation
1200 Cypress Creek Rd., Cedar Park, TX 78613 USA
(512) 331-0411 (tel)
(512) 331-2219 (fax)
Manufactures load cells and load/force systems for professional industrial applications.

The Means of Production
(818) 988-8667 (tel)
(818) 988-8600 (fax)
TMOP5@aol.com
Sells new and used audio equipment for professional studio recording and live music performances.

Measurement & Control Products
16027 Brookhurst St., #G-224, Fountain Valley, CA 92708 USA
(714) 775-7991 (tel)
Manufactures battery gauges for radio-control model airplanes. Offers a 90-day money-back guarantee.

MEC Computer Express
50 Marcus Blvd., Hauppage, NY 11788 USA
(800) 632-6913 (tel)
(516) 435-1199 (tel)
(516) 435-2113 (fax)
Sells personal home and laptop computers and

MEC Computer Express

printers; also sells parts and boards, such as monitors, terminals, video capture cards, modems, CD-ROM drives, etc.

Mechem Electronics
P.O. Box 7846, Fredricksburg, VA 22404 USA
3605 Loren Whitney Dr., Massaponax Bus. Pk., Fredricksburg, VA 22408 USA
(703) 891-0569 (tel)
(703) 891-0538 (fax)
Visa, MasterCard
Sells base stations, repeaters, power cords, speaker microphones, battery chargers, test equipment, etc. for cellular microwave radio applications.

MECI
340 E. First St., Dayton, OH 45402 USA
(800) 344-4465 (tel)
(800) 344-6324 (fax)
(513) 461-3525 (tel)
(513) 461-3391 (fax)
http://www.meci.com
Visa, MasterCard, Discover, American Express
Sells surplus electronics, including parts, batteries, computer cables, wires, cellular phone accessories, wall transformers, etc. Mininum $20 order. Free 32-page catalog.

Media Factory
1930 Junction Ave., San Jose, CA 95131 USA
(800) 879-9536 (tel)
(408) 456-8848 (tel)
(408) 456-9298 (fax)
sales@mediafactoryinc.com
http://www.mediafactoryinc.com
Visa, MasterCard, Discover, American Express
Sells a variety of computer data storage media, including floppy diskettes, CD-Rs, QIC tapes, optical disks, removable cartridges, etc.

Media Logic ADL
4999 Pearl E. Circle, Boulder, CO 80301 USA
(303) 939-9780 (tel)
(303) 939-9745 (fax)
Manufactures data storage systems for a wide variety of professional computer network applications.

Media Management
4405 International Blvd., Norcross, GA 30093 USA
(800) 849-8978 (tel)
(404) 564-3299 (tel)
(404) 564-0221 (fax)
Sells computerized video editing software and hardware, including a professional video editing suite.

Media Mation
2461 W. 205th St., B100, Torrance, CA 90501 USA
(310) 320-0696 (tel)
(310) 320-0699 (fax)
Visa, MasterCard, Discover, American Express
Manufactures MIDI lighting dimmer for professional and personal audio and musical applications.

Media Source
2197 Canton Rd., Ste. 210, Marietta, GA 30066 USA
(800) 241-8857 (tel)
(770) 919-9228 (fax)
http://www.mediasource
Visa, MasterCard, Discover, American Express
Sells computer diskettes, data carts, optical disks, CD-ROMs, and accessories. Offers a 60-day money-back guarantee.

Mediastore
2238 N. Glassell Unit A., Orange, CA 92665 USA
(800) 555-5551 (tel)
(714) 997-5551 (tel)
(714) 997-5553 (fax)
http://www.mediastore.com
Sells a wide variety of computer data recording media, including optical cartridges, CD-ROMs, etc.

Medicine Man CB
P.O. Box 37, Clarksville, AR 72830 USA
Sells equipment for citizens band radio applications.

Megabytes
3070 Bristol Pike, Bensalem, PA 19020 USA
(800) 214-5455 (tel)
(215) 244-9392 (tech support tel)
(215) 244-8189 (inter order tel)
(215) 244-9056 (fax)

(215) 244-7785 (BBS)
Visa, MasterCard, Discover, American Express
Sells computer parts and boards, such as hard drives, motherboards, memory chips, cables, floppy drives, multimedia upgrades, modems, etc.

MegaBytes
1224 Executive Blvd., #101, Chesapeake, VA 23320 USA
(804) 548-3148 (tel)
Sells personal desktop computers and parts, such as hard drives, motherboards, CD-ROM drives, etc.

Megacomp
261 NE 1 St., #200, Miami, FL 33132 USA
(800) 946-2459 (tel)
Sells personal desktop computers; also sells computer parts and boards, such as floppy drives, hard drives, video cards, modems, memory chips, etc.

MegaDyne Communications
P.O. Box 1899, Auckland, New Zealand
0800-663-367 (tel)
09-379-0551 (tel)
09-373-2679 (fax)
sales@megadyne.co.nz
Sells modems, cellular telephone equipment, ISDN, etc. for personal and professional computer data communications applications.

Mega Electronics
21 S. Main St., Winter Garden, FL 34787 USA
(800) 676-6342 (tel)
Visa, MasterCard, COD
Sells cable TV descramblers and converters. Offers a free catalog.

Mega Haus Hard Drives
2201 Pine Dr., Dickinson, TX 77539 USA
(800) 786-1185 (tel)
(713) 534-3919 (tel)
(713) 534-2630 (tech support tel)
(800) 786-1192 (order inquiry tel)
(800) 786-1146 (corporate, etc. sales tel)
(713) 534-6580 (fax)
(800) 473-0972 (TTY)

Visa, MasterCard, Discover, American Express, COD
Sells computer hard drives, tape drives, CD-ROM drives, CD-ROM recorders, controllers, etc.

Megahertz
(801) 320-7777 (tech support tel)
(800) LAPTOPS (fax back)
techsupport@mhz.com
Manufactures modems for personal computer applications.

Mega Hertz
6940 S. Holly, #200, Englewood, CO 80112 USA
(800) 525-8366 (tel)
(303) 779-1717 (tel)
(303) 779-1749 (fax)
(800) 962-5966 (Atlanta tel)
(800) 821-6800 (St. Louis tel)
(800) 922-9200 (Florida tel)
Manufactures digital video noise reducers, modulators, processors, demodulators, etc. for professional video applications.

Megasound
Forchhammersvej 11, 1920 Frederiksberg C, Denmark
3325-8909 (tel)
Sells high-end home audio and video equipment, such as speakers, cassette decks, amplifiers, preamplifiers, etc.

Megatech
3070 Bristol Pike, Bensalem, PA 19020 USA
(800) 214-5455 (tel)
(215) 244-8189 (international order tel)
(215) 244-9392 (tech support tel)
(215) 244-9056 (fax)
(215) 244-7785 (BBS)
Visa, MasterCard, Discover, American Express
Sells computer parts and boards, such as floppy drives, hard drives, monitors, video cards, modems, motherboards, memory chips, keyboards, etc.

Megatel
125 Wendell Ave., Weston, ON M9N 3K9 Canada
(416) 245-2953 (tel)

Megatel

(416) 245-6505 (fax)
Manufactures miniature computer boards for professional and industrial computer applications.

Megatronics
165 Whitney Pl., Fremont, CA 94539 USA
(800) 898-6283 (tel)
Visa, MasterCard, American Express
Manufactures and sells joystick control systems for personal computer applications. Offers a 30-day money-back guarantee.

MEI
1100 Steelwood Rd., Columbus, OH 43212 USA
(800) 634-3478 (tel)
(614) 481-4417 (tel)
(614) 487-6417 (fax)
Visa, MasterCard, Discover
Sells computer parts and boards, such as tape drives, CD drives, backup power supplies, diskettes, keyboards, etc.

Melcher
187 Billerica Rd., Chelmsford, MA 01824 USA
(800) 828-9712 (tel)
(508) 256-4642 (tel)
Manufactures power regulators for a wide variety of electronics applications.

Melcor
1040 Spruce St., Trenton, NJ 08648 USA
(609) 393-4178 (tel)
(609) 393-9461 (fax)
Manufactures thermoelectrics for electro-optics applications.

MEM Electronics
3119 Burn Brae Dr., Dresher, PA 19025 USA
(215) 657-3119 (fax)
Sells discounted Radio Shack equipment for a wide variety of electronics applications.

Memocom Electronics
3rd Fl., 303, Fu Hsing N. Rd., Taipei, Taiwan
886-2-5475959 (tel)

886-2-5463199 (fax)
Manufactures uninterruptible power banks for a wide variety of computer and electronics applications.

Memory Conversion Products
428 NW 9th St., Corvallis, OR 97330 USA
(800) 809-1379 (tel)
(503) 753-7918 (tel)
(503) 758-5331 (fax)
Manufactures memory adapter boards for computer applications.

Memory and CPU Warehouse
8361 E. Evans Rd., #105, Scottsdale, AZ 85260 USA
(800) RAM-7091 (tel)
(602) 443-0696 (tel)
(602) 443-0918 (fax)
Visa, MasterCard, American Express
Sells memory and CPU chips for personal desktop computers.

Memory Depot
7460 Girard Ave., Ste. 4, La Jolla, CA 92037 USA
(800) 755-0586 (tel)
memdepot@cts.com
Buys, sells, and trades memory chips for personal computer applications.

Memory, Etc.
1751 E. Broadway Rd., #1, Tempe, AZ 85282 USA
(602) 784-2223 (tel)
(602) 784-2275 (fax)
Sells memory and CPU chips, motherboards, etc. for personal computer applications.

The Memory Exchange Wholesalers
7946 Ivanhoe Ave., #220, La Jolla, CA 92037 USA
(800) 501-2770 (tel)
(619) 454-9263 (tel)
(619) 454-9076 (fax)
Visa, MasterCard, Discover, American Express, COD
Sells memory chips for personal computer applications.

Memory Express
15140 Valley Blvd., City of Industry, CA 91744 USA
(800) 877-8188 (tel)
(818) 333-6389 (customer service tel)
(818) 369-1236 (fax)
Sells memory chips for a wide variety of computer applications.

Memory 4 Less
2622 W. Lincoln, #104, Anaheim, CA 92801 USA
(800) 821-3354 (tel)
(714) 826-5981 (tel)
(714) 821-3361 (fax)
Visa, MasterCard, Discover, American Express, COD
Sells printers and memory chips for home and laptop computers.

Memory Source
8204 NW 12th, Miami, FL 33126 USA
(800) 559-CHIP (tel)
Sells memory chips for a wide variety of computer applications.

Memory World
3392 Progess Dr., Units B & C, Bensalem, PA 19020 USA
(800) 272-9578 (tel)
(215) 244-7930 (tel)
(215) 244-7932 (fax)
Buys and sells memory chips for a wide variety of computer applications.

Memphis Amateur Electronics
1465 Wells Station Rd., Memphis, TN 38108 USA
(800) 238-6168 (tel)
(901) 682-7165 (fax)
Visa, MasterCard, COD
Sells a wide variety of amateur radio equipment and accessories, including amateur radio transceivers, receivers, antenna tuners, antennas, amplifiers, etc.

Meniscus
2575 28th St. Unit 2, Wyoming, MI 49509 USA
(616) 534-9121 (tel)
(616) 534-7676 (fax)
Sells loudspeakers, crossovers, inductors, capacitors, etc. for high-end home audio and professional audio applications.

M2 Enterprises
7560 N. Del Mar Ave., Fresno, CA 93711 USA
(209) 432-8873 (tel)
(209) 432-3059 (fax)
Manufactures and sells antennas for shortwave and VHF/UHF amateur radio and other two-way radio applications.

MentorPlus
P.O. Box 356, 22781 Airport Rd. NE, Aurora, OR 97002 USA
(800) 628-4640 (tel)
(503) 678-1431 (tel)
(503) 678-1480 (fax)
Manufactures GPS systems for professional aviation applications.

Merchant Data Systems
Rt. 5 Box 131, Willspoint, TX 75169 USA
(903) 873-3999 (tel)
(903) 873-3434 (fax)
Buys and sells printers, terminals, controllers, keyboards, etc. for a wide variety of computer applications.

Meridian America
3800 Camp Creek Pkwy., Bldg. 2400, #112, Atlanta, GA 30331 USA
(404) 344-7111 (tel)
Manufactures a digital sound processor for high-end home theater applications.

Merisel
200 Continental Blvd., El Segundo, CA 90245 USA
(800) 622-4786 (tel)
(310) 615-3080 (tel)
(310) 416-9600 (literature library tel)
Sells Sun Microsystems computer equipment.

MetaStor
1036 Elkton Dr., Colorado Springs, CO 80907 USA
(800) 86-ARRAY (tel)

MetaStor

saligent@salignet.com
Manufactures RAID and rack-mount data storage systems for professional computer applications.

Methode Electronics
7444 W. Wilson Ave., Chicago, IL 60656 USA
(800) 323-6858 (tel)
(708) 867-9130 (fax)
Manufactures fiberoptic connectors for a wide variety of electronics applications.

Metric Equipment Sales
351A Foster City Blvd., Foster City, CA 94404 USA
(800) 432-3424 (tel)
(415) 341-8874 (fax)
http://www.metricsales.com
metricsale@aol.com
metric1@ix.netcom.com
Rents, buys, and sells new and used test equipment, including oscilloscopes, spectrum analyzers, frequency counters, etc. Offers a six-month warranty and a five-day free trial.

Metrix Instruments
Chemin de la Croix-Rouge, B.P. 30, F74010 Annecy, Cedex, France
50-52-82-02 (tel)
50-51-70-93 (fax)
Manufactures DMMs for professional electronics testing applications.

Metrocom
10702 Fall Stone Rd., Houston, TX 77099 USA
(800) 364-8838 (tel)
(713) 783-8838 (tel)
(713) 783-8832 (fax)
metrocom@metrocominc.com
http://www.metrocominc.com
Visa, MasterCard
Buys and sells new and used modems, multiplexers, channel banks, etc. for a wide variety of computer applications.

Metronome Music
Sondermarksvej 16, 2500 Valby, Denmark
36-46-77-55 (tel)

Sells high-end audio amplifiers for home audio applications.

Metroplex Mobile Data
1640 W. Oakland Park Blvd., #300, Ft. Lauderdale, FL 33311 USA
(800) 328-2626 (tel)
Manufactures wireless mobile computing systems for professional two-way radio applications.

MetroWest
822 N. Spring Ave., LaGrange Park, IL 60525 USA
(708) 354-2125 (tel)
Manufactures and sells scanner chargers, antennas, cords, etc. for VHF/UHF scanners.

Metrum Peripheral Products
4800 E. Dry Creek Rd., Littleton, CO 80122 USA
(303) 773-4700 (tel)
(303) 773-4909 (fax)
Manufactures data storage systems for a wide variety of professional computer network applications.

Meyer Hill Lynch
352 Tomahawk Dr., Maumee, OH 43537 USA
(800) 343-1764 (tel)
(419) 897-9700 (tel)
(419) 897-9710 (fax)
Buys and sells new and refurbished UNISYS computer equipment.

MF Electronics
10 Commerce Dr., New Rochelle, NY 10801 USA
(914) 576-6570 (tel)
(914) 576-6204 (fax)
Manufactures miniature oscillator chips for a wide variety of computer applications.

MFJ
P.O. Box 494, Mississippi State, MS 39762 USA
(800) 647-1800 (tel)
(601) 323-5869 (tel)
(601) 323-6551 (fax)
76206.1763@compuserve.com
Manufactures and sells a number of shortwave and

amateur radio products, including antennas, DSP filters, antenna tuners, coax switches, dummy loads, digital decoders, speaker/microphones, SWR meters, clocks, iambic paddles, etc.

MGC Technologies
16605 Gale Ave., City of Industry, CA 91745 USA
(818) 968-6798 (tel)
Manufactures monitors for personal desktop computer systems.

M & G Electronics
2 Aborn St., Providence, RI 02903 USA
(800) 886-8699 (tel)
Visa, MasterCard, Discover, COD
Offers project kits to descramble scrambled satellite television broadcasts. Descramblers are illegal for personal or business use, so these kits are sold for educational or entertainment use only.

MGR Industries
450B Industrial Dr., Ft. Collins, CO 80524 USA
(970) 221-2201 (tel)
(970) 484-4078 (fax)
Manufactures keypads for a wide variety of electronics applications.

MHz Electronics
3802 N. 27th Ave., Phoenix, AZ 85017 USA
(602) 278-4062 (tel)
(602) 269-1737 (fax)
Buys and sells test equipment for a wide variety of electronics applications.

Miami Radio Center
5590 W. Flagler St., Miami, FL 33134 USA
(305) 264-8406 (tel)
Sells scanners and amateur and citizens band radio equipment and accessories, including amateur radio transceivers, receivers, antenna tuners, antennas, amplifiers, etc.

Mian Data Systems
2350 Centerline Industrial Dr., St. Louis, MO 63146 USA
(800) 645-2296 (tel)
(314) 432-5900 (tel)
(314) 432-7742 (fax)
Manufactures file servers for a wide variety of professional computer network applications.

Mic Heaven
1032 Washington St., Dept. 5., Hoboken, NJ 07030 USA
(201) 656-3936 (tel)
(201) 963-4764 (fax)
Sells tube, condenser, and vintage microphones for professional studio recording and live music performances.

Michigan Radio
23040 Schoenherr, Warren, MI 48089 USA
(800) TRU-HAMM (tel)
(810) 771-4711 (tel)
(810) 771-4712 (serv tel)
(810) 771-6546 (fax)
Visa, MasterCard, Discover, American Express, COD
Sells a wide variety of amateur radio equipment and accessories, including amateur radio transceivers, receivers, antenna tuners, towers, antennas, radio teletype equipment, amplifiers, etc.

Micom
2353 Eastman Ave., Bldg. 1 & 2, Ventura, CA 93003 USA
(805) 658-0800 (tel)
(805) 658-0822 (fax)
Visa, MasterCard
Sells new and refurbished personal desktop computer equipment.

Micom
4100 Los Angeles Ave., Simi Valley, CA 93063 USA
(800) 642-6687 (tel)
(805) 583-8600 (tel)
(800) 343-0329 (fax)
Manufactures routers for professional computer network applications.

Micrel Semiconductor
1849 Fortune Dr., San Jose, CA 95131 USA

Micrel Semiconductor

(408) 944-0800 (tel)
(408) 944-0970 (fax)
Manufactures integrated circuits for a wide variety of electronics applications.

Micro-Assist

145 Hudson St., 4th Fl., New York, NY 10013 USA
(800) 390-7428 (tel)
(212) 343-7953 (customer service tel)
(212) 274-9193 (fax)
Sells computer parts and boards, such as motherboards, hard drives, multimedia kits, etc.

Microbus

10849 Kinghurst, #105, Houston, TX 77099 USA
(800) 688-4405 (tel)
(713) 568-4744 (tel)
(713) 568-4604 (fax)
Manufactures single-board computers for professional and industrial computer applications.

MicroCache

(713) 946-1221 (tel)
(713) 946-2429 (fax)
Sells computer parts and boards, such as hard drives, memory chips, floppy drives, cooling fans, etc.

Micro City

17707 Valley View Ave., Cerritos, CA 90703 USA
(800) 567-APOG (tel)
gst@primenet.com
http://www.primenet.com/_7Egst/
Manufactures and sells personal desktop computers; also sells computer printers and scanners.

MicroCom Industries

(800) 665-2002 (tel)
Visa, MasterCard, American Express, COD
Sells cable TV converters and descramblers. Offers a free brochure.

Micro Computer Center

4406 E. Independence Blvd., Charlotte, NC 28205 USA
(704) 532-0206 (tel)
5219 Buford Hwy., Atlanta, GA 30340 USA
(404) 454-0371 (tel)
Manufactures and sells personal desktop computers; also sells computer parts and boards, such as floppy drives, hard drives, monitors, motherboards, modems, cases, etc.

Micro Control Specialties

23 Elm Park, Groveland, MA 01834 USA
(508) 372-3442 (tel)
(508) 373-7304 (fax)
Manufactures repeaters for amateur radio applications.

Microcraft

P.O. Box 513, Thiensville, WI 53092 USA
(414) 241-8144 (tel)
Visa, MasterCard
Manufactures and sells several different Morse code and radioteletype readers.

Micro Design International

6985 University Blvd., Winter Park, FL 32792 USA
(800) 920-3316 (tel)
(407) 677-8333 (tel)
(407) 677-8365 (fax)
Aztec Centre, Aztec West, Almondsbury, Bristol BS1 24TD UK
+44-1-454-614182 (tel)
+44-1-454-618206 (fax)
Manufactures data storage systems for a wide variety of professional computer network applications.

Microelectronics Technology

130 Rose Orchard Way, San Jose, CA 95134 USA
(408) 954-1818 (tel)
(408) 954-0908 (fax)
No. 1 Innovation Rd. II, Hsinchu Science Based Industrial Park, Hsinchu 300, Taiwan
886-35-779-855 (tel)
886-35-773-335 (tel)
886-35-777-121 (fax)
Imperial House, 14 Rue Dicks, L-1417 Luxembourg
352-408-583 (tel)
352-408-584 (tel)
352-408-582 (fax)

Manufactures television and CATV links for professional broadcasting applications.

Microenergy
745 State Rd. 434 W., Longwood, FL 32750 USA
(407) 831-2000 (tel)
Manufactures power supplies for a wide variety of electronics applications.

Micro 2000 Europe
P.O. Box 2000, Letchworth, Herts SG6 1UT UK
+44-462-483483 (tel)
+44-462-481484 (fax)
Manufactures and sells toolkits for computer diagnostic applications.

Micro Exchange
682 Passaic Ave., Nutley, NJ 07110 USA
(800) 478-8723 (tel)
(214) 553-6770 (tel)
(201) 284-1200 (tel)
microexch@aol.com
Visa, MasterCard
Sells used personal desktop computers. Advertises more than 100,000 parts and 5,000 systems/peripherals ready to ship.

The Micro Group
333-C W. State Rd., Island Lake, IL 60042 USA
(800) 560-1195 (tel)
(708) 487-4600 (tech support tel)
(708) 487-4637 (fax)
Visa, MasterCard, American Express
Sells computer parts and boards, such as tape drives, hard drives, monitors, video cards, modems, motherboards, memory chips, keyboards, etc. Offers a 30-day return policy.

Microintel
20814 Aurora Rd., Cleveland, OH 44146 USA
(800) 739-9137 (tel)
(216) 662-1910 (tel)
(216) 662-4761 (fax)
Visa, MasterCard, Discover, American Express
Manufactures and sells personal desktop computers; also sells computer parts and boards, such as floppy drives, hard drives, monitors, video cards, modems, keyboards, etc.

Micro International
10850 Seaboard Loop, Houston, TX 77099 USA
(800) 967-5667 (tel)
(713) 495-7791 (fax)
Manufactures and sells laptop computers.

Micro Linear
2092 Concourse Dr., San Jose, CA 95131 USA
(408) 433-5200 (tel)
Manufactures integrated circuits for a wide variety of electronics applications.

MicroLogic Systems
3707 Westway Dr., Tyler, TX 75703 USA
(903) 561-0007 (tel)
(903) 561-4974 (fax)
Sells printers and parts for a wide variety of computer applications.

Micro Mart
159 State Rd., Buzzards Bay, MA 02532 USA
(508) 833-2225 (tel)
(508) 888-2157 (fax)
Sells IBM personal desktop computers and parts.

Micro Media Computers
6026 State Rd., #B, Parma, OH 44134 USA
(800) 645-7670 (tel)
(216) 885-1421 (customer service tel)
(216) 885-1422 (fax)
Visa, MasterCard, Discover, COD
Sells computer parts and boards, such as hard drives, motherboards, memory chips, controllers, sound cards, etc.

Micro Memory Bank
60 James Way, Unit 8, Southampton, PA 18966 USA
(215) 396-1000 (tel)
(215) 396-1001 (fax)
Buys and sells memory chips for a wide variety of computer applications.

MicroModule Systems
10500-A Ridgeview Ct., Cupertino, CA 95014 USA
(800) 297-8849 (tel)
(408) 864-5950 (fax)
moreinfo_c@mms.com
Manufactures SRAM boards for a wide variety of computer applications.

Micron Electronics
915 E. Karcher Rd., Nampa, ID 83687 USA
Visa, MasterCard, American Express
(800) 347-3490 (tel)
(208) 465-1740 (fax)
(800) 708-1758 (Canada tel)
(208) 465-8970 (International tel)
(208) 465-8993 (International fax)
95-800-708-1755 (Mexico tel)
(800) 708-1756 (Puerto Rico tel)
0800-22-000-3 (New Zealand tel)
009-489-2890 (New Zealand tel)
Technology Division
8000 S. Federal Way, Boise, ID 83707 USA
(800) 932-4992 (tel)
(208) 368-5800 (data fax)
(208) 368-3342 (fax)
prodmktg@micron.com
http://www.micro.com
Manufactures and sells complete home computer systems. Also manufactures and sells memory chips with a 30-day money-back guarantee.

Micropace
P.O. Box 648, Northville, MI 48167 USA
(810) 347-7654 (tel/fax)
Visa, MasterCard
Manufactures and sells microprocessor-controlled NiCd systems for radio-control model airplanes. Offers a 30-day money-back guarantee.

Micro Parts & Supplies
308 Cary Point Dr., Cary, IL 60013 USA
(800) 336-2131 (tel)
Sells computer parts for a wide variety of computer applications.

The Microphone Company
5805 Chimney Rock Rd., Houston, TX 77081 USA
(800) 346-3642 (tel)
Sells a variety of high-level audio equipment for audio enthusiasts and for professional studio recording.

Microplex Systems
8525 Commerce Ct., Burnaby, BC V5A 4N3 Canada
(800) 665-7798 (tel)
(604) 444-4232 (tel)
(604) 444-4239 (fax)
Manufactures print servers for professional computer network applications.

Micro Pneumatic Logic
2890 NW 62nd St., Ft. Lauderdale, FL 33309 USA
(305) 973-6166 (tel)
(305) 973-6339 (fax)
Manufactures miniature switches and sensors for professional industrial applications. Offers a free data package.

Micropolis
21211 Nordhoff St., Chatsworth, CA 91311 USA
(800) 395-3748 (tel)
(818) 718-5264 (tel)
Manufactures computer hard disk drives and arrays.

Micro Pro
5400 Brookpark Rd., Cleveland, OH 44129 USA
(800) 353-3003 (tel)
(800) 809-6483 (tech support tel)
(216) 661-2454 (fax)
Visa, MasterCard, Discover, American Express, COD
Manufactures and sells personal desktop computers; also sells computer parts and boards, such as floppy drives, hard drives, monitors, motherboards, modems, keyboards, etc. Offers a 30-day money-back guarantee.

Microprocessors Unlimited
24000 S. Peoria Ave., Beggs, OK 74421 USA
(918) 267-4961 (tel)
(918) 267-3879 (fax)
Visa, MasterCard, Discover
Sells memory chips for a wide variety of computer applications.

MicroProfessionals
19261 Burnham, #100, Lansing, IL 60438 USA
(800) 800-8300 (tel)
(800) 800-8308 (tech support tel)
(703) 339-8398 (tel)
(708) 339-8333 (fax)
Visa, MasterCard, Discover, American Express
Manufactures and sells personal desktop and laptop computers. Offers a 30-day money-back guarantee.

MicroSat
4343 Shallowford Rd., #F3, Marietta, GA 30062 USA
(800) 438-0812 (tel)
(404) 643-0899 (fax)
Sells equipment for professional cable TV applications.

Micro Sense
370 Andrew Ave., Leucadia, CA 92024 USA
(800) 544-4252 (tel)
(619) 632-8621 (tel)
(619) 752-6133 (fax)
Visa, MasterCard
Sells laptop hard drives for a wide variety of computer applications.

MicroSolutions
132 W. Lincoln Hwy., DeKalb, IL 60115 USA
(800) 295-1214 (tel)
(815) 756-3411 (tel)
(815) 756-2928 (fax)
Manufactures CD-ROM drives for personal computer multimedia applications.

Microsource
801 Clanton Rd., #108, Charlotte, NC 28217 USA
(800) 796-6590 (tel)
(704) 529-1031 (tel)
(704) 529-1899 (tech support tel)
(704) 529-1035 (fax)
questions@msi-online.com
http://www.msi-online.com
Manufactures and sells personal desktop computers; also sells computer parts and boards, such as floppy drives, hard drives, monitors, motherboards, modems, cases, etc.

Microstar Laboratories
2265 116th Ave. NE, Bellevue, WA 98004 USA
(206) 453-2345 (tel)
(206) 453-3199 (fax)
info@mstarlabs.com
http://www.mstarlabs.com/mstarlabs/
Manufactures data-acquisition personal computer boards, etc. for professional industrial applications.

Micro/sys
3447 Ocenview Blvd., Glendale, CA 91208 USA
(818) 244-4600 (tel)
(818) 244-4246 (fax)
Manufactures single-board computers for professional and industrial computer applications. Offers a free 284-page catalog.

MicroTech
7304 15th Ave. NE, Seattle, WA 98115 USA
(800) 521-9035 (tel)
(206) 526-7989 (service tel)
(206) 522-6727 (fax)
Visa, MasterCard
Sells memory chips, motherboards, cases, etc. for personal computer applications.

Micro Technologies International
117 E. Main St., Round Rock, TX 78664 USA
(800) 288-1487 (tel)
(512) 244-1838 (fax)
http://www.mticom.com
Sells terminals for professional computer applications.

Micro Technology
P.O. Box 21061, Raleigh, NC 27619 USA
(919) 870-0344 (tel)
(919) 870-7163 (fax)
Manufactures digital audio workstations for professional studio recording.

Microtek
3715 Doolittle Dr., Redondo Beach, CA 90278 USA
(800) 654-4160 (tel)
(310) 297-5050 (fax)
(310) 297-5101 (fax back)

Microtek

http://www.mteklab.com
Manufactures image scanners for a wide variety of personal and professional desktop computer applications.

Microtek
RR#3 Box 4361, Bumpass, VA 23024 USA
(540) 872-7020 (tel)
Manufactures microcontrollers for Ramsey amateur radio kits.

Microtest
4747 N. 22nd St., Phoenix, AZ 85016 USA
(800) 526-9675 (tel)
(602) 952-6400 (tel)
(602) 952-6401 (fax)
2A Kingfisher House, Northwood Park, Gatwick Rd., Crawley, W. Sussex, RH10 2XN UK
44-1293-894000 (tel)
44-1293-894008 (fax)
Manufactures CD servers for professional computer network applications.

Microtext Systems
P.O. Box 4020-116, Mississauga, ON L5A 3A4 Canada
(905) 276-1257 (tel)
(905) 276-7236 (fax)
Manufactures tape controller boards for a wide variety of professional computer network applications.

Micro Thinc
P.O. Box 63/6025, Nargate, FL 33063 USA
(305) 752-9202 (tel)
Sells TV notch filters. Offers a free brochure.

Micro Time
172 Madison Ave., #303, New York, NY 10016 USA
(212) 725-5542 (tel)
(212) 725-5549 (fax)
Sells personal desktop and laptop computers.

MicroTools
P.O. Box 624, 714 Hopmeadow St., #14, Simsbury, CT 06070 USA

(800) 651-6170 (tel)
(203) 651-0019 (fax)
Manufactures intermittent detectors for computer testing applications.

Microtouch Systems
300 Griffin Brook Park Dr., Methuen, MA 01844 USA
(508) 659-9000 (tel)
Manufactures touch screens for a wide variety of personal computer applications.

Micro Trade
1666 Garnet Ave., #103, San Diego, CA 92109 USA
(800) 934-1701 (tel)
(619) 483-1703 (fax)
Visa, MasterCard, Discover, American Express
Buys and sells memory chips, hard drives, CPUs, motherboards, etc. for a wide variety of computer applications.

Microwave Data Systems
175 Science Pkwy., Rochester, NY 14620 USA
(716) 442-4000 (tel)
(716) 242-9600 (tel)
(716) 242-9620 (fax)
sales@mdsroc.com
http://www.mdsroc.com
Manufactures point-to-point radio links for microwave radio applications.

Microwave Filter
6743 Kinne St., E. Syracuse, NY 13057 USA
(800) 448-1666 (tel)
(315) 437-3953 (tel)
(315) 463-1467 (fax)
mfc@ras.com
http://www.ras.com/mwfilter/mwfilter.htm
Manufactures bandpass filters, notch filters, pay TV traps, etc. for professional cable TV applications.

Microwave Networks
10795 Rockley Rd., Houston, TX 77099 USA
(713) 495-7123 (tel)
(713) 879-4728 (fax)
525-273-1629 (Mexico tel)
525-515-6333 (Mexico fax)

65-538-8583 (Singapore tel)
65-538-5636 (Singapore fax)
Manufactures networks for professional cellular microwave radio applications.

MicroXperts

6230 Cochran Rd., Solon, OH 44139 USA
(800) 73MICRO (tel)
(800) 214-4029 (home computers tel)
(800) 214-4031 (business computers tel)
(800) 215-5213 (repeat customer tel)
(800) 215-5214 (government and education tel)
(800) 557-TECH (tech support tel)
(216) 498-3330 (tel)
(216) 498-3522 (tech tel)
(216) 498-3520 (sales tel)
(216) 498-3521 (BBS)
Spanish/Italian/French, ask for X230
Russian, ask for X232
German, ask for X245
ktavalli@salamander.net
http://www.microx.com
Visa, MasterCard, Discover
Manufactures and sells personal desktop computers and other computer components, such as motherboards, hard drives, cases, CPUs, monitors, etc. Offers a 30-day money-back guarantee.

Micro X-Press

5646-48 W. 73rd. St., Indianapolis, IN 46278 USA
(800) 875-9737 (tel)
(317) 328-5780 (tel)
(317) 328-5787 (tech support tel)
(317) 328-5782 (fax)
Manufactures and sells personal desktop computers; also sells computer parts and boards, such as floppy drives, hard drives, monitors, video cards, modems, keyboards, etc. Offers a 30-day money-back guarantee.

Midian Electronics

2302 E. 22nd St., Tucson, AZ 85713 USA
(800) MIDIANS (tel)
(602) 884-7981 (tel)
(602) 884-0422 (fax)
Manufactures voice scrambler boards for a variety of two-way radio and telephone applications.

The Midi Station

(914) 534-7683 (tel)
(914) 534-9220 (fax)
smidijack@aol.com
Sells a variety of professional music technology products for professional recording studios and live music performances.

Midland ComputerMart

8268 Lehigh Ave., Morton Grove, IL 60053 USA
(800) 407-0700 (tel)
(708) 967-0700 (tel)
(708) 967-0760 (customer service tel)
(708) 967-0703 (tech support tel)
(708) 967-0710 (fax)
Visa, MasterCard, Discover
Sells personal desktop and laptop computers, printers, and scanners; also sells computer parts and boards, such as floppy drives, hard drives, monitors, motherboards, video cards, modems, keyboards, etc.

Midland International

1690 North Topping, Kansas City, MO 64120 USA
(800) 669-4567 (tel)
(800) MIDLAND (tel)
(816) 241-8500 (tel)
(816) 920-1144 (fax)
(905) 420-4102 (Canada tel)
(905) 420-5848 (Canada fax)
(800) 561-5951 (Ontario, Canada tel)
(800) 561-6190 (Quebec, Canada tel)
(905) 839-7411 (Ontario, Canada fax)
(514) 923-2003 (Quebec, Canada fax)
Manufactures commercial two-way radio communications and transceivers for citizens band, amateur, and VHF marine radio applications.

Midwest Cable Services

P.O. Box 96, Argos, IN 46501 USA
(219) 892-5537 (tel)
(219) 892-5624 (fax)
Buys and sells scrap TV cable and used line equipment.

Midwest Computer Brokers

648 W. Mulberry St., W. Baraboo, WI 53913 USA

Midwest Computer Brokers
(608) 356-9604 (tel)
(608) 356-9607 (fax)
Visa, Discover
Sells personal laptop computers, printers, terminals, and monitors for a wide variety of computer applications.

Midwest Computer Electronics
180 Lexington Dr., Buffalo Grove, IL 60089 USA
(800) 869-6757 (tel)
(708) 459-9410 (customer service tel)
(708) 459-6933 (fax)
Visa, MasterCard, Discover, COD
Sells personal desktop and laptop computers, printers, and scanners; also sells computer parts and boards, such as floppy drives, hard drives, monitors, motherboards, video cards, modems, keyboards, etc. More than 30,000 different products.

Midwest Electronics
(800) 648-3030 (tel)
Sells cable TV converters and descramblers. Offers a free catalog.

Midwest Electronics
124 12th Ave. S., Minneapolis, MN 55415 USA
(800) 895-0336 (tel)
(612) 339-9533 (tel)
(612) 305-0964 (fax)
Sells laptop computers; also sells computer parts and boards, such as floppy drives, hard drives, monitors, motherboards, modems, cases, etc.

Midwestern Diskette
509 W. Taylor, Creston, IA 50801 USA
(800) 221-6332 (tel)
(515) 782-4166 (fax)
Visa, MasterCard, COD ($4.50)
Sells professional computer diskette equipment, including diskette labelers, duplication equipment, blank diskettes, etc.

Midwest MAC
P.O. Box 22532, Minneapolis, MN 55422 USA
(612) 529-4829 (tel)
Sells Apple and MacIntosh parts for a wide variety of personal computer applications.

Midwest Micro
6910 U.S. Rt. 36 E., Fletcher, OH 45326 USA
(800) 235-8940 (tel)
(800) 262-6622 (tech support tel)
(800) 728-8584 (customer service tel)
(800) 883-1461 (dealers tel)
(800) 562-6622 (fax)
(800) 445-2025 (info fax)
Visa, MasterCard, Discover
Manufactures and sells personal desktop and laptop computers, printers, and scanners; also sells computer parts and boards, such as floppy drives, hard drives, monitors, processor upgrades, video cards, modems, keyboards, etc. Offers a 30-day money-back guarantee and a free catalog

Midwest Micro-Tek
2308 E. 6th St., Brookings, SD 57006 USA
(605) 697-8521 (tel)
(605) 692-5112 (fax)
Manufactures embedded controller boards for a wide variety of computer applications.

MIE Systems
451 Johannah Pl., Lilburn, GA 30247 USA
(404) 921-6167 (tel)
Manufactures PC video encoder boards for a wide variety of computer applications.

Migration Solutions
1274 Eagan Industrial Rd., Eagan, MN 55121 USA
(612) 888-8099 (tel)
(612) 888-8188 (fax)
Manufactures file servers for a wide variety of professional computer network applications.

Mike's Electronics
1001 NW 52nd St., Ft. Lauderdale, FL 33309 USA
(305) 491-7110 (tel)
(305) 491-7011 (fax)
Sells amateur radio equipment and accessories, including amateur radio transceivers, receivers, antenna tuners, antennas, amplifiers, etc.

Milam Audio
1470 Valle Vista Blvd., Pekin, IL 61554 USA

(800) 334-8187 (tel)
(309) 346-3161 (tel)
(309) 346-6431 (fax)
Sells audio equipment, including mixing consoles, DAT decks, microphones, multitrack recorders, monitors, amplifiers, etc., for professional studio recording and live music performances.

Milan Technology
894 Ross Dr., #101, Sunnyvale, CA 94089 USA
(408) 752-2770 (tel)
Manufactures modular hubs for professional computer network applications.

Milbert
18 Warrior Brook, Germantown, MD 20874 USA
(301) 963-9355 (tel)
(301) 428-0831 (fax)
Manufactures amplifiers for car stereo applications.

Milford Instruments
+44-01977-683665 (tel)
+44-01977-681465 (fax)
Manufactures miniature computer boards for a wide variety of electronics applications.

Miller & Kreisel Sound
10391 Jefferson Blvd., Culver City, CA 90232 USA
(310) 204-2854 (tel)
(310) 202-8782 (fax)
Manufactures loudspeakers for high-end home audio and home theater applications.

Millimeter Wave Technology
1395 S. Marietta Pkwy., Bldg. 800, #104, Marietta, GA 30067 USA
Manufactures microwave reflectometers for professional aviation applications.

Milo Associates
5736 N. Michigan Rd., Indianapolis, IN 46208 USA
(317) 257-6811 (tel)
(317) 257-1590 (fax)
Sells test equipment and electronic components, including semiconductors, electron tubes, transformers, batteries, relays, etc. for a wide variety of electronics applications.

Miltronix
P.O. Box 80041, Toledo, OH 43608 USA
Sells entirely refurbished military R-390A shortwave receivers and parts. Offers information for a SASE.

Milwaukee Resistor
8920 W. Heather Ave., Milwaukee, WI 53224 USA
(414) 362-8900 (tel)
(414) 362-9876 (fax)
Manufactures resistors for a wide variety of electronics applications.

Minco Products
7300 Commerce Ln., Minneapolis, MN 55432 USA
(612) 571-3121 (tel)
(612) 571-0927 (fax)
Manufactures temperature probes for professional industrial applications.

Mini-Circuits
P.O. Box 350166, Brooklyn, NY 11235 USA
(718) 934-4500 (tel)
(718) 332-4661 (fax)
Manufactures laser-sealed miniature voltage-controlled oscillators.

Minnesota Computers
2475 Xenium Ln. N., Minneapolis, MN 55441 USA
(612) 544-7900 (tel)
(612) 544-8031 (fax)
Sells new and used IBM and Sun computer systems.

Minnesota Western Visual Presentation Systems
5828 Vallejo St., Oakland, CA 94608 USA
(800) 999-8590 (tel)
Manufactures LCD projection panels for computer multimedia applications.

Minolta
101 Williams Dr., Ramsey NJ 07446 USA
(800) 9-MINOLTA (tel)

Minolta

(201) 825-4000 (tel)
Manufactures cameras, fax machines, digital systems, imagers, camcorders, photocopiers, etc. for a wide variety of photographic, office, video, etc. applications.

Minuteman

1455 LeMay Dr., Carollton, TX 75007 USA
(800) 238-7272 (tel)
(214) 446-7363 (tel)
(214) 446-9011 (fax)
Manufactures uninterruptible power supplies and power-line filters for a wide variety of personal computer applications.

Mirage

3641 McNicoll Ave., Scarborough, ON M1X 1G4 Canada
(416) 231-1800 (tel)
(416) 321-1500 (fax)
Manufactures high-end speakers for audio and home theater applications.

Mirage Communications

P.O. Box 1000, Morgan Hill, CA 95038 USA
(408) 779-7363 (tel)
(408) 779-8845 (fax)
Manufactures amplifiers and antennas for amateur radio applications.

Mirage Communications Equipment

300 Industrial Park Rd., Starkville, MS 39759 USA
(800) 647-1800 (tel)
(601) 323-8287 (tel)
(601) 323-6551 (fax)
Manufactures and sells VHF, UHF, repeater, and amateur TV amplifiers and accessories for amateur radio applications. Offers a free catalog.

Miro Computer Products

955 Commercial St., Palo Alto, CA 94303 USA
(800) 249-6476 (tel)
(415) 855-0955 (tel)
http://www.miro.com
Manufactures video capture boards for personal computer multimedia applications.

MIS

45395 Northport Loop W., Fremont, CA 94538 USA
(800) 733-9188 (tel)
(510) 226-0230 (fax)
Manufactures and sells personal desktop computers.

Mis-Cyn Electronics

7211 Olde Salem Cr., Hanmore than Park, IL 60103 USA
(800) 487-4549 (tel)
(708) 837-6052 (fax)
Visa, MasterCard
Sells switches, LEDs, and power supplies for a wide variety of electronics applications.

Mitra

20504 Earlgate St., Walnut, CA 91789 USA
(800) 324-1441 (tel)
(909) 468-2891 (fax)
Visa, MasterCard, Discover, American Express
Manufactures and sells personal desktop and laptop computers.

Mitsubishi Cable America

520 Madison Ave., New York, NY 10022 USA
(800) 262-6200 (tel)
(212) 888-2270 (tel)
(212) 888-2276 (fax)
Manufactures LEDs for a wide variety of electronics applications.

Mitsubishi Electronics America

6100 Atlantic Blvd., Norcross, GA 30071 USA
5757 Plaza Dr., P.O. Box 6007, Cypress, CA 90630 USA
(800) 828-6372 (tel)
(800) 553-7278 (tel)
(800) 843-2515 (Display Products tel)
(800) 937-0000 X435 (tel)
(800) 387-9630 (Canada Display Products tel)
(714) 220-2500 (tel)
(800) 937-2094 (tel)
(800) 825-6655 (fax)
(714) 220-4792 (fax)
Semiconductor Division
1050 E. Arques Ave., Sunnyvale, CA 94086 USA

(408) 730-5900 (tel)
http://www.directnet.com/mitsubishi
Visa, MasterCard
Manufactures VCRs for home, amateur, and professional video editing, recording, and viewing applications; manufactures integrated circuits for a wide variety of electronics applications.

Mitsumi Electronics
(800) 648-7864 (tel)
(415) 691-4465 (tech support tel)
(415) 691-4469 (BBS)
Manufactures and sells peripherals (such as floppy disk drives, keyboards, mice, and CD-ROM drives) for personal desktop computers.

Mizar
2410 Luna Rd., Carrollton, TX 75006 USA
(800) 635-0200 (tel)
(214) 277-4600 (tel)
(214) 277-4666 (fax)
Manufactures DSP computer boards for professional and industrial computer applications.

M & K
10391 Jefferson Blvd., Culver City, CA 90232 USA
(310) 204-2854 (tel)
Mobile Expo Tel
(800) 228-8972 (tel)
Sells pager and telephone accessories for cellular telephone applications.

M2 Electronics
3526 Jasmine, #44, Los Angeles, CA 90034 USA
(310) 837-7818 (tel)
(310) 841-6050 (BBS/fax)
loving@cs.ucla.edu
http://www.cs.ucla.edu/csd-grads-gs3/loving/
www/m2l.html
Manufactures EEPROM programmers for a wide variety of electronics applications.

MMATS Professional Audio
863 W. 13th Ct., Riviera Beach, FL 33404 USA
(800) 767-0034 (tel)
Manufactures amplifiers and loudspeakers for mobile audio applications.

Mobile Authority
1900 W. Artesia Blvd., Compton, CA 90220 USA
(310) 223-0400 (tel)
(310) 223-0412 (fax)
Manufactures head units, amplifiers, equalizers, and loudspeakers for mobile audio applications.

Mobile Mark Communications Antennas
3900B River Rd., Schiller Park, IL 60176 USA
(800) 648-2800 (tel)
(708) 671-6690 (tel)
(708) 671-6715 (fax)
Manufactures mobile and base antennas for a wide variety of radio bands.

Mobilex
120 Constitution Dr., Menlo Park, CA 94025 USA
(800) 269-7678 (tel)
(415) 462-4510 (tel)
(415) 462-1051 (fax)
(800) 746-6245 (fax)
http://www.mobilex.com
Manufactures antennas, battery chargers, power cords, etc. for cellular telephone applications.

Mobius Computer
5627 Stoneridge Dr., Bldg. 312, Pleasonton, CA 94588 USA
(800) MOBIUS-1 (tel)
(510) 460-5252 (tel)
(510) 460-5249 (fax)
Manufactures workstations for a wide variety of professional computer network applications.

Model Control Devices
P.O. Box 173, Bobcaygeon, ON K0M 1A0 Canada
(705) 738-1335 (tel/fax)
Manufactures switching systems, signal monitor safety systems, and switch accessory control systems for radio-control model airplanes and boats.

Model Rectifier
200 Carter Dr., Edison, NJ 08817 USA
Manufactures digital sound-switch control panels for model railroading systems.

Modern Electronics
2125 S. 156th Circle, Omaha, NE 68130 USA
(800) 906-6664 (tel)
Visa, MasterCard, Discover, American Express, COD
Sells cable TV converters, descramblers, video stabilizers, filters, etc. Offers a money-back guarantee, a 30-day trial, and a free catalog.

Modern Gold Enterprises
Block J, 16/F, Gold King Industrial Bldg., 35-41, Tai Lin Pai Rd., Kwai Chung, Hong Kong
852-2489-8187 (tel)
852-2489-1366 (fax)
Manufactures system boards, video cards, etc. for a wide variety of computer applications.

Modular Communication Systems
13309 Saticoy St., N. Hollywood, CA 91605 USA
(818) 764-1333 (tel)
(818) 764-1992 (fax)
1824 Geneva Ln., Plano, TX 75075 USA
(214) 509-8144 (tel)
Manufactures computerized workstations for a variety of communications applications. Offers a free literature package and demo programming disk.

Molex
2222 Wellington Ct., Lisle, IL 60532 USA
(800) 78MOLEX (tel)
(708) 969-4550 (tel)
49-89-413092-0 (Germany tel)
81-427-21-5539 (Japan tel)
65-660-8555 (Singapore tel)
Manufactures connectors for computers.

Monarchy Audio
380 Swift Ave., #21, S. San Francisco, CA 94080 USA
(415) 873-3055 (tel)
(415) 588-0335 (fax)
Manufactures a D/A converter for high-end home audio and home theater applications.

Mondial Design Limited
20 Livingstone Ave., Dobbs Ferry, NY 10522 USA
(914) 693-8008 (tel)
(914) 693-7199 (fax)
Manufactures audio amplifiers for high-end home audio and home theater applications.

Mondo-tronics
524 San Anselmo Ave., #107-62, San Anselmo, CA 94960 USA
(800) 374-5764 (tel)
(415) 455-9330 (tel)
(415) 455-9333 (fax)
Visa, MasterCard
Sells wires that contract under the presence of electricity for robots, model planes, model railroads, etc. Offers a free brochure.

Moneke
7205 Carmelita Ave., Atascadero, CA 93422 USA
(805) 481-2858 (tel)
(805) 481-4478 (fax)
Sells parts for computer tape drives.

Monroe Electronics
Lyndonville, NY 14098 USA
(800) 821-6001 (tel)
(716) 765-2254 (tel)
(716) 765-9330 (fax)
Manufactures trunk switches for professional cable TV applications.

Monster Cable
274 Wattis Way, S. San Francisco, CA 94080 USA
(415) 871-6000 (tel)
Manufactures power and audio cabling for car audio applications.

Monterey Bay Communications
1010 Fair Ave., Santa Cruz, CA 95060 USA
(408) 429-6144 (tel)
(408) 429-1918 (fax)
Sells refurbished Hewlett-Packard computer workstations and printers; also sells mass storage systems, monitors, memory, interfaces, etc. for computer applications.

Monument Computer
8400 E. Iliff, Ste. 10, Denver, CO 80231 USA
(303) 696-0905 (tel)
(303) 696-0941 (fax)
Buys and sells personal desktop computer systems and WYSE terminals for a wide variety of computer applications.

M.O.O.
809 W. Estes, Schaumburg, IL 60193 USA
(800) 323-8962 (tel)
Sells surplus components, speakers, microphones, etc. for a wide variety of electronics applications.

Moore Diversified Products
1441 Sunshine Ln., Lexington, KY 40505 USA
(800) 769-1441 (tel)
(606) 299-6288 (tel)
(606) 299-6653 (fax)
Manufactures and sells fiberoptic equipment for professional cable TV applications.

Moore Products
Sumneytown Pike, Spring House, PA 19477 USA
(215) 646-7400 (tel)
(215) 283-6358 (fax)
Manufactures automated batch management systems for industrial applications.

Morehouse Instrument
1742 Sixth Ave., York, PA 17403 USA
(717) 843-0081 (tel)
(717) 846-4193 (fax)
Manufactures digital readout ring force gauges for industrial applications.

Morel Acoustics USA
414 Harvard St., Brookline, MA 02146 USA
(617) 277-6663 (tel)
(617) 277-2415 (fax)
11 Foxtail Rd., Nacton Rd., Industrial Estate, Ipswitch, IP3 9RT UK
0473-719212 (tel)
0473-716206 (fax)
Manufactures loudspeakers for professional and home audio applications.

Morley
185 Detroit St., Cary, IL 60013 USA
(708) 639-4646 (tel)
Manufactures electric guitar pedals and switches for professional and personal musical applications.

Morning Star Technologies
3518 Riverside Dr., Colombus, OH 43221 USA
(800) 558-7827 (tel)
(614) 451-1883 (tel)
sales@morningstar.com
Manufactures Internet routers for professional network computer applications.

Morris Hobbies
(800) 826-6054 (tel)
(502) 451-0901 (tel)
(502) 451-6602 (fax)
Sells sport and computer radios, batteries, and servos for radio-control model airplanes.

Moseley Associates
111 Castilian Dr., Santa Barbara, CA 93117 USA
(805) 968-9621 (tel)
(805) 685-9638 (fax)
Manufactures digital encoders and decoders for professional broadcasting applications.

Mosel Vitelic
3910 N. 1st St., San Jose, CA 95134 USA
(408) 433-6000 (tel)
(408) 433-0952 (fax)
886-2-545-1213 (Taiwan tel)
852-2665-4883 (Hong Kong tel)
81-3-3365-2851 (Japan tel)
Manufactures memory chips for computer applications.

Mosley Electronics
1344 Baur Blvd., St. Louis, MO 63132 USA
(800) 966-7539 (tel)
(800) 325-4016 (tel)
(314) 994-7872 (tech tel)
Manufactures and sells antennas and accessories for amateur, commercial, and military radio applications.

Most
11205 Knott Ave., #B, Cypress, CA 90630 USA
(714) 898-9400 (tel)
(714) 373-9960 (fax)
Manufactures data storage systems for a wide variety of professional computer network applications.

Mo-Tech
(810) 566-7262 (tel)
(810) 566-7253 (fax)
Sells cable TV converters.

Motherboard Discount Center
3701 E. Baseline Rd., #107, Gilbert, AZ 85234 USA
(800) 486-2026 (tel)
(602) 813-6547 (tel)
(602) 813-8441 (fax)
http://www.homepage.com/mail/mrupgrad.html
Visa, MasterCard, Discover, American Express
Sells computer parts and boards, such as motherboards, etc.

Motherboards International
8361 E. Evans Rd., #107, Scottsdale, AZ 85260 USA
(800) 499-3970 (tel)
(800) 499-3970 (tech support tel)
(602) 596-5226 (tel)
(602) 596-1818 (inter tel)
(602) 596-1554 (fax)
http://www.motherboards.com/
Visa, MasterCard, Discover, American Express
Sells computer parts and boards, such as motherboards, video cards, controllers, etc.

Motorola
Computer division
2900 S. Diablo Way, Tempe, AZ 85282 USA
(602) 438-3287 (tel)
(800) 766-4883 (modem info tel)
(800) 894-7353 (PDA info tel)
(708) 632-3409 (cable info tel)
Electronics division
5005 E. McDowell Rd., Phoenix, AZ 85018 USA
(602) 244-4911 (tel)
Test equipment division
POB 1417 80201 E. McDowell Rd., Scottsdale, AZ 85252 USA
(800) 505-TEST (tel)
(818) 365-5742 (fax)
Microprocessor and memory technologies division
6501 William Cannon Dr. W., Austin, TX 78735 USA
(512) 891-2142 (tel)
http://www.mot.com/MIMS/ISG/
Manufactures a variety of semiconductors and consumer electronics, such as wireless PDAs, two-way radios, etc.; also test sets for professional communication applications.

MoTron Electronics
310 Garfield St., #4, P.O. Box 2748, Eugene, OR 97402 USA
(503) 687-2118 (info tel)
(800) 338-9058 (tel orders)
(503) 687-2492 (fax)
motron.info@emerald.com
Visa, MasterCard, Discover, American Express, COD ($5 fee, cash or money order only)
Sells DTMF and rotary test decoders, including kits, cables, and power adapters.

MountainGate Data Systems
9393 Gateway Dr., Reno, NV 89511 USA
(800) 556-0222 (tel)
(702) 851-9393 (tel)
(702) 851-5533 (fax)
sales@mntgate.lockheed.com
Manufactures high-capacity data storage systems for professional computer applications.

Mountain International
501 4th St. SE, Brandon-by-the-Sea, OR 97411 USA
(503) 347-4700 (tel)
(503) 347-4163 (fax)
Sells digital sampler memory chips for professional and personal musical applications.

Mouser Electronics
958 N. Main St., Mansfield, TX 76063 USA
(800) 992-9943 (tel)
Sells more than 50,000 different new electronic components. Offers a free catalog and same-day shipping.

MPC Technologies
2915 Damier St., Santa Ana, CA 92705 USA
(800) 672-8808 (tel)
(800) 672-9919 (tel)
(800) 672-6686 (tech support tel)
(714) 724-9000 (tel)
(800) 672-8882 (fax back)
(714) 724-9648 (fax)
help@mpctech.com
Visa, MasterCard, Discover
Manufactures and sells personal laptop computers. Offers a 30-day money-back guarantee.

MPR Taltech
8999 Nelson Way, Burnaby, BC V5A 4B5 Canada
Postfach 1855, Ob. Haupstr. 52, D-85354 Freising Germany
(800) 555-7700 (US tel)
(604) 294-1471 (Canada tel)
(604) 293-5787 (Canada fax)
+49-8161-42400 (Germany tel)
+49-8161-41234 (Germany fax)
Manufactures real-time MPEG audio codecs for audio transmission and recording.

MPS
2235 First St., Unit 120, Simi Valley, CA 93065 USA
(805) 581-9592 (tel)
(805) 581-9594 (fax)
Visa, MasterCard
Sells data storage systems, such as hard drives and tape drives, for a wide variety of computer applications.

MPS Multimedia
379 Oyster Pt., Unit 7, South San Francisco, CA 94080 USA
(888) 237-6688 (tel)
(415) 583-4677 (information tel)
(415) 583-0270 (fax)
mpscdrom@aol.com
Visa, MasterCard, Discover
Sells computer data storage, such as CD-ROM drives, recorder, towers, etc.

MQP Electronics
Unit 2, Park Road Centre, Malmesbury, Wiltshire SN16 0BX UK
+44-01666-825146 (tel)
+44-01666-825141 (fax)
Visa, Access
Manufactures EEPROM programmers for a wide variety of electronics applications.

Mr. Mike's Music
816 A. East St., Carbondale, IL 62901 USA
(618) 529-3444 (tel)
(618) 457-6605 (fax)
mrmikes@midewest.net
Sells synthesizer keyboards for professional and personal musical applications.

Mr. NICD (E.H. Yost)
2211D Parview Rd., Middleton, WI 53562 USA
(608) 831-3443 (tel)
(608) 831-1082 (fax)
Manufactures and sells batteries for computers, cellular phones, camcorders, amateur radio applications, etc. Offers a free catalog.

M-Systems
(408) 654-5820 (California tel)
(408) 654-9107 (California fax)
972-3-647-7776 (Israel tel)
972-3-647-6668 (Israel fax)
Manufactures flash disk boards for professional and industrial computer applications.

MTX
The Pointe at South Mountain, 4545 E. Baseline Rd., Phoenix, AZ 85044 USA
(602) 438-4545 (tel)
(602) 438-8692 (fax)
(604) 942-9865 (British Columbia tel)
(905) 890-0298 (Ontario tel)
809-768-0660 (Puerto Rico tel)
+61-02-674-7171 (Australia tel)
+61-02-674-7175 (Australia tel)
Manufactures amplifiers and loudspeakers for mobile audio applications.

Mueller
1583 E. 31st. St., Cleveland, OH 44114 USA

Mueller

(800) 955-2629 (tel)
(216) 771-5225 (tel)
(216) 771-3068 (fax)
Manufactures test leads and accessories for a variety of professional and personal electronics applications. Offers a free 72-page catalog.

Multi-Band Antennas
7131 Owensmouth Ave., #363C, Canoga Park, CA 91303 USA
(818) 341-5460 (tel)
Manufactures and sells mobile multiband shortwave transceiving antennas for amateur radio applications.

Multicom
(800) 423-2593 (tel)
(407) 331-7779 (tel)
(407) 339-0204 (fax)
Sells a wide variety of equipment for professional cable TV applications.

Multi-Contact USA
5560 Skylane Blvd., Santa Rosa, CA 95403 USA
(800) 592-4585 (tel)
(707) 575-7575 (tel)
Manufactures test accessories for professional electronics testing applications.

MultiFAX
143 Rollin Irish Rd., Milton, VT 05468 USA
(802) 893-7006 (tel)
Manufactures external demodulators, TVRO downconverters, etc. for receiving data transmissions via satellite.

MultiMedia Direct
101 Reighard Ave., Williamsport, PA 17701 USA
(800) 386-3342 (tel)
Visa, MasterCard
Sells personal computer parts and boards, such as sound cards, speakers, modems, etc.

Multimedia Effects
40 Wynford Dr., #106, Don Mills, ON M3C 1J5 Canada
(800) 367-3954 (tel)
(416) 444-2324 (tel)
(416) 444-0465 (fax)
Sells recordable CD-ROM drives, optical hard drives, etc.

Multimedia Enterprises
P.O. Box 221, Carlisle, PA 17013 USA
(717) 249-0220 (tel)
(717) 249-0242 (fax)
Visa, MasterCard
Sells discounted personal desktop and laptop computers that were purchased at bankruptcy auctions. Offers a 30-day money-back guarantee.

Multi-Micro Systems
2124 Zanker Rd., San Jose, CA 95131 USA
(800) 875-0333 (tel)
(408) 437-1583 (fax)
Visa, MasterCard
Manufactures ruggedized and rack-mount equipment for professional and industrial computer applications.

Multiplier Industries
135 Radio Circle, P.O. Box 630, Mt. Kisco, NY 10549 USA
(914) 241-9510 (tel)
(914) 241-1103 (fax)
Manufactures batteries for professional cellular microwave radio applications.

Multi-Tech Systems
2205 Woodale Dr., Mounds View, MN 55112 USA
(800) 328-9717 (tel)
(612) 785-3500 (tel)
(612) 785-9874 (fax)
(612) 785-3702 (BBS)
Manufactures communications servers and routers for professional computer network applications.

Multi-Vision
(800) 833-2915 (tel)
Sells cable TV and electronic equipment, including descrambler/converters, protectors, voice-mail systems, digital on-hold announcers, laser pens, etc.

Multi-Vision Electronics
(800) 835-2330 (tel)
Visa, MasterCard, COD
Sells a variety of cable TV converters and descramblers. Offers a free catalog.

Multiwave Technology
15318 Valley Blvd., City of Industry, CA 91746 USA
(800) 234-3358 (tel)
(800) 587-1730 (customer service tel)
(800) 861-0050 (tech support tel)
(800) 595-6908 (corporate customers tel)
Visa, MasterCard, Discover, American Express
Manufactures and sells personal desktop computers.

Murata Electronics
2200 Lake Park Dr., Smyrna, GA 30080 USA
(800) 831-9172 (tel)
Manufactures surface-mount components for a wide variety of electronics applications.

Murrietta Circuits
4761 E. Hunter Ave., Anaheim, CA 92807 USA
(714) 970-2430 (tel)
(714) 970-2406 (fax)
Manufactures PC board prototypes for a wide variety of professional electronics applications.

Musical Concepts
5749 Westwood Dr., St. Charles, MO USA
(314) 447-0040 (tel)
Manufactures driver boards for amplifiers, phono/line preamplifier boards, and interconnects for high-end audio applications.

Music Industries
99 Tulip Ave., Floral Park, NY 11001 USA
(800) 431-6699 (tel)
(516) 352-4110 (tel)
(516) 352-0754 (fax)
Manufactures keyboard controllers for a wide variety of home and professional musical applications.

MusiCraft
1280 E. Dundee Rd., Palatine, IL 60067 USA
(708) 991-7220 (tel)
1800 S. Halsted, Homewood, IL 60430 USA
(708) 799-2400 (tel)
Sells home and amateur audio and video equipment, such as camcorders, VCRs, speakers, cassette decks, TVs, home theater equipment, etc.

Mustek
1702 McGaw Ave., Irvine, CA 92714 USA
(714) 250-8855 (tel)
Manufactures computer image scanners for amateur and professional desktop publishing applications.

MVS
P.O. Box 850, Merrimack, NH 03054 USA
(508) 792-9507 (tel)
Manufactures single-board computers for a wide variety of professional computer applications.

MWK Industries
1269 W. Pomona, Corona, CA 91720 USA
(800) 356-7714 (tel)
(909) 278-0563 (tel)
(909) 278-4887 (fax)
Visa, MasterCard, Discover, American Express, COD, OAC
Sells new and surplus laser and optical supplies, including books, power supplies, complete laser assemblies, laser pointers, image intensifiers, light show equipment, mirrors, optics, and safety materials. Offers a free 36-page catalog

MWS Wire Industries
31200 Cedar Valley Dr., Westlake Village, CA 91362 USA
(800) 423-5097 (tel)
(818) 991-8553 (tel)
(818) 706-0911 (fax)
Manufactures wire for a wide variety of electronics applications.

Mx-Com
4800 Bethania Station Rd., Winston-Salem, NC 27105 USA
(800) 638-5577 (tel)
Manufactures scrambler/descrambler boards for two-way radio communications.

MYAT

MYAT
380 Chestnut St., P.O. Box 425, Norwood, NJ 07648 USA
(201) 767-5380 (tel)
(201) 767-4147 (fax)
Manufactures rigid-line coaxial cable and components for professional broadcast transmitting.

Mylex
34551 Ardenwood Blvd., Fremont, CA 94555 USA
(800) 77-MYLEX (tel)
(510) 796-6100 (tel)
(510) 745-8016 (fax)
Manufactures disk controller boards and interface cards for a wide variety of professional computer network applications.

Mytek Digital
P.O. Box 1023, New York, NY 10276 USA
(212) 388-2677 (tel)
(212) 686-4948 (fax)
Manufactures a digital AES/EBU meter for professional studio recording.

NAD
200 Williams Dr., Ramsey, NJ 07446 USA
(800) 265-4NAD (tel)
633 Granite Ct., Pickering, ON L1W 3K1 Canada
(800) 263-4641 (tel)
(416) 831-6333 (tel)
(416) 831-6936 (fax)
Manufactures audio equipment, such as receivers, cassette decks, etc., for high-end home audio and home theater applications.

Nadel Enterprises
80 Galaxy Blvd., #13, Rexdale, ON M9W 4Y8 Canada
(416) 798-2622 (tel)
Distributes Sima video editing equipment in Canada.

Nadine's Music & Pro Audio
6251 Santa Monica Blvd., Hollywood, CA 90038 USA
(800) 918-4800 (tel)
(213) 464-7550 (tel)
(213) 464-2897 (fax)
Sells keyboards, electric guitars, audio equipment, etc. for a wide variety of professional musical applications.

Nady Systems
6701 Bay St., Emeryville, CA 94608 USA
(510) 652-2411 (tel)
(510) 652-5075 (fax)
Manufactures a variety of home and professional audio and video equipment, including stereo A/V mixers, camcorder microphones, wireless video microphones, professional wireless microphones, etc.

Nagra
240 Gt. Circle Rd., #326, Nashville, TN 37228 USA
Manufactures reel-to-reel tape decks for high-end home audio and broadcasting applications.

NAi (National Advantages)
470 Cloverleaf Dr., Units A/B, Baldwin Park, CA 91706 USA
(800) 777-0439 (tel)
(818) 855-1820 (tel)
(818) 855-1825 (fax)
Visa, MasterCard
Manufactures and sells personal desktop and laptop computers. Offers a 30-day money-back guarantee.

Nakamichi America
19701 S. Vermont Ave., Torrance, CA 90502 USA
(310) 538-8150 (tel)
(310) 324-7614 (fax)
Manufactures cassette decks, amplifiers, receivers, CD players, etc. for home audio and car audio applications.

Nanao
23535 Telo Ave., Torrance, CA 90505 USA
(800) 800-5202 (tel)
(310) 325-5202 (tel)
(310) 530-1679 (fax)
(800) 416-FLEX (fax back)
http://www.traveller.com/nanao
Manufactures monitors for personal desktop computer systems.

Narco Avionics
270 Commerce Dr., Ft. Washington, PA 19034 USA
(800) 223-3636 (tel)
(215) 643-2905 (tel)
Manufactures GPS and Loran systems for professional aviation applications.

National Cable Brokers
(219) 935-4128 (tel)
Sells a wide variety of cable TV equipment.

National Computer Resources
9750 230th St. E., Lakeville, MN 55044 USA
(612) 461-2872 (tel)
(612) 461-2874 (fax)
Buys and sells new and used terminals for professional computer applications.

National Data Mux
21634 Lassen St., Chatsworth, CA 91311 USA
(818) 772-1591 (tel)
(818) 772-6854 (fax)
Visa, MasterCard, Discover, American Express
Sells modems and multiplexers for a wide variety of computer applications.

National Instruments
6504 Bridgepoint Pkwy., Austin, TX 78730 USA
(800) 433-3488 (tel)
(512) 794-0100 (tel)
(512) 794-8411 (fax)
http://www.natinst.com
info@natinst.com
Manufactures plug-in boards, PCMCIA cards, I/O modules, etc. for professional temperature-measurement applications. Offers a free catalog.

National LAN Exchange
1403 W. 820 N., Provo, UT 84601 USA
(801) 377-0074 (tel)
(801) 377-0078 (fax)
Sells new and used network equipment for professional computer applications.

National Peripherals
1111 Pasquinelli Dr., #200, San Jose, CA 95117 USA
(800) 441-5758 (tel)
(408) 985-7100 (tel)
(408) 246-3127 (fax)
Manufactures data storage systems for a wide variety of professional computer network applications.

National Peripheral Services
1220 N. Simon Circle., Ste. C, Anaheim, CA 92806 USA
(800) 628-9012 (tel)
(714) 630-4005 (fax)
Sells tape drives for a wide variety of computer applications.

National Recording Supplies
764 5th Ave., Brooklyn, NY 11232 USA
(800) 538-2336 (tel)
(718) 369-8273 (tel)
(718) 369-8275 (fax)
Visa, MasterCard, American Express
Sells a variety of recording supplies, including bulk audio and video cassettes, multitrack tape, etc., for professional recording studios. Offers a free catalog.

National Semiconductor
P.O. Box 5100, Crawfordsville, IN 47933 USA
(800) 272-9959 (tel)
+49-0-89-247-11-222 (Europe fax)
81-43-299-2300 (Japan tel)
852-2737-1600 (Southeast Asia tel)
2900 Semiconductor Dr., Santa Clara, CA 95052 USA
Manufactures semiconductors for a wide variety of electronics applications. Offers free designing and applications literature.

National Sound and Video
6500 McDonough Dr., Norcross, GA 30093 USA
(404) 447-0101 (tel)
Sells a variety of audio and video equipment for professional broadcasting, professional studio recording, and consumer audio applications.

National Tower
P.O. Box 15417, Shawnee Mission, KS 66285 USA
(800) 762-5049 (tel)

National Tower

Visa, MasterCard
Sells a wide variety of amateur radio equipment and accessories, including amateur radio transceivers, receivers, towers, antennas, amplifiers, etc.

Nationwide Computers Direct
110 McGaw Dr., Edison, NJ 08837 USA
(800) 747-6923 (tel)
(908) 417-4455 (tel)
(800) 329-6923 (fax)
Visa, MasterCard, Discover, American Express
Sells personal desktop and laptop computers, monitors, fax machines, and printers.

Nationwide Computer Distributor
P.O. Box 7AQ, Jersey City, NJ 07307 USA
(800) 777-1054 (tel)
(201) 659-6074 (tel)
(201) 659-2977 (tech support tel)
(201) 659-3345 (fax)
Visa, MasterCard, Discover
Sells computer parts and boards, such as hard drives, motherboards, memory chips, cables, floppy drives, multimedia upgrades, modems, etc.

Naval Electronics
5417 Jetview Circle, Tampa, FL 33634 USA
(813) 885-6091 (tel)
(813) 885-3789 (fax)
Manufactures and sells audio boosters and squelches for two-way radio applications.

NBO Distributors
5631 Palmer Way, Carlsbad, CA 92008 USA
(800) 346-6466 (tel)
Sells a variety of satellite TV reception equipment, including receivers, dish antennas, cabling, etc.

NCA Computer Products
1202 Kifer Rd., Sunnyvale, CA 94086 USA
(800) NCA-9444 (tel)
(800) NCA-1115 (tel)
(408) 567-1777 (international sales)
(800) 961-1200 (corp tel)
(408) 739-2420 (corp fax)
(800) NCA-1666 (fax)
(408) 567-1915 (fax)
Visa, MasterCard, Discover, American Express, COD
Manufactures and sells personal desktop and laptop computers, printers, and scanners; also sells computer parts and boards, such as floppy drives, hard drives, monitors, video cards, modems, keyboards, etc. Offers a 30-day money-back guarantee.

NCG Companies
1275 N. Grove St., Anaheim, CA 92806 USA
(800) 962-2611 (tel)
(714) 630-4541 (tel)
(714) 630-7024 (fax)
Sells UHF transceivers and antennas for two-way radio applications.

NCI
100 Brainford Place, Newark, NJ 07102 USA
(800) 879-6240 (tel)
Sells power dialers and software for professional telecommunications applications.

NCS Industries
2255E Wyandotte Rd., Willow Grove, PA 19090 USA
(800) 523-2342 (tel)
(215) 657-0840 (fax)
Cebu, Phillipines
633-246-1651 (tel/fax)
Sells a wide variety of equipment for professional cable TV applications.

NEC
475 Ellis St., Mountain View, CA 94039 USA
(800) 366-9782 (tel)
(800) 729-9288 (fax)
(800) NEC-INFO (tel)
(800) 374-8000 (tel)
(800) 366-9782 (tel)
(508) 264-8000 (tel)
(800) 366-0476 (info fax)
Electronics and Computer divisions
1255 Michael Dr., Wood Dale, IL 60191 USA
(800) 366-3632 (fax)
(708) 860-9500 (tel)
(800) 356-2415 (fax)
(708) 860-5593 (fax)

http://www.nec.com
Manufactures and sells memory chips, personal desktop and laptop computers, CD-ROM drives, file servers, printers, monitors. Offers a 30-day money-back guarantee and a free price list.

NECX
Four Technology Dr., Peabody, MA 01960 USA
(800) 961-9208 (tel)
http://www.necx.com
Visa, MasterCard, Discover, American Express
Sells personal desktop and laptop computers, printers, and scanners; also sells computer parts and boards, such as floppy drives, hard drives, monitors, processor upgrades, video cards, modems, keyboards, etc.

Needham's Electronics
4630 Beloit Dr., #20, Sacramento, CA 95838
(916) 924-8037 (tel)
(916) 924-8065 (fax)
(916) 924-8094 (BBS)
ftp://ftp.crl.com/users/ro/needhams
Visa, MasterCard, COD
Manufactures EEPROM programmers for a wide variety of computer and electronics applications.

Needle Doctor
419 14th Ave. SE, Minneapolis, MN 55414 USA
(800) 229-0644 (tel)
(612) 378-0543 (tel)
(612) 378-9024 (fax)
Sells a wide variety of record player needles and cartridges.

N.E. Engineering
11 Safford St., Quincy, MA 02171
(800) 926-4030 (tel)
(617) 770-3830 (tel)
(617) 770-2305 (fax)
Visa, MasterCard, COD
Sells cable TV test chips.

NEFF Instrument
700 S. Myrtle Ave., Monrovia, CA 91016 USA
(800) 423-7151 (tel)
(818) 357-2281 (tel)
(818) 303-2286 (tel)
Manufactures VXI interface controllers for professional industrial applications.

Nemal Electronics
12240 NE 14th Ave., N. Miami, FL 33161 USA
(305) 893-3924 (tel)
(305) 895-8178 (fax)
Manufactures and sells coaxial cable, RF transmission line, and accessories for a variety of electronics applications.

Neopost
1345 Valwood Pkwy., Carrollton, TX 75006 USA
(800) 624-7892 (tel)
Manufactures electronic mailers for a wide variety of office applications.

NESI
61 Holyoke Rd., Richboro, PA 18954 USA
(800) 216-NESI (tel)
(215) 357-1056 (tel)
(215) 355-5739 (fax)
Sells floppy drives for a wide variety of computer applications.

Net Computers International
13725-27 Gamma Rd., Dallas, TX 75244 USA
(800) 252-9855 (tel)
(214) 404-9877 (tel)
(214) 404-0610 (fax)
http://www.netcomputers.com/
Visa, MasterCard, Discover, American Express
Manufactures and sells personal desktop computers; also sells computer parts and boards, such as motherboards, hard drives, monitors, sound cards, multimedia packages, modems, etc. Offers a 30-day money-back guarantee.

NetFRAME Systems
1545 Barber Ln., Milpitas, CA 95035 USA
(800) 852-3726 (tel)
(408) 944-0600 (tel)
(408) 434-4190 (fax)
Manufactures file servers for a wide variety of professional computer network applications.

Network Appliance
295 N. Bernardo Ave., Mountain View, CA 94043 USA
(800) 952-5005 (tel)
(415) 428-5100 (tel)
(415) 428-5151 (fax)
Manufactures file servers for a wide variety of professional computer network applications.

Network Computing Devices
350 N. Bernardo Ave., Mountain View, CA 94043 USA
(415) 366-4400 (tel)
(415) 366-5675 (fax)
Manufactures terminals for a wide variety of professional computer network applications.

Network Equipment Technologies
800 Saginaw Dr., Redwood City, CA 94063 USA
(415) 366-0690 (tel)
(415) 366-7711 (fax)
Manufactures routers for professional computer network applications.

Network Express
1720 Oak St., P.O. Box 301, Lakewood, NJ 08701 USA
(800) 374-9899 (tel)
(941) 359-2876 (office tel)
(941) 355-5841 (fax)
netexp@netline.net
Visa, MasterCard, Discover, American Express, Diner's Club
Sells computer network equipment, such as complete computer networks, printers, hard drives, modems, file servers, WAN kits, CD-ROM recorders, etc.

Net Works
511 Montague Expressway, Milpitas, CA 95035 USA
(800) 345-2377 (tel)
(408) 942-6160 (fax)
Visa, MasterCard
Sells computer parts and boards, such as network cards, CD-ROM drives, etc.

Neumann
6 Vista Dr., P.O. Box 987, Old Lyme, CT 06371 USA
(203) 434-5220 (tel)
(203) 434-3148 (tel)
221 LaBrosse Ave., Pte-Claire, PQ H9R 1A3 Canada
(514) 426-3013 (tel)
(514) 426-3953 (fax)
Manufactures a variety of microphones for professional studio recording.

Neutrik
195 Lehigh Ave., Lakewood, NJ 08701 USA
(908) 901-9488 (tel)
(908) 901-9608 (fax)
(514) 344-5220 (tel)
(514) 344-5221 (fax)
+75-237-24-24 (Liechtenstein tel)
+75-232-53-93 (Liechtenstein fax)
+0941-98041 (Germany tel)
+0941-99-97-72 (Germany fax)
+0983-811-441 (UK tel)
+0983-811-439 (UK fax)
+01-734-0400 (Switzerland tel)
+01-734-3891 (Switzerland fax)
Manufactures equipment for a variety of electronics testing applications.

Newbridge Microsystems
603 March Rd., Kanata, ON K2K 2M5 Canada
(800) 267-7231 (tel)
(613) 599-3600 (tel)
(613) 592-3120 (fax)
Manufactures WAN access cards for professional computer network applications.

Newmar
P.O. Box 1306, Newport Beach, CA 92663 USA
(800) 854-3906 (tel)
(714) 957-1621 (fax)
Manufactures and sells integrated power systems for professional two-way radio applications.

New Micros (NMI)
1601 Chalk Hill Rd., Dallas, TX 75212 USA
(214) 339-2204 (tel)
(214) 339-1585 (fax)
(214) 339-2234 (fax back)
general@mewmicros.com

http://www.newmicros.com
Manufactures single-board computers for a wide variety of professional computer applications.

Newport Computer Services
1 Lafayette Rd., Hampton, NH 03842 USA
(603) 926-4300 (tel)
(603) 926-4646 (fax)
Sells hard drives, memory chips, and network equipment for a wide variety of computer applications.

Newport Digital
(800) 383-3642 (tel)
Sells memory chips for a wide variety of computer applications.

Newport Technology
6321 Angus Dr., Raleigh, NC 27613 USA
(919) 571-9405 (tel)
(919) 571-9262 (fax)
Manufactures dc-to-dc power converter modules for a wide variety of electronics applications. Offers a free data book.

New Sensor
20 Cooper Sq., New York, NY 10003 USA
(800) 633-5477 (tel)
(212) 529-0466 (tel)
(212) 529-0486 (fax)
sovtek@emedia.net
http://www.emedia.net/sovtek
Sells guitar amplifiers and electron tubes for professional and personal musical applications.

New Voice
1893 Preston White Dr., Ste. 120, Reston, VA 22091 USA
(703) 648-0585 (tel)
(703) 648-9430 (fax)
newvoice@planetcom.com
Manufactures voice boards for a wide variety of professional computer applications.

New West Electronics
4120 Meridian, Bellingham, WA 98226 USA
(800) 488-8877 (tel)
(360) 734-3314 (fax)
Visa, MasterCard, Discover, American Express, COD
Sells high-end home, amateur, and professional audio and video equipment, such as camcorders, VCRs, video editors, monitors, TVs, home theater equipment, etc.

New York Music
7144 Market St., Boardman, OH 44512 USA
(800) 241-6330 (tel)
(216) 758-9432 (tel)
Sells a wide variety of musical instruments and equipment for home and professional musical applications.

Next Destination
7 Barnack Business Centre, Blakey Rd., Salisbury, Wiltshire SP1 2LP UK
01722-410800 (tel)
Sells GPS equipment for a wide variety of navigational applications.

Next Generation
35 Cherry Hill Dr., Danvers, MA 01923 USA
(800) 896-7530 (tel)
(800) 896-7530 X230 (Spanish\Italian\French tel)
(800) 896-7530 X232 (Russian tel)
(800) 896-7530 X245 (German tel)
Manufactures and sells personal desktop computers; also sells computer parts and boards, such as tape drives, hard drives, monitors, video cards, modems, keyboards, etc. Offers a 30-day money-back guarantee.

Next International
13622 Neutron Rd., Dallas, TX 75244 USA
(800) 730-6398 (tel)
(214) 404-8260 (tel)
(214) 386-9834 (tech support tel)
(214) 404-8263 (fax)
Visa, MasterCard, Discover, American Express, COD
Manufactures and sells personal desktop computers; also sells computer parts and boards, such as UPS systems, hard drives, monitors, video cards, modems, keyboards, etc.

NHT (Now Hear This)
535 Getty Ct. Bldg. A, Benecia, CA 94510 USA
(800) NHT-9993 (tel)
(707) 747-1252 (fax)
74101.1754@compuserve.com
http://www.nthifi.com
Manufactures loudspeakers for home audio applications.

The NiCad Lady
P.O. Box 654, Wildomar, CA 92595 USA
(909) 698-1901 (tel)
(909) 678-0065 (fax)
Sells rebuilt rechargeable batteries for laptop computers.

Nichi Standard Enterprises
Flat A, 13/f, International Industrial Center, 2-8 Kwei Tei St., Fo Tan, Shatin, NT, Hong Kong
852-2603-2853 (tel)
852-2606-2800 (fax)
4th Fl., Block 1, Yin Peng 1st Industrial Zone, 26th District, Bao An County, Shenzhen, China
Manufactures floppy disks, joysticks, data switches, loudspeakers, keyboards, etc. for a wide variety of computer applications.

Nico Systems
103 Geiger Rd., Philadelphia, PA 19115 USA
(800) 989-9897 (tel)
(215) 969-1969 (tel)
(215) 969-6133 (tech support tel)
(215) 969-6058 (fax)
Visa, MasterCard, Discover, American Express
Manufactures and sells personal desktop computers; also sells computer parts and boards, such as floppy drives, hard drives, monitors, motherboards, modems, controller cards, etc. Offers a 30-day money-back guarantee.

NICS (New Interactive Concept Systems)
11861 E. Telegraph Rd., Santa Fe Springs, CA 90670 USA
(800) 289-2345 (corp sales tel)
(800) 948-8330 (educ sales tel)
(310) 948-5970 (fax)
(310) 948-1059 (BBS)
nics@primenet.com
Visa, MasterCard, Discover, American Express
Manufactures and sells personal desktop computers.

NIE International
3000 E. Chambers, Phoenix, AZ 85040 USA
(800) 448-9713 (tel)
(800) 633-2869 (tel)
(602) 470-1500 (tel)
(602) 470-1540 (fax)
Visa, MasterCard, Discover, American Express
Sells personal home and laptop computers, and parts and accessories.

Nikon
1300 Walt Whitman, Melville, NY 11747 USA
(800) 645-6678 (tel)
(800) NIKON-35 (brochure tel)
Manuafactures cameras and other optical electronics for a wide variety of applications.

Niles Audio
P.O. Box 160818, Miami, FL 33116 USA
(800) 289-4434 (tel)
(305) 238-4373 (tel)
Manufactures loudspeakers for indoor/outdoor audio applications.

NKK Switches
7850 E. Gelding Dr., Scottsdale, AZ 85260 USA
(602) 991-0942 (tel)
(602) 998-1435 (fax)
Manufactures switches for a wide variety of electronics applications. Offers a free, 456-page design guide.

NKX
1814 Hancock St., Gretna, LA 70053 USA
(800) 237-6519 (tel)
(504) 361-5525 (tel)
(504) 361-5526 (fax)
Manufactures and sells crystals for pager and two-way radio applications.

NMB Marketing
London House, 100 New Kings Rd., London SW6 4LX UK
0171-731-8199 (tel)
0171-731-8312 (fax)
Sells single-board computers for a wide variety of hobbyist computer applications.

No. 1 Battery Specialists
74 W. 38th St. 2nd Fl., New York, NY 10018 USA
(800) 275-2879 (tel)
(212) 768-7595 (tel)
(212) 768-9530 (fax)
Sells a wide variety of specialty batteries, battery packs, and battery belts, especially for camcorders, computers, cellular telephones, etc.

Noble House
22230-6 Germaine, Chatsworth, CA 91311 USA
(800) 260-5053 (tel)
(818) 709-5053 (tel)
(818) 709-3503 (fax)
Sells CD-ROM recorders for professional computer applications.

Noise Com
E. 49 Midland Ave., Paramus, NJ 07652 USA
(201) 261-8797 (tel)
(201) 261-8339 (fax)
Manufactures amplifier testers for CDMA applications.

Nokia Display Products
1505 Ridgeway Blvd., Sausilito, CA 94965 USA
(800) 296-6542 (tel)
(415) 331-6622 (tel)
nokia@aol.com
Manufactures monitors for personal desktop computer systems.

Nokia Mobile Phones
2300 Tall Pines Dr., #120, Largo, FL 34641 USA
(800) 666-5553 (tel)
(813) 530-3599 (fax)
Manufactures telephones for cellular telephone applications.

Norad A/S
Specialelektronik, Frederikshavnsvej 74, DK-9800 Hjorring, Denmark
45-98-909-999 (tel)
45-98-909-988 (fax)
Sells receivers, antennas, and accessories for shortwave radio-listening applications.

Norcomm
12438 Loma Rica Dr., Grass Valley, CA 95945 USA
(800) 874-8663 (tel)
(916) 477-8403 (fax)
Manufactures decoders and encoders for remote-control DTMF applications.

NorComp Interconnect Devices
P.O. Box 3867, Gastonia, NC 28054 USA
(800) 849-4450 (tel)
(704) 868-3581 (tel)
(704) 867-0577 (fax)
Manufactures connectors and accessories for a wide variety of computer and electronics applications. Offers a free catalog.

Nordisk Systems
1116 W. Collins Ave., Orange, CA 92667 USA
(800) 929-4789 (tel)
(805) 644-6195 (tel)
(805) 644-5146 (fax)
Sells hard drives and tape drives for a wide variety of computer applications.

Norham Radio
4373 Steeles Ave. W., N. York, ON M3N 1V7 Canada
(416) 667-1000 (tel)
(416) 667-9995 (fax)
http://www2.westonia.com/shortwave.html
Visa, MasterCard
Sells receivers, antennas, and accessories for shortwave listening applications.

Noritake
919 Atlanta Gift Mart, 230 Spring St. NW., Atlanta, GA 30303 USA
(800) 837-4727 (tel)

Noritake

(414) 529-9468 (tel)
(414) 529-9455 (fax)
945 Concord St., #118, Framingham, MA 01701 USA
(508) 626-0811 (tel)
(508) 626-0429 (fax)
2635 Clearbrook Dr., Arlington Heights, IL 60005 USA
(708) 439-9020 (tel)
(708) 593-2285 (fax)
2454 Dallas Trade Mart, Dallas, TX 75207 USA
(214) 742-9389 (tel)
(214) 747-5065 (fax)
2050 E. Vista Bella Way, Compton, CA 90220 USA
(310) 603-9770 (tel)
(310) 603-9810 (fax)
75 Seaview Dr., Secaucus, NJ 07094 USA
(201) 319-0600 (tel)
(201) 319-1962 (fax)
Manufactures fluorescent displays for a wide variety of electronics applications.

Nortel (Northern Telecom)

2100 Lakeside Blvd., Richardson, TX 75082 USA
(800) 4-NORTEL (tel)
(800) 262-9334 (tel)
http://www.nortel.com
Manufactures a wide variety of equipment for professional cable TV and telephone network applications.

North American CAD

4A Hillview Dr., Barrington, IL 60010 USA
(800) 619-2199 (tel)
(708) 381-8834 (tech support tel)
(708) 381-7374 (fax)
shopping@nacad.com (sales)
plotters@nacad.com (questions)
http://nacad.com
Visa, MasterCard, Discover
Specializes in the sales of CAD design products, including printers, plotters, monitors, digitizing tablets, and scanners. Offers a free catalog.

North American Capacitor

7545 Rockville Rd., Indianapolis, IN 46214 USA
(317) 273-0090 (tel)
Manufactures capacitors for a wide variety of electronics applications.

North Atlantic Instruments

170 Wilbur Pl., Bohemia, NY 11716 USA
(516) 567-1100 (tel)
(516) 567-1823 (fax)
Manufactures test equipment for a wide variety of professional electronics applications.

North Atlantic Radio Service

P.O. Box 1, Marion Bridge, NS B0A 1P0 Canada
Distributes Lowe shortwave receivers to North America.

North Country Radio

P.O. Box 53, Wykagyl Station, New Rochelle, NY 10804 USA
Manufactures and sells radio/audio/video kits, including amateur TV transmitters and receivers, etc.

North Creek Music Systems

P.O. Box 1120, Old Forge, NY 13420 USA
(315) 369-2500 (tel/fax)
Sells loudspeaker kits for high-end home audio applications.

North East Memory

1051 County Line Rd., #198, Huntington Valley, PA 19006 USA
(800) 797-2447 (tel)
(215) 396-2860 (tel)
(215) 396-2870 (fax)
Buys and sells memory chips for a wide variety of computer applications.

Northeast Electronics

P.O. Box 3310N, N. Attleboro, MA 02761 USA
(800) 886-8699 (tel)
Visa, MasterCard, COD ($7.50)
Sells cable TV descrambler kits.

Northeast Minicomputer

55 High St., #6, Billerica, MA 01862 USA
(800) 343-8302 (tel)
(508) 663-2550 (tel)
(508) 667-0718 (fax)

Sells keyboards, etc. for a wide variety of computer applications.

Northern Exposure Office Products
119 E. 19th Ave., Torrington, WY 82240 USA
(800) 952-7372 (tel)
Visa, MasterCard
Sells office supply products, including photocopiers.

North Star Avionics
30 Sudbury Rd., Acton, MA 01720 USA
(800) 628-4487 (tel)
(508) 897-6600 (tel)
Manufactures GPS and Loran systems for professional aviation applications.

Northwest DevTech
10220 N. Nevada, #170, Spokane, WA 99218 USA
(800) 877-1979 (tel)
(509) 468-0411 (tel)
(509) 468-0414 (fax)
Manufactures file servers for a wide variety of professional computer network applications.

Nott
4001 La Plata Hwy., Farmington, NM 87401 USA
(505) 327-5646 (tel)
(505) 325-1142 (fax)
Manufactures antenna systems for professional broadcasting, military, maritime, and amateur radio applications.

Novacomm
6409 Independence Ave., Woodland Hills, CA 91367 USA
(800) 757-3006 (tel)
(818) 340-3180 (fax)
Manufactures cellular telephones, laptop computers, alarms, etc. for communications, computer, security, and communications applications.

Nova Computers
22052 D. Lakeshore Blvd., Euclid, OH 44123 USA
(800) 461-5535 (tel)
(216) 731-8298 (tech support tel)
(216) 731-9150 (fax)
Visa, MasterCard, Discover, American Express
Manufactures and sells personal desktop computers; also sells computer parts and boards, such as floppy drives, hard drives, monitors, video cards, modems, keyboards, etc. Offers a 30-day money-back guarantee.

Novak Electronics
18910 Teller Ave., Irvine, CA 92715 USA
(714) 833-8873 (tel)
Sells speed controls for radio-control models.

Novaplex
8818 Bradley Ave., Sun Valley, CA 91352 USA
(800) 644-6682 (tel)
(818) 504-4007 (tel)
(818) 504-6522 (fax)
Visa, MasterCard, COD
Sells converters, remote controls, splitters, surge protectors, and components for cable TV applications. Minimum order of $10.

Now! Recording Systems
32 W. 39th St., 9th Fl., New York, NY 10018 USA
(800) 859-2579 (tel)
(212) 768-7800 (tel)
(212) 768-9740 (fax)
Visa, MasterCard, American Express
Sells blank audio and video recording cassettes, etc. Offers a free catalog; minimum $25 order.

N & G Electronics
1950 NW 94th Ave., Miami, FL 33126 USA
(305) 592-9685 (tel)
(305) 592-2330 (fax)
Sells amateur radio equipment and accessories, including amateur radio transceivers, receivers, antenna tuners, antennas, amplifiers, etc.

NRG Research
840 Rogue River Hwy., Bldg., #144, Grants Pass, OR 97527 USA
(800) 753-0356 (tel)
Sells a variety of amateur and professional audio and video equipment, including on-camera lights, line conditioner/converters, power packs, cabling, etc.

NSA
100 Lowderbrook Dr., Ste. 2400, Westwood, MA 02090 USA
(800) 441-4832 (tel)
(617) 461-8300 (tel)
Manufactures monitors for personal desktop computer systems.

NS Electronics Service
3610 Dekalb Technology Pkwy., #110/#111, Atlanta, GA 30340 USA
(770) 451-3264 (tel)
(770) 458-8785 (fax)
Sells new and used equipment for professional two-way radio applications.

NSM
40 Cain Ave., Woodbury, NY 11797 USA
(800) 423-1122 (tel)
(516) 273-4240 (fax)
Manufactures high-capacity CD player/changers for professional broadcasting.

NSM
P.O. Box 326, Garden City, NY 11530 USA
(516) 486-8285 (tel)
(516) 538-0933 (fax)
Sells loudspeaker kits for high-end home audio applications.

N.T.E. Electronics
44 Ferrand St., Bloomfield, NJ 07003 USA
Manufactures high-voltage diodes for industrial and microwave ovens.

Numerous Complaints Music
1537B Howell Mill Rd., Atlanta, GA 30318 USA
(404) 351-4422 (tel)
(404) 351-4442 (fax)
Visa, MasterCard, American Express
Sells new and used synthesizer keyboards, electric pianos, etc. for professional and personal musical applications.

Nupon Computing
1391 Warner Ave., #A, Tustin, CA 92680 USA

Manufactures routers and print servers for professional computer network applications.

NuReality
2907 Daimler St., Santa Ana, CA 92705 USA
(800) 501-8086 (tel)
(714) 442-1080 (tel)
(714) 852-1059 (fax)
sales@nureality.com
http://www.nureality.com
Manufactures sound enhancement systems for a wide variety of computer applications.

Nu-Tek Electronics
3250 Hatch Cedar Park, TX 78613 USA
(800) 228-7404 (tel)
Visa, MasterCard, Discover, American Express, COD
Sells a variety of cable TV equipment. Offers a free catalog.

M.M. Newman
P.O. Box 615, Marblehead, MA 01945 USA
(617) 631-7100 (tel)
(617) 631-8887 (fax)
Manufactures quick-heating soldering irons.

nView
860 Omni Blvd., Newport News, VA 23606 USA
(804) 873-1354 (tel)
Manufactures LCD projection panels for computer multimedia applications.

NYCE
790 Suma Ave., Woodbury, NY 11590 USA
(516) 997-7197 (tel)
(516) 997-7170 (tel)
(516) 997-7072 (fax)
(516) 997-7237 (fax)
Buys, sells, and trades personal home and laptop computers, printers, memory chips, power supplies, tape drives, monitors, etc.

NYE Engineering
4020 Galt Ocean Dr., #606, Ft. Lauderdale, FL 33308 USA

(305) 566-3997 (tel)
(305) 537-3534 (fax)
Visa, MasterCard
Manufactures and sells digital field strength meters for amateur radio applications.

Oar
10447 Roselle St., San Diego, CA 92121-1503 USA
(619) 453-4014 (tel)
(619) 546-8739 (fax)
Manufactures direction finders for professional radio applications.

Oasis Technologies
P.O. Box 267, Tracyton, WA 98393 USA
(360) 373-3818 (tel)
(360) 613-1737 (BBS)
Sells printers, video cards, hard drives, keyboards, memory chips, I/O cards, etc.

Ocean Electronics
4905 34th St. S., #344, St. Petersburg, FL 33711 USA
(800) 225-5550 (tel)
(813) 571-1508 (fax)
Sells audio accessories and systems for mobile audio and security applications.

Océ-Bruning
1800 Bruning Dr W., Itasca, IL 60143 USA
(800) 445-3526 (tel)
(708) 351-7549 (tel)
Manufactures plotter/scanner/copier systems for professional computer applications.

Odyssey Pro Sound
11 Eden St., Salem, MA 01970 USA
(508) 744-2001 (tel)
(508) 744-7224 (fax)
Visa, MasterCard
Buys and sells new and used consoles, tape machines, microphones, etc. for professional studio recording.

OFS WeatherFAX
6404 Lakerest Ct., Raleigh, NC 27612 USA

(919) 847-4545 (tel)
Manufactures a board with software that plugs into a computer to receive radio weather faxes.

O'Gara Associates
21 Canal St., Lawrence, MA 01840 USA
(508) 683-8899 (tel)
(508) 686-4229 (fax)
Visa, MasterCard, Discover, American Express
Sells remanufactured printers for a wide variety of computer applications.

Ohio Automation
7840 Angel Ridge Rd., Athens, OH 45701 USA
(614) 592-1810 (tel)
Visa, MasterCard
Sells a computer program for designing printed circuit boards.

Ohm Acoustics
241 Taaffe Pl., Brooklyn, NY 11205 USA
(718) 783-1111 (tel)
Sells a wide variety of components (such as loudspeakers, repair kits, crossovers, etc.) for high-end home audio and professional audio applications.

Okidata
532 Fellowship Rd., Mount Laurel, NJ 08054 USA
(800) OKIDATA (tel)
(800) OKI-TEAM (tel)
(609) 235-2600 (tel)
(800) 654-6651 (fax back)
comments@okidata.com
http://www.okidata.com
Manufactures printers for personal computers.

OKI Semiconductor
785 Mary Ave., Sunnyvale, CA 94086 USA
(800) 654-6388 (tel)
(408) 720-1918 (fax)
Manufactures integrated circuits for a wide variety of electronics applications.

Oklahoma Comm Center
13424 Railway Dr., Oklahoma City, OK 73114 USA

Oklahoma Comm Center

(800) 765-4267 (tel)
(405) 748-3066 (tel)
(405) 748-3066 (fax)
Visa, MasterCard, Discover, COD
Sells a wide variety of amateur radio equipment and accessories, including amateur radio transceivers, receivers, antenna tuners, antennas, amplifiers, etc.

Old Colony Sound Lab

P.O. Box 243, Peterborough, NH 03458 USA
(603) 924-6371 (tel)
(603) 924-6526 (tel)
(603) 924-9467 (fax)
Visa, MasterCard, Discover
Manufactures a variety of high-end audio kits, including amplifiers, crossovers and preamplifiers.

Old Tyme Radio

2445 Lyttonsville Rd., #317, Silver Springs, MD 20910 USA
(301) 585-8776 (tel)
Sells parts to repair antique radios.

Omar Electronics

2130 GA Hwy. 81 SW, Loganville, GA 30249 USA
(404) 466-3241 (tel)
(404) 466-1952 (tel)
(404) 466-9013 (fax)
Sells amateur radio equipment and accessories, including amateur radio transceivers, receivers, antenna tuners, antennas, etc.

Omation

253 Polaris Ave., Mountain View, CA 94043 USA
(415) 966-1396 (tel)
(415) 966-8205 (fax)
Manufactures envelope openers for a wide variety of office applications.

Omega Technologies

1 Omega Dr., P.O. Box 4047, Stamford, CT 06907 USA
(800) 826-6342 (tel)
(203) 359-1660 (tel)
(800) 848-4271 (fax back)
(203) 359-7700 (fax)
(412) 967-5363 (BBS)
http://www.industry.net/omega
Manufactures and sells signal conditioners, plug-in computer boards, temperature/humidity data loggers, data acquisition systems, etc. for a wide variety of professional and industrial applications.

Omnicron Electronics

581 Liberty Hwy., P.O. Box 623, Putnam, CT 06260 USA
(203) 928-0377 (tel)
(203) 928-6477 (fax)
Manufactures and sells voice logging recorders for professional two-way radio and telephone applications.

Omni Data Communications

906 N. Main, #3, Witchita, KS 67203 USA
(800) 922-2329 (tel)
(316) 264-5589 (tel)
odcom@aol.com
Visa, MasterCard, COD ($4.75)
Sells modems for computer data applications.

Omni Data Systems

(800) 725-2345 (tel)
(713) 469-4365 (tel)
(713) 469-5841 (fax)
Buys and sells computer systems, printers, terminals, etc. for a wide variety of computer applications.

Omni Electronics

1007 San Dario, Laredo, TX 78040 USA
(800) 827-6664 (tel)
(512) 722-5195 (tel)
(512) 722-6664 (tel)
(512) 722-8184 (fax)
Sells amateur radio equipment and accessories, including amateur radio transceivers, receivers, antenna tuners, antennas, amplifiers, etc.

Omni Electronics

174 Dalkeith Rd., Edinburgh EH16 5DX UK
0131-667-2611 (tel)
Visa, Access
Sells a wide variety of electronics components. Offers a catalog for £2.

OmniModels
P.O. Box 708, Mahomet, IL 61853 USA
(800) 342-6464 (tel)
(217) 398-7738 (tel)
(217) 398-7731 (fax)
Visa, MasterCard, Discover, COD ($5)
Sells radios, batteries, battery chargers, speed controls, etc. for radio-control models.

Omnionics
P.O. Box 684, 2970 Main St., Peru, NY 12972 USA
(800) 533-9222 (tel)
Manufactures a sound and light effects box for radio-control models of warplanes.

OmniTester
423 Commerce Ln., Berlin, NJ 08009 USA
(800) 523-2283 (tel)
(609) 767-7617 (tel)
(609) 767-4289 (fax)
Manufactures cable/harness testers for a variety of professional industrial applications.

Oneac
27944 N. Bradley Rd., Libertyville, IL 60048 USA
(800) 327-8801 (tel)
(708) 816-6000 (tel)
(708) 680-5124 (fax)
Manufactures uninterruptible power supplies for a wide variety of personal computer applications.

One Call West
418 W. Riverside, Spokane, WA 99201 USA
(800) 633-5255 (tel)
(509) 838-4387 (fax)
onecall@iea.com
Visa, MasterCard, Discover, American Express, COD
Sells high-end home and amateur audio and video equipment, such as camcorders, VCRs, speakers, car audio equipment, TVs, home theater equipment, etc.

One Stop Micro
524 Prospect Ave., Little Silver, NJ 07739 USA
(800) 248-9666 (tel)
(908) 741-8888 (tel)
Visa, MasterCard, Discover, American Express
Sells personal home and laptop computers, printers, etc.

Onkyo
200 Williams Dr., Ramsey, NJ 07446 USA
(201) 825-7950 (tel)
(201) 934-1845 (fax)
Manufactures equipment for home stereo applications.

Onset Computer
P.O. Box 3450, 536 MacArthur Blvd., Pocasset, MA 02559 USA
(508) 563-9000 (tel)
(508) 563-9477 (fax)
Manufactures data logger/controller boards, etc. for professional industrial applications.

Ontario Surplus
585 Ridge Rd., Ontario, NY 14519 USA
(716) 2665-9096 (tel)
Sells surplus electronics, such as portable radio towers and hard-line cable for a wide variety of radio and electronics applications.

Opcode
3950 Federal Way, #100, Palo Alto, CA 94303 USA
(415) 856-3333 (tel)
(415) 856-0777 (fax)
Manufactures MIDI cards for a wide variety of home and professional computerized musical applications.

Open Connect Systems
2711 LBJ Freeway, Dallas, TX 75234 USA
(214) 484-5200 (tel)
(214) 888-0688 (fax)
Manufactures servers for professional computer network applications.

Operating Technical Electronics
850 Greenview Dr., Grand Prarie, TX 75050 USA
(214) 988-6828 (tel)
(214) 641-7089 (fax)
Manufactures power supplies for a wide variety of electronics applications.

Optec Sales
18221 Edison Ave., Chesterfield, MO 63005 USA
(315) 530-9170 (tel)
(315) 530-9169 (fax)
Manufactures magnet wire for a wide variety of electronics applications.

Optibase
5000 Quorum Dr., #700, Dallas, TX 75240 USA
(214) 774-3800 (tel)
Manufactures digital serial interfaces for computer audio/video applications.

Optical Data Systems
1101 E. Arapaho Rd., Richardson, TX 75081 USA
(214) 234-6400 (tel)
(214) 301-3893 (fax)
Manufactures routers for professional computer network applications.

Optical Media International
51 E. Campbell Ave., Ste. 170, Campbell, CA 95008 USA
(408) 376-3511 (tel)
(408) 376-3519 (fax)
sales@optmedia.com
Manufactures and sells CD-ROM recording systems for personal computer applications.

Optik Technology
1215 W. Crosby Rd., Carrollton, TX 75006 USA
(800) 481-7819 (tel)
(214) 323-2200 (tel)
(214) 323-2396 (fax)
Manufactures sensors for a wide variety of electronics applications.

Optima
P.O. Box 1366, Covina, CA 91722 USA
(818) 855-1848 (fax)
Sells discount PCs, hard drives, CD-ROMs, fax modems, etc.

Optima Batteries
770 W. 17th St., Costa Mesa, CA 92627 USA
(800) 311-3591 (tel)
Visa, MasterCard, Discover, American Express
Manufactures batteries for mobile audio applications.

Optimal Computing Systems
2100 Hwy. 360 N., #800, Grand Prarie, TX 75050 USA
(800) 211-0383 (tel)
(214) 641-1898 (tel)
(214) 641-2010 (fax)
Visa, MasterCard
Manufactures and sells personal desktop computers; also sells computer parts and boards, such as tape drives, hard drives, monitors, motherboards, modems, video adapters, controllers, etc. Offers a 30-day money-back guarantee.

OPTIM Electronics
Middlebrook Tech Park, 12401 Middlebrook Rd., Germantown, MD 20874 USA
(301) 428-7200 (tel)
(301) 353-0129 (fax)
Manufactures data acquisition systems for a wide variety of professional industrial applications.

Optimum Data
5018 Levenworth St., Omaha, NE 68106 USA
(402) 575-3000 (tel)
(813) 826-0017 (East Coast tel)
(402) 575-2011 (fax)
jturco@optimumdata.com
Manufactures modems, multiplexers, packet switches, hubs, CSU/DSUs, remote access servers, routers, interface converters, data compressors, etc. for professional computer network applications.

Optoelectronics
5821 NE 14th Ave., Ft. Lauderdale, FL 33334 USA
(800) 327-5912 (tel)
(305) 771-2050 (tel)
(305) 771-2052 (fax)
Visa, MasterCard, COD (cash or money order only)
Manufactures and sells electronic equipment (primarily test equipment), including a variety of frequency counters, computer control interfaces, tone decoders, and Reaction Tuners for use with scanners.

Optronics
7 Stuart Rd., Chelmsford, MA 01824 USA
(508) 256-4511 (tel)
Manufactures computer image scanners for professional desktop publishing applications.

Opus Systems
3000 Coronado Dr., Santa Clara, CA 95054 USA
(408) 562-9340 (tel)
(408) 562-9341 (fax)
Manufactures workstations for a wide variety of professional computer network applications.

ORA Electronics
9410 Owensmouth Ave., Chatsworth, CA 91311 USA
(818) 772-2700 (tel)
(818) 718-8626 (fax)
Manufactures activation centers, etc. for professional cellular microwave radio applications.

Orbacom Systems
1704 Taylors Ln., Cinnaminson, NJ 08077 USA
(609) 829-4455 (tel)
(609) 829-6980 (fax)
Manufactures dispatch consoles for professional communication applications.

Orban
1525 Alvarado St., San Leandro, CA 94577
(510) 351-3500 (tel)
(510) 351-0500 (fax)
custserv@orban.com
Manufactures audio processors for professional broadcast stations.

Orbit Semiconductor
1215 Bordeaux Dr., Sunnyvale, CA 94089 USA
(800) 331-4617 (tel)
(800) 647-0222 (Calif tel)
(408) 744-1800 (tel)
(408) 747-1263 (fax)
Manufactures semiconductors for a wide variety of electronics applications.

Orca
1531 Lookout Dr., Agoura, CA 91301 USA
(818) 707-1629 (tel)
(818) 991-3072 (fax)
Manufactures speaker damping material and audio cable for high-end home audio and professional audio applications.

Orchid Technology
232 E. Warren Ave., Fremont, CA 94539 USA
(800) 7ORCHID (tel)
(510) 651-2300 (tel)
(510) 651-5612 (fax)
Manufactures sound cards for home computer applications.

Orion
118 W. Julie Dr., Tempe, AZ 85283 USA
(602) 730-8200 (tel)
(602) 831-8101 (fax)
Manufactures amplifiers and loudspeakers for car stereo applications.

Orion Telecom
1925 E. Beltline Rd., #405, Carrollton, TX 75006 USA
(800) 669-8088 (tel)
(800) 371-1760 (govern and corp order tel)
(214) 416-3720 (tel)
(214) 416-4141 (fax)
(214) 416-6248 (fax back)
Visa, MasterCard, American Express, COD
Manufactures and sells fax switches, modems, and voice mailbox systems for computer applications.

Ortel
2015 W. Chestnut St., Alhambra, CA 91803 USA
(818) 281-3636 (tel)
(818) 281-8231 (fax)
mktbba@ortel.com
Manufactures fiberoptic systems for professional cable TV applications.

Ortofon
65 E. Bethpage Rd., Plainview, NY 11803 USA

Ortofon

(516) 454-6570 (tel)
(516) 454-6515 (fax)
Manufactures phono cartridges for home stereo applications.

Allen Osborne Associates
756 Lakefield Rd., Westlake Village, CA 91361 USA
(805) 495-8420 (tel)
(805) 373-6067 (fax)
Manufactures a variety of GPS systems for professional aviation applications.

O.S. Computers
58 Second St., 5th Fl., San Francisco, CA 94105 USA
(800) 938-6722 (tel)
(415) 543-6082 (cust serv tel)
(415) 543-6095 (fax)
Visa, MasterCard, Discover, American Express, COD ($5)
Sells hardware (such as scanners, plotters, printers, video cards, monitors, etc.) and software for computerized art applications.

Ositech Communications
679 Southgate Dr., Guelph, ON N1G 4S2 Canada
(800) 563-2386 (tel)
(519) 836-8063 (tel)
(519) 836-6156 (fax)
http://www.ositech.com
Manufactures laptop modem cards for a wide variety of computer applications.

Oslo Switch
328 Industrial Ave., Cheshire, CT 06410 USA
(203) 272-2794 (tel)
Manufactures switches for a wide variety of electronics applications. Offers a free catalog.

Otari
378 Vintage Park Dr., Foster City, CA 94404 USA
(415) 341-5900 (tel)
(415) 341-7200 (fax)
(818) 972-3687 (Los Angeles sales tel)
+81-4-2481-8626 (Japan tel)
+81-4-2481-8633 (Japan fax)
+49-2159-50861 (Germany tel)
+49-2159-50862 (Germany tel)
+49-2159-1778 (Germany fax)
+65-284-7211 (Singapore tel)
+65-284-4727 (Singapore fax)
Manufactures digital audio recorders and open-reel tape decks for professional studio recording and professional broadcasting.

Ott Machine Services
P.O. Box 1701, Lombard, IL 60148 USA
(708) 889-8030 (tel)
(708) 964-0587 (tel)
(708) 719-0114 (fax)
Manufactures and sells digital command controls for model railroads.

Otto
2 E. Main St., Carpentersville, IL 60110 USA
(708) 428-7171 (tel)
(708) 428-1956 (fax)
Manufactures headsets, headphones, microphones, and surveillance kits for a wide variety of professional two-way radio communications.

Outbacker Antenna Sales
330 Cedar Glen Circle, Chattanooga, TN 37412 USA
(615) 899-3390 (tel)
Imports amateur radio antennas into North America from Terlin Aerials, Australia.

Outdoor Outfitters
705 Elm Ct., Waukesha, WI 53186 USA
(414) 542-7772 (tel)
(414) 542-4435 (fax)
(800) 558-2020 (tel)
Manufactures metal detectors for hobbyists. Offers free information.

Ovation
P.O. Box 507, Bloomfield, CT 06002 USA
http://www.kamanmusic.com
Manufactures electric guitars for professional and personal musical applications.

Overcity International
Flat 1, 21/F, Wing Hing Industrial Bldg., 83-93 Chai

Wan Kok St., Tsuen Wan, Hong Kong
852-2413-9388 (tel)
852-2413-1920 (fax)
Manufactures motherboards for a wide variety of computer applications.

Overland Data

8975 Balboa Ave., San Diego, CA 92123 USA
(619) 571-5555 (tel)
(619) 571-3664 (fax)
+44-1734-891891 (UK tel)
+44-1734-891897 (UK fax)
odisales@ovrland.com
ftp://ovrland.com/pub/odisales
http://www.ovrland.com/_7Eodisales
Manufactures tape drives and controllers for professional and personal computer applications.

Oxley

P.O. Box 814, Branford, CT 06405 USA
(203) 488-4135 (tel)
(203) 488-1033 (tel)
Manufactures surface-mount test points for a wide variety of electronics applications.

PacComm Packet Radio Systems

4413 N. Hesperides St., Tampa, FL 33614 USA
(800) 486-7388 (tel)
(813) 874-2980 (tel)
(813) 875-6417 (tech tel)
(813) 876-0351 (purchase tel)
(813) 872-8696 (fax)
(813) 874-3078 (BBS)
pico@paccomm.com
http://www.paccomm.com
ftp.paccomm.com (Internet FTP site)
Manufactures controllers, modems, digital transceivers, and adapters for amateur packet radio applications.

Pacific Avionics

8640 154th Ave. NE, Redmond, WA 98052 USA
(800) PAC-4629 (tel)
http://www.pacav.com
Manufactures VXI/VME modules for professional electronics applications.

Pacific Coast Micro

4901 Morena Blvd., #1111, San Diego, CA 92117 USA
(800) 581-6040 (tel)
(619) 581-1439 (cust serv tel)
(619) 581-0125 (fax)
Visa, MasterCard, Discover, American Express
Sells computer parts and boards, such as hard drives, motherboards, memory chips, controllers, video cards, modems, etc.

Pacific Coast Parts Distributor

15014 Staff Ct., Gardena, CA 90248 USA
(800) 421-5080 (tel)
(800) 262-1312 (tel)
(310) 515-0207 (tel)
(310) 538-4919 (fax)
Sells replacement parts and accessories for audio equipment.

Pacific Computer Exchange

1031 SE Mill, #B, Portland, OR 97214 USA
(503) 236-2949 (tel)
Visa, MasterCard, Discover, COD
Buys and sells Tandy personal desktop computers and parts and boards.

Pacific Computer Products

1255 Birchwood Dr., Sunnyvale, CA 94089 USA
(800) 733-9730 (tel)
(408) 734-4800 (fax)
pcp@best.com
http://www.pcp.com/pcp
Visa, MasterCard, Discover, American Express
Sells computer parts and boards, such as hard drives, motherboards, memory chips, cables, video cards, multimedia upgrades, modems, etc.

Pacific Electronics

1580 Oakland Rd., C112, San Jose, CA 95131 USA
(408) 436-8080 (tel)
(408) 436-8883 (fax)
Buys, sells, and trades personal computer systems, terminals, and disk drives for a wide variety of computer applications.

Pacific Power Source
15122 Bolsa Chica St., Huntington Beach, CA 92649 USA
(800) 854-2433 (tel)
(714) 898-2691 (tel)
(714) 898-8076 (fax)
Manufactures power controllers for a variety of professional industrial applications.

Pacific Recorders & Engineering
2070 Las Palmas Dr., Carlsbad, CA 92009 USA
(619) 438-3911 (tel)
(619) 438-9277 (fax)
Manufactures mixing consoles for professional studio recording and professional broadcasting.

Packard Bell
8285 W. 3500 S., Magna, UT 84044 USA
(800) 733-4411 (tel)
(801) 579-0161 (tel)
(801) 579-0092 (tech support fax)
(801) 250-1600 (BBS)
http://www.packardbell.com
Manufactures complete computer systems for a wide variety of applications.

Page Computer
4665 Melrose Ave., Los Angeles, CA 90029 USA
(800) 886-0055 (tel)
(213) 665-7777 (tel)
(213) 660-0444 (fax)
Visa, MasterCard, Discover, American Express
Sells personal desktop and laptop computers and printers; also sells IBM computer parts and boards.

Pagecorp Industries (PCI)
366 San Miguel, Ste. 211, Newport Beach, CA 92660 USA
(800) 957-8700 (tel)
Manufactures and sells crystals for pager and two-way radio applications.

PageTap
(800) 735-3650 (tel)
(303) 337-4811 (tel)
(303) 337-3084 (fax)
Manufactures remote control switches for professional cellular microwave radio applications.

Pagine
1961A Concourse Dr., San Jose, CA 95131 USA
(408) 944-9200 (tel)
(408) 944-9728 (fax)
Manufactures terminals for a wide variety of professional computer network applications.

PairGain Technologies
14404 Franklin Ave., Tustin, CA 92680 USA
(800) 638-0031 (tel)
(714) 832-9922 (tel)
(714) 832-9924 (fax)
sales@pairgain.com
http://www.pairgain.com
Manufactures Ethernet bridges for professional computer network applications.

Palomar Engineers
P.O. Box 462222, Escondido, CA 92046 USA
(619) 747-3343 (tel)
(619) 747-3346 (fax)
Visa, MasterCard
Manufactures and sells a number of different antennas, baluns, and preamplifiers for shortwave listeners, amateur radio operators, and other radio hobbyists. Offers a free catalog.

Panamax
150 Mitchell Blvd., San Rafael, CA 94903 USA
(800) 472-5555 (tel)
(415) 499-3900 (tel)
Manufactures surge protectors/line conditioners for electronic equipment. Offers a lifetime product and a $5,000 connected equipment warranty.

Pana-Pacific
541 Division St., Campbell, CA 95008 USA
(800) 726-2333 (tel)
1007 Commerce Ct., Buffalo Grove, IL 60089 USA
(800) 726-2777 (tel)
10340 SW Nimbus Ave., #N-A, Portland, OR 97223 USA

(800) 457-2333 (tel)
8375 Camino Santa Fe, #D, San Diego, CA 92121 USA
(800) 275-2333 (tel)
6333 Chalet Dr., Commerce, CA 90040 USA
(800) 726-2689 (tel)
1901 E. Carnegie Ave., #1C, Santa Ana, CA 92705 USA
(800) 726-2234 (tel)
Sells telephones and accessories for cellular telephone applications; also sells two-way radio equipment.

Panasonic
1 Panasonic Way, Seacaucus, NJ 07094 USA
(800) 742-8086 (tel)
(800) 365-1515 X333 (tel)
(201) 348-7000 (tel)
(201) 392-4500 (tel)
Manufactures laptop computers, computer monitors and printers, televisions, and a wide variety of portable stereo equipment and radios for home stereo and shortwave radio listening applications.

Pan-Com International
P.O. Box 130, Paradise, CA 95967 USA
Sells hundreds of kits, plans, and books for licensed and unlicensed broadcasting; ham, CB, shortwave, and longwave radio; surveillance, phone devices, etc. Offers a catalog for $1.

The Panda Project
5201 Congress Ave., Ste. C100, Boca Raton, FL 33487 USA
(407) 994-2300 (tel)
(407) 994-0191 (fax)
sysinfo@archistrat.com
Manufactures computer servers and personal workstations.

Panduit
1333 Schoolhouse Rd., New Lenox, IL 60451 USA
(800) 777-3300 (tel)
(708) 538-1811 (fax)
1819 Atlanta Hwy., Cumming, GA 30130 USA
Manufactures wire disconnects for a wide variety of electronics applications.

Panelight Display Systems
101 The Embarcadero, #100-A, San Francisco, CA 94105 USA
P.O. Box 190940, San Francisco, CA 94119 USA
(800) 726-3599 (tel)
(415) 772-5800 (tel)
(415) 986-3817 (fax)
Sells LCD projectors, projection panels, monitors, scan converters, etc. for a wide variety of video and multimedia applications. Offers a free catalog.

Panson Electronics
I-80 & New Maple Ave., Pine Brook, NJ 07058 USA
Sells replacement parts and accessories for audio equipment.

Pappy John's R/C Warehouse
RR #2, Box 1885, S.R. 7, Newport, ME 04953 USA
(800) 973-7244 (tel)
(207) 257-2305 (tel)
(800) 279-6532 (fax)
Sells radio-control systems, etc. for radio-control model airplanes.

Paradigm Electronics
P.O. Box 2410, Niagra Falls, NY 14302 USA
(905) 632-0180 (tel)
101 Hanlan Rd., Woodbridge, ON L4L 3P5 Canada
(905) 850-2889 (tel)
Manufactures and sells high-end speakers for audio and home theater applications.

Paralan
7875 Convoy Ct., San Diego, CA 92111 USA
(619) 560-7266 (tel)
(619) 560-8929 (fax)
Manufactures SCSI bus regenerators for professional computer network applications.

Parallax
3805 Atherton Rd. #102, Rocklin, CA 95765 USA
(916) 624-8333 (tel)
(916) 624-8003 (fax)
(916) 624-1869 (fax back)
(916) 624-7101 (BBS)

Parallax

info@parallaxinc.com
ftp.parallaxinc.com
http://www.parallaxinc.com
Manufactures BASIC stamp modules and PIC development boards for a wide variety of computer and electronics applications.

Paramax

251 Jeanelle Dr., Ste. 3, Carson City, NV 89703 USA
(800) 473-8080 (tel)
http://www.digitalage.com/vm/paramax
Manufactures stepper motor controller boards for a wide variety of electronics applications.

Paramount Audio

12737 Moore St., Cerritos, CA 90703 USA
(310) 483-8111 (tel)
(310) 483-8106 (fax)
Manufactures amplifiers and loudspeakers for car stereo applications.

Paramount Computer

1607 Cuernavaca #400, Austin, TX 78733 USA
(512) 263-7010 (tel)
(512) 263-7018 (fax)
Sells personal desktop and network computer systems and printers; also sells computer parts and boards, such as hard drives, memory chips, multiplexers, monitors, etc.

Paramount Video & Audio

110 Lincoln Hwy., Box 15, Langhorne, PA 19030 USA
(800) 477-0330 (tel)
(215) 949-2860 (tel)
(215) 949-1085 (fax)
Visa, MasterCard, Discover, American Express, Canon, COD
Sells high-end home, amateur, and professional audio and video equipment, such as camcorders, VCRs, video editors, monitors, TVs, CD and cassette decks, etc.

Parasound Products

950 Battery St., San Francisco, CA 94111 USA
(800) 822-8802 (tel)
(415) 397-7100 (tel)
(415) 397-0144 (fax)
Manufactures and sells CD changers and high-end amplifiers for audio and home theater applications.

Par Electronics

6869 Bayshore Dr., Lantana, FL 33462 USA
(407) 586-8278 (tel)
(407) 582-1234 (fax)
Manufactures a notch filter for VHF scanners.

Parity Systems

110 Knowles Dr., Los Gatos, CA 95030 USA
(800) 514-4080 (tel)
(408) 378-1000 (tel)
(408) 378-1022 (fax)
Manufactures workstations and data storage systems for a wide variety of professional computer network applications.

Parker Guitars

89 Frost St., Westbury, NY 11590 USA
Manufactures electric guitars for professional and personal musical applications.

Parker Hannefin

77 Dragon Ct., Woburn, MA 01888 USA
(617) 935-4850 (tel)
Manufactures EMI-suppressing ferrites for a wide variety of electronics applications.

The Parts Connection

2790 Brighton Rd., Oakville, ON L6H 5T4 Canada
(800) 769-0747 (tel)
(905) 829-5858 (tel)
(905) 829-5388 (fax)
Visa, MasterCard, American Express
Sells a wide variety of high-end audio kits and electronics, including cable, jacks, capacitors, tubes, transformers, etc. Offers a free catalog.

Parts Express

340 E. First St., Dayton, OH 45402 USA
(800) 338-0531 (tel)
Sells a variety of electronics components for con-

sumer electronics, and high-end home audio equipment, such as speakers, cassette decks, amplifiers, preamplifiers, etc. Offers a free 188-page catalog.

PartStock
2920 Talmage Ave. SE, Minneapolis, MN 55414 USA
(612) 378-3996 (tel)
(612) 378-7299 (fax)
Visa, MasterCard
Sells IBM computer parts, including system boards, hard drives, memory chips, power supplies, adapters, etc. for a wide variety of computer applications.

Paul Reed Smith Guitars
1812 Virginia Ave., Annapolis, MD 21401 USA
Manufactures electric guitars for professional and personal musical applications. Offers a brochure for $5.

PC America
60 N. Harrison Ave., Congers, NY 10920 USA
(800) PC-AMERICA (tel)
(914) 267-3500 (tel)
(914) 267-3550 (fax)
Visa, MasterCard, American Express
Manufactures computerized cash registers, UPC scanners, etc. for business applications.

PCB Piezotronics
3425 Walden Ave., Depew, NY 14043 USA
(716) 684-0001 (tel)
(716) 684-0987 (fax)
Manufactures pressure sensors for professional industrial applications.

PC Computer Solutions
130 W. 32nd Ave., New York, NY 10001 USA
(800) 573-6245 (tel)
(212) 629-8300 (tel)
(212) 629-0798 (fax)
14 W. 45th St., New York, NY 10036 USA
(212) 869-4411 (tel)
(212) 869-4486 (fax)
Sells personal desktop and laptop computers, printers, and scanners; also sells computer parts and boards, such as memory chips, cases, monitors, fax modems, etc.

PC Connection
6 Mill St., Marlow, NH 03456 USA
(800) 800-0003 (tel)
(800) 600-9145 (tel)
(800) 600-9243 (tel)
(800) 600-9245 (tel)
(800) 600-9251 (tel)
(800) 600-8253 (tel)
(800) 600-8252 (tel)
(603) 446-0004 (tel)
(603) 446-7791 (fax)
Visa, MasterCard, Discover, American Express
Sells image scanners, printers, hard drives, modems, and software for a wide variety of computer applications.

PC Direct
Corner Cook & Nelson Sts., P. O. Box 38-713, Freepost 4300, Auckland, New Zealand
0800-377-677 (tel)
09-356-5300 (tel)
09-378-9917 (fax)
Sells personal desktop and laptop computers, printers, and scanners; also sells computer parts and boards, such as memory chips, cases, monitors, fax modems, etc.

P.C. Electronics
2522 S. Paxson Ln., Arcadia, CA 91007 USA
(818) 447-4565 (tel)
Visa, Master Card, COD
Manufactures and sells amateur television (ATV) downconverters, transceivers, and transmitters. Offers a free 10-page catalog.

PC Importers
8295 Darrow Rd., Twinsburg, OH 44087 USA
(800) 619-9119 (tel)
(800) 886-5155 (tel)
(800) 698-3820 (tech support tel)
(216) 487-5242 (fax)
Visa, MasterCard, Discover, American Express, COD
Manufactures and sells personal desktop computers; also sells motherboards, hard disk drives, controllers, keyboards, video cards, modems, monitors, memory chips, etc. Offers a 30-day money-back guarantee.

PC Interactive
8295 Darrow Rd., Twinsburg, OH 44087 USA
(800) 347-3807 (tel)
(800) 698-3820 (tech support tel)
(216) 487-5242 (fax)
Manufactures and sells personal desktop computers; also sells computer parts and boards, such as floppy drives, hard drives, monitors, video cards, motherboards, keyboards, etc. Offers a 30-day money-back guarantee.

PC International
8295 Darrow Rd., Twinsburg, OH 44087 USA
(800) 458-3133 (tel)
(800) 698-3820 (tech support tel)
(216) 487-5242 (fax)
Visa, MasterCard, Discover, American Express
Manufactures and sells personal desktop computers; also sells computer parts and boards, such as floppy drives, hard drives, monitors, video cards, modems, keyboards, etc. Offers a 30-day money-back guarantee.

PC Parts
1800 Paxton St., Harrisburg, PA 17104 USA
(800) 666-9373 (tel)
(800) 288-9373 (fax)
(717) 233-6650 (tel)
(717) 233-2774 (fax)
Visa, MasterCard, Discover, American Express
Sells personal desktop and laptop computers, printers, monitors, and computer parts and boards. Offers a free catalog.

PC Peripherals
3401 Industrial Ln., Unit E, Broomfield, CO 80020 USA
(303) 439-8650 (tel)
(303) 439-8651 (fax)
Sells disk drives for a wide variety of computer applications.

PC Power
309 Broadway, Newmarket, Auckland, New Zealand
0-9-529-0200 (tel)
0-9-529-0164 (fax)
62 Hurstmere Rd., Takapuna, Auckland, New Zealand
0-9-489-3100 (tel)
0-9-489-3101 (fax)
Manufactures personal computers.

PCRC
651 Ocean Ave., Bohemia, NY 11716 USA
(516) 563-4715 (tel)
(516) 563-4716 (fax)
(516) 286-1518 (BBS)
Visa, MasterCard, Discover
Manufactures and sells digital audio delays for repeaters for amateur radio applications.

PCs Compleat
34 St. Martin Dr., Marlboro, MA 01752 USA
(800) 578-4727 (tel)
(508) 624-6400 (tel)
sales@pcscompleat.com
Sells personal desktop and laptop computers, printers, and scanners. Offers a free product guide and a 30-day money-back guarantee.

PC Service Source
2350 Valley View Ln., Dallas, TX 75234 USA
(800) 727-2787 (tel)
(214) 406-8583 (tel)
(214) 406-9081 (fax)
Sells parts for a wide variety of computer applications.

PC Video Conversion
1340 Tully Rd., #309, San Jose, CA 95122 USA
(408) 279-2442 (tel)
(408) 279-6105 (fax)
Manufactures computer converter boards to produce NTSC, PAL, S-VHS, RGB, etc. video from a computer.

The PC Zone
15815 SE 37th St., Bellevue, WA 98006 USA
(800) 258-2088 (tel)
(206) 603-2500 (fax)
Visa, MasterCard, Discover, American Express
Sells personal desktop and laptop computers, printers, software, and scanners; also sells computer parts

and boards, such as floppy drives, hard drives, monitors, processor upgrades, video cards, modems, keyboards, etc.

Peach State Photo
1706 Chantilly Dr. NE, Atlanta, GA 30324 USA
(800) 766-9653 (tel)
(404) 633-2699 (customer service tel)
(404) 321-6316 (fax)
Visa, MasterCard, Discover, American Express
Sells high-end home, amateur, and professional photographic, audio, and video equipment, such as cameras, lenses, camcorders, VCRs, video editors, monitors, tripods, etc.

Peak Computing
244 W. Main St., Goshen, NY 10924 USA
(800) 732-5908 (tel)
(914) 294-6287 (fax)
Visa, MasterCard, Discover, American Express
Manufactures and sells personal desktop and laptop computers. Offers a 30-day money-back guarantee.

The Peak Technologies Group
8990 Old Annapolis Rd., Columbia, MD 20871 USA
(800) 950-6372 (tel)
(410) 992-9922 (tel)
(410) 992-4520 (fax)
Manufactures data storage systems for a wide variety of professional computer network applications.

Pearl Electronics
312 Dexter Ave. N., Seattle, WA 98109 USA
(206) 622-6200 (tel)
Sells receivers, antennas, and accessories for shortwave-listening applications.

Peavey Electronics
711 A St., Meridian, MS 39301 USA
(601) 483-5365 (tel)
(601) 486-1278 (fax)
peavey@aol.com
http://www.peavey.com
Manufactures synthesized sound modules, keyboards, amplifiers, mixers, microphones, PA speakers, etc. for professional studio recording and live music performances.

Pembleton Electronics
1222 Progress Rd., Ft. Wayne, IN 46808 USA
(219) 484-1812 (tel)
(219) 484-0163 (fax)
Sells new and surplus components for a wide variety of electronics applications.

Peninsula Computer
5336 George Washington Hwy., Yorktown, VA 23692 USA
(804) 875-0215 (tel)
(804) 874-1737 (fax)
Sells personal desktop computers and parts, such as hard drives, motherboards, CD-ROM drives, etc.

J.C. Penney
National Parts Center
6840 Barton Rd., Morrow, GA 30260 USA
(800) 933-7115 (tel)
(404) 961-8408 (fax)
Manufactures and sells a wide variety of consumer electronics products.

Penril Datability Networks
1300 Quince Orchard Blvd., Gaithersburg, MD 20878 USA
(800) 4-PENRIL (tel)
(301) 921-8600 (tel)
(301) 921-8376 (fax)
Manufactures routers for professional computer network applications.

Penstock
10015 Old Columbia Rd., Ste. D235, Columbia, MD 21046 USA
(800) 736-7862 (tel)
(408) 730-0300 (tel)
(613) 592-6088 (Canada tel)
http://www.penstock.avnet.com
Sells coaxial attenuators, terminators, adapters, test cables, etc. for a wide variety of electronics testing applications. Sells more than 10,000 different products.

PEP Modular Computers
750 Holiday Dr., Bldg. 9, Pittsburgh, PA 15220 USA

PEP Modular Computers

(800) 228-1737 (tel)
(412) 921-3356 (tel)
5010 E. Shea Blvd., #C226, Scottsdale, AZ 85254 USA
(602) 483-7100 (tel)
Manufactures single-board computers for a wide variety of computer applications.

Lee Pepper Sound Studios

3983 Renate Dr., Las Vegas, NV 89103 USA
(702) 247-8929 (tel)
Sells a variety of equipment for professional studio recording and live music performances.

Percomp Microsystems

1111 Digital Dr., Richardson, TX 75081 USA
(800) 856-6688 (tel)
(214) 234-8658 (fax)
http://www.percomp.com
Visa, MasterCard, Discover, American Express
Manufactures and sells personal desktop computers.

Michael Percy

P.O. Box 526, Inverness, CA 94937 USA
(415) 669-7181 (tel)
(415) 669-7558 (fax)
Sells a variety of high-end audio parts and components, including diodes, inductors, switches, plugs, attenuators, tubes, tools, ICs, etc. Offers a free catalog.

Performance Cable TV Products

P.O. Box 9947, Roswell, GA 30077 USA
(800) 2799-6330 (tel)
jud.williams@industry.net
Manufactures standby power supply systems for a variety of professional electronics applications.

Performance Hobby

1293 NW Wall #1492, Bend, OR 97701 USA
(800) 308-6160 (tel)
(541) 389-6160 (tel)
Manufactures batteries and battery protectors for radio-control models.

Performance Motion Devices

97 Lowell Rd., Corcord, MA 01742 USA
(508) 369-3302 (tel)
(508) 369-3819 (fax)
Manufactures motion control system boards for a wide variety of computer applications.

Performance Technology

800 Lincoln Center, 7800IH10 W., San Antonio, TX 78230 USA
(800) 327-8526 (tel)
(210) 349-2000 (tel)
http://www.perftech.com
Manufactures router/modems for professional network computer applications.

Periphex

300 Centre St., Hollbrook, MA 02343 USA
(800) 634-8132 (tel)
(617) 767-5516 (tel)
(617) 767-4599 (fax)
Manufactures and sells batteries for amateur radio transceivers, camcorders, cordless phones, NiCads, lithium cells, and custom battery packs.

Peripherals Unlimited

240 Mayfield Dr., #103, Smyrna, TN 37167 USA
(615) 459-3639 (tel)
(615) 355-6386 (fax)
Sells printers and terminals for a wide variety of computer applications.

Peripheral Systems (PSI)

150 Wright Brothers Dr., #550, Salt Lake City, UT 84116 USA
(801) 521-0366 (tel)
(801) 521-0383 (fax)
Manufactures handheld processing terminals for a wide variety of professional computer applications.

Persoft

465 Science Dr., Madison, WI 53711 USA
(800) EMULATE (tel)
(608) 273-6000 (tel)
(608) 273-8227 (fax)

Manufactures bridges for professional computer network applications.

Personal Security
P.O. Box 579361, Modesto, CA 95357 USA
Sells stun guns. Offers a free catalog.

A.K. Peters
289 Linden St., Wellesley, MA 02181 USA
(617) 235-2210 (tel)
(617) 235-2404 (fax)
kpeters@geom.umn.edu
Manufactures and sells "The Rug Warrior" mobile robot kit.

Petras
Petras Industrial Plaza, 1313 Knight St., Arlington, TX 76015 USA
(817) 468-5780 (tel)
Manufactures crossovers and loudspeakers for mobile audio applications.

Phase Linear
25 Tri-State International Office Ctr., #400, Lincolnshire, IL 60069 USA
(800) 753-6736 (tel)
(708) 317-3826 (fax)
Manufactures amplifiers for car stereo applications.

Phase X Systems
19545 NW Von Neumann Dr., #210, Beaverton, OR 97006 USA
(800) 845-4064 (tel)
(503) 531-2400 (tel)
(503) 531-2401 (fax)
Manufactures terminals for a wide variety of professional computer network applications.

Phelps Instruments
2631 Hillside Ave., Norco, CA 91760 USA
(909) 279-7347 (tel)
Visa, MasterCard, Discover, American Express
Sells test equipment, including oscilloscopes, power supplies, signal generators, voltmeters, frequency counters, spectrum analyzers, etc. for a wide variety of electronics testing applications.

Philips Consumer Electronics
1 Philips Dr., Knoxville, TN 37914 USA
(800) 822-4788 (tel)
(800) 235-7373 (tel)
(615) 521-4316 (tel)
(615) 521-4391 (tel)
(800) 522-7464 (tel)
(315) 682-9006 (fax)
Service Department
401 Old Andrew Johnson Hwy., Jefferson City, TN 37760 USA
(800) 851-8885 (replacement parts tel)
(615) 475-8869 (tech tel)
(800) 535-3715 (fax)
Broadband Networks division
(800) 448-5171 (tel)
(800) 522-7464 (tel)
(315) 682-9006 (fax)
http://pps.philips.com
Manufactures and sells a variety of consumer audio and video electronics, including TVs, VCRs, home stereo equipment (including amplifiers, cassette decks, CD players, DCC decks, etc.), car stereo equipment (including head units, amplifiers, speakers, etc.), cable TV equipment, computer equipment (including CD-ROM recorders), and shortwave radios.

Philips Power Systems
Bd. de l'Europe 131, B 1301, Wavre, Belgium
+32-10-43-85-02 (tel)
+32-10-43-82-13 (fax)
Manufactures power supplies for a wide variety of electronics applications.

Phillips-Tech Electronics
P.O. Box 8533, Scottsdale, AZ 85252 USA
(602) 947-7700 (tel)
Visa, MasterCard, American Express, COD
Sells a variety of microwave TV antennas. Offers a free catalog and a lifetime guarantee.

Philly's Camera
27 S. 11th St., Philadelphia, PA 19107 USA
(800) 923-2022 (tel)
(215) 922-5130 (tel)
(215) 922-5135 (fax)

Philly's Camera

Visa, MasterCard, Discover, American Express, Diner's Club, COD
Sells a variety of optical, audio, and video equipment, including cameras, camcorders, CD players, receivers, cassette decks, video editors, VCRs, etc.

Philtec

P.O. Box 359, Arnold, MD 21012 USA
(410) 757-4404 (tel)
Manufactures fiberoptic displacement sensors for a wide variety of professional industrial applications.

Phoenix Contact

P.O. Box 4100, Harrisburg, PA 17111 USA
(800) 322-3225 (tel)
(717) 944-1625 (fax)
Manufactures terminal blocks for a wide variety of electronics applications.

Phoenix Gold

9300 N. Decatur, Portland, OR 97203 USA
(503) 288-2008 (tel)
(503) 978-3380 (fax)
Manufactures loudspeakers for car stereo applications.

PhoneMate

P.O. Box 2914, Torrance, CA 90509 USA
20665 Manhattan Pl., Torrance, CA 90501 USA
(800) 322-9995 (tel)
(310) 618-9910 (tel)
(310) 533-0187 (fax)
(410) 686-3015 (literature request tel)
Manufactures telephones and answering machines.

Photocomm

7681 E. Gray Rd., P.O. Box 14670, Scottsdale, AZ 85267 USA
(800) 223-9580 (tel)
(602) 948-8003 (tel)
(602) 483-6431 (fax)
Sells photovoltaic (solar) power systems for a wide variety of professional electronics applications.

Pico Electronics

453 N. MacQuesten Pkwy., Mt. Vernon, NY 10552 USA
(800) 431-1064 (tel)
(914) 699-5514 (tel)
(914) 699-5565 (fax)
Manufactures surface-mount transformers and inductors and power modules for a wide variety of electronics applications. Offers a free catalog.

Pico Macom

12500 Foothill Blvd., Lakeview Terrace, CA 91342 USA
(800) 421-6511 (tel)
(818) 899-1165 (fax)
Manufactures distribution amplifiers for professional cable TV applications.

Pico Products

6315 Fly Rd., E. Syracuse, NY 13057 USA
(800) 822-7420 (tel)
(315) 437-1711 (tel)
(315) 437-7525 (fax)
Manufactures line amplifiers, power supplies, and power inserters for professional cable TV applications.

Pico Technology

Broadway House, 149-151 St. Neots Rd., Hardwick, Cambridge, CB3 7QJ UK
44-0-1954-211716 (tel)
44-0-1954-211880 (fax)
100073.2365@compuserve.com
Visa, Access
Manufactures thermocouple-to-PC converters and logic analyzers for computerized electronics testing applications.

Picotronic

Zollamtstr. 48, 67663 Kaiserslautern, Germany
49-631-29187 (tel)
49-631-29579 (fax)
49-631-21048 (BBS)
Manufactures microwave audio/video transmitters and receivers for hobby and experimental applications.

P.I.E. Engineering
2296 Williamston Rd., Williamston, MI 48895 USA
(800) 628-3185 (tel)
(517) 655-5523 (tel)
Visa, MasterCard, Discover
Manufactures dual mice adapters. Offers a 30-day money-back guarantee.

Pika Technologies
155 Terence Matthews Cres., Kanata, ON K2M 2A8 Canada
(613) 591-1555 (tel)
(613) 591-1488 (fax)
pika.info@pika.ca
Manufactures voice boards for a wide variety of professional computer applications.

Pikul & Associates
101 Glenfield Dr., Festus, MO 63028 USA
(314) 937-0335 (tel)
Sells refurbished monitors for a wide variety of computer applications.

Pilot Avioncs
10015 Muirlands Blvd., Unit G, Irvine, CA 92718 USA
(800) 874-1140 (tel)
Manufactures a variety of headsets and intercoms for professional aviation applications.

Pinnacle Data Systems
2155 Dublin Rd., Columbus, OH 43228 USA
(800) 882-8282 (tel)
(614) 487-1150 (tel)
(614) 487-8568 (fax)
Manufactures file servers for a wide variety of professional computer network applications.

Pinnacle Micro
19 Technology, Irvine, CA 92718 USA
(800) 553-7070 (tel)
(714) 789-3000 (tel)
(714) 789-3150 (fax)
(714) 789-3110 (BBS)
(714) 789-3048 (BBS)
Manufactures several different recordable computer CD drives. Offers free the Guide to Compact Disc Recordable (CD-R).

Pioneer Data
1515 N. Pacific Hwy., Woodburn, OR 97071 USA
(503) 982-5115 (tel)
Sells image scanners for a wide variety of computer applications.

Pioneer Electronics
2265 E. 220th St., P.O. Box 1720, Long Beach, CA 90801 USA
(800) 745-3271 (tel)
(800) 421-1404 (tel)
(800) 227-1693 (tel)
(800) 444-6784 (tel)
(213) PIONEER (tel)
(310) 952-2247 (fax)
Service Department
1925 E. Dominguez St., P.O. Box 1760, Long Beach, CA 90801 USA
(800) 457-2881 (parts tel)
(310) 746-6337 (tel)
(310) 816-0412 (fax)
(310) 952-2247 (fax)
Manufactures head units, amplifiers, cassette decks, CD changers, loudspeakers, etc. for home and car audio applications.

Pipo Communications
P.O. Box 2020, Pollock Pines, CA 95726 USA
(916) 644-5444 (tel)
(916) 644-7476 (fax)
Manufactures DTMF encoders for professional cellular microwave radio applications.

Pirelli Cable
700 Industrial Dr., Lexington, SC 29072 USA
(800) 669-0808 (tel)
(803) 951-4800 (fax)
Manufactures fiberoptic cables for professional cable TV applications.

Pixel Express
516 Valley Way, Milpitas, CA 95035 USA

Pixel Express
(800) 749-3966 (tel)
Visa, MasterCard
Sells Relisys scanners and transparency adapters.

Pixelworks
7 Park Ave., Hudson, NH 03051 USA
(800) 247-2476 (tel)
(603) 880-1322 (tel)
(603) 880-6558 (fax)
Manufactures terminals for a wide variety of professional computer network applications.

Plainview Batteries
23 Newtown Rd., Plainview, NY 11803 USA
(516) 249-2873 (tel)
(516) 249-2876 (fax)
Manufactures battery packs for a wide variety of electronics applications.

Planet Electronics
8418 Lilley Rd., Canton, MI 48187 USA
(800) 247-4663 (tel)
(313) 453-4750 (tel)
(313) 453-4750 (fax)
Visa, MasterCard, Discover
Sells high-end home, amateur, and professional audio and video equipment, such as camcorders, VCRs, video editors, monitors, TVs, home theater equipment, etc. Offers a free 64-page catalog.

Plexsys International
607 Herndon Pkwy., #201, Herndon, VA 22070 USA
(703) 709-3044 (tel)
Manufactures entire systems for professional cellular microwave radio applications.

Plextor
4255 Burton Dr., Santa Clara, CA 95054 USA
(800) 886-3935 (tel)
(800) 475-3986 (tel)
http://www.plextor.com
Manufactures CD-ROM drives for personal computer multimedia applications.

PMC
57 Harvey Rd., Londonderry, NH 03053 USA
(603) 432-9473 (tel)
(800) 639-8701 (fax)
Manufactures wire and cable for a wide variety of professional industrial sensor applications.

PMC/BETA Limited Partnership
9 Tech Circle, Matick, MA 01760 USA
(617) 237-6920 (tel)
(508) 651-9762 (fax)
Manufactures vibration switches for industrial applications.

PMC-Sierra
8501 Commerce Ct., Burnaby, BC V5A 4N3 Canada
(604) 668-7300 (tel)
(604) 668-7301 (fax)
info@pmc-sierra.bc.ca
Manufactures broadband network equipment for a wide variety of professional computer applications.

Polaris Industries
141 W. Wieuca Rd. 300B, Atlanta, GA 30342 USA
(800) 752-3571 (tel)
(404) 252-3340 (tech info tel)
(404) 252-8929 (fax)
Visa, MasterCard, Discover, American Express
Manufactures and sells radio programmers for cellular microwave radio applications; also sells miniature video cameras.

Polaroid
565 Technology Sq., Cambridge, MA 02139 USA
(800) 343-5000 (tel)
(617) 386-2000 (tel)
Manufactures a variety of instant cameras, including "talking cameras" and also LCD projection panels for computer multimedia applications.

Polk Audio
5601 Metro Dr., Baltimore, MD 21215 USA
(800) 377-7655 (tel)
(410) 358-3600 (tel)
(416) 847-8888 (Canada tel)
Manufactures loudspeakers for high-end home audio and home theater applications.

Polotec
2372 Walsh Ave., Santa Clara, CA 95051 USA
(408) 986-8417 (tel)
(408) 986-8721 (fax)
Manufactures equipment for professional cable TV applications.

Polydax Speaker Corp.
10 Upton Dr., Wilmington, MA 01887 USA
(508) 658-0700 (tel)
(508) 658-0703 (fax)
Sells a wide variety of Audax loudspeakers for home and car audio applications. Offers a free catalog.

Polyfusion Audio
30 Ward Rd., Lancaster, NY 14086 USA
(716) 681-3040 (tel)
(716) 681-2763 (fax)
Manufactures audio amplifiers for high-end home audio and home theater applications.

PolyPhaser
2225 Park Pl., P.O. Box 9000, Minden, NV 89423 USA
(800) 325-7170 (tel)
(702) 782-2511 (tel)
(702) 782-4476 (fax)
Manufactures cable ground straps for a variety of electrical applications.

Polywell Computers
1461-1 San Mateo Ave., S. San Francisco, CA 94080 USA
(800) 999-1278 (tel)
Manufactures and sells personal desktop computers.

Popless Voice Screens
P.O. Box 1014, New Paltz, NY 12561 USA
(800) 252-1503 (tel)
(914) 255-3367 (tel/fax)
Manufactures variable acoustic compressors and microphone screens for professional recording studios and live music performances. Offers a free brochure.

Portage Electric Products
7700 Freedom Ave. NW, N. Canton, OH 44720 USA
(216) 499-2727 (tel)
(216) 499-1853 (fax)
Manufactures thermal controls for a wide variety of electronics applications.

Portland Radio Supply
234 SE Grand Ave., Portland, OR 97214 USA
(503) 233-4904 (tel)
Sells a variety of amateur radio equipment and accessories, including amateur radio transceivers, receivers, antenna tuners, antennas, amplifiers, etc.

Portrait Display Labs
6665 Owens Dr., Pleasanton, CA 94588 USA
(800) 858-7744 (tel)
(510) 227-2700 (tel)
(510) 227-2721 (tech support tel)
http://www.portrait.com
Manufactures monitors for personal desktop computers.

P.O.S. International
5300 N. Federal Hwy., Ste. 200, Ft. Lauderdale, FL 33308 USA
(800) 646-4767 (tel)
pos@pointl.com
Manufactures computerized cash registers, UPC scanners, etc. for business applications.

Potomac Instruments
932 Philadelphia Ave., Silver Spring, MD 20910 USA
(301) 589-2662 (tel)
(301) 589-2665 (fax)
Manufactures audio analyzers for professional broadcasting applications.

Potter & Brumfield
200 S. Richland Creek Dr., Princeton, IN 47671 USA
(812) 386-2561 (tel)
(812) 386-2072 (fax)
Manufactures relays and circuit breakers for a wide variety of electronics applications.

Power Acoustic
5920 E. Slauson Ave., Commerce, CA 90040 USA

Power Acoustic

(213) 722-3333 (tel)
(213) 722-1122 (fax)
Manufactures head units, amplifiers, and loudspeakers for mobile audio applications.

Power Convertibles

3450 S. Broadmont Dr., Tucson, AZ 85713 USA
(520) 628-8292 (tel)
(520) 770-9369 (fax)
Manufactures dc-to-dc converters for a wide variety of electronics applications.

Power and Environment International

10521 Laramie Ave., Chatsworth, CA 91311 USA
(800) 324-6367 (tel)
(818) 775-0909 (fax)
Buys and sells uninterruptible power systems for computer backup applications.

Power Guard

801 Fox Trail, P. O. Box 2796, Opelika, AL 36803 USA
(800) 288-1507 (tel)
(205) 742-0055 (tel)
(205) 742-0058 (fax)
Manufactures powering systems for professional cable TV broadcasting applications.

Power-Lab

7578 El Cajon Blvd., La Mesa, CA 91941 USA
(619) 589-0444 (tel)
(619) 589-9014 (fax)
Manufactures dc-to-dc power converters for a wide variety of electronics applications.

Power-One Power Supplies

740 Calle Plano, Camarillo, CA 93012 USA
(800) 765-7448 (tel)
(805) 388-0476 (fax)
Manufactures power supplies for a wide variety of electronics applications. Offers a free catalog.

Power Pros

105 Cromwell Ct., Raleigh, NC 27614 USA
(800) 788-0070 (tel)
(919) 782-9210 (tel)
Visa, MasterCard, American Express
Sells power converters and protectors for a variety of electronics products.

Power-Sonic

P. O. Box 5243, Redwood City, CA 94063 USA
(415) 364-5001 (tel)
Manufactures batteries for a wide variety of electronics applications.

Power Trends

1101 N. Raddant Rd., Batavia, IL 60510 USA
(708) 406-0900 (tel)
Manufactures dc-to-dc power converter modules for a wide variety of electronics applications.

PowerTronics

P.O. Box 735, 143 Raymond Rd., Candia, NH 03034 USA
(800) 746-9586 (tel)
Manufactures power line monitors for a variety of professional industrial applications.

Powervideo

6808 Hornwood Dr., Houston, TX 77074 USA
(800) 877-2900 (tel)
(713) 772-4400 (tel)
(713) 772-6500 (fax)
Visa, MasterCard, Discover, American Express, COD
Sells home, amateur, and professional audio and video equipment, such as camcorders, video editors, speakers, VCRs, cables, etc.

Practical Peripherals

P.O. Box 921789, Norcross, GA 30092 USA
(770) 840-9966 (tech support tel)
(770) 734-4600 (BBS)
(770) 734-4601 (fax)
(800) 225-4774 (fax back)
http://practinet.com
Manufactures modems for personal computer communications applications.

Prarie Digital

846 17th St., Prarie du Sac, WI 53578 USA

(608) 643-8599 (tel)
(608) 643-6754 (fax)
Manufactures RS-232 interfaces for a wide variety of professional computer applications.

PRC
9233 Eton Ave., Chatsworth, CA 91311 USA
(800) 627-DISK (tel)
(818) 700-8482 (tech supp tel)
(818) 700-0533 (fax)
Visa, MasterCard, American Express
Buys, sells, and trades hard drives for a wide variety of computer applications.

Precision Interconnect
16640 SW 72nd Ave., Portland, OR 97224 USA
(503) 620-9400 (tel)
(503) 620-7131 (fax)
Manufactures cable, connectors, and interconnect systems for a wide variety of electronics and computer applications.

Precision Micro Dynamics
1961 Ferndale Rd., Victoria, BC V8N 2Y4 Canada
(604) 472-7249 (tel)
(604) 472-1830 (fax)
pmd@islandnet.com
http://pmdi.com
Manufactures mobile robots for a wide variety of electronics applications.

Precision Power
4829 S. 38th St., Phoenix, AZ 85040 USA
(800) 62-POWER (tel)
Manufactures amplifiers for mobile audio applications.

Premier Communications
20277 Valley Bl., #J, Walnut, CA 91789 USA
(909) 869-5709 (tel)
(909) 869-5710 (fax)
Manufactures VHF handheld and mobile transceivers for amateur radio applications. Offers free information.

Premier Technologies
2005 Shaughnessy Circle., P.O. Box 159, Mong Lake, MN 55256 USA
(800) 466-8642 (tel)
sales@premtech.com
Manufactures digital recorders for professional studio recording and live music performances.

Premio Express
1300-A John Reed Ct., City of Industry, CA 91745 USA
(800) 834-4558 (tel)
(818) 336-2184 (tel)
(818) 336-4374 (fax)
Visa, MasterCard
Manufactures and sells personal desktop computers. Offers a 30-day money-back guarantee.

Prem Magnetics
3521 N. Chapel Hill Rd., McHenry, IL 60050 USA
(815) 385-2700 (tel)
(815) 385-8578 (fax)
Manufactures power transformers for a wide variety of computer and electronics applications.

Pre-Owned Electronics
205 Burlington Rd., Bedford, MA 01730 USA
(800) 274-5343 (tel)
(617) 275-4600 (tel)
(617) 275-4848 (fax)
Sells Apple computer parts and boards, monitors, printers, etc.

PreSonus Audio Electronics
501 Government St., Baton Rouge, LA 70802 USA
(504) 344-7887 (tel)
(504) 344-8881 (fax)
Manufactures audio processors for professional studio recording applications.

Primax Electronics
254 E. Hacienda Ave., Campbell, CA 95008 USA
(800) 774-6291 (tel)
(408) 364-2800 (tel)
Manufactures personal text scanners and

Primax Electronics

telephone/fax switches for a wide variety of personal computer applications.

Prime Electronic Components
150 W. Industry Ct., Deer Park, NY 11729 USA
(516) 254-0101 (tel)
(516) 242-8995 (fax)
http://www.imsworld.com/prime
Visa, MasterCard
Sells surplus modems, hard drives, image scanners, CD-ROM drives, RAM chips, etc. for a wide variety of personal computer applications.

PrimeTime Video and Cameras
1104 Chestnut St., Philadelphia, PA 19107 USA
(800) 477-8445 (tel)
(215) 627-1080 (tel)
(215) 627-5886 (fax)
Visa, MasterCard, Discover, American Express
Sells home, amateur, and professional audio and video equipment, such as camcorders, laser disc players, computer editing equipment, VCRs, accessories, etc.

Princeton Graphic Systems
2801 S. Yale St., Suite 110, Santa Ana, CA 92704 USA
(800) 747-6249 (tel)
(714) 751-8405 (tel)
(714) 751-5736 (fax)
Manufactures and sells monitors for personal desktop computers.

Princo Instruments
11020 Industrial Hwy., Southampton, PA 18966 USA
(215) 355-1500 (tel)
(215) 355-7766 (fax)
Manufactures level controls, etc. for a wide variety of professional industrial applications. Offers a free 6-page guide.

Printec Enterprises (PEI)
1271 Rankin Dr., Troy, MI 48083 USA
(800) 346-2618 (tel)
(810) 588-2800 (tel)
(810) 588-1288 (fax)
Sells terminals for professional computer applications.

The Printer Connection
P. O. Box 927061, San Diego, CA 92192 USA
(800) 479-6090 (tel)
(800) 479-6091 (tel)
Visa, MasterCard, Discover, American Express
Sells computer printers and accessories.

Printer Plus
940B Ann St., Stroudsburg, PA 18360 USA
(717) 421-7138 (tel)
(610) 558-2600 (tech support tel)
(717) 424-7334 (fax)
Sells printers, plotters, and accessories for a wide variety of electronics applications.

The Printer Works
3481 Arden Rd., Hayward, CA 94545 USA
(800) 225-6116 (tel)
Visa, MasterCard
Buys and sells laser printers and printer parts for a wide variety of computer applications.

Priority Computer Parts
125 Spring Hill Blvd., Unit 1, Grass Valley, CA 95945 USA
(800) 994-1329 (tel)
(916) 272-8084 (fax)
Sells NEC computer parts.

Proac USA
112 Swanhill Ct., Baltimore, MD 21208 USA
(410) 486-5975 (tel)
Manufactures loudspeakers for home audio and home theater applications.

ProBoard Circuits
100 Market St., #16, Galveston, TX 77550 USA
(409) 762-5436 (tel)
(409) 762-4167 (fax)
Visa, MasterCard, COD
Manufactures logic analyzer boards for computerized electronics testing applications. Offers a 30-day money-back guarantee.

Processor Technology
P.O. Box 192, Sayville, NY 11782 USA
(516) 567-7850 (tel)
Sells test equipment, including oscilloscopes, spectrum analyzers, IC testers, etc. for a wide variety of electronics testing applications.

Procomm
1372 Harmony Ct., Thousand Oaks, CA 91362 USA
(800) 497-2394 (tel)
(805) 497-2397 (info tel)
(805) 497-3430 (info tel)
(805) 494-5078 (info tel)
(805) 494-3115 (fax)
Manufactures and sells VHF/UHF transceivers for professional two-way radio applications.

ProCom Technology
2181 Dupont Dr., Irvine, CA 92715 USA
(800) 800-8600 (tel)
(714) 852-1000 (tel)
(714) 261-7380 (fax)
info@procom.com
http://www.procom.com
Manufactures CD-ROM drive arrays for professional computer applications.

Pro-Disk Service
706 Charcot Ave., San Jose, CA 95131 USA
(408) 955-0929 (tel)
(408) 955-9890 (fax)
Sells hard drives for a wide variety of computer applications.

Products International
8931 Brookville Rd., Silver Spring, MD 20910 USA
(301) 587-7824 (tel)
(800) 545-0058 (fax)
(800) 654-9838 (tech support tel)
Visa, MasterCard, American Express
Sells test equipment from AMREL, including digital multimeters, LCR meters, frequency counters, power supplies, E/EEPROM programmers, etc. Offers a free 84-page catalog.

Profeel Marketing
42 Main St., #109, Monsey, NY 10952 USA
(800) 776-3335 (tel)
(914) 425-2070 (tel)
(914) 425-0221 (fax)
Visa, MasterCard, Discover, American Express, COD
Sells high-end home, amateur, and professional audio and video equipment, such as camcorders, video editors, monitors, VCRs, cables, etc.

Professional Audio Design
357 Liberty St., Rockland, MA 02370 USA
(617) 982-2600 (tel)
(617) 982-2610 (fax)
Visa, MasterCard
Buys, sells, and trades new and used audio equipment for professional studio recording and live music performances.

Professional and Leisure Computers
12085 University City Blvd., Harrisburg, NC 28075 USA
(704) 455-2927 (tel)
Visa, MasterCard, Discover, American Express
Sells computer parts and boards, such as hard drives, memory chips, cases, multimedia upgrades, etc.

Professional Technologies
21038 Commerce Pointe Dr., Walnut, CA 91789 USA
(800) 949-5018 (tel)
(909) 468-3730 (tel)
(909) 468-1372 (fax)
Visa, MasterCard, Discover, American Express, COD
Manufactures and sells complete home and laptop computer systems. Offers a 30-day money-back guarantee.

Profile Consumer Electronics
17211 S. S. Valley View Ave., Cerritos, CA 90703 USA
(310) 802-9007 (tel)
Manufactures amplifiers for mobile audio applications.

Proflex Lighting
333 Encinal St., Santa Cruz, CA 95060 USA
(408) 454-9100 (tel)

Proflex Lighting

Manufactures lighting systems for amateur and professional video production.

ProGen Systems
15501 Redhill Ave., Tustin, CA 92680 USA
(800) 848-8777 (tel)
(800) 848-8767 (tech support tel)
(714) 566-9200 (tel)
(714) 566-9267 (fax)
progen@ix.netcom.com
Visa, MasterCard, Discover, American Express
Manufactures and sells personal desktop computers.

Programmable Designs
41 Enterprise Dr., Ann Arbor, MI 48103 USA
(313) 769-7540 (tel)
(313) 769-7242 (fax)
Manufactures probes for electronics testing applications. Offers a 30-day money-back guarantee.

Progressive Concepts
1434 N. Mills Ave., Ste B., Claremont, CA 91711 USA
(909) 626-4969 (tel)
(909) 626-4FAX (fax)
Visa, MasterCard
Sells plans and kits for Part 15 FM transmitters, SWR meters, dc power supplies, frequency counters, audio limiters, etc. Offers a catalog for $2.

Project One AV
10627 Burbank Blvd., N. Hollywood, CA 91601 USA
(800) 818-8986 (tel)
(818) 753-8273 (tel)
(213) 464-2285 (tel)
01@projectone.com
Sells a wide variety of recording formats, including DATs, open reel tapes, CD-ROM discs, magneto optical disks, ADATs, etc. for professional studio recording.

Pro-Log
12 Upper Ragsdale Dr., Monterey, CA 93904 USA
(800) 252-3279 (tel)
(408) 646-3649 (tel)
Manufactures single-board computers for professional and industrial computer applications.

Prologic Designs
P.O. Box 19026, Baltimore, MD 21204 USA
(410) 661-5950 (tel)
Visa, MasterCard
Manufactures microcontroller boards for a wide variety of computer applications.

Pro-Match
1850 McCulloch Blvd., #B2, Lake Havasu City, AZ 86403 USA
(800) 770-4643 (tel)
(520) 855-2226 (tel)
Sells batteries, etc. for radio-control models.

ProMedia Computer Supplies
432 N. Canal St., #1, S. San Francisco, CA 94080 USA
(800) 583-5833 (tel)
(415) 583-9932 (fax)
Sells floppy diskettes for a wide variety of computer applications. Offers a free catalog.

Promethus Products
9524 Tualatin Sherwood Rd., Tualatin, OR 97062 USA
(800) 477-3473 (tel)
(503) 692-9600 (tel)
Manufactures voice mail modems for a wide variety of computer applications.

Pro-Rec Synth Sounds
106 W. 13th St., #13, New York, NY 10011 USA
(212) 675-5606 (tel)
(212) 627-3148 (demos tel and fax)
prorec@aol.com
http://www.users.aol.com/prorec/prorec.html
Sells synthesizers for a wide variety of home and professional musical applications.

Pro Sound & Stage Lighting
11711 Monarch St., Garden Grove, CA 92641 USA
(800) 945-9300 (tel)
(714) 530-6760 (tel)
Visa, MasterCard, Discover, American Express
Sells a wide variety of professional sound, lighting, and video equipment. Offers a free 72-page catalog.

ProSource Power
107 Whitman Dr., Schaumburg, IL 60173 USA
(800) 949-4797 (tel)
Visa, MasterCard
Manufactures backup power systems for computer applications.

ProStar
4102 W. Valley Blvd., Walnut, CA 91789 USA
(800) 576-1134 (tel)
(909) 869-6018 (tel)
(909) 869-6026 (fax)
Manufactures and sells personal laptop computers. Offers a 30-day money-back guarantee.

ProSystems
65 36th St., Wheeling, WV 26003 USA
(304) 233-2223 (tel)
(304) 233-2258 (fax)
Manufactures loudspeakers for professional audio applications.

Protek
154 Veterans Dr., Northvale, NJ 07647 USA
(201) 767-7242 (tel)
(201) 767-7343 (fax)
Manufactures digital multimeters (DMMs).

Proteon
9 Technology Dr., Westboro, MA 01581 USA
(800) 830-1300 (tel)
(508) 898-2800 (tel)
(508) 366-8901 (fax)
Manufactures routers for professional computer network applications.

Proteus Industries
340 Pioneer Way, Mountain View, CA 94041 USA
(415) 964-4163 (tel)
(415) 965-9355 (fax)
(415) 968-5840 (fax back)
Manufactures flow sensors for industrial applications.

Proton
16826 Edwards Rd., Cerritos, CA 90703 USA
(310) 404-2222 (tel)
(310) 404-2322 (fax)
Manufactures loudspeakers and power amplifiers for car stereo applications.

Proven Technology Group
20 Keyland Ct., Bohemia, NY 11716 USA
(516) 567-1867 (tel)
(516) 567-2453 (fax)
Buys, sells, and leases new and used computer systems for professional computer applications.

Proxima
9440 Carroll Park Dr., San Diego, CA 92121 USA
(619) 457-5500 (tel)
(619) 457-9647 (fax)
Horsterweg 24, 6191 RX Beek, The Netherlands
+31-43-650-248 (tel)
+31-43-649-220 (fax)
Manufactures LCD projection panels for computer multimedia applications.

PRV Sales
894 Green Pl., Woodmere, NY 11598 USA
(800) 670-6555 (tel)
Buys and sells new and lightly used video equipment.

PSB Speakers
P.O. Box 33018, Detroit MI 48232 USA
(800) 263-4641 (tel)
Manufactures loudspeakers for home audio and home theater applications.

Psitech Plus
531 Gordon Ct., Room A, Benicia, CA 94510 USA
(707) 745-4804 (tel)
(707) 747-5277 (fax)
Sells RF wattmeters for a wide variety of radio transmitting applications.

Ptek
1814 Schooldale Dr., San Jose, CA 95124 USA
(408) 448-3342 (tel)
(510) 279-5912 (fax)
Sells repeater and pager power amplifiers for cellular microwave radio applications.

PTL Cable Service
612 N. Orange Ave., Unit D8, Jupiter, FL 33458 USA
(407) 747-3647 (tel)
(407) 575-4635 (fax)
Sells test equipment and satellite antennas, particularly for professional cable TV applications.

Publisher's Toolbox
8845 S. Greenview Dr., #8, Middleton, WI 53562 USA
(800) 204-9935 (tel)
(800) 233-3898 (Canada tel)
(608) 828-1100 (tel)
(608) 828-1112 (fax)
0130813266 (Germany tel)
078119498 (Belgium tel)
1553247 (Switzerland tel)
006633800144 (Japan toll-free fax)
Visa, MasterCard, Discover, American Express, COD (add $5)
Sells a variety of computer desktop publishing software and hardware, including monitors, removable storage media, video cards, printers, tablets, etc.

Publishing Perfection
W134 N5490 Campbell Dr., P.O. Box 307, Menomonee Falls, WI 53052 USA
(800) 716-5000 (tel)
(414) 252-5000 (tel)
(414) 252-2502 (fax)
Visa, MasterCard, Discover, American Express
Sells on-line publishing equipment, including scanners, monitors, graphic cards, printers, LCD projectors, video cards, modems, keyboards, etc.

Pulsar Barco
1000 Cobb Place Blvd., Kennesaw, GA 30144 USA
(404) 590-7900 (tel)
(404) 590-8836 (fax)
Manufactures modulators for professional cable TV applications.

Pulse Metric
6190 Cornerstone Ct. E., #103, San Diego, CA 92121 USA
(800) 92-PULSE (tel)
(619) 546-9461 (tel)
(619) 546-9470 (fax)
Manufactures and sells cardiovascular monitors that plug into computers. Offers a 30-Day money-back guarantee.

Purchase Radio Supply
327 E. Hoover Ave., Ann Arbor, MI 48104 USA
(313) 668-8696 (tel)
(313) 668-8802 (fax)
Sells electron tubes, electronic components, hardware, etc., particularly for amateur radio applications.

Purdy Electronics
720 Palomar Ave., Sunnyvale, CA 94086 USA
(408) 523-8230 (tel)
(408) 733-1287 (fax)
Manufactures LEDs for a wide variety of electronics applications.

Pure Audio
(818) 887-6600 (tel)
(818) 887-4700 (fax)
Sells a variety of audio equipment, including compressors, mixing consoles, recorders, microphones, etc., for professional recording studio and live music performances.

PWI/MobileFax
2533 N. Carson St., #1956, Carson City, NV 89706 USA
(800) 910-4099 (tel)
(805) 379-9447 (fax)
Visa, MasterCard, American Express
Manufactures sweep/function generators for a wide variety of electronics testing applications.

P&W International
1501 Monroe St., #114, Toledo, OH 43624 USA
(800) 779-1723 (tel)
Manufactures high-end home loudspeakers.

Pyle
501 Center St., Huntington, IN 46750 USA

(219) 356-1200 (tel)
(219) 356-2830 (fax)
Manufactures loudspeakers for car stereo applications.

Pyramid
1501-60th St., Brooklyn, NY 11219 USA
(718) 436-1616 (tel)
(718) 436-1401 (fax)
Manufactures amplifiers and loudspeakers for mobile audio applications.

Pyramid Communications
1198 Pacific Coast Hwy., #D-286, Seal Beach, CA 90740 USA
(310) 430-5892 (tel)
Manufactures vehicular repeaters for professional two-way radio applications.

Pyramid Technology
3860 N. First St., San Jose, CA 95734 USA
(408) 428-9000 (tel)
(408) 428-5058 (fax)
Manufactures workstations for a wide variety of professional computer network applications.

Q Components
638 Colby Dr., Waterloo, ON N2V 1A2 Canada
(519) 884-1142 (tel)
Sells a variety of speaker components. Offers a free catalog.

QEI
One Airport Dr., P.O. Box 805, Williamstown, NJ 08094 USA
(800) 334-9154 (tel)
(609) 728-2020 (service hotline)
(609) 629-1751 (fax)
Manufactures a variety of solidstate FM transmitters and exciters for broadcast stations.

Qintar
700 Jefferson Ave., Ashland, OR 97520 USA
(503) 488-5500 (tel)
(503) 488-5567 (fax)
Manufactures splitters, taps, switches, amplifiers, etc. for professional cable TV applications.

QMS
1 Magnum Pass, Mobile, AL 36618 USA
(800) 972-6695 (tel)
(800) 762-8894 (tel)
(800) 633-7213 (fax)
(800) 991-2000 (Canada tel)
(334) 633-4300 (tel)
(334) 633-4866 (fax)
http://www.qms.com
Manufactures color laser printers for professional computer applications.

QRV Electronics
503 Main St., P.O. Box 330, Crawford, GA 30630 USA
(706) 743-3344 (tel/fax)
Buys, sells, and trades new and used Azden amateur radio equipment.

QSC Audio
1675 MacArthur Blvd., Costa Mesa, CA 92626 USA
(714) 754-6175 (tel)
(714) 754-6174 (fax)
Manufactures professional amplifiers for studio recording.

QSTAR Technologies
600 E. Jefferson St., Rockville, MA 20852 USA
(800) 568-2578 (tel)
(301) 762-9800 (tel)
(301) 762-9829 (fax)
Manufactures data storage systems for a wide variety of professional computer network applications.

QT Optoelectronics
16775 Addison Rd., Ste. 200, Dallas, TX 75248 USA
(800) 533-6786 (tel)
322-466-3540 (Europe tel)
603-735-2417 (Asia/Pacific tel)
Manufactures optoelectronics components for a wide variety of electronics applications. Offers a free product data book.

Qualimetrics
1165 National Dr., Sacramento, CA 95834 USA

Qualimetrics

(916) 928-1000 (tel)
(916) 928-1165 (fax)
Manufactures sensors and systems for a wide variety of professional industrial applications. Offers a 160-page catalog.

Quality Cable & Electronics

1950 NW 44th St., Pompano Beach, FL 33064 USA
(800) 978-8845 (tel)
(305) 978-8845 (tel)
(305) 978-8831 (fax)
Sells new and used professional cable TV equipment, including converters, headends, line equipment, etc.

Quality Power Products

5500 Oakbrook Pkwy., #250, Norcross, GA 30093 USA
(800) 525-7502 (tel)
(404) 242-1901 (tel)
(404) 242-3069 (fax)
Visa, MasterCard
Sells power systems for operating a variety of electronics products.

Quality RF Services

850 Park Way, Jupiter, FL 33477 USA
(800) 327-9767 (tel)
(407) 747-4998 (tel)
(407) 744-4618 (fax)
Sells replacement equalizers for equipment from a variety of manufacturers, including Philips, Magnavox, Scientific Atlanta, and RCA.

Quality Semiconductor

851 Martin Ave., Santa Clara, CA 95050 USA
(408) 450-8000 (tel)
(408) 496-0773 (fax)
Ste. B, Unit 4A, Mansfield Park, Four Marks, Hampshire, GU34 5PZ UK
44-0-420-563333 (tel)
44-0-420-561142 (fax)
Manufactures semiconductors for a wide variety of electronics applications.

Qualix Group

1900 S. Norfolk St., #224, San Mateo, CA 94403 USA
(415) 572-0200 (tel)
(415) 572-1300 (fax)
info@qualix.com
http://www.qualix.com
55 Madison Ave., #200, Morristown, NJ 07960 USA
(201) 538-8772 (tel)
(201) 538-8826 (fax)
markk@qualix.com
Sells network systems for professional computer applications.

Qualstar

6709 Independence Ave., Canoga Park, CA 91303 USA
(800) 468-0680 (tel)
(818) 592-0116 (tel)
(818) 592-0161 (fax)
Sells tape drives for a wide variety of computer applications.

Quanta Micro

12110 E. Slauson Ave., Santa Fe Springs, CA 90670 USA
(310) 907-1180 (tel)
(310) 907-1186 (fax)
http://www.qmicro.com
Visa, MasterCard
Sells computer parts and boards, such as motherboards, video cards, controllers, hard drives, keyboards, memory chips, etc.

Quantex Microsystems

400B Pierce St., Somerset, NJ 08873 USA
(800) 644-4680 (tel)
(800) 864-9022 (customer service tel)
(800) 896-4898 (tel)
(908) 563-4166 (tel)
(908) 563-0407 (fax)
http://www.qtx.com
Visa, MasterCard, Discover, American Express
Manufactures and sells personal desktop computers. Offers a 30-day money-back guarantee.

Quantum

500 McCarthy Blvd., Milpitas, CA 95035 USA

(800) 624-5545 (tel)
(408) 894-4000 (tel)
(408) 894-5088 (fax)
Manufactures data storage systems for a wide variety of professional computer network applications.

Quantum Composers
P.O. Box 4316, Bozeman, MT 59772 USA
(406) 586-3159 (tel)
(406) 586-3220 (fax)
Manufactures pulse generators for electro-optical applications.

Quantum Data
2111 Big Timber Rd., Elgin, IL 60123 USA
(708) 888-0450 (tel)
(708) 888-2802 (fax)
Manufactures video test generator boards for professional electronics testing applications.

Quark Technology
7900 Plaza Blvd., Unit #182, Mentor, OH 44060 USA
(800) 443-8807 (tel)
(216) 974-3287 (tech support tel)
(216) 974-8857 (fax)
Visa, MasterCard, Discover, American Express
Manufactures and sells personal desktop computers; also sells computer parts and boards, such as floppy drives, hard drives, monitors, motherboards, modems, keyboards, etc.

Quasar
1707 N. Randall Rd., Elgin, IL 60123 USA
(708) 468-5600 (tel)
Manufactures televisions for home video applications.

Quasar Electronics
Unit 14, Summingdale, Bishop's Stortford, Hertfordshire, CM23 2PA UK
Manufactures and sells a wide variety of electronics kits, including surveillance kits, motion detectors, radios, amplifiers, flashers, etc. Offers a catalog for two first-class stamps.

Quement Communications
1000 S. Bascom Ave., San Jose, CA 95128 USA
(408) 998-2355 (tel)
(408) 292-9920 (fax)
Sells base, mobile, and handheld scanners, CB transceivers, antennas, battery packs, etc. Offers free shipping.

Quest Electronics
5715 W. 11th Ave., Denver, CO 80214 USA
(303) 274-7545 (tel)
(303) 274-2317 (fax)
Visa, MasterCard
Sells electron tubes and semiconductors for a wide variety of electronics applications.

QuickLogic
2933 Bunker Hill, Ste. 100A, Santa Clara, CA 95054 USA
(800) 842-3742 (tel)
+49-89-899-143-28 (Europe tel)
+49-89-857-77-16 (Europe fax)
qlogicinfo@net.com
Manufactures integrated circuits for a wide variety of computer applications.

Qume
3475A N. First St., San Jose, CA 95134 USA
(800) 457-4447 (tel)
(408) 473-1500 (tel)
(408) 473-1510 (fax)
Manufactures terminals for a wide variety of professional computer network applications.

Quorum Communications
8304 Esters Blvd., #850, Irving, TX 75063 USA
(214) 915-0256 (tel)
(214) 915-0270 (fax)
(214) 915-0346 (BBS)
Manufactures and sells a Wefax receiver and scan converter for amateur radio applications.

Racal-Datacom
1601 N. Harrison Pkwy., Sunrise, FL 33323 USA
(800) RACAL-55 (tel)
http://www.racal.com

Racal-Datacom

Manufactures digital access multiplexers, fiber data transporters, etc. for professional computer applications.

Racal Recorders
15375 Barranca Pkwy., #H-101, Irvine, CA 92718 USA
(800) 847-1226 (tel)
(714) 727-3444 (tel)
(714) 727-1774 (fax)
72056@compuserve.com
Manufactures portable data recorders for professional data recording applications.

Radar Sales
5485 Pineview Ln., Plymouth, MN 55442 USA
(612) 557-6654 (tel)
(612) 550-0454 (fax)
Sells a variety of new and refurbished radar guns for a variety of applications. Offers a free catalog.

Rader Video Products
340 E. 9th, Freemont, NE 68025 USA
(402) 721-9121 (tel)
Sells remote video camera systems for professional video production.

Radio Accessories
PO Box 168, Melvin Village, NH 03850 USA
Sells a brush-on coating that is used to prevent computer-generated radio and television interference.

Radio Adventures Corp. (RAC)
Main St., Box 339, Seneca, PA 16346 USA
(814) 677-7221 (tel)
(814) 677-6456 (fax)
rac@usa.net.au2
Sells receiver kits and CMOS iambic keyer/controller chips for amateur radio applications.

Radio Astronomy Supplies
190 Jade Cove Dr., Roswell, GA 30075 USA
(404) 992-4959 (tel)
Sells radio telescopes, discrete components, modules, and RF amplifiers for amateur radio astronomy systems. Offers a free catalog.

Radio Center
525 E. 70th Ave., #1W, Denver, CO 80229 USA
(800) 227-7373 (tel)
(303) 288-7373 (tel)
Sells a wide variety of amateur radio equipment and accessories, including amateur radio transceivers, receivers, antenna tuners, antennas, amplifiers, etc.

Radio Center USA
630 NW Englewood Rd., Kansas City, MO 64110 USA
(800) 821-7323 (tel)
(816) 741-8118 (tel)
Sells a wide variety of amateur radio equipment and accessories, including amateur radio transceivers, receivers, antenna tuners, antennas, amplifiers, etc.

Radio Central
P.O. Box 1365, Robertsdale, AL 36567 USA
(800) 923-6872 (tel)
(205) 438-6152 (fax)
Sells telephones and accessories for cellular telephone applications.

Radio City
2663 County Rd. I, Mounds View, MN 55112 USA
(800) 426-2891 (tel)
(612) 786-4475 (tel)
(612) 786-6513 (fax)
radiocty@skypoint.com
http://www.radioinc.com
Visa, MasterCard, Discover
Sells a wide variety of amateur radio equipment and accessories, including amateur radio transceivers, receivers, antenna tuners, towers, antennas, radio teletype equipment, amplifiers, etc. Offers a free catalog on hardcopy or send a SASE for a catalog on computer diskette.

Radio Communications
P.O. Box 212, Royal Oak, MI 48068 USA
(800) 551-1955 (tel)
(810) 546-2010 (fax)

Sells amateur and two-way radio equipment and accessories, including antenna tuners, antennas, etc.

Radio Communications Center
Amsterdamsestraatweg 516, NL-3553 Utrecht, The Netherlands
030-433838 (tel)
Sells equipment for shortwave radio receiving applications.

Radio Communications of Charleston
102 Farm Rd., Goose Creek, SC 29445 USA
(803) 553-4101 (tel)
(803) 553-3564 (fax)
Sells two-way radio equipment for a wide variety of radio applications.

Radio Communications Services
1007 Eastfield, Lansing, MI 48917 USA
Sells antennas for CB, scanner, cellular, two-way, etc. communications. Offers a price list for $1.

Radio Communications Wholesalers
9635 Girard Ave. S., Bloomington, MN 55341 USA
(800) 726-9015 (tel)
(612) 884-8352 (tel)
(612) 884-8356 (fax)
Sells transceivers, etc. for two-way radio communications.

Radio Component Specialists
337 Whitehorse Rd., Croydon, Surrey CR0 2HS UK
0181-684-1665 (tel)
Visa, Access
Sells radio components, including power supplies, transformers, capacitors, tubes, etc.

Radio Control Systems (RCSI)
8125G Ronson Rd., San Diego, CA 92111 USA
(800) 560-7234 (tel)
(619) 560-7008 (tech supp tel)
(619) 292-6955 (fax)
rcsi@cts.com
http://www.cts.com/browse/rcsi
Visa, MasterCard, COD
Manufactures and sells an expansion box for several Kenwood transceivers for amateur radio applications.

Radio Design Labs (RDL)
P.O. Box 1286, Carpinteria, CA 93014 USA
(800) 281-2683 (tel)
(805) 684-5415 (tel)
+31-20-6238-983 (Europe tel)
+31-20-6225-287 (Europe fax)
Manufactures a variety of interfaces: audio or video DAs, mic preamps, line amplifiers, audio meters, and video attenuators for broadcast stations. Offers a free catalog.

Radio Electric Supply
P.O. Box 2790, Rt. 2, Melrose, FL 32666 USA
(904) 475-1950 (tel/fax)
Sells a wide variety of used and new/old stock tubes. Offers a free price list.

Radio Engineers
7969 Engineer Rd., #102, San Diego, CA 92111 USA
(619) 565-1319 (tel)
(619) 571-5909 (fax)
Manufactures UHF/VHF radio direction finders.

Radio Express
P.O. Box 632, Centreville, VA 22020 USA
(800) 545-7748 (tel)
(703) 266-1928 (tel)
(703) 830-8710 (fax)
http://www.internext.com/radio
Visa, MasterCard, Discover, American Express
Sells transceivers, etc. for two-way radio communications.

RadioMate
4030-A Pike Ln., Concord, CA 94520 USA
(800) 346-6442 (tel)
(510) 676-3376 (tel)
(510) 676-3387 (fax)
Manufactures and sells headsets for professional two-way radio applications.

Radio One
(716) 661-9964 (tel)
(716) 763-0371 (fax)
Sells used service monitors for professional cellular microwave radio applications.

The Radio Place
5675A Power Inn Rd., Sacramento, CA 95824 USA
(916) 387-0730 (tel)
(916) 387-0744 (fax)
Visa, MasterCard, Discover
Sells a wide variety of amateur radio equipment and accessories, including amateur radio transceivers, receivers, antenna tuners, antennas, amplifiers, etc.

Radio Shack
One Tandy Center, Ft. Worth, TX 76102 USA
(817) 390-3011 (tel)
(817) 878-4852 (tel)
Business Products Division
1801 S. Beach St., Ft. Worth, TX 76105 USA
(817) 390-3011 (tel)
Consumer Parts Center
7439 Airport Freeway, Ft. Worth, TX 76118 USA
(800) 243-1311 (tel)
National Parts Division
900 E. Northside Dr., Ft. Worth, TX 76102 USA
(800) 442-2425 (tel)
(817) 870-5600 (tel)
Visa, MasterCard
Manufactures and sells electronics components, TVs, VCRs, amplifiers, speakers, turntables, computers, scanners, antennas, amateur radio and citizens band transceivers, wire and cable, coaxial cable, etc.

Radio South
3702 N. Pace, Pensacola, FL 32505 USA
(800) 962-7802 (tel)
(904) 434-0909 (tel)
Visa, MasterCard
Sells radio equipment for radio-control model airplanes.

Radio-Tech
Overbridge House, Weald Hall Ln., Thornwood, Epping, Essex CM16 6NB UK
+44-0181-368-8277 (sales tel)
+44-0181-361-3434 (fax)
+44-01992-576107 (administration tel)
+44-01992-561994 (fax)
Visa, MasterCard
Manufactures data transmitters and receivers for a wide variety of radio applications.

Radioware
P.O. Box 1478, Westford, MA 01886 USA
(800) 950-WARE (tel)
(508) 452-5555 (tel)
Visa, Master Card, Discover, American Express
Manufactures and sells scanner antennas and accessories. Offers a free catalog.

Radio Wholesale Marketing
3132 Mercury Dr., Columbus, GA 31906 USA
(800) 537-2346 (tel)
(706) 561-7000 (tel)
(706) 568-4504 (fax)
Sells amateur and two-way radio equipment and accessories.

Radio Works
P.O. Box 6159, Portsmouth, VA 23703 USA
(800) 280-1873 (tel)
(804) 484-0140 (tel)
(804) 483-1873 (fax)
jim@radioworks.com
Manufactures and sells wire antennas and baluns; sells wire, coaxial cable, guy wire, insulators, etc. for amateur radio applications.

Radius
215 Moffett Dr., Sunnyvale, CA 94089 USA
(800) 572-3487 (tel)
(408) 541-6100 (tel)
support@radius.com
Manufactures monitors for personal desktop computer systems.

Radlinx
900 Corporate Dr., Mahwah, NJ 07430 USA
(201) 529-1100 (tel)
(201) 529-5777 (fax)

Manufactures terminal servers for professional computer network applications.

Radstone Technology
50 Craig Rd., Montvale, NJ 07645 USA
(800) 368-2378 (tel)
(201) 391-2700 (tel)
Water Lane, Towcester, Northants, N12 6JN, UK
sales@radstone.com
Manufactures FDDI controller boards for a wide variety of computer applications.

RA Enterprises
2260 De La Cruz Blvd., Santa Clara, CA 95050 USA
(800) 801-0230 (tel)
(408) 986-8286 (tel)
(408) 986-1009 (fax)
Visa, MasterCard, Discover, COD ($5)
Sells surplus electronics, including cables, controller cards, batteries, plotters, earphones, etc. for a wide variety of electronics applications.

RAG Electronics
2450 Turquiose Circle, Newbury Park, CA 91320 USA
(800) 923-3457 (tel)
(805) 498-9933 (tel)
(805) 498-3733 (fax)
Buys and sells a wide variety of new and used electronics test equipment, including RF and microwave equipment, spectrum analyzers, synthesizers, oscilloscopes, power supplies, etc. Offers a free catalog.

Railcom
P. O. Box 38881, Germantown, TN 38183 USA
(901) 755-1514 (tel/fax)
Sells scanners, two-way radios for railroad communications, mobile antennas, and accessories.

Railway Depot
124 King St. E., Oshawa, ON L1H 1B6 Canada
(800) 422-7962 (tel)
(905) 433-0507 (tel)
(905) 433-3863 (fax)
Manufactures infrared direction detectors for model railroad applications.

Raindirk
34 Nelson St., Oakville, ON L6L 3H6 Canada
(800) 300-4756 (tel)
Manufactures handmade mixing consoles for professional studio recording applications.

RAm
229C E. Rollins Rd., Round Lake Beach, IL 60073 USA
(847) 740-8726 (tel)
(847) 740-8727 (fax)
Sells lighting systems, sound effects electronics, and radios for radio-control model airplanes and boats. Offers free info for a SASE.

Ramfixx Technologies
5213 Fallingbrook Dr., Mississauga, ON L5V 1N7 Canada
(905) 542-2347 (tel)
(905) 542-2659 (fax)
73447.3315@compuserve.com
http://www.ttrains.com/ramtraxx
Manufactures digital command control systems for model railroad applications.

RAM International Computer
4750 NE 11th Ave., Ft. Lauderdale, FL 33334 USA
(305) 776-3450 (tel)
(305) 776-2218 (fax)
(305) 771-FAXU (fax)
Visa, MasterCard
Buys, sells, and trades memory chips for a wide variety of computer applications.

Ramsey Electronics
793 Canning Pkwy., Victor, NY 14564 USA
(716) 924-4560 (tel)
Manufactures and sells a variety of electronics kits, including frequency counters, amateur radio amplifiers, oscilloscopes, shortwave receiver kits, Part 15 AM and FM transmitters, etc.

The Ranch Pit Stop
1655 E. Mission Blvd., Pomona, CA 91766 USA
(800) 285-7223 (tel)

The Ranch Pit Stop

(909) 623-1506 (tel)
(909) 623-7933 (fax)
Visa, MasterCard, COD ($5.95)
Sells radios, battery chargers, etc. for radio-control models.

Randall Amplifiers
255 Corporate Woods Pkwy., Vernon Hills, IL 60061 USA
(708) 913-5511 (tel)
(708) 913-7772 (fax)
Manufactures amplifiers for professional and personal musical applications.

Rane
10802 47th Ave. W., Mukilteo, WA 98275 USA
(206) 355-600 (tel)
(296) 347-7757 (fax)
Manufactures an 18- to 24-bit to 16-bit digital converter for professional studio recording.

Raritan Computer (RCI)
10-1 Ilene Ct., Belle Mead, NJ 08502 USA
(908) 874-4072 (tel)
(908) 874-5274 (fax)
31-10-458-6673 (Europe tel)
31-10-451-9610 (Europe fax)
886-2-218-1117 (Taiwan tel)
886-2-218-1221 (Taiwan fax)
sales@raritan.com
http://www.raritan.com
Manufactures computer server controllers for a wide variety of professional computer applications. Offers a 30-day money-back guarantee.

Ray's Amateur Electronics
1701-G N. Main St., High Point, NC 27262 USA
(910) 883-6038 (tel)
(910) 883-1464 (fax)
Sells a wide variety of amateur radio equipment and accessories, including amateur radio transceivers, receivers, antenna tuners, antennas, amplifiers, etc.

Raytheon Electronic Systems
465 Centre St., Quincy, MA 02169 USA
(617) 984-4104 (tel)
(617) 984-8515 (fax)
Manufactures semiconductors for a wide variety of electronics applications.

RBH Sound
4042 Pacific Ave., Riverdale, UT 84405 USA
(801) 399-4900 (tel)
(801) 543-3300 (fax)
Manufactures loudspeakers for home audio applications.

RCA
600 N. Sherman Dr., Indianapolis, IN 46201 USA
(800) 336-1900 X112 (catalog tel)
(317) 267-5000 (tel)
(502) 491-8110 (manual info tel)
Manuafactures a wide variety of consumer electronics, including TVs, VCRs, and DSS systems for home audio applications.

RCC Electronics
11B Princess Rd., Lawrenceville, NJ 08648 USA
(800) 733-4722 (tel)
(609) 895-0505 (fax)
Manufactures portable memory recorders for industrial electronics applications.

RC Distributing
PO Box 552, South Bend, IN 46624 USA
(219) 236-5776 (tel)
Visa, Master Card
Manufactures and sells video sync generators, satellite antennas, and satellite systems. Offers free information packets.

RCK
465 Croft Dr., Idaho Falls, ID 83401 USA
(208) 522-2839 (tel)
Visa, MasterCard
Manufactures and sells bandpass filters for amateur radio applications. Offers a brochure for a SASE.

RCR
318 Cooper Center, N. Park Dr., Pennsauken, NJ 08109 USA

(800) 526-7368 (tel)
(203) 265-5052 (fax)
Sells used personal computer systems and printers; also sells computer parts and boards, such as hard drives, modems, floppy drives, etc.

R/C Skydivers
P. O. Box 662, St. Croix Falls, WI 54024 USA
Sells equipment to radio-control model parachutists from radio-controlled kit airplanes. Offers a catalog for $1.

RC Systems
2513 Highway 646, Santa Fe, TX 77510 USA
(409) 925-7808 (tel)
(409) 925-1078 (fax)
Manufactures transmitters/controllers/alarms/RTUs for industrial applications.

R-Cubed
11126 Shady Trail, Ste. 101, Dallas, TX 75229 USA
(800) 672-8233 (tel)
(214) 243-3830 (Dallas tel)
(713) 683-6824 (Houston tel)
Visa, MasterCard
Sells new and refurbished terminals for professional computer applications.

R & D Computers
6767 Peachtree Industrial Blvd., #C, Norcross, GA 30092 USA
(770) 416-0103 (tel)
(770) 416-0155 (fax)
Sells Data General computer systems for professional computer applications.

RDL Acoustics
26 Pearl St., #15, Bellingham, MA 02019 USA
(800) 227-0390 (tel)
Sells loudspeakers for high-end home audio and professional audio applications.

RDP Electrosense
2216 Pottstown Pike, Pottstown, PA 19465 USA
(800) 334-5838 (tel)

(610) 469-0852 (fax)
Manufactures sealed displacement transducers for a wide variety of professional industrial applications.

Reach Electronics
P.O. Box 308, Lexington, NE 68850 USA
(800) 445-0007 (tel)
Manufactures pagers for telecommunications applications.

Reamp
3559 Monterey Blvd., Oakland, CA 94619 USA
(510) 530-2402 (tel)
(510) 530-2404 (fax)
reamp@aol.com
Manufactures impedance and level-adjustment boxes for professional recording studios and live music performances.

Reckon Data Systems
733 Elmont Rd., Elmont, NY 11003 USA
(516) 285-1000 (tel)
(516) 285-0809 (fax)
American Express
Sells IBM personal desktop computers and parts and accessories, including memory chips, system boards, hard drives, network equipment, monitors, etc.

Recording Systems
32 W. 39th St., 9th Fl., New York, NY 10018 USA
(800) 859-3579 (tel)
(212) 768-7800 (tel)
(212) 768-9740 (fax)
Visa, MasterCard, American Express, COD
Sells unlabelled and unboxed blank cassettes of variable lengths for professional recording. Offers a free catalog. Mininum order: $25.

Recortec
1290 Lawrence Station Rd., Sunnyvale, CA 94089 USA
(800) 729-7654 (tel)
(408) 734-1290 (tel)
(408) 734-2140 (fax)
(408) 734-9374 (fax back)
recortec@ix.netcom.com

Recortec

http://www.industry.net/recortec
Visa, MasterCard
Manufactures and sells single-board computers and rack-mount computer equipment for professional computer applications. Offers a free catalog.

Recoton

46-23 Crane St., LIC, NY 11101 USA
(718) 392-6442 (tel)
2950 Lake Emma Rd., Lake Mary, FL 32746 USA
(800) 223-6009 (tel)
Manufactures antennas, blank audio tapes, connectors, microphones, speakers, surge protectors, voltage converters, etc. for a wide variety of audio, video, office, and computer applications.

Red Castle Services

43 Enfield St., Pemberton, Wigan WN5 8DY UK
01942-227568 (tel)
01942-709957 (fax)
Sells a wide variety of electronics components, including resistors, capacitors, terminals, transistors, etc. Offers a list for a SAE.

Reed Brothers

593 King, Charelston, SC 29403 USA
(803) 723-7276 (tel)
Visa, MasterCard, Discover, American Express
Sells a wide variety of high-end home audio equipment.

Reel-Talk

4790 Irvine Blvd., #105-406, Irvine, CA 92720 USA
(800) 766-8255 (tel)
(714) 544-6725 (fax)
Manufactures radio talk show recorders for professional broadcasting applications.

Reference Audio Video

18214 Dalton Ave., Gardena, CA 90248 USA
(310) 517-1700 (tel)
(310) 517-1732 (fax)
Visa, MasterCard, Discover, American Express
Sells a wide variety of high-end home audio and video equipment.

Reliable Computer Parts

7401-R Fullerton Rd., Springfield, VA 22153 USA
(800) 569-5300 (tel)
(800) 569-5301 (fax)
Sells personal desktop and laptop computer parts and accessories, including memory chips, system boards, hard drives, power supplies, keyboards, etc.

Relisys

320 S. Milpitas Blvd., Milpitas, CA 95035 USA
(800) 783-2333 (tel)
(408) 945-9000 (tel)
(408) 945-0587 (fax)
Manufactures terminals for a wide variety of professional computer network applications.

Relm Communications

7707 Records St., Indianapolis, IN 46226 USA
(800) 821-2900 (tel)
Manufactures and sells VHF/UHF transceivers for professional two-way radio applications.

Reltec

11333 Edison St., Franklin Park, IL 60131 USA
(847) 455-8010 (tel)
Manufactures broadband enclosures for professional cable TV applications.

Renco Electronics

60 Jefryn Blvd. E., Deer Park, NY 11729 USA
(800) 645-5828 (Engineering tel)
(516) 586-5566 (tel)
(516) 586-5562 (fax)
Manufactures toroids for a wide variety of electronics applications.

Renu International

3315 West Catalina Dr., Phoenix, CA 85017 USA
(800) 473-3778 (tel)
1221 Halifax St., Regina, Saskatchewan, S4R 1T7 Canada
(306) 525-1721 (tel)
Manufactures batteries for a wide variety of applications.

Resurrection Electronics
3504 King St., Austin, TX 78705 USA
(512) 451-5900 (tel)
(512) 323-5152 (fax)
bobo@bga.com
Sells replacement synthesizer keyboard circuit boards for professional and personal musical applications.

Results Computer
2 NW 6th St., Oklahoma City, OK 73102 USA
(800) 730-2589 (tel)
(405) 341-2589 (tel)
(405) 330-1930 (fax)
Sells refurbished printers and terminals for a wide variety of computer applications.

Retix
2401 Colorado Ave., Santa Monica, CA 90404 USA
(800) 255-2333 (tel)
(310) 828-3400 (tel)
(310) 828-9107 (fax)
Manufactures routers for professional computer network applications.

Reveal Computer Products
(800) 326-2222 (tel)
(800) 4-REVEAL (tech support tel/fax back)
Manufactures multimedia upgrade kits for a wide variety of personal computer applications.

Revered Technology
4750 Calle Quetzal, Camarillo, CA 93012 USA
(800) 220-8919 (tel)
Manufactures LCD projection panels for computer multimedia applications.

Reynolds Radio
1134 Clara Ave., Ft. Wayne, IN 46805 USA
(219) 482-5630 (tel/fax)
Sells transceivers and other equipment for commercial and amateur two-way radio applications.

RF Applications
9310 Little Mountain Rd., Kirtland Hills, OH 44060 USA
(800) 423-7252 (tel)
(216) 974-1961 (tel)
(216) 974-9506 (fax)
Visa, MasterCard
Manufactures digital wattmeters for amateur radio applications.

RF Concepts
P. O. Box 11039, Reno, NV 89510 USA
(702) 324-3290 (tel)
(702) 324-3289 (fax)
Manufactures and sells VHF and UHF amplifiers for amateur radio applications.

The RF Connection
213 N. Fredrick Ave., #11, Gaithersburg, MD 20877
(800) 783-2666 (tel)
(301) 840-5477 (tel)
(301) 869-3680 (fax)
Sells a variety of RF connectors, coaxial cable, and relays, especially for amateur radio applications.

RF Engineering and Systems
8884 Wagner St., Westminster, CO 80030 USA
(303) 430-8281 (tel)
(303) 430-4023 (fax)
Manufactures power controllers to provide power conditioning and surge protection for a variety of electronic equipment.

RF Industries
7610 Miramar Rd., San Diego, CA 92126 USA
(800) 233-1728 (tel)
(619) 549-6340 (tel)
(619) 549-6345 (fax)
Manufactures mobile-mount whip antennas for cellular telephone applications.

RF Limited
P. O. Box 1124, Issaquah, WA 98027 USA
(206) 222-4295 (tel)
(206) 222-4294 (fax)
Manufactures and sells antennas, accessories, and microphones for amateur radio and citizens band radio applications.

RF Monolithics (RFM)
4441 Sigma RD., Dallas, TX 75244 USA
(214) 233-2903 (tel)
44-1483-747077 (UK tel)
33-1-39-16-42-89 (France tel)
Manufactures SAW components for a wide variety of electronics applications.

RF Parts
435 S. Pacific St., San Marcos, CA 92069 USA
(800) 737-2787 (tel)
(619) 744-0700 (tel)
(619) 744-1943 (fax)
Visa, MasterCard, COD
Sells a wide variety of power tubes, transistors, capacitors, heatsinks, tube sockets, etc. for amateur radio and broadcast transmitters. Mininum order of $25 for telephone and credit card orders; $40 limit for personal checks.

Rhetorex
200 Hacienda Ave., Campbell, CA 95008 USA
(408) 370-0881 (tel)
(408) 370-1171 (fax)
rhet.info@octel.com
Manufactures voice-processing boards for a wide variety of professional computer applications.

Rhythm City
1485 NE Expressway, Atlanta, GA 30329 USA
(404) 320-7253 (tel)
Visa, MasterCard, Discover, American Express
Sells new and used open-reel and cassette decks, mixers, microphones, special effects units, reference monitor loudspeakers, and digital recorders for professional recording studios.

RIA Electronic
15 Meridian Rd., P. O. Box 447, Eatontown, NJ 07724 USA
(908) 389-1300 (tel)
(908) 389-9066 (fax)
Manufactures connectors for a wide variety of electronics applications.

Rich Music
1007 Ave. C, Denton, TX 76201 USA
(800) 795-8493 (tel)
(817) 898-8659 (fax)
Visa, MasterCard, Discover, American Express
Sells recording equipment, mixers, special effects, amplifiers, and musical instruments for professional recording studios, institutions, and live performances.

Rickenbacker International
3895 S. Main St., Santa Ana, CA 92707 USA
Manufactures electric guitars for professional and personal musical applications.

Ricoh
5 Dedrick Pl., W. Caldwell, NJ 07006 USA
(800) 63-RICOH (tel)
(201) 882-2000 (tel)
3001 Orchard Pkwy., San Jose, CA 95134 USA
(408) 432-8800 (tel)
Manufactures photocopiers for a wide variety of office applications.

Riegl
8516 OWG Rd., #101, Orlando, FL 32835 USA
(407) 294-2799 (tel)
(407) 294-3215 (fax)
Manufactures laser range finders for a wide variety of applications.

Rigel
P. O. Box 90040, Gainesville, FL 32607 USA
(904) 373-4629 (tel)
(904) 377-4435 (BBS)
Manufactures embedded controller boards for professional computer applications.

The Rip-Tie Company
P. O. Box 77394, San Francisco, CA 94107 USA
(415) 543-0170 (tel)
(415) 777-9868 (fax)
Manufactures Velcro cable ties for audio and radio cables.

Riser Bond Instruments
(880) 688-8377 (tel)
(402) 466-0933 (tel)
(402) 466-0967 (fax)
Manufactures test equipment for professional cable TV applications.

Rivendell Electronics
8 Londonderry Rd., Derry, NH 03038 USA
(603) 434-5371 (tel)
(603) 432-3674 (fax)
Sells a wide variety of amateur radio equipment and accessories, including amateur radio transceivers, receivers, antenna tuners, antennas, amplifiers, etc.

R&L Electronics
1315 Maple Ave., Hamilton, OH 45011 USA
(800) 221-7735 (tel)
(513) 868-6399 (tel)
(513) 868-6574 (fax)
8524 E. Washington St., Indianapolis, IN 46219 USA
(800) 524-4889 (tel)
(317) 897-7362 (tel)
(317) 898-3027 (fax)
Sells a wide variety of amateur radio equipment and accessories, including amateur radio transceivers, receivers, antenna tuners, antennas, amplifiers, etc.

RMR Electronics
32 Mountain Home Rd., Londonderry, NH 03053 USA
(603) 434-7445 (tel)
Sells a variety of satellite communications equipment, scanners, shortwave receivers, and antennas.

RMS Communications Group
5100 W. Silver Springs Blvd., Ocala, FL 34482 USA
(800) 627-2022 (tel)
Buys and sells pagers for cellular telephone applications.

RMSE
900 Leidy Rd., Westminster, MD 21157 USA
(410) 857-8402 (tel)
(410) 857-8403 (fax)
Sells remanufactured printers and printer parts for a wide variety of computer applications.

RMS Electronics
41-51 Hartz Way, Secaucus, NJ 07094 USA
(800) 223-8312 (tel)
(201) 601-9191 (tel)
(201) 601-0011 (fax)
+44-0256-881-525 (UK tel)
+44-0256-882-866 (UK fax)
Manufactures a wide variety of equipment for professional cable TV applications. Offers a free catalog.

RO Associates
246 Caspian Dr., Sunnyvale, CA 94088 USA
(800) 443-1450 (tel)
(408) 744-1450 (tel)
(408) 744-1521 (fax)
Manufactures power modules for a wide variety of electronics applications.

Robotech
43 E. 7200 S., Midvale, UT 84047 USA
(800) 533-0633 (tel)
(800) 533-0633 X157 (fax back)
Manufactures and sells personal desktop computers.

ROBOT Research
5636 Ruffin Rd., San Diego, CA 92123 USA
(619) 278-9430 (tel)
(619) 279-7931 (fax)
Manufactures and sells color SSTV converters for amateur radio applications.

Rockford Fosgate
P. O. Box 1860, Tempe, AZ 85280 USA
(602) 967-3565 (tel)
(800) 238-7011 (fax)
http://www.rockfordcorp.com
Manufactures head units, CD changers, amplifiers, and loudspeakers for mobile audio applications.

Rock'n Rhythm
19880 State Line Rd., S. Bend, IN 46637 USA
(800) 348-5003 (tel)

Rock'n Rhythm

Sells musical instruments, multitrack recorders, DAT decks, signal processors, etc. for professional studio recording and live performances. Offers a free catalog.

Rocksonics
P. O. Box 442, Los Alamitos, CA 90720 USA
(714) 229-0840 (tel)
Manufactures compressors for professional audio applications.

Rocktek America
P. O. Box 4241, Warren, NJ 07059 USA
(908) 469-2828 (tel)
(908) 469-2882 (fax)
Manufactures guitar special effects pedals for professional and personal musical applications.

Rockwell International
7402 Hollister Ave., Santa Barbara, CA 93117 USA
(800) 262-8023 (tel)
(805) 968-4262 (tel)
(805) 968-6478 (fax)
Manufactures equipment for professional computer network applications.

Rockwood
16001 Gothard St., Huntington Beach 92647 USA
(714) 841-3200 (tel)
Manufactures head units, equalizers, crossovers, amplifiers, and loudspeakers for mobile audio applications.

Rocky Mountain Instrument
P. O. Box 683, Thermopolis, WY 82443 USA
(307) 864-9300 (tel)
Manufactures altitude encoder systems for professional aviation applications.

Rocky Mountain Jumper Cables
P. O. Box 9707, Helena, MT 59604 USA
(406) 458-6563 (tel)
Manufactures jumper cables for professional cable TV applications. Offers free samples.

Roger's Systems Specialist
21021-504 Soledad Canyon Rd., Santa Clarita, CA 91351 USA
(800) 366-0579 (tel)
(805) 288-2288 (tel)
(805) 288-2293 (fax)
Visa, MasterCard, Discover, American Express
Sells surplus electronic and computer equipment, parts, and assemblies. Mininum order of $10.

Rogue Music
251 W. 30th St., New York, NY 10001 USA
(212) 629-3708 (tel)
(212) 947-0027 (fax)
roguemus@pipeline.com
Sells musical instruments and recording gear for personal and professional audio and musical applications.

Rogus Electronics
250 Meriden-Waterbury Tnpke., Southington, CT 06489 USA
(203) 621-2252 (tel)
Sells amateur radio equipment and accessories, including antenna tuners, antennas, amplifiers, etc.

Rohde & Schwarz
Mühldorfstrasse 15, D-81671 München, Germany
+4989-4129-3115 (tel)
Manufactures rod antennas for amateur radio applications and receivers for professional monitoring applications.

ROHN
P. O. Box 2000, Peoria, IL 61656 USA
(309) 697-4400 (tel)
(309) 697-7931 (fax)
Manufactures and sells antenna towers, particularly for amateur radio applications.

Roland
7200 Dominion Circle, Los Angeles, CA 90040 USA
(800) 386-7575 (tel)
(213) 685-5141 (tel)
(213) 722-0911 (fax)

5480 Parkwood Way, Richmond, BC V6V 2M4 Canada
(604) 270-6626 (tel)
(604) 270-6552 (fax)
go roland@compuserve.com
http://www.rolandus.com
Manufactures sound cards, music software, multi-track disc recorders for professional studio recording, and keyboards.

Rolls
5143 S. Main St., Salt Lake City, UT 84107 USA
Manufactures guitar pedals for professional and personal musical applications.

Rolls Music Center
1065 W. Broad St, Falls Church, VA 22046 USA
(800) 336-0199 (tel)
(703) 533-9500 (tel)
Sells electric guitars, amplifiers, signal processors, lighting, and audio equipment for a wide variety of professional audio recording and musical applications.

Rolyn Optics
706 Arrowgrand Circle, Covina, CA 91722 USA
(818) 915-5707 (tel)
(818) 915-5717 (tel)
(818) 915-1379 (fax)
Manufactures optical components for a wide variety of industrial and electronics applications. Offers a free 120-page catalog.

Rondo Music
1597 Rt. 22 W., Union, NJ 07083 USA
(800) 845-1947 (tel)
(908) 687-2250 (fax)
Sells electric guitars, synthesizers, amplifiers, PA systems, DJ equipment, etc. for a wide variety of professional and personal musical applications.

Rorke Data
9700 W. 76th St., Eden Prarie, MN 55344
(800) 328-8147 (tel)
(612) 829-0300 (tel)
(612) 829-0988 (fax)
Manufactures a direct-to-disk video recorder for professional studio recording.

Rose Electronics
P. O. Box 742571, Houston, TX 77274 USA
(800) 333-9343 (tel)
(713) 933-7673 (tel)
(713) 933-0044 (fax)
Manufactures print servers, data switches, and keyboard/video controls for a wide variety of personal and professional computer applications. Offers a free catalog.

Rose Enclosures
7330 Executive Way, Fredrick, MD 21704 USA
(301) 696-9800 (tel)
(301) 696-9494 (fax)
Manufactures enclosures for a wide variety of electronics applications.

Rosen's Electronics
104 E. 2nd Ave., Williamson, WV 25661 USA
(304) 235-3677 (tel)
(304) 235-8038 (fax)
Sells amateur radio and computer equipment.

Rosewood
P. O. Box 229, Elko, SC 29826 USA
(800) 875-7762 (tel)
(803) 266-7900 (tel)
Sells transceivers, antennas, power supplies, and accessories for amateur radio applications.

Ross Distributing
78 S. State St., Preston, ID 83263 USA
(208) 852-0830 (tel)
Sells a wide variety of amateur radio equipment and accessories, including amateur radio transceivers, receivers, antenna tuners, antennas, amplifiers, etc.

Ross Systems
P. O. Box 2344, Ft. Worth, TX 76113 USA
(817) 336-5114 (tel)
Manufactures a variety of studio monitor speakers for professional studio recording.

Rotal Telecommunication
+972-3-5596480 (Israel tel)
+972-3-5596484 (Israel fax)
Sells a wide variety of equipment for professional cable TV applications.

Rotel of America
54 Concord St., N. Reading, MA 01864 USA
(800) 370-3741 (tel)
(508) 664-3820 (tel)
(508) 664-4109 (fax)
Manufactures high-end amplifiers that are designed specifically for home theater applications.

Roxy Music
1012 Lincolnway, Laporte, IN 46350 USA
(800) 535-7699 (tel)
Sells electric guitars, amplifiers, etc. for professional and personal musical applications. Offers a free catalog.

Royal Computer
1208 John Reed Ct., Industry, CA 91745 USA
(800) 486-0008 (tel)
(800) 486-2708 (tech support tel)
(818) 855-5077 (tel)
(818) 330-2717 (fax)
Visa, MasterCard, Discover, American Express
Manufactures and sells personal desktop computers. Offers a 30-day money-back guarantee.

Royal Copystar
(800) 824-STAR (tel)
Manufactures photocopiers and fax machines for a wide variety of office applications.

RP Electronics
(800) 304-3604 (tel)
Sells cable TV descramblers. Offers a money-back guarantee.

R & R Communications
2312 Carpenter Rd., Wilmington, DE 19810 USA
(302) 475-1351 (tel)
(302) 475-1352 (tel)
Sells a variety of citizens band radio equipment, including base, mobile, and handheld transceivers, antennas, microphones, etc.

RSP Technologies
2870 Technology Dr., Rochester Hills, MI 48309 USA
(810) 853-3055 (tel)
Manufactures rack-mount encoders, decoders, and controllers for professional studio recording applications.

R & S Surplus
1050 E. Cypress St., Covina, CA 91724 USA
(818) 967-0846 (tel)
(818) 967-1999 (fax)
Visa, MasterCard
Sells surplus test equipment, including frequency counters, DMMs, spectrum analyzers, pulse generators, signal generators, power supplies, oscilloscopes, voltmeters, LCR meters, etc.

RT Aerospace
P. O. Box 170938, Hialeah, FL 33017 USA
(305) 688-5803 (tel)
Sells electric retracts for radio-control model airplanes.

R&T Enterprises
1310 S. 13th, P. O. Box 883, Kelso, WA 98686 USA
(800) 451-9779 (tel)
(360) 423-1840 (tel)
(206) 423-3396 (fax)
Manufactures amplifiers and loudspeakers for mobile audio applications.

RT Systems Amateur Radio Supply
8207 Stephanie Dr., Huntsville, AL 35802 USA
(800) 723-6922 (tel)
(205) 882-9292 (tel)
Visa, MasterCard, Discover, American Express
Sells a wide variety of amateur radio equipment and accessories, including amateur radio transceivers, receivers, antenna tuners, antennas, amplifiers, etc.

RTVC
376 Edgware Rd., London W2 1EB UK

+44-0171-723-8432 (tel)
+44-0171-723-3462 (tel)
+44-0171-723-3467 (fax)
Manufactures amplifiers kits for home audio applications.

Runco
26203 Production Ave., #8, Hayward, CA 94545 USA
(510) 293-9154 (tel)
(510) 293-0201 (fax)
Manufactures laser disc players, scan converters, digital video projectors, etc. for home theater applications.

Russian Dragon
(800) 880-8776 (tel)
(210) 344-3299 (fax)
Manufactures timing accuracy meters for professional recording studios. Offers a two-week money-back guarantee.

Rustrak
1900 S. County Trail, E. Greenwich, RI 02818 USA
(800) 332-3202 (tel)
Manufactures data loggers for professional industrial applications. Offers free information.

Rutland Arrays
109 Finch Ct., Stephens City, VA 22655 USA
(800) 566-4530 (tel)
(540) 869-4530 (tel)
(540) 869-5116 (fax)
Manufactures and sells beam antennas for VHF and UHF amateur radio applications.

Ryzex Re-Marketing
805 W. Orchard Dr., Ste. 3, Bellingham, WA 98225 USA
(800) 941-4004 (tel)
(604) 599-8570 (tel)
(604) 599-8940 (fax)
http://www.Ryzex.com
Buys and sells surplus and reconditioned bar-code equipment for business applications. Offers a 90-day money-back guarantee.

Sabine
4637 NW 6th St., Gainesville, FL 32609 USA
(904) 371-3829 (tel)
(904) 371-7441 (fax)
Manufactures guitar auto tuners for a wide variety of music applications.

S.A.C.
5403 Van Horne, Montreal, PQ H3X 1G2 Canada
(514) 739-2524 (tel)
(514) 343-9599 (fax)
Manufactures dc/ac power inverters for a wide variety of electronics applications.

Sadelco
75 W. Forest Ave., Englewood, NJ 07631 USA
(800) 569-6299 (tel)
(201) 569-3323 (tel)
(201) 569-6285 (fax)
Manufactures calibrators for professional cable TV applications.

The Saelig Company
1193 Mosely Rd., Victor, NY 14564 USA
(716) 425-3753 (tel)
(716) 425-3835 (fax)
Manufactures bus card kits for computer education applications.

Safari Radio
4710 Eisenhower Blvd., Ste. C-4, Tampa, FL 33634 USA
(800) RADIO-80 (tel)
Sells transceivers and accessories for two-way radio applications.

SAFT America
2155 Paseo de las Americas, #31, San Diego, CA 92173 USA
(619) 661-5069 (tel)
(619) 661-7992 (tel)
(619) 661-5096 (fax)
Manufactures batteries for a wide variety of electronics applications.

SAG Electronics
867 Tnpke., N. Andover, MA 01845 USA
(800) 989-3475 (tel)
(508) 682-0055 (tel)
(508) 689-0180 (fax)
Manufactures and sells file servers, RAID systems, and computer storage devices, including CD-ROMs, recordable CD-ROMs, hard drives, tape drives, etc. Offers 30-day money-back guarantee.

Sager
18005 Cortney Ct., City of Industry, CA 91748 USA
(800) 669-1624 (tel)
(818) 964-8682 (tel)
(818) 964-2381 (fax)
Visa, MasterCard, Discover, American Express
Manufactures and sells personal laptop computers. Offers a 30-day money-back guarantee.

Saguaro Computer Services
(602) 952-9675 (tel)
(602) 952-8446 (fax)
Sells new, used, and refurbished parts, peripherals, and systems for a wide variety of computer applications.

Sales Systems
632 Ocean Rd., #8, Point Pleasant, NJ 08742 USA
(800) 277-1308 (sales tel)
(908) 899-6663 (purchase tel)
(908) 899-7238 (fax)
Sells terminals for professional computer applications.

Sam Ash Music Stores
P.O. Box 9047, Hicksville, NY 11802 USA
(800) 472-6274 (tel)
(800) 726-2740 (Canada tel)
(516) 333-8700 (New York tel)
(718) 347-7757 (New York tel)
(609) 667-6696 (New Jersey tel)
(201) 843-0119 (New Jersey tel)
(908) 572-5595 (New Jersey tel)
(516) 931-3881 (fax)
Sells musical instruments, recording equipment, sound and lighting gear, DJ equipment, and computers for a wide variety of professional and personal musical, recording, and computer applications.

Samman's Electronics
1166 Hamburg Tnpke., Wayne, NJ 07470 USA
(800) 937-3537 (tel)
(201) 696-6531 (tel)
Sells high-end, amateur, and home audio and video equipment, such as camcorders, VCRs, video editors, receivers, monitors, TVs, home theater equipment, car audio equipment, etc.

Samson Technologies
P.O. Box 9031, Syosset, NY 11791 USA
(516) 364-2244 (tel)
(516) 364-3888 (fax)
Manufactures graphic equalizers for a wide variety of professional audio applications.

Samsung
Electronics Division
105 Challenger Rd., Ridgefield, NJ 07660 USA
(800) 656-3036 (tel)
(800) 933-4110 (tel)
(800) 229-2239 (fax back)
(201) 229-4000 (tel)
Service Division
1 Samsung Pl., Ledgewood, NJ 07852 USA
(800) SAMSUNG (tech service tel)
(201) 691-6200 (tel)
(201) 347-8650 (tech service fax)
(800) 248-0498 (fax)
Semiconductor Division
3655 N. 1st St., San Jose, CA 95134 USA
(800) 446-2760 (tel)
Visa, MasterCard, Discover, American Express
Manufactures and sells personal laptop computers; also manufactures memory chips for a wide variety of computer applications.

Samtec
P.O. Box 1147, New Albany, IN 47151 USA
(800) SAMTEC-9 (tel)
(812) 944-6733 (tel)
(812) 948-5047 (fax)
http://www.smatec.com

Satellite Systems

Manufactures SMT and microinterconnects for a wide variety of electronics applications.

Samtron
18600 Broadwick St., Rancho Dominguez, CA 90220 USA
(800) SAMTRON (tech support tel)
(310) 537-7000 (tel)
(310) 537-1055 (fax)
Manufactures video monitors for personal computer applications.

Sanar Systems
3350 Scott Blvd., Bldg. 65-01, Santa Clara, CA 95054 USA
(800) 80-SANAR (tel)
(408) 982-0288 (tel)
(408) 982-9369 (fax)
Manufactures workstations for a wide variety of professional computer network applications.

Sangean America
2651 Troy Ave., S. El Monte, CA 91733 USA
(818) 579-1600 (tel)
(818) 579-6806 (fax)
Manufactures a variety of portable receivers for shortwave radio applications.

Sangoma Technologies
7170 Werden Ave., #2, Markhem, ON L5C 347 Canada
(800) 388-2475 (tel)
(905) 474-1990 (tel)
(905) 474-9223 (fax)
Manufactures cards for professional computer network applications.

Sansui
1290 Wall St. W., Lyndhurst, NJ 07071 USA
(201) 460-9710 (tel)
17150 S. Margay Ave., Carson, CA 90746 USA
(800) 421-1500 (tel)
(310) 604-7300 (fax)
(310) 604-1664 (fax)
Manufactures a wide variety of equipment for home and car audio applications.

Santa Fe Distributing
9640 Legler Rd., Lenexa, KS 66219 USA
(800) 255-6595 (tel)
(913) 492-8288 (tel)
(800) 255-6596 (fax)
Sells antennas, duplexers/combiners, coaxial cable, connectors, etc. for professional cellular microwave telephone applications.

Sanyo Fisher USA
Consumer Sales Division
21350 Lassen St., P.O. Box 2329, Chatsworth, CA 91313 USA
(818) 998-7322 (tel)
Service Corporation
1200 W. Artesia Blvd., Compton, CA 90220 USA
(800) 421-5013 (tel)
(310) 537-5830 (tel)
(310) 605-6741 (tel)
(310) 605-6699 (fax)
(310) 605-6744 (fax)
600 Supreme Dr., Bensenville, IL 60106 USA
(708) 350-1505 (tel)
(708) 350-1621 (fax)
210 Riser Rd., Little Ferry, NJ 07643 USA
(201) 641-3000 (tel)
(201) 641-2207 (fax)
1790 Corporate Dr., #340, Norcross, GA 30093 USA
(404) 925-8900 (tel)
(404) 925-9308 (fax)
Manufactures home appliances, vacuum cleaners, heat pumps, home and car stereo equipment, etc. for a wide variety of applications.

Sanyo Technica USA
5420 W. Southern Ave., Indianapolis, IN 46241 USA
(800) 528-0116 (tel)
Manufactures radar detectors for mobile avoidance applications.

Satellite Systems
615 E. Brookside, Colorado Springs, CA 80906 USA
(719) 634-6319 (tel)
(719) 635-8151 (fax)
Manufactures Ariel digital satellite receivers for broadcast stations.

Satman
6310 N. University #3798, Peoria, IL 61612 USA
(800) 472-8626 (tel)
Sells a variety of satellite TV reception equipment, including receivers, dish antennas, cabling, etc.

Satter
P.O. Box 7234, Denver, CO 80207 USA
(303) 399-7493 (tel)
Manufactures lighting systems for amateur and professional video production.

Savant Audio
287 Clarksville Rd., Princeton Junction, NJ 08550 USA
(800) 628-0627 (tel)
(609) 799-9664 (tel)
(609) 799-8480 (fax)
pp000792@interramp.com
Sells home and amateur audio and video equipment, such as camcorders, VCRs, speakers, cassette decks, TVs, home theater equipment, etc.

SavOn Hobbies
P.O. Box 421220, Miami, FL 33242 USA
(305) 634-8959 (tel)
(305) 634-3950 (fax)
Visa, MasterCard, American Express, COD ($5)
Sells temperature probes, radios, batteries, battery chargers, etc. for radio-control models.

SBE
4550 Norris Canyon Rd., San Ramon, CA 94583 USA
(800) 214-4SBE (tel)
(510) 355-2000 (tel)
(510) 355-2020 **(fax)**
+49-8062-8889 (Europe tel)
+44-0-1753-710442 (UK tel)
Manufactures EISA bus controllers for professional computer applications.

Scala
P.O. Box 4580, Medford, OR 97501 USA
(503) 779-6500 (tel)
(503) 779-3991 (fax)
Manufactures antennas for professional cellular microwave radio applications. Offers a free catalog.

Scanner World, USA
10 New Scotland Ave., Albany, NY 12208 USA
(518) 436-9606 (tel)
(518) 465-2945 (fax)
Sells a wide variety of scanner equipment, including base, mobile, and handheld scanners, antennas, two-way radios, CB transceivers, telephones, etc.

SC Engineering
1961 E. Whittier Blvd., La Habra, CA 90631 USA
(800) 546-2529 (tel)
(714) 870-4237 (tel/fax)
Visa, MasterCard
Sells I/O boards, terminal boards, relay boards, etc. for a wide variety of computer applications.

Sceon Lighting Systems
7805 NE 135th Pl., Kirkland, WA 98034 USA
Manufactures fiberoptic lighting systems for hard-to-reach locations.

Schaefer Radio
130 E. Fayette St., P.O. Box 395, Denver, IA 50622 USA
(319) 984-6115 (tel)
(319) 984-6220 (fax)
Sells transceivers, etc. for cellular telephone applications.

Scharff Weisberg
599 11th Ave., New York, NY 10036 USA
(212) 582-2345 (tel)
Sells and rents wireless audio equipment, including microphones, intercoms, digital fiberoptic snakes, walkie-talkies, custom antennas, wireless IFB systems, and professional audio gear.

Schott
1000 Parkers Lake Rd., Wayzata, MN 55391 USA
(612) 475-1786 (tel)
Manufactures transformers for a wide variety of electronics applications.

Schurter
1016 Clegg Ct., Petaluma, CA 94975 USA
(800) 848-2600 (tel)
(707) 778-6311 (tel)
Manufactures fuse clips for a wide variety of electronics applications.

Sciemetric
27 Northside Rd., Nepean, ON K2H 8S1 Canada
(613) 596-3995 (tel)
(613) 820-3746 (fax)
Manufactures turnkey test systems for a wide variety of professional industrial applications.

Science First/Morris & Lee
95 Botsford Place, Buffalo, NY 14216 USA
Sells Van de Graaff generators and accessories for experimental electric applications.

Scientific Atlanta
4291 Communications Dr., Norcross, GA 30093 USA
(800) 722-2009 (tel)
(800) 433-6222 (tel)
(404) 903-6701 (tel)
(404) 903-5375 (fax)
+61-2-452-3388 (Australia tel)
+61-2-451-4432 (Australia fax)
(604) 420-5322 (Canada tel)
(604) 420-5941 (Canada fax)
+86-21-475-0770 (China tel)
+86-21-475-0132 (China fax)
+852-2-522-5059 (Hong Kong tel)
+852-2-522-5624 (Hong Kong fax)
+81-3-3497-9711 (Japan tel)
+81-3-3497-0582 (Japan fax)
(305) 592-3948 (Miami tel)
(305) 592-9646 (Miami fax)
+65-733-9805 (Singapore tel)
+65-738-5806 (Singapore fax)
+34-1-415-6843 (Spain tel)
+34-1-415-6809 (Spain fax)
+44-923-266-133 (UK tel)
+44-923-269-018 (UK fax)
Manufactures transmission equipment for professional cable TV applications.

Scientific Dimensions (SDI)
P.O. Box 26778 Albuquerque, NM 87125 USA
(800) 523-6180 (tel)
(505) 345-8874 (tel)
(505) 345-2812 (fax)
Manufactures vehicle console/slide mounts and docking stations for professional voice/data communication applications.

Scientific Stereo
128 Main St., Brattleboro, VT 05301 USA
(800) 456-4434 (tel)
(802) 257-5855 (tel)
Visa, MasterCard, Discover, American Express
Sells high-end home audio equipment, such as speakers, cassette decks, TVs, home theater equipment, etc. Offers five-year warranties on all new equipment, 100% financing, and free installation in New York City.

Scientific Technologies (STI)
6550 Dumbargon Circle, Fremont, CA 94555 USA
(800) 221-7060 (tel)
(510) 471-9752 (fax)
Manufactures optical scanners for a wide variety of professional industrial applications.

SciNet
3255-2 Scott Blvd., #102, Santa Clara, CA 95054 USA
(800) 653-1010 (tel)
(408) 492-9365 (tel)
(408) 492-9379 (fax)
Manufactures CD-ROM servers for professional computer applications.

Sci Tran Products
1734 Emery Dr., Allison Park, PA 15101 USA
(412) 367-7063 (tel/fax)
3-9-2 Imaizumi, Fujishi, Shizuoka 417 Japan
0545-53-8965 (tel)
0545-53-8978 (fax)
Manufactures data acquisition cards for a wide variety of professional computer applications.

H.H. Scott
5601 Westside Ave., North Bergen, NJ 07047 USA
(201) 952-0077 (tel)
Parts and Literature division
State Rt. 41 & County Rd. 100 W., Princeton, IN 47670 USA
See Emerson for more information.

Scott Communications
P.O. Box 248, St. Ignatius, MT 59865 USA
(406) 745-3218 (tel)
Sells new and used repeaters, base stations, transceivers, etc. for professional cellular microwave telephone applications.

R. Scott Communications
(604) 642-2859 (tel)
(604) 642-7742 (fax)
Manufactures and sells low-power FM broadcast transmitters for hobbyist applications.

SC&T International
(800) 408-4084 (tel)
Manufactures loudspeakers for personal computer multimedia applications.

SDI Technologies
800 Federal Bldg., Carteret, NJ 07008 USA
1299 Main St., Rahway, NJ 07065 USA
(800) 925-5679 (tel)
(800) 888-4491 (customer service tel)
(908) 574-9000 (tel)
(908) 855-0220 (tel)
(908) 855-0331 (traffic dept. tel)
(908) 382-2954 (fax)
(908) 855-0224 (fax)
Manufactures Soundesign audio products, including clock radios, cassette-player clock radios, mini stereo systems, and karaoke systems.

SEA
1769 Leyburn Ct., Jacksonville, FL 32223 USA
(904) 260-3738 (tel)
(904) 260-7553 (fax)
Manufactures loudspeakers for car stereo applications.

Sea Change
6695 Millcreek Dr., Unit 1, Mississauga, ON L5N 5R8 Canada
(800) 661-7274 (tel)
(905) 542-9484 (tel)
(905) 542-9479 (fax)
info@seachange.com
http://www.seachange.com
Manufactures Internet servers for professional network computer applications.

Seagate
920 Disk Dr., Scotts Valley, CA 95066 USA
(408) 438-8111 (tel)
(408) 626-6637 (tel)
http://www.seagate.com
Manufactures hard drives, and personal desktop and laptop computers.

Sears
333 Beverly Rd., Hoffman Estates, IL 60195 USA
(708) 286-5222 (tel)
Sells a wide variety of electronics products, including those for personal computer, home audio, home video, shortwave radio, office supply, etc. applications.

Seco-Larm
17811 Sky Park Circle, #D & E, Irvine, CA 92714 USA
(800) 662-0800 (tel)
(714) 261-2999 (tel)
(714) 261-7326 (fax)
Manufactures systems for car security applications.

Secondhand Sounds
3867 W. Market St., #188, Akron, OH 44333 USA
(216) 665-4755 (tel)
(216) 666-2903 (fax)
70701.1132@compuserve.com
Buys and sells used synthesizer keyboards for professional and personal musical applications.

Second Source Cellular
1403 Alexandria Pike, Ft. Thomas, KY 41075 USA

Select Solutions

(606) 781-8156 (tel)
(606) 781-5574 (fax)
Buys and sells telephones and accessories for cellular telephone applications.

Security Call
15425 Los Gatos Blvd., Los Gatos, CA 95032 USA
(800) 334-0074 (tel)
Visa, MasterCard, American Express
Manufactures a wiretap alarm system for telephones and computers. Offers a one-year warranty.

Security Dynamics
1 Alewife Center, Cambridge, MA 02140 USA
(617) 547-7820 (tel)
(617) 354-8836 (fax)
Manufactures access-control hardware for professional computer network applications.

Security Trade
6065 Hillcroft, #413, Houston, TX 77081 USA
(713) 774-1000 (tel)
(713) 774-6222 (fax)
Sells tape recorders, microphones, bugging/phone tap detectors, scramblers, etc. for a wide variety of security and surveillance applications.

SEEK Systems
11014 120th Ave. NE, Kirkland, WA 98033 USA
(800) 790-7335 (tel)
(206) 822-7400 (tel)
(206) 822-3898 (fax)
Manufactures data storage systems for a wide variety of professional computer network applications.

Seewhy Computer Systems
1424B S. Azusa Ave., W. Covina, CA 91791 USA
(800) 275-1668 (tel)
(818) 916-1088 (tech support tel)
(818) 916-1080 (fax)
seewhy88@aol.com
Visa, MasterCard, Discover, American Express, COD ($5)
Sells computer parts and boards, such as floppy drives, hard drives, monitors, sound cards, modems, motherboards, keyboards, etc.

S.E.G. Technologies
2910 Franks Rd., Huntingdon Valley, PA 19006 USA
(800) 372-7666 (tel)
(215) 947-7669 (international tel)
(215) 947-7666 (tech support tel)
(215) 947-7440 (fax)
Visa, MasterCard, American Express, COD
Sells computer parts and boards, such as hard drives, motherboards, memory chips, etc.

Seiko Instruments
2990 W. Lomita Blvd., Torrance, CA 90505 USA
(310) 517-7771 (tel)
(800) 688-0817 (tel)
(408) 922-5900 (tel)
Manufactures label printers and LCD displays for a wide variety of computer applications.

Sejus
2618 Palisades Crest Dr., Lake Oswego, OR 97034 USA
(503) 638-9000 (tel)
(503) 638-9009 (fax)
Manufactures VXI plug-in computer data storage systems for a variety of professional industrial applications.

Selectra
P.O. Box 5497, Walnut Creek, CA 94596 USA
(800) 874-9889 (tel)
(510) 284-3320 (tel)
Visa, MasterCard
Sells a turnkey A/B-roll editing system for professional video production applications.video production.

Select Solutions
P.O. Box 6512, Champaign, IL 61826 USA
(800) 4-SNAPPY (tel)
(800) 322-1261 (tel)
(217) 355-2785 (customer service tel)
(217) 356-4312 (fax)
solutions@dvideo.com
Visa, MasterCard, Discover, COD ($5)
Specializes in software and equipment for computer video applications, such as sound and video cards, hard drives, Bernoulli drives, etc.

Self-Reliance
P.O. Box 306, Florissant, MO 63032 USA
Sells solar, wind, and hydroelectric systems, inverters, electric boat and car kits, pool heaters, battery chargers, heating stystems, ac/dc refridgerators, etc. Offers a 150+ page catalog for $5.75.

Semaphore Communications
2040 Martin Ave., Santa Clara, CA 95050 USA
(408) 980-7750 (tel)
(408) 980-7769 (fax)
Manufactures encryption sytems for professional computer network applications.

Semtech
625 Mitchell Rd., Newbury Park, CA 91320 USA
(805) 498-3804 (tel)
Manufactures dc-to-dc power converters for a wide variety of electronics applications.

Sencore
3200 Sencore Dr., Sioux Falls, SD 57107 USA
(800) 736-2673 (tel)
(605) 339-0100 (tel)
(605) 339-0317 (fax)
Manufactures test equipment, such as cable TV monitors and analyzers, for a wide variety of applications.

Senix
52 Maple St., Bristol, VT 05443 USA
(800) 677-3649 (tel)
(802) 453-2549 (fax)
Manufactures ultrasonic sensors for a wide variety of electronics applications.

Sennheiser
6 Vista Dr., P.O. Box 987, Old Lyme, CT 06371 USA
(203) 434-9190 (tel)
(203) 434-1759 (fax)
4116 W. Magnolia Blvd., #100, Burbank, CA 91505 USA
(818) 845-7366 (tel)
(818) 845-7140 (fax)
Palenque No. 663, Col. Narvarte, 03600 Mexico DF, Mexico
+5-605-7686 (tel)
+5-605-6473 (tel)
221 Labrosse Ave., Pte-Claire. PQ H9R 1A3 Canada
(514) 426-3013 (tel)
(514) 426-3953 (tel)
Manufactures a variety of RF wireless systems for professional broadcasting and live performances and headphones for high-end home audio applications; also manufactures headsets for professional aviation applications.

SensorData Technologies
43626 Utica Rd., Sterling Heights, MI 48314 USA
(810) 739-4254 (tel)
(810) 739-5689 (fax)
Manufactures torque transducers for a wide variety of professional electronics applications. Offers a free product brochure.

Sensotec
1200 Chesapeake Ave., Columbus, OH 43212 USA
(800) 848-6564 (tel)
(614) 486-0506 (fax)
http://www.sensotec.com
Manufactures pressure transducers, load cells, LVDTs, accelerometers, etc. for a wide variety of professional industrial applications. Offers a free 195-page catalog.

Sentex
1129 Dell Ave., Campbell, CA 95008 USA
(408) 364-0112 (tel)
(408) 364-2038 (fax)
http://www.sentex.com
Manufactures color scanners for a variety of professional applications.

Sentry Manufacturing
P.O. Box 250, Chickasha, OK 73023 USA
(800) 252-6780 (tel)
Manufactures pager and radio crystals for cellular telephone applications.

Sequent Computer Systems
15450 SW Koll Pkwy., Beaverton, OR 97006 USA
(800) 346-2683 (tel)

(503) 626-5700 (fax)
Manufactures workstations for a wide variety of professional computer network applications.

Sequoia Systems
400 Nickerson Rd., Marlborough, MA 01752 USA
(800) 562-0011 (tel)
(508) 480-0800 (tel)
(508) 480-0184 (fax)
http://www.sequoia.com
Manufactures computer systems for professional and industrial computer applications.

Sermax
1751 Second Ave., New York, NY 10128 USA
(800) 209-7126 (tel)
(212) 410-1597 (tel)
(212) 410-0452 (fax)
http://www.simmstack.com
Manufactures memory chip adapters for computer applications.

Serpent
P.O. Box 180, 2100 AD Heemstede, The Netherlands
31-0-23-5-28-49-50 (tel)
http://www.serpent.nl
team@serpent.nl
Manufactures and sells computer trainers for radio-control model applications.

Service Trading
57 Brigman Rd., Chiswick, London W4 5BB UK
0181-995-1560 (tel)
0181-995-0549 (fax)
Visa, Access
Sells a wide variety of electronic equipment and parts, including transformers, EPROM erasure kits, strobe kits, lights, fans, etc.

Servis-Elekto-Praha
K Hrncirum 17, CR-14900 Prague 4, Czech Republic
Sells head units, loudspeakers, amplifiers, receivers, cassette decks, CD players, etc. for home audio and car audio applications.

Sescom
2100 Ward Dr., Henderson, NV 89015 USA
(800) 634-3457 (tel)
(702) 565-3400 (tel)
(702) 565-3993 (tech tel)
(800) 551-2749 (fax)
(702) 565-4828 (fax)
Visa, MasterCard
Manufactures rack and chassis boxes for electronic projects. Offers a free catalog.

Sessions Music
10 Congress St., Portsmouth, NH 03801 USA
(800) 328-0069 (tel)
(603) 431-6251 (tel)
(603) 427-0679 (fax)
Visa, MasterCard, Discover, American Express
Sells a wide variety of home audio and video equipment, including entire home stereo and car stereo systems.

Sextant Avionique
Aérodrome de Villacoublay, BP 59-78141 Vélizy-Villacoublay Cedex, France
33-1-46-29-70-00 (tel)
33-1-46-29-78-60 (fax)
Manufactures Global Navigation Satellite Systems for professional aviation applications.

Seymour Duncan
5427 Hollister Ave., Santa Barbara, CA 93111 USA
(805) 964-9610 (tel)
(805) 964-9749 (fax)
sdpickups@aol.com
Manufactures guitar pickups for professional and personal musical applications.

S.G.
1780 E. Chase Ave., El Cajon, CA 92020 USA
(800) 431-9002 (tel)
Visa, MasterCard
Manufactures and sells radio-control systems for model balloons, boats, and cars.

SGC

The SGC Bldg., P.O. Box 3526, Bellevue, WA 98009 USA
(800) 259-7331 (tel)
(206) 746-6310 (tel)
(206) 746-6384 (fax)
Manufactures amateur, marine, and government transceivers, RF amplifiers, antennas, antenna tuners, etc.

SGS-Thomson Microelectronics

55 Old Bedford Rd., Lincoln, MA 01773 USA
(617) 259-0300 (tel)
(617) 259-4420 (fax)
Manufactures integrated circuits, voltage regulators, etc. for a wide variety of electronics applications.

Shadin

6831 Oxford St., Minneapolis, MN 55426 USA
(800) 328-0584 (tel)
(612) 927-6500 (tel)
(612) 924-1111 (fax)
Manufactures control-panel instruments for professional aviation applications.

Shaffstall

7901 E. 88th St., Indianapolis, IN 46256 USA
(800) 248-3475 (tel)
(317) 594-5457 (international tel)
(317) 842-2077 (tel)
(317) 842-8294 (fax)
sales@shaffstll.com
http://sam.on-net:80/info
Manufactures data transfer and format converters for a wide variety of professional computer applications.

Shakespeare

P.O. Box 733, Newberry, SC 29108 USA
(803) 276-5504 (tel)
Manufactures antennas for citizens band, VHF, and cellular applications.

Sharp

Sharp Plaza, Mahwah, NJ 07430 USA
(800) BE-SHARP (tel)
(800) 642-0261 (tel)
(800) 833-9437 (tel)
(201) 529-8200 (tel)
(201) 512-0055 (tel)
(201) 529-0362 (tel)
(201) 529-8731 (LCD Product Group tel)
(201) 529-9636 (fax)
(201) 512-3456 (fax)
(201) 529-9284 (fax)
Manufactures a wide variety of consumer electronics, such as computer monitors, TVs, VCRs, LCD projection panels for computer multimedia applications, etc.

Sharp Communication

3403 Governors Dr., Huntsville, AL 35805 USA
(800) 548-2484 (tel)
(205) 539-1663 (fax)
Visa, MasterCard, Discover, American Express
Sells equipment for cellular microwave telephone and two-way radio applications.

Sheffield Electronics

P.O. Box 377940B, Chicago, IL 60637 USA
(312) 324-2196 (tel)
Sells partially assembled surveillance transmitter kits. Offers a catalog for $2.

Sheldon's Hobbies

2135 Old Oakland Rd., San Jose, CA 95131 USA
(800) 822-1688 (tel)
(408) 943-0872 (tel)
(408) 943-0904 (fax)
rcs95131@aol.com
http://www.btown.com/sheldon.html
Visa, MasterCard, Discover, COD ($5.50)
Sells radios and servos for radio-control model airplanes. Offers a free catalog.

Shenandoah Equipment

P.O. Box 104, Churchville, VA 24421 USA
(800) 864-0079 (tel)
(540) 337-7582 (tel)
(540) 337-6294 (fax)
http://www.cfw.com/~shenan/home.html
Sells new and refurbished printers, plotters, and accessories for a wide variety of computer applications.

Sherwood
14830 Alondra Blvd., La Mirada, CA 90638 USA
(714) 521-6100 (tel)
Manufactures a wide variety of equipment for home audio, car audio, and car security applications.

Sherwood Engineering
1268 S. Ogden St., Denver, CO 80210 USA
(303) 722-2257 (tel)
(303) 744-8876 (fax)
Sells modified receivers and accessories for shortwave-listening applications. Offers a catalog for $1.

Sherwood Technologies
4181 Business Center Dr., Fremont, CA 94568 USA
(800) 777-8755 (tel)
(510) 623-8900 (tel)
(510) 623-8945 (fax)
Manufactures terminals for a wide variety of professional computer network applications.

A.T. Shindler Communications
101-21 Antares Dr., Nepean, ON K2E 7T8 Canada
(613) 723-1103 (tel)
firlan@fox.nstn.ca
Manufactures and sells wireless LAN links for professional computer networking applications.

Shivelbine Music & Sound
533 Broadway, Cape Girardeau, MO 63701 USA
(314) 334-5216 (tel)
(314) 334-8559 (fax)
Sells new and used synthesizer keyboards and electronic gear for professional and personal musical applications.

Shogyo International
45 Executive Dr., Plainview, NY 11803 USA
(516) 349-5200 (tel)
(516) 349-7744 (fax)
Manufactures transformers, lamps, and bulbs for a wide variety of electronics applications.

Shoppers Market
10875 Washington Blvd., Culver City, CA 90232 USA
(800) 607-2329 (tel)
(310) 842-3700 (sales tel)
(800) 607-2253 (fax)
(310) 842-8130 (fax)
Sells personal desktop and laptop computers, monitors, and printers.

Shredex
49 Natcom Dr., Shirley, NY 11967 USA
(516) 345-0300 (tel)
(516) 345-0791 (fax)
Manufactures paper shredders for a wide variety of office applications.

Shreve Systems
1200 Marshall St., Shreveport, LA 71101 USA
(800) 227-3971 (tel)
(318) 424-9791 (customer service tel)
(318) 424-9771 (fax)
(318) 424-9713 (BBS)
Sells Macintosh and Apple computer parts and boards, such as logic board upgrades, modems, printers, mice, etc.

Shure
222 Hartrey Ave., Evanston, IL 60202 USA
(800) 25-SHURE (tel)
(708) 866-2200 (tel)
(708) 866-2553 (customer service tel)
(708) 866-5732 (service tel)
(708) 866-5725 (fax)
(708) 866-2279 (fax)
(800) 488-3297 (data fax)
Visa, MasterCard, American Express
Manufactures and sells microphones for two-way radio, consumer, and professional audio applications.

Shurter
1016 Clegg Ct., Petaluma, CA 94954 USA
(800) 848-2600 (tel)
(707) 778-6401 (fax)
73024.2314@compuserve.com
http://www.shurterinc.com
Manufactures fuse holders, fuses, and power-entry modules for a wide variety of electronics applications.

Sidney Aircraft Radio Industries
131 Leeder Hill Dr., Hamden, CT 06517 USA
(203) 281-0563 (tel)
Sells unused military and commercial aircraft radio equipment from the 1940s, 1950s, and 1960s.

Siecor
P.O. Box 489, Hickory, NC 28603 USA
(800) SIECOR-4 (tel)
Manufactures cable for a wide variety of electronics applications.

Siemens
10950 N. Tantau Ave., Cupertino, CA 95014 USA
(800) 77-SIEMENS (tel)
http://www.sci.siemens.com/
Manufactures ICs and other components for a wide variety of electronics applications.

Sierra Systems
3510 Frutas Ave., El Paso, TX 79905 USA
(800) 339-4018 (tel)
(915) 577-0049 (fax)
Visa, MasterCard, Discover, American Express
Sells computer parts and boards, such as hard drives, motherboards, memory chips, floppy drives, monitors, modems, etc.

Sigmatest
Im Strickmann 210, D-78344 Gottmadingen, Germany
+49-7731-977001 (tel)
+49-7731-73541 (fax)
Manufactures data acquisitions and analysis systems for electronics troubleshooting applications.

Signal Engineering & Communications
155 San Lazaro Ave., Sunnyvale, CA 94086 USA
(408) 733-1580 (tel)
(408) 732-4456 (fax)
Visa, MasterCard, Discover, American Express
Manufactures quad antennas for CB radio applications.

Signal Enterprises
8265 Sierra College Blvd., Ste. 308, Rosehill, CA 95661 USA
(800) 332-1121 (tel)
(916) 791-8802 (tel)
(916) 791-8803 (fax)
sigent@ix.netcom.com
Sells modems for a wide variety of computer applications.

Signal Transformer
500 Bayview Ave., Inwood, NY 11096 USA
(516) 239-5777 (tel)
(516) 239-7208 (fax)
Manufactures transformers for a wide variety of electronics applications.

Signatec
355 N. Sheridan St., #117, Corona, CA 91720 USA
(909) 734-3001 (tel)
(909) 734-4356 (fax)
Manufactures waveform capture and DSP boards for professional computer applications.

Signature Computer
(702) 387-1922 (tel)
Visa, MasterCard
Sells personal desktop and laptop computers, printers, and scanners.

Signum Systems
200 Brown Rd., Ste. 206, Fremont, CA 94539 USA
(800) 838-8012 (tel)
http://www.signum.com
sales@signum.com
Manufactures in-circuit emulators for professional debugging and development applications. Offers a 10-day free trial.

Sigtronics
822 N. Dodsworth Ave., Covina, CA 91724 USA
(818) 915-1993 (tel)
Manufactures automatic squelch, VOX, pilot isolate systems, etc. for professional aviation applications.

Silent Sam
1627 Basil Dr., Columbus, OH 43227 USA

(800) 398-5605 (tel)
Visa, MasterCard
Manufactures and sells complete and kit vehicle turn-signal reminders.

Silicon Back-Up Solutions
2355 Oakland Rd., #47, San Jose, CA 95131 USA
(408) 434-6697 (tel)
(408) 434-6659 (fax)
Sells data storage systems, such as tape drives, for a wide variety of computer applications.

Silicon Designs
1445 NW Mall St., Issaquah, WA 98027 USA
(206) 391-8329 (tel)
(206) 391-0446 (fax)
Manufactures accelerometers for a variety of professional instrumentation applications.

Silicon Graphics
2011 N. Shoreline Blvd., MS 915, Mountain View, CA 94043 USA
(800) 800-7441 (tel)
(415) 960-1980 (tel)
(415) 961-0595 (fax)
100 W. Cyress Creek Rd., #975, Ft. Lauderdale, FL 33309 USA
(800) 800-4561 (tel)
(305) 938-8388 (tel)
(305) 938-9269 (fax)
Manufactures computer systems for a wide variety of professional computer network applications.

Siliconix
P.O. Box 54951, Santa Clara, CA 95056 USA
(800) 554-5565 (tel)
Manufactures integrated circuits for a wide variety of electronics applications.

Siliconrax
1050 E. Duane Ave. Bldg. I, Sunnyvale, CA 94086 USA
(800) 700-8560 (tel)
(408) 720-1090 (tel)
(408) 245-2570 (fax)
Visa, MasterCard, American Express
Sells equipment for rack-mount personal computers, including chassis, hard disk drives, keyboards, etc.

Silicon Sound
P.O. Box 371694, Reseda, CA 91337 USA
(818) 996-5073 (tel)
Manufactures light-show kits for hobby electronics applications.

Silicon Systems
14351 Myford Rd., Tustin, CA 92680 USA
(800) 624-8999 (tel)
(714) 573-6000 (tel)
(714) 573-6914 (fax)
cpd@ssil.com
Manufactures semiconductors for a wide variety of electronics applications.

Silicon Valley Power Amplifiers
529 Foreman Dr., Unit B, Campbell, CA 95008 USA
(800) 986-9700 (tel)
Manufactures FM RF power amplifiers for broadcast stations.

Sima Products
6153 Mulford St., Niles, IL 60714 USA
(708) 966-0300 (tel)
Manufactures video editors for professional and amateur video production applications.

SimmSaver Technology
228 N. Pennsylvania, Witchita, KS 67214 USA
(800) 636-7281 (tel)
(316) 264-2244 (tel)
(316) 264-4445 (fax)
Manufactures memory adapter boards for computer applications.

Simpson Electric
853 Dundee Ave., Elgin, IL 60120 USA
(708) 697-2260 (tel)
(708) 697-2272 (fax)
Manufactures digital panel indicators for a wide variety of electronics applications.

Sipex
22 Linnell Circle, Billerica, MA 01821 USA
(508) 667-8700 (tel)
(508) 670-9001 (fax)
Manufactures semiconductors for a wide variety of electronics applications.

SIRIO Antenna
Strada dei Collo Sud, 1/Q, 46049 Volta Mantovana, Mantova, Italy
011-39-376-801515 (tel)
011-39-376-801254 (fax)
Manufactures antennas for two-way radio applications.

6th Avenue Electronics
1030 6th Ave., New York, NY 10018 USA
(800) 394-6283 (tel)
(201) 467-0100 (tel)
331 Rt. 4W, Paramus, NJ 07652 USA
22 Rt. 22W, Springfield, NJ 07081 USA
Sells a wide variety of home video and audio equipment, including camcorders, VCRs, TVs, CD/laser disc players, karaoke, home theater systems, etc.

SK America
3858 Carson St., #105, Torrance, CA 90503 USA
(800) SK-USA95 (tel)
(310) 540-6424 (fax)
Distributes Furkawa batteries in North America.

Sky Computers
27 Industrial Ave., Chelmsford, MA 01824 USA
(800) 486-3400 (tel)
(508) 250-1920 (tel)
(508) 250-0036 (fax)
Manufactures computer boards for professional and industrial computer applications.

Skycraft Parts & Surplus
2245 W. Fairbanks Ave., Orlando, FL 32853 USA
(407) 647-4831 (tel)
(407) 628-5634 (fax)
Sells surplus electronics components, wires, batteries, panel meters, circuit boards, etc. for a wide variety of electronics applications.

SkyMall
P.O. Box 20426, Phoenix, AZ 85036 USA
(800) 759-6255 (tel)
(602) 254-5711 (tel)
(800) 986-6255 (fax)
(602) 528-3290 (fax)
Sells a wide variety of products for travelers, including cordless telephones, headsets, portable stereos, laser pointers, electronic thermometers, Caller ID, etc.

Sky1 Technologies
437 Chestnut St., Philadelphia, PA 19106 USA
(800) 294-5640 (tel)
(215) 922-2904 (tel)
(215) 922-6920 (fax)
Visa, MasterCard, Discover, American Express, COD
Sells computer parts and boards, such as hard drives, memory chips, processor upgrades, CD-ROM drives, motherboards, etc.

Skyvision
1050 Frontier Dr., Fergus Falls, MN 56537 USA
(800) 334-6455 (tel)
(218) 739-5231 (tel)
(218) 739-4879 (fax)
Sells satellite TV systems, antennas, receivers, etc. Offers a free 64-page catalog.

Slaughter
801 Hailey St., Ardmore, OK 73401 USA
(800) 421-1921 (tel)
(405) 223-4773 (tel)
Manufactures and sells hipot testers for professional electronics testing applications.

Slep Electronics
P.O. Box 100, Otto, NC 28763 USA
(704) 524-7519 (tel)
Sells a wide variety of amateur radio equipment and accessories, including amateur radio transceivers, receivers, antenna tuners, antennas, amplifiers, military surplus equipment, antique radios, etc.

Smart Choice
1220 Champion Circle, Ste. 100, Carrollton, TX

75006 USA
(800) 444-6278 (tel)
(214) 280-9380 (tel)
Sells baby audio/video monitors.

Smart Modular Technologies
4305 Cushing Pkwy., Fremont, CA 94538 USA
(800) 956-7627 (tel)
(510) 623-1434 (fax)
http://www.smartm.com
Manufactures memory chips for a wide variety of computer applications.

Smile Photo & Video
29 W. 35th St., New York, NY 10001 USA
(800) 516-4202 (tel)
(212) 967-5900 (tel)
(800) 699-2836 (fax)
(212) 967-5912 (fax)
Visa, MasterCard, Discover, American Express
Sells home, amateur, and professional video equipment, such as camcorders, VCRs, video editors, monitors, TVs, PAL equipment, LCD projectors, etc.

Smiley Antenna
408 La Cresta Heights Rd., El Cajon, CA 92021 USA
(800) 527-5439 (tel)
(619) 579-8598 (fax)
Manufactures whip antennas for radio-control model airplanes.

R.E. Smith
4311 Tylersville Rd., Hamilton, OH 45011 USA
(513) 874-4796 (tel)
(513) 874-1236 (fax)
Manufactures communications converter boards for professional computer applications.

Smith Corona
Rt. 13, P.O. Box 2020, Cortland, NY 13045 USA
(607) 753-6011 (tel)
(800) 523-2881 (fax)
Manufactures equipment for office communications applications.

Smith Design
207 E. Prospect Ave., N. Wales, PA 19454 USA
(215) 661-9107 (tel/fax)
Manufactures PC-board RF current measurers for professional electronics applications.

Smith-Victor
301 N. Colfax St., Griffith, IN 46319 USA
(800) 348-9862 (tel)
(219) 924-7356 (tel)
Sells lighting kits, stands, mounts, umbrellas, etc. for professional video production.

S-MOS
2460 N. 1st St., San Jose, CA 95131 USA
(408) 922-0200 (tel)
Manufactures integrated circuits for a wide variety of electronics applications.

SMS Data Products
1501 Farm Credit Dr., McLean, VA 22102 USA
(800) 331-1767 (tel)
(800) 333-7717 (fax)
http://www.sms.com
Manufactures CD-ROM drive towers for professional computer applications.

S & N Cellular
934 University Dr., #152, Coral Springs, FL 33071 USA
(954) 974-6161 (tel)
Sells phone programmers and ESN readers for professional cellular telephone applications.

Snell Acoustics
143 Essex St., Haverhill, MA 01832 USA
(508) 373-6114 (tel)
(508) 373-6172 (fax)
(800) 88-SNELL (fax back)
support@snell.com
Manufactures loudspeakers for home audio and home theater applications.

SNI Innovation
39 Green St., Waltham, MA 02154 USA

SNI Innovation

(617) 736-9007 (tel)
http://www.sniinc.com
Manufactures voice-mail compatible telephones.

SOAT

1155 NW 76th St., Miami, FL 38126 USA
(305) 477-5596 (tel)
(305) 477-3575 (fax)
Manufactures amplifiers and loudspeakers for car stereo applications.

Soborghus Hi-Fi

Frederiksborgvej 221, 2860 Soborg, Denmark
31-67-16-33 (tel)
Sells high-end home audio equipment, such as speakers, cassette decks, amplifiers, preamplifiers, etc.

Softhard Systems Enterprises

3580 Sheridan Dr., Amherst, NY 14226 USA
(716) 834-2125 (tel)
(716) 834-2079 (fax)
sales@soft-hard.com
6301 DeSoto Ave., Woodland Hills, CA 91367 USA
(818) 999-9531 (tel)
(818) 999-9683 (fax)
tilevi@aol.com
http://www.soft-hard.com
Visa, MasterCard, COD
Sells computer parts and boards, such as hard drives, motherboards, controllers, video cards, monitors, cases, etc.

The Software Group

642 Welham Rd., Barrie, ON L4M 6E7 Canada
(800) 4NETCOM (tel)
(705) 725-9999 (tel)
(705) 725-9666 (fax)
Manufactures adapter boards for professional computer network applications.

Sola

1717 Burse Rd., Elk Grove Village, IL 60007 USA
(800) 289-7652 (tel)
(708) 439-2800 (tel)
(800) 626-6269 (fax)
Manufactures uninterruptible power supplies and voltage regulators for a wide variety of personal computer applications.

SolarCon

P.O. Box 176, Holland, OH 43528 USA
Manufactures and sells citizens band radio mobile and vertical antennas.

Solar Electric

4901 Morena Blvd., #305, San Diego, CA 92117 USA
(800) 842-5678 (tel)
(619) 581-0051 (tel)
(619) 581-6440 (fax)
solar@cts.com
Visa, Master Card, Discover, American Express
Sells a wide variety of solar-electricity generating products, including solar panels and dc power inverters.

Solar Electric Specialties

P.O. Box 537, Willits, CA 95490 USA
(800) 344-2003 (tel)
(707) 459-9496 (tel)
(707) 459-5132 (fax)
Sells photovoltaic (solar) power systems for a wide variety of professional electronics applications.

Soldano Custom Amplification

1537 NW Ballard Way, Seattle, WA 98107 USA
(206) 781-4636 (tel)
Manufactures amplifiers for professional and personal musical applications.

Solen

4470 Thibault Ave., St. Hubert, QC J3Y 7T9 Canada
(514) 656-2759 (tel)
(514) 443-4949 (fax)
Sells a variety of components for high-end audio applications, including capacitors, inductors, crossovers, speaker parts, etc. Offers a catalog for $6.

Solflower Computer

1821 Zanker Rd., San Jose, CA 95112 USA

(408) 436-0906 (tel)
(408) 436-0814 (fax)
Manufactures workstations and data storage systems for a wide variety of professional computer network applications.

SoLiCo (Sorenson Lighted Controls)
75 Locust St., Hartford, CT 06114 USA
(800) 275-7089 (tel)
(203) 527-3092 (tel)
(203) 527-6047 (fax)
Manufactures LEDs for a wide variety of electronics applications.

Solid State Logic
Begbroke, Oxford OX5 1RU England
(800) 343-0101 (tel)
+44-01865-842300 (UK tel)
+1-34-60-46-66 (France tel)
+2-612-17-20 (Italy tel)
+3-54-74-11-44 (Japan tel)
(212) 315-1111 (New York tel)
(213) 463-4444 (Los Angeles tel)
Manufactures a full-scale digital film dubbing console (direct from hard disk) with up to 24 tracks for professional studio recording.

Solunet
1591 Robert J. Conlan Blvd., #100, Palm Bay, FL 32905 USA
(800) 795-2814 (tel)
(407) 676-0809 (fax)
mike@solunet.com
Manufactures Internet routers for professional network computer applications.

Somerset Electronics
1290 Hgwy. A1A, Satellite Beach, FL 32937 USA
(800) 678-7388 (tel)
(407) 773-8097 (fax)
Manufactures and sells decoder interfaces for amateur radio applications.

Sonance
961 Calle Negocio, San Clemente, CA 92672 USA
(800) 538-5151 (tel)

(714) 492-7777 (tel)
(714) 361-5151 (fax)
Manufactures loudspeakers, cable, interconnects, and connectors for high-end home audio and home theater applications.

Songtech International
46560 Fremont Blvd., #106, Fremont, CA 94538 USA
(510) 770-9051 (tel)
(510) 770-9060 (fax)
Visa, MasterCard
Buys, sells, and trades memory chips for personal computer applications.

Sonic Communications
4 Colonial Ctr., Box 287, New Ipswich, NH 03071 USA
(800) 688-1944 (tel)
(603) 878-1944 (tel)
(603) 878-1773 (fax)
Manufactures headsets, surveillance kits, ear microphones, motorcycle accessories, etc. for a wide variety of electronics applications.

Sonic Frontiers
2790 Brighton Rd., Oakville, ON L6H 5T4 Canada
(905) 829-3838 (tel)
(905) 829-3033 (fax)
Manufactures high-end digital stereo components, including CD transports, digital processors, and digital-to-analog converters.

Sonic Solutions
101 Rowland Way, #110, Novato, CA 94945 USA
(415) 893-8000 (tel)
(415) 893-8008 (fax)
Manufactures several digital sound processors for professional studio recording.

Sony Electronics
1 Sony Dr., Park Ridge, NJ 07656 USA
(800) 352-7669 (tel)
(201) 930-1000 (tel)
Sony Semiconductor
3300 Zanker Rd., San Jose, CA 95134 USA

Sony Electronics

(800) 288-7669 (tel)
(408) 955-5116 (fax)
National Parts Center
8281 NW 107th Terrace, Kansas City, MO 64153 USA
(816) 891-7550 (tel)
Service company
120 Interstate N. Pkwy. E., #414, Atlanta, GA 30339 USA
(404) 953-5862 (tel)
Service company
1200 N. Arlington Heights Rd., Itasca, IL 60143 USA
(708) 773-7500 (tel)
Service company
3201 Premier Dr., #100, Irving, TX 75063 USA
(214) 550-5270 (tel)
Service company
2729 Bristol St., Costa Mesa, CA 92626 USA
(714) 549-0294 (tel)
Publications department
P.O. Box 20407, Kansas City, MO 64195 USA
(816) 891-7550 (tel)
(800) 821-5662 (fax)
Manufactures digital signal processors for professional studio recording; integrated circuits for a wide variety of electronics applications; audio equipment (including tape decks, turntables, portable stereos, loudspeakers, car stereo equipment, etc); video equipment (including TVs, VCRs, video disc players, etc.); computer equipment (monitors, multimedia equipment, data storage devices, etc.); office equipment (including telephones, etc.), etc.

Sony Mangnescale America

137 W. Bristol Ln., Orange, CA 92665 USA
(714) 921-0630 (tel)
(714) 921-1162 (fax)
Manufactures DAT data recorders, probes, etc. for a wide variety of professional data recording applications.

The Sound Approach

6067 Jerico Tnpke, Commack, NY 11725 USA
(800) 368-2344 (tel)
Sells head units, cassette decks, speakers, receivers, CD players, amplifiers, etc. for home and car stereo applications.

Sound City

Meadtown Shopping Center, Rt. 23 S., Kinnelon, NJ 07405 USA
(800) 542-7283 (order tel)
(800) 233-4010 (service tel)
(201) 838-9100 (tel)
Visa, MasterCard, Discover, American Express
Sells high-end home, amateur, and professional audio and video equipment, such as camcorders, VCRs, video editors, monitors, TVs, home theater equipment, car audio equipment, etc. Offers a free 100-page catalog.

Sound Connections International

203 Flagship Dr., Lutz, FL 33549 USA
(813) 948-2707 (tel)
(813) 948-2907 (fax)
Manufactures and sells high-end audio cable and connectors.

Sound Deals

250 Old Towne Rd., Birmingham, AL 35216 USA
(800) 822-6434 (tel)
(205) 823-4888 (tel)
(205) 979-1811 (fax)
72662.135@compuserve.com
Sells synthesizers, digital samplers, computer hardware, and analog/digital recording gear for a wide variety of professional audio and musical applications.

Sound Dynamics

3641 McNicoll Ave., Scarborough, ON M1X 1G5 Canada
(416) 321-1800 (tel)
Manufactures loudspeakers for home audio and car audio applications.

Sound Electro Flight (SEF)

4545 Industrial St. 5N, Simi Valley, CA 93063 USA
(800) 777-3475 (tel)
(805) 527-9953 (tel)
(805) 527-0046 (tech support tel)
(805) 527-0705 (fax)
electrosef@aol.com
Visa, MasterCard, Discover

Sells computer parts and boards, such as tape drives, hard drives, CD-ROM drives, removable disk drives, adapter boards, etc. Offers a 30-day money-back guarantee.

Sound Ethix
3087 12th St., Riverside, CA 92502 USA
(909) 274-3747 (tel)
Manufactures guitar special effects for professional and personal musical applications.

Sound Ideas
P.O. Box 124, Commack, NY 11725 USA
(516) 864-6548 (tel)
(516) 864-6437 (tel)
http://www.olinsys.com/soundc/soundc.html
Sells head units, loudspeakers, amplifiers, CD changers, cassette decks, etc. for home audio and car audio applications.

Sound Investment
3586 Pierce Dr., Chamblee, GA 30341 USA
(800) 659-8273 (tel)
(404) 458-1679 (tel)
(404) 458-0276 (fax)
Specializes in selling a variety of blank audio/video recording tapes for home, amateur, and professional applications.

Soundnorth Electronics
1802 Hwy. 53, International Falls, MN 56679 USA
(800) 932-3337 (tel)
(218) 283-9290 (fax)
Sells amateur radio equipment and accessories, including amateur radio transceivers, receivers, antenna tuners, antennas, amplifiers, etc.

Sound Quest
2950 Lake Emma Rd., Lake Mary, FL 32746 USA
Manufactures cables, connectors, fuse panels, etc. for mobile audio applications.

Soundscape
705A Lakefield Rd., Westlake Village, CA 91361 USA
(805) 495-7375 (tel)
(805) 379-2648 (fax)
Manufactures digital recording systems for professional audio applications.

The Sound Seller
2808 Cahill Rd., P.O. Box 224, Marinette, WI 54143 USA
(800) 826-0520 (tel)
(715) 735-9002 (tel)
Visa, MasterCard, Discover, American Express
Sells a wide variety of high-end audio equipment.

Sound Source
525-590-0791 (tel)
Sells a wide variety of equipment for mobile audio applications.

SoundSpace
845 Dayton St., Yellow Springs, OH 45387 USA
(800) 767-7353 (tel)
Sells bulk audio cassettes and boxes for professional recording studios.

Soundstream Technologies
120 Blue Ravine Rd., Folsom, CA 95630 USA
(800) 35STREAM (tel)
(916) 351-0414 (fax)
Manufactures amplifiers, preamplifiers, and D/A converters for high-end home audio and home theater applications.

SoundTech
255 Corporate Woods Pkwy., Vernon Hills, IL 60061 USA
(800) 877-6863 (tel)
(708) 913-5511 (tel)
(708) 913-7772 (fax)
http://www.washburn.com
Manufactures amplified audio mixers and loudspeakers for live music performances.

Soundwave
1150 University Ave., Bldg. 5, Rochester, NY 14607 USA
(800) 318-6731 (tel)

Soundwave

(716) 271-8799 (fax)
Manufactures loudspeakers for high-end home audio and home theater applications.

Source Graphics
1540 E. Katella Ave., Anaheim, CA 92805 USA
(800) 553-5285 (tel)
(714) 939-0114 (tech support tel)
(714) 939-0124 (printer service tel)
(714) 939-0525 (fax)
Visa, MasterCard
Sells a variety of computer desktop publishing software and hardware, including digitizers, printers, etc.

Source Innovations
3780 Kilroy Airport Way, #200, Long Beach, CA 90806 USA
(310) 988-6574 (tel)
(310) 988-6570 (fax)
tom2265@ix.netcom.com
http://www.iwsc.com/shoppers/electronics/modula/index.html
Manufactures and sells desk pad touch-screens for personal and professional desktop computer applications.

Southern Audio Services Inc. (SAS)
15049 Florida Blvd., Baton Rouge, LA 70819 USA
(800) 843-8823 (tel)
(504) 272-9844 (fax)
Manufactures loudspeakers for car stereo applications.

Southern California Pro Audio
4603 Reforma Rd., Woodland Hills, CA 91367 USA
(818) 222-4522 (tel)
(818) 222-2248 (fax)
Sells a wide variety of equipment for professional music and studio recording applications.

Southern Computer Repair
2990 Gateway Dr., Ste 500A, Norcross, GA 30071 USA
(770) 441-4980 (tel)
(770) 734-0200 (fax)
Sells Apple and MacIntosh parts for a wide variety of computer applications.

Southern Electronics
10097 Cleary Blvd., #285, Plantation, FL 33324 USA
(954) 723-0557 (tel)
Sells cable TV descramblers.

Southern Music Supplies
100 Woodruff Rd., Greenville, SC 29607 USA
(800) 450-8273 (tel)
Sells a variety of recording supplies, including bulk audio and video cassettes, DAT, ADAT, reel-to-reel tape, etc., for professional recording studios. Offers a free catalog. No mininum order.

Southern Technical
1467 Story Ave., Louisville, KY 40205 USA
(502) 585-5635 (tel)
(502) 584-6008 (fax)
Buys and sells printers and printer parts for a wide variety of computer applications.

South Essex Communications
191 Francis Rd., Leyton, London E10 6NQ UK
081-5580854 (tel)
Sells equipment for shortwave radio receiving applications.

Southside Hobbys
29 Southland Center, Terre Haute, IN 47802 USA
(800) 782-4164 (tel)
(812) 235-7097 (tel)
(812) 877-3618 (fax)
Visa, MasterCard, COD ($5.95)
Sells batteries, radios, speed controls, power supplies, etc. for radio-control models.

Southwest Computers
21429 N. 2nd Ave., #700, Phoenix, AZ 85027 USA
(602) 581-6670 (tel)
(602) 581-6850 (fax)
Sells Hewlett-Packard computer equipment, including printers, disk drives, memory chips, workstations, terminals, tape drives, etc., for a wide variety of computer applications.

South World
12F, 146, Section 5, Min Sheng E. Rd., Taipei, Tai-

wan
886-2-767-6768 (tel)
886-2-765-8620 (fax)
Manufactures splitters, filters, attenuators, amplifiers, etc. for professional cable TV applications.

Sparkomatic
P.O. Box 277, Milford, PA 18337 USA
(800) 233-8831 (tel)
(800) 592-8891 (PA tel)
(717) 296-1230 (tel)
(717) 296-6444 (tel)
Manufactures car stereo equipment.

Sparton Technology
4901 Rockaway Blvd. SE, Rio Rancho, NM 87124 USA
(800) 772-7866 (tel)
Visa, MasterCard, Discover, American Express
Manufactures and sells call routers for professional telecommunications applications.

Speaker City, USA
115 S. Victory Blvd., Burbank, CA 91501 USA
(818) 846-9921 (tel)
(818) 846-1009 (fax)
Sells loudspeakers and components for high-end home audio and professional audio applications.

Speakers Etc.
1828 W. Peoria, Phoenix, AZ 85029 USA
(602) 944-1878 (tel)
(602) 371-0605 (fax)
Sells speaker components, kits, crossovers, and components for high-end audio applications. Offers a free catalog.

Speakerworks
4931 A-1 S. Mingo, Tulsa, OK 74146 USA
(800) 526-8879 (tel)
Visa, MasterCard
Sells a wide variety of components (such as loudspeakers, repair kits, crossovers, etc.) for high-end home audio and professional audio applications.

Speaker World
535 Tabor Rd., Morris Plains, NJ 07950 USA
(201) 984-5200 (tel)
(201) 538-2578 (fax)
Sells loudspeakers for home and car audio applications.

Special Economies
3 Margit St., Wilmington, DE 19810 USA
Sells surplus electronics for a wide variety of applications.

Specialix
745 Camden Ave., #A, Campbell, CA 95008 USA
(800) 423-5364 (tel)
(408) 378-5364 (tel)
(408) 378-0786 (fax)
+44-932-349128 (UK tel)
+65-749-2003 (Asia tel)
Manufactures I/O boards and terminal servers for professional computer applications.

Specialized Products
3131 Premier Dr., Irving, TX 75063 USA
(800) 866-5353 (tel)
(800) 234-8286 (fax)
Sells tools, tool kits, cases, and test equipment for a wide variety of electronics applications.

Spectra
Anaheim Corporate Center, 5101 E. La Palma Ave., 2nd Fl., Anaheim, CA 92807 USA
(800) 745-1233 (tel)
(714) 970-7000 (tel)
(714) 970-7095 (fax)
Buys, sells, rents, and leases new and used computers and peripherals for professional computer applications.

Spectra Logic
1700 N. 55th St., Boulder, CO 80301 USA
(800) 833-1132 (tel)
(303) 449-7759 (tel)
(303) 939-8844 (fax)
Manufactures data storage systems for a wide variety of professional computer network applications.

Spectra Test Equipment
3263 Kifer Rd., Santa Clara, CA 95051 USA
(800) 716-2421 (tel)
(408) 738-2421 (tel)
(408) 738-1503 (fax)
paulhall@spectratest.com
http://spectratest.com
Buys and sells test equipment for a wide variety of electronics applications.

Spectral
18800 142 Ave. NE, Woodinville, WA 98072 USA
(800) 407-5832 (tel)
(206) 487-2931 (tel)
(206) 487-3431 (fax)
Manufactures digital audio workstations for professional studio recording.

Spectron
595 Old Willets Path, P.O. Box 13368, Hauppauge, NY 11788 USA
(516) 582-5600 (tel)
(516) 582-5671 (fax)
Manufactures tilt sensors and inclinometers for a wide variety of professional industrial applications.

Spectron Systems
1425 Hacienda Ave., Campbell, CA 95008 USA
(800) 424-8898 (tel)
(800) 541-0009 (tech support tel)
Visa, MasterCard, Discover
Sells printers for personal computer applications.

Spectronics
1009 Garfield St., Oak Park, IL 60304 USA
(708) 848-6777 (tel)
(708) 848-3398 (fax)
Sells VHF/UHF two-way radio equipment and accessories.

Spectrum Cellular
2611 Cedar Springs Rd., Dallas, TX 75201 USA
(214) 999-6000 (tel)
(214) 880-0151 (fax)
Manufactures computer interfaces for cellular telephones.

Spectrum Computer
5601 Butler St., Pittsburgh, PA 15201 USA
(412) 784-8600 (tel)
(412) 784-6609 (fax)
Visa, MasterCard
Sells personal home computer and printer parts and accessories.

Speir Music
1207 S. Buckner, Dallas, TX 75217 USA
(800) 219-3281 (tel)
Sells synthesizer keyboards, microphones, recording equipment, etc. for professional and personal audio and musical applications.

A.W. Sperry Instruments
245 Marcus Blvd., Hauppauge, NY 11788 USA
(516) 231-7050 (tel)
Manufactures microwave leakage detectors for professional electronics testing applications.

Spica
3425 Bryn Mawr NE, Albuquerque, NM 87107 USA
(415) 397-7100 (tel)
Manufactures loudspeakers for high-end home audio and home theater applications.

Spirit
8760 S. Sandy Pkwy., Sandy, UT 84070 USA
(801) 566-9135 (tel)
(801) 566-2086 (fax)
Manufactures a rack-mount portable mixing console for professional studio recording and live music performances.

Spi-Ro Mfg.
P.O. Box 2800, Dept. 106, Hendersonville, NC 28793 USA
(800) 728-7594 (tel)
(704) 693-1001 (tel)
(704) 693-3002 (fax)
Visa, MasterCard
Manufactures wire antennas, surge protectors, baluns, antenna traps, etc. for amateur radio applications.

Sports-Communications Dist.
P.O. Box 36, Scotts Mills, OR 97375 USA
(800) 573-2256 (tel)
(503) 873-2256 (tel)
(503) 873-2051 (fax)
Sells GMRS radios and accessories.

SpringBoard Technology
1 Federal St., Springfield, MA 01105 USA
(413) 747-2804 (tel)
(413) 747-2438 (fax)
Sells data storage systems, such as hard drives and tape drives, for a wide variety of computer applications.

Spy Emporium
6065 Hillcroft, #414, Houston, TX 77081 USA
Sells devices for surveillance, privacy, and security. Offers a catalog for $5.

Spy Outlet
P.O. Box 337, Buffalo, NY 14226 USA
(716) 691-3476 (tel)
Sells mini FM transmitters, micro recorders, covert video cameras, voice changers, microphones, etc. Offers a catalog for $5.

Spy Supply
1212 Boylston St., #120, Chestnut Hill, MA 02167 USA
(617) 327-7272 (tel)
Sells cellular software, cables, phones, and modification guides.

Square 1 Electronics
P.O. Box 501, Kelseyville, CA 95451 USA
(800) 220-3916 (tel)
(707) 279-8881 (tech support tel)
(707) 279-8883 (fax)
Visa, MasterCard, COD
Manufactures microcontroller trainer kits for a wide variety of electronics applications.

SR Batteries
P.O. Box 287, Bellport, NY 11713 USA
(516) 286-0901 (tel)
(516) 286-0079 (fax)
74167.751@compuserve.com
Manufactures battery chargers, etc. for radio-control model airplanes.

SRS
2044 N. Hwy. 360, Grand Prarie, TX 75050 USA
(800) 335-3789 (tel)
(214) 606-0307 (tel)
(214) 660-1682 (fax)
5915 Coopers Ave., Mississauga, ON L4Z 1R9 Canada
(800) 387-3262 (tel)
(905) 507-4333 (tel)
(905) 507-2325 (fax)
Sells personal desktop computers, parts, and accessories.

SRS
Fallvindsgatan 3-5, S-65106 Karlstad, Sweden
054-850340 (tel)
Sells equipment for shortwave radio receiving applications.

SSB Electronic
124 Cherrywood Dr., Mountaintop, PA 18707 USA
(717) 868-5643 (tel)
Visa, MasterCard
Manufactures and sells preamplifiers, linear amplifiers, transverters, antennas, etc. for amateur radio applications. Offers a catalog for two first-class stamps.

S&S Engineering
14102 Brown Rd., Smithsburg, MD 21783 USA
(301) 416-0661 (tel)
(301) 416-0963 (fax)
Visa, MasterCard
Manufactures frequency counter and transceiver kits for amateur radio applications.

S.S.R.
1025 Eden Way N., Chesapeake, VA 23320 USA
(804) 547-7170 (tel)
Sells computer parts and boards, such as floppy

S.S.R.

drives, hard drives, motherboards, modems, cases, etc.

Stable It
P.O. Box 613, Fairhaven MA 02719 USA
(800) 622-0815 (tel)
Manufactures and sells camcorder stabilizers.

Stabo Ricofunk
Muenchewiese 14-16, 31137 Hildesheim, Germany
49-5121-7620-0 (tel)
49-5121-516646 (fax)
Sells receivers, antennas, and accessories for shortwave radio-listening applications.

Stallion Technologies
60 Penny Ln., Watsonville, CA 95076 USA
(800) 347-7979 (tel)
(408) 761-9499 (tel)
(408) 761-3288 (fax)
Manufactures terminal servers for professional computer network applications.

Standard Amateur Radio Products
P.O. Box 48480, Niles, IL 60714 USA
(312) 763-0081 (tel)
(312) 763-3377 (fax)
Manufactures and sells a variety of UHF and VHF transceivers and accessories for amateur radio use. Offers free information.

Standard Communications
P.O. Box 92151, Los Angeles, CA 90009 USA
(800) 745-2445 (tel)
(310) 532-2329 (tel)
(310) 532-0397 (tel)
(310) 532-5300 (tel)
(800) 722-2329 (fax)
(310) 532-0397 (CA and Int'l fax only)
41 Industrial Pkwy. S., Units 5 & 6, Aurora, ON L4G 3Y5 Canada
(800) 638-4741 (sales tel)
(905) 841-7557 (tel)
(905) 841-3639 (fax)
Manufactures stereo generators, descramblers, modulators, etc. for professional cable TV broadcasting applications.

Standard Microsystems
80 Arkay Dr., Hauppage, NY 11788 USA
(800) 443-SEMI (tel)
Manufactures integrated circuits for a wide variety of computer applications.

Stanford Research Systems
1290D Reamwood Ave., Sunnyvale, CA 94089 USA
(408) 744-9040 (tel)
(408) 744-9049 (fax)
Manufactures thermocouple monitor/scanners for a wide variety of professional industrial applications.

Standford Telecom
480 Java Dr., Sunnyvale, CA 94089 USA
(408) 541-9031 (tel)
(408) 541-9030 (fax)
tpg_marketing@stelhq.com
Manufactures modulators and demodulators for professional cable TV applications.

Stanlite
602 House Wren Circle, Palm Harbor, FL 34683 USA
(800) 772-9373 (tel)
(813) 781-7362 (fax)
Manufactures transceivers for two-way radio applications.

Stanton Video Services
2902 E. Charleston Ave., Phoenix, AZ 85032 USA
(602) 224-6162 (tel)
(602) 493-2468 (fax)
Manufactures a camcorder crane for amateur and professional video production.

Star Circuits
P.O. Box 94917, Las Vegas, NV 89193 USA
(800) 535-STAR (tel)
Sells cable TV notch filters. Offers a free brochure.

Star Computer
70-01 Queens Blvd., Woodside, NY 11377 USA
(800) 853-7827 (tel)

(718) 507-9629 (tel)
(718) 507-7840 (fax)
Visa, MasterCard, American Express, COD
Manufactures and sells personal desktop computers; also sells printers and computer parts.

Star Electronics
(800) 282-4336 (tel)
Sells a cable TV "magic box" and device to "unjam" video tapes. Offers a free catalog.

Stargate Microsystems
47 Meadowlark Rd., Enfield, CT 06082 USA
(800) 462-1800 (tel)
Visa, MasterCard
Manufactures and sells personal desktop computers; also sells computer parts and boards, such as CD-ROM drives, hard drives, monitors, motherboards, memory chips, video cards, etc. Offers a 30-day money-back guarantee.

StarKom
6200 N. Middleville Rd., Middleville, MI 49333 USA
(800) 307-8275 (tel)
(616) 795-2025 (tel)
(616) 795-4323 (fax)
Visa, MasterCard, COD
Sells computer parts and boards, such as hard drives, motherboards, memory chips, multimedia upgrades, monitors, graphic cards, etc.

Starled
1050 E. Dominguez St., #A, Carson, CA 90746 USA
(310) 603-0403 (tel)
(310) 603-1335 (fax)
Manufactures incandescent replacement lights for a wide variety of electronics applications.

Starlite Electronic
1161 Chess Dr., #D, Foster City, CA 94404 USA
(800) 552-9880 (tel)
(415) 372-9289 (tel)
(415) 372-0689 (service tel)
(415) 372-9778 (fax)
Visa, MasterCard, American Express
Manufactures and sells personal desktop computers; also sells computer parts and boards, such as hard drives, video cards, CD-ROM drives, memory chips, etc. Offers a 30-day money-back guarantee.

Starnet Development
404 W. Ironwood Dr., Salt Lake City, UT 84115 USA
(801) 464-1600 (tel)
(801) 464-1699 (fax)
Manufactures digital inserters for professional cable TV applications.

Starquest Computers
4491 Mayfield Rd., S. Euclid, OH 44121 USA
(800) 945-0202 (tel)
(216) 691-9966 (tel)
(216) 691-9967 (tech support tel)
http://www.cybergate.net/_7Esquest
Visa, MasterCard, Discover, American Express
Manufactures and sells personal desktop computers; also sells computer parts and boards, such as floppy drives, hard drives, monitors, video cards, modems, controllers, etc.

StarrSwitch
1717 Fifth Ave., San Diego, CA 92101 USA
(619) 233-6715 (tel)
Manufactures audio processing boards for professional and personal audio and musical applications.

StarTech Computer Products
175 Stonach Cresent, London, ON N5V 3G5 Canada
(800) 265-1844 (tel)
(519) 438-8529 (tel)
(519) 438-6555 (fax)
0-590-3661 (France tel)
0-800-96-7710 (UK tel)
startech.computer@onlinesys.com
Visa, MasterCard, American Express
Manufactures computer server controllers for a wide variety of professional computer applications. Offers a money-back guarantee.

Star Technologies
23162 La Cadena, Laguna Hills, CA 92653 USA
(800) 486-2533 (tel)
(714) 768-6460 (tel)

Star Technologies

(714) 768-8763 (fax)
Manufactures workstations for a wide variety of professional computer network applications.

Startek International
398 NE 38th St., Ft. Lauderdale, FL 33334 USA
(305) 561-2211 (tel)
(800) 638-8050 (tel orders)
(305) 561-9133 (fax)
Visa, MasterCard, Discover, COD ($5 fee)
Manufactures and sells a variety of frequency counters and accessories (bandpass filters, rubber duck antennas, probes, cases, cables, etc.).

Star-Tronics
P.O. Box 98102, Las Vegas, NV 89193 USA
(702) 795-7151 (tel/fax)
Sells surplus electronic components. Offers a free monthly flyer.

The Starving Musician
3427 El Camino, Santa Clara, CA 95051 USA
(408) 554-9041 (tel)
(408) 554-9598 (fax)
Buys, sells, and trades used keyboard synthesizers for professional and personal musical applications.

STB Systems
1651 N. Glenville, Ste. 210, Richardson, TX 75081 USA
(800) 234-4334 (tel)
(214) 234-8750 (tel)
http://www.stb.com
Manufactures video cards for a wide variety of computer applications.

STD Entertainment
10945 McCormick Rd., Hunt Valley, MD 21031 USA
(800) 929-7831 (tel)
Manufactures speakers for personal computer multimedia applications.

Steadi Systems
30 W. 21st St. New York, NY 10010 USA
1014 N. Highland Ave., Hollywood, CA 90038 USA
(800) 626-0946 (NY tel)
(212) 647-0900 (NY tel)
(800) 451-6920 (CA tel)
(213) 461-6868 (CA tel)
(800) 990-8273 (Miami tel)
(305) 375-8273 (Miami tel)
Sells a full line of 3M recording media, including professional open reel tapes, video tapes, cassettes, DATs, computer diskettes, etc.

Stedman
4167 Stedman Dr., Richland, MI 49083 USA
(800) 873-0544 (tel)
(616) 629-5930 (tel)
(616) 629-4149 (fax)
Manufactures microphones for professional studio recording and live music performances.

Steinmetz
7519 Maplewood Ave., Hammond, IN 46324 USA
Sells a variety of electron tubes. Offers a tube list for a SASE.

Stepp Audio Technologies
P.O. Box 1088, Flat Rock, NC 28731 USA
(800) 747-3692 (tel)
(704) 697-9001 (tel)
Visa, MasterCard
Sells surround replacement kits for hi-fi speakers.

Stereo Classics
75 Church St., New Brunswick, NJ 08901 USA
(908) 221-1144 (tel)
(908) 220-1284 (fax)
Sells high-end home audio equipment, such as speakers, cassette decks, amplifiers, preamplifiers, etc.

The Stereo Trading Outlet
320 Old York Rd., Jenkintown, PA 19046 USA
(215) 886-1650 (tel)
Buys and sells a variety of used home audio and video equipment.

Stereo World
P.O. Box 596, Monroe, NY 10950 USA

(914) 782-6044 (tel)
Sells cassette decks, receivers, CD players, loudspeakers, head units, crossovers, etc. for home stereo and car stereo applications.

Sterling Communications
203 N. Chestnut St., McKinney, TX 75069 USA
(800) 786-2199 (tel)
(214) 562-7957 (fax)
Buys and sells site equipment, paging transmitters, trunking systems, etc. for cellular telephone applications.

H. Stewart Designs
P.O. Box 643, Oregon City, OR 97045 USA
(503) 654-3350 (tel)
Manufactures mobile antennas for amateur radio applications.

Stewart Electronics
11460 Sunrise Gold Circle, Rancho Cordova, CA 95742 USA
(800) 316-7626 (tel)
Manufactures a variety of rack-mount amplifiers for professional studio recording and live music performances. Offers a five-year warranty.

Stewart Electronics
1411C First Capital Dr., St. Charles, MO 63303 USA
(314) 949-8890 (tel)
Sells transceivers, antennas, microphones, sound processors, antenna tuners, frequency counters, power supplies, antenna switches, etc. for CB radio applications; also sells marine transceivers and scanner receivers.

Stewart Filmscreen
1161 W. Sepulveda Blvd., Torrance, CA 90502 USA
(310) 326-1422 (tel)
(310) 326-1422 (fax)
Manufactures projection screens for high-end home theater applications.

Stewart of Reading
110 Wykeham Rd., Reading, Berks RG6 1PL UK
01734-268041 (tel)
01734-351696 (fax)
Visa, Access
Sells new and used test equipment, including oscilloscopes, signal generators, spectrum analyzers, noise meters, color bar generators, frequency counters, etc.

STG International
7, Derech Hashalom, Tel-Aviv 67892, Israel
972-3-696-5231 (tel)
972-3-696-5141 (fax)
Sells receivers, antennas, and accessories for shortwave radio-listening applications.

Stick Enterprises
6011 Woodlake Ave., Woodland Hills, CA 91367 USA
(818) 884-2001 (tel)
(818) 883-0668 (fax)
Manufactures electric stringed instruments for professional and personal musical applications.

STI-CO Industries
11 Cobham Drive, Orchard Park, NY 14127 USA
(800) 685-1122 (fax)
(716) 662-2680 (tel)
(716) 662-5150 (fax)
Manufactures mobile antennas for professional two-way radio applications.

Stillwater Designs
P.O. Box 459, Stillwater, OK 74076 USA
(800) 256-5425 (tel)
(405) 624-8510 (tel)
Manufactures amplifiers and loudspeakers for mobile audio applications.

Stirling Connectors
211 Telson Rd., Markham, ON L3R 1E7 Canada
(800) 285-3940 (tel)
(905) 475-6920 (tel)
(905) 475-7228 (fax)
Manufactures connectors for a wide variety of electronics applications.

Stoneman Guitars
20 Russell Blvd., Bradford, PA 16701 USA
(814) 362-8820 (tel)
Manufactures and sells electric guitars for professional and personal musical applications.

Storage Computer
11 Riverside St., Nashua, NH 03062 USA
(603) 880-3005 (tel)
Manufactures data storage systems for a wide variety of professional computer network applications.

Storage Concepts
2652 McGaw Ave., Irvine, CA 92714 USA
(800) 525-9217 (tel)
(714) 852-8511 (tel)
(714) 557-5064 (fax)
Manufactures data storage systems for a wide variety of professional computer network applications.

Storage Dimensions
1656 McCarthy Blvd., Milpitas, CA 95035 USA
(800) 765-7895 (tel)
(408) 954-0710 (tel)
(408) 944-1200 (fax)
Manufactures data storage systems for a wide variety of professional computer network applications.

Storage Solutions
2440 Fulton Ave., #6, Sacramento, CA 95825 USA
(800) 777-3309 (tel)
(916) 484-3475 (tech support tel)
(916) 484-1590 (fax)
Visa, MasterCard, Discover, COD
Sells computer data storage equipment, such as hard drives, tape drives, CD-ROM drives, etc.

Storage Solutions (SSI)
550 West Ave., Stamford, CT 06902 USA
(203) 325-0035 (tel)
Manufactures and sells storage systems for professional computer applications.

StorageTek
801 Warrenville Rd., Lisle, IL 60532 USA
(800) 323-3289 (tel)
(708) 434-1200 (tel)
(708) 434-1254 (fax)
Manufactures data storage systems for a wide variety of professional computer network applications.

Storage USA
101 Reighard Ave., Williamsport, PA 17701 USA
(800) 538-3475 (tel)
(717) 327-9200 (tel)
(717) 327-1217 (fax)
http://www.storageusa.com
Visa, MasterCard
Sells hard disk drives, CD-ROM drives, tape drivers, etc. for personal desktop computers.

Stormer Racing
P.O. Box 126, Hwy. 2 East, Glasgow, MT 59230 USA
(800) 255-7223 (tel)
(800) 255-3133 (cust serv tel)
(406) 228-4569 (tel)
(800) 255-3225 (fax)
(406) 228-8152 (fax)
Visa, MasterCard, Discover, American Express, COD ($5.95)
Sells radios, batteries, battery chargers, speed controls, etc. for radio-control models.

Storm Power Sound
Vesterbrogade 179, 1800 Fredriksberg C, Denmark
31-21-66-20 (tel)
Manufactures amplifiers for high-end home audio applications.

Storm Software
521 Almanor Ave., Sunnyvale, CA 94086 USA
(800) 275-5734 (tel)
http://www.easyphoto.com/storm
Manufactures image scanner systems for a wide variety of personal computer applications.

Strings & Things
1492 Union Ave., Memphis, TN 38104 USA
(901) 278-0500 (tel)
Sells a variety of equipment for professional studio recording.

STS Electronics
1001 NW 62nd St., #108, Ft. Lauderdale, FL 33309 USA
(800) 226-STS1 (tel)
(305) 491-9740 (tel)
(305) 491-0278 (fax)
Visa, MasterCard, Discover, American Express, COD
Sells amateur, broadcast, and professional audio and video equipment, such as character generators, video editors, tripods, etc.

Studer Editech
1947 Leslie St., Toronto, ON M3B 2M3 Canada
(415) 326-7030 (tel)
(404) 257-8829 (Atlanta tel)
(818) 703-1100 (Los Angeles tel)
(212) 626-6734 (New York tel)
(416) 510-1347 (Toronto tel)
(416) 510-1294 (Toronto fax)
Manufactures a digital audio workstation with up to 24 disk channels, 48 assignable mixer strips, and 72 inputs/outputs for professional studio recording.

Studio Audio Digital Equipment (SADiE)
1808 W. End Ave., #119, Nashville, TN 37203 USA
(615) 327-1140 (tel)
(615) 327-1699 (fax)
Manufactures SADiE professional digital audio editors for broadcast stations and for professional studio recording.

Studio Consultants
321 W. 44th St., New York, NY 10036 USA
(212) 586-7376 (tel)
Sells a variety of components for professional studio recording, live music performances, and broadcasting, including consoles, compressor/limiters, rack systems, preamplifiers, etc.

Studio Electronics
18034 Ventura Blvd., #169, Encino, CA 91316 USA
(818) 776-8104 (tel)
(818) 776-1733 (fax)
Manufactures MIDI programmers for professional musical applications.

Studio 1 Productions
1524 County Line Rd., York Springs, PA 17327 USA
(800) 788-0068 (tel)
(717) 528-8374 (fax)
http://members.gnn.com/studio1/home.htm
Visa, MasterCard, Discover, American Express, COD
Sells a variety of professional video equipment, including video and audio distribution amps, video cables, wireless microphones, etc. Offers a free catalog.

Studio Techniques
100 Mill Plain Rd., 3FL, Danbury, CT 06811 USA
(203) 791-3919 (tel)
(203) 791-3918 (fax)
Manufactures an adjustable microphone stand for professional studio recording.

SubTech
112 Primrose Ln., Brick, NJ 08724 USA
(908) 206-1573 (tel)
Manufactures automatic pitch controls and a safety control for radio-control model submarines.

Sullivan Products
P.O. Box 5166, Baltimore, MD 21224 USA
(410) 732-3500 (tel)
(410) 327-7443 (fax)
Sells electric starters for radio-control model airplanes.

Suma Designs
The Workshops, 95 Main Rd., Baxterley Near Atherstone, Warwickshire CV9 2LE UK
01827-714476 (tel/fax)
Manufactures and sells a wide variety of electronics kits, including transmitters, surveillance kits, bug detectors, radios, subcarrier decoders, etc. Offers a catalog for two first-class stamps.

Summit Instruments
2236 N. Cleveland-Massillion Rd., Akron, OH 44333 USA
(216) 659-3312 (tel)
(216) 659-3286 (fax)
Manufactures accelerometers for a wide variety of professional electronics applications.

Suncoast Technologies
P.O. Box 5835, Spring Hill, FL 34606 USA
(904) 596-7599 (tel/fax)
Manufactures single-board computers for professional and industrial computer applications.

Suncom Technologies
6400 W. Gross Point Rd., Niles, IL 60714 USA
(708) 647-4040 (tel)
Manufactures joysticks for computer gaming applications.

Sun Data
1300 Oakbrook Dr., Norcross, GA 30093 USA
(800) 241-9882 (tel)
(770) 449-6116 (tel)
(770) 448-4536 (fax)
http://www.sundata.com
Sells personal desktop computer systems.

SunDisk
3270 Jay St., Santa Clara, CA 95054 USA
(408) 562-3400 (tel)
(503) 967-0072 (fax back)
(408) 986-1186 (BBS)
Manufactures interface boards for a wide variety of personal computer applications.

Sunfire
P.O. Box 1589, Snohomish, WA 98290 USA
(206) 335-4748 (tel)
Manufactures and sells high-end stereo power amplifiers.

Sun Microsystems
2550 Garcia Ave, Mountain View, CA 94043 USA
(800) 681-8845 (tel)
(800) 786-0785 (tel)
(415) 960-1300 (tel)
(415) 969-9131 (fax)
http://www.sun.com/sparc/net.engine
Manufactures microprocessors and Internet servers for computer, multimedia, and network applications.

Sun Moon Star
1941 Ringwood Ave., San Jose, CA 95131 USA
(408) 452-7811 (tel)
(408) 452-1411 (fax)
Sells CD-ROM drives and software for personal computer multimedia applications.

Sunny Bright Electronic
5-1 Fl. No. 490, Fu Shing N. Rd., Taipei, Taiwan
Sells head units, loudspeakers, amplifiers, etc. for car audio applications.

Sunpak
300 Webro Ave., Parsippany, NJ 07054 USA
(201) 428-9800 (tel)
Manufactures lighting systems for amateur and professional video production.

Sunrise Micro Systems
540 Weddell Dr., #1, Sunnyvale, CA 94089 USA
(408) 541-9496 (tel)
(408) 541-9497 (fax)
Sells motherboards, controller cards, memory chips, etc. for a wide variety of computer applications.

SunRiver
2600 McHale Ct., #125, Austin, TX 78754 USA
(512) 835-8001 (tel)
(512) 835-8026 (fax)
Manufactures terminals for a wide variety of professional computer network applications.

Sunset Computer Services
32 Rome St., Farmingdale, NY 11735 USA
(516) 579-8152 (tel)
(516) 420-1251 (fax)
Sells personal desktop computers and printers; also sells parts and accessories, including motherboards, hard drives, power supplies, etc.

Sunshine Computers
1240 E. Newport Center Dr., Deerfield Beach, FL 33442 USA
(800) 828-2992 (tel)
(305) 422-9680 (tel)
(305) 422-5894 (fax)
sergio@gate.net

http://www.sunshinec.com
Visa, MasterCard, Discover, American Express
Sells personal desktop and laptop computers, printers, and scanners; also sells computer parts and boards, such as removable drives, hard drives, monitors, memory upgrades, video cards, modems, etc.

Sunstar

900 W. Hyde Park Blvd., Inglewood, CA 90302 USA
(800) 663-5523 (tel)
(310) 330-2900 (tel)
(310) 330-2910 (fax)
sunstar@ix.netcom.com
Visa, MasterCard
Sells data storage systems, such as hard drives and tape drives, for a wide variety of computer applications.

Super Cable

541-433-2427 (Argentina tel)
541-433-0297 (Argentina fax)
Sells a wide variety of equipment for professional cable TV applications.

Supercircuits

1 Supercircuits Plaza, Leander, TX 78641 USA
(800) 335-9777 (tel)
http://www.scx.com
Manufactures miniature video cameras and transmitters for a wide variety of electronics applications. Offers a 50-page catalog for $5.

Superior Audio

55-56/1 Sukhumvit 63 Rd., Prakanong Bangkok 10110 Thailand
Sells head units, loudspeakers, amplifiers, receivers, cassette decks, CD players, etc. for home audio and car audio applications.

Superior Cellular Products (SCP)

738 Arrow Grand Circle, Covina, CA 91722 USA
(800) 522-4SCP (tel)
(818) 339-1213 (tel)
(818) 339-2223 (fax)
Manufactures battery chargers for cellular telephone applications.

Superior Electric

383 Middle St., Bristol, CT 06010 USA
(800) 787-3532 (tel)
(203) 585-4500 (tel)
(203) 582-3784 (fax)
Manufactures the Stabiline series of uninterruptible power supplies with a two-year warranty.

Superior Electronics Group

6432 Parkland Dr., Sarasota, FL 32243 USA
(941) 756-6000 (tel)
(941) 758-3800 (fax)
Manufactures network management and monitoring systems for professional cable TV applications.

Support Measures

P.O. Box 1279, 20 Main St., #8, Manotick, ON K4M 1B1 Canada
(613) 692-2915 (tel)
(613) 692-2878 (fax)
Sells receivers, antennas, and accessories for shortwave radio-listening applications.

Supra

7101 Supra Dr. SW, Albany, OR 97321 USA
(800) 727-8772 (tel)
(503) 967-2410 (tel)
(503) 967-2401 (fax)
Carl-Friedrich Gauss Strasse 7, D-5024 Pulheim Brauweiler Germany
49-22-34-98590 (tel)
Manufactures modems for a wide variety of personal computer applications.

Supreme

1562 Coney Island Ave., Brooklyn, NY 11230 USA
(800) 332-2661 (tel)
(800) 241-2661 (Canada tel)
(718) 692-4140 (tel)
(718) 692-4760 (fax)
Visa, MasterCard, Discover, American Express
Sells high-end home, amateur, and professional audio and video equipment, such as camcorders, VCRs, video editors, monitors, TVs, home theater equipment, etc.

Surplus Center
P.O. Box 82209, Lincoln, NE 68501 USA
Sells surplus 12-volt electric motors, transformers, rectifiers, etc. for a variety of electronics applications.

Surplus Sales of Nebraska
1502 Jones St., Omaha, NE 68102 USA
(800) 244-4567 (tel)
(402) 346-4750 (tel)
(402) 346-2939 (fax)
Visa, MasterCard, American Express, COD
Sells a variety of surplus electronics components-- especially rare parts for amateur radio and military transmitters and receivers.

Surplus Supply
1446 Sutton Bridge Rd., Rainbow City, AL 35906 USA
(205) 442-7360 (tel)
(205) 442-7343 (fax)
Sells used personal desktop computers and parts and accessories.

Surplus Traders
P.O. Box 276, Alburg, VT 05440 USA
(514) 739-9328 (tel)
(514) 345-8303 (fax)
Sells a wide variety of surplus electronics. Offers a free catalog.

Surplus Trading
P.O. Box 1082, 2700 No. M-63, Benton Harbor, MI, 49022 USA
(616) 849-1800 (tel)
(616) 849-2995 (fax)
Sells surplus parts and equipment for a wide variety of electronics applications.

Surround Sound (SSI)
11836 Clark St., Arcadia, CA 91006 USA
(800) 845-4774 (tel)
(818) 305-0508 (tel)
(818) 305-0201 (fax)
Manufactures and sells loudspeakers and decoder/ amplifiers for home theater applications.

Sutter Buttes 2-Way
598 Garden Hwy., Ste. 16, Yuba City, CA 95991 USA
(916) 674-7532 (tel)
(916) 674-1941 (fax)
Sells remotes, consoles, repeaters, amplifiers, microphones, etc. for professional cellular microwave radio applications.

Sutton Designs
215 N. Cayuga St., Ithaca, NY 14850 USA
(800) 326-8119 (tel)
(607) 277-4301 (tel)
(607) 277-6983 (fax)
Manufactures uninterruptible power supplies for a wide variety of personal computer applications.

Suzalex Technologies
218 Roswell St., Marietta, GA 30060 USA
(800) 526-1669 (tel)
(770) 426-1669 (tel)
(770) 426-1146 (fax)
+234-1-439-6644 (Nigeria tel)
+234-1-439-6661 (Nigeria fax)
Sells uninterruptible power supplies for a wide variety of computer applications.

Svetlana Electron Devices
8200 S. Memorial Pkwy., Hunstville, AL 35802 USA
(800) 239-6900 (tel)
(205) 882-1344 (tel)
(205) 880-8077 (fax)
3000 Alpine Rd., Portola Valley, CA 94028 USA
(800) 578-3852 (tel)
(415) 233-0429 (international tel)
(415) 233-0439 (fax)
Manufactures electron power tubes for amateur radio and professional broadcast transmitters. Offers a free listing.

SVS
1273 Industrial Pkwy. W., #460, P.O. Box 55125, Hayward, CA 94545 USA
(510) 582-6602 (tel)
Sells a laser light kit and a computerized motor kit. Offers a free flyer.

Swagur Enterprises
P.O. Box 620035, Middleton, WI 53562 USA
(608) 592-7409 (tel/fax)
Visa, MasterCard, Discover, American Express
Sells a variety of satellite radio/video receiving equipment, including feed horns and dishes.

Swan Technologies
313 Boston Post Rd., Ste. 200, Marlboro, MA 01752 USA
(800) 645-7789 (tel)
(800) 382-4919 (federal customer tel)
(800) 468-7926 (customer service tel)
(814) 237-4450 (fax)
(814) 237-6145 (BBS)
http://www.tisco.com/swan
Manufactures and sells personal desktop and laptop computers. Offers a 30-day money-back guarantee and a free 24-page catalog.

S&W Computers & Electronics
31 W. 21 St., New York, NY 10010 USA
(800) 874-1235 (tel)
(212) 463-8330 (tel)
(212) 463-8335 (fax)
Visa, MasterCard, Discover, COD
Sells home audio and video, office, and computer equipment, such as camcorders, VCRs, cordless telephones, TVs, answering machines, etc.

Sweetwater Sound
5335 Bass Rd., Ft. Wayne, IN 46808 USA
(219) 432-8176 (tel)
(219) 432-1758 (fax)
sales@sweetwater.com
http://www.sweetwater.com
Visa, MasterCard, Discover, American Express
Sells a variety of components and instruments for professional studio recording and high-end consumer audio applications. Offers a free newsletter.

Swiss Global Informatica
1500 NW 49th St., #500, Ft. Lauderdale, FL 33309 USA
(800) 733-8885 (tel)
Visa, MasterCard
Sells multimedia kits for personal desktop computers. Offers a 30-day money-back guarantee.

SWS Security
(410) 879-4035 (tel)
Sells transceivers to governments for cellular microwave telephones and other two-way radio applications.

Sycard Technology
1180-F Miraloma Way, Sunnyvale, CA 94086 USA
(408) 749-0130 (tel)
(408) 749-1323 (fax)
Visa, MasterCard
Manufactures PC card development tools for a wide variety of professional computer applications.

Symbios Logic
2001 Danfield Ct., Ft. Collins, CO 80525 USA
(800) 334-5454 (tel)
(800) 856-3093 (tel)
http://www.symbios.com
Manufactures computer logic chips for a wide variety of computer applications.

Synapse Micro
15230 Surveyor Blvd., Addison, TX 75244 USA
(800) 460-7357 (tel)
(214) 417-3959 (tel)
(214) 418-9245 (fax)
Sells scanners, CD-ROM towers, and multimedia upgrade kits for a wide variety of computer applications.

Sync-A-Link Communications
P.O. Box 4, Locust Grove, OK 74352 USA
(918) 479-6451 (tel)
Manufactures universal video sync generators for satellite TV receiving applications.

Synchronous Marketing
77 Las Colinas Ln., San Jose, CA 95119 USA
(800) 659-6750 (tel)
(408) 362-4800 (tel)
(408) 362-4826 (fax)

Synchronous Marketing

Manufactures transmitters and amplifiers for professional cable TV applications.

Synctronics
980 Buenos Ave., #C2, San Diego, CA 92110 USA
(800) 444-5397 (tel)
(619) 275-3525 (tel)
(619) 275-3520 (fax)
Manufactures and sells keyboards for a wide variety of computer applications.

Synergetics Surplus
P.O. Box 809, Thatcher, AZ 85552 USA
(520) 428-4073 (tel)
Visa, MasterCard
Sells telephone interfaces for a wide variety of electronics applications.

Synergistic Solutions
1023 Calle Sombia, Unit K, San Clemente, CA 92673 USA
(714) 369-7856 (tel)
Sells new and used printers and printer parts for a wide variety of computer applications.

Synergy Semiconductor
3450 Central Expressway, Santa Clara, CA 95051 USA
(800) 788-3297 (tel)
(408) 730-1313 (tel)
info@synergysemi.com
http://www.synergysemi.com
Manufactures semiconductors for a wide variety of electronics applications.

Synergy Microsystems
9605 Scranton Rd., San Diego, CA 92121 USA
(619) 452-0020 (tel)
(619) 452-0060 (fax)
Manufactures single-board computers for professional and industrial applications.

Syntax Retail Systems
(800) 279-6829 (tel)
(814) 262-0210 (fax)
Visa, MasterCard, Discover, American Express
Manufactures computerized cash registers, UPC scanners, credit card readers, receipt printers, etc. for business applications.

Systech
10505 Sorrento Valley Rd., San Diego, CA 92121 USA
(800) 800-0610 (tel)
(619) 453-8970 (tel)
(800) 205-0611 (fax)
Manufactures Ethernet cards for professional computer network applications.

Systems Beyond
26851 Zemen Ave., Euclid, OH 44132 USA
(216) 261-8846 (tel)
(216) 731-5549 (fax)
Sells NCR and Data General systems, peripherals, upgrades, and parts for professional computer applications.

Systems Design Group
1310 Kingsdale Ave., Redondo Beach, CA 90278 USA
(310) 370-8575 (tel)
Visa, MasterCard, Discover, American Express
Sells high-end home, amateur, and professional audio equipment, such as amplifiers, cables, speakers, preamplifiers, etc.

Systems & Software International
4639 Timber Ridge Dr., Dumfries, VA 22026 USA
(703) 680-3559 (tel)
(703) 878-1460 (fax)
74065.1140@compuserve.com
Manufactures remote scanning systems for computerized VHF/UHF radio scanning.

Systems With Reliability (SWR)
P.O. Box 856, Ebensburg, PA 15931 USA
(814) 472-5436 (tel)
(814) 472-5552 (fax)
Manufactures a variety of customized FM antennas for broadcast stations.

Systran
4126 Linden Ave., Dayton, OH 45432 USA
(800) 252-5601 (tel)
(513) 252-5601 (tel)
(513) 258-2729 (fax)
info@systran.com
Manufactures network adapter cards and switches for a wide variety of professional computer applications.

Tadiran Electronic Industries
2 Seaview Blvd., Port Washington, NY 11050 USA
(800) 537-1368 (tel)
(516) 621-4980 (tel)
(516) 621-4517 (fax)
Manufactures batteries for a wide variety of electronics applications.

Tad Radio of Canada
3101 29 Ave., Vernon, BC V1T 1Z2 Canada
(604) 545-1150 (tel)
Sells VHF/UHF transceivers for professional two-way radio applications.

Tagram System
1451-B Edinger Ave., Tustin, CA 92680 USA
(800) 824-7267 (tel)
(714) 258-3222 (tel)
(714) 258-3220 (fax)
Visa, MasterCard, Discover, American Express
Manufactures and sells personal desktop computers. Offers a 30-day money-back guarantee.

Tait Electronics
9434 Old Katy Rd., #110, Houston, TX 77055 USA
(800) 222-1255 (tel)
Manufactures transceivers for professional two-way radio applications.

Taiwan Cable Connection
3 Fl., No. 2, Alley 2, Ln. 222, Lien-Cheng Rd., Chung Ho City, Taipei, Taiwan
886-2-2480502 (fax)
Manufactures taps for professional cable TV applications.

Talema Electronic
3 Industrial Park Dr., St. James, MO 65559 USA
(314) 265-5500 (tel)
Crolley, Donegal, Ireland
353-75-48666 (tel)
Riesstrasse 8, W-8034 Germering, Germany
089-84-1000 (tel)
1160 #02-05/07, Depot Rd., Singapore 0410
65-2719927-8 (tel)
Manufactures toroidal transformers, coils, and chokes for a wide variety of electronics applications.

Tandem Computers
14321 Tandem Blvd., Austin, TX 78728 USA
(512) 244-8000 (tel)
(512) 244-8588 (fax)
Manufactures workstations for a wide variety of professional computer network applications.

Tandy Electronics
(see Radio Shack)

A.G. Tannenbaum
P.O. Box 386, Ambler, PA 19002 USA
(215) 540-8055 (tel)
(215) 540-8327 (fax)
Visa, MasterCard
Sells vintage electronics components for radio collectors and experimenters. Offers a free catalog.

Tannoy/TGI
300 Gage Ave., Kitchener, ON K2M 2C8 Canada
(519) 745-1158 (tel)
(519) 745-2364 (fax)
Manufactures reference monitor loudspeakers for professional studio recording.

Tape Connection
Jaegergårdsgade 63, Århus C, Denmark
86760806 (tel)
Kalkvaerksvej 12A, Frederikshavn, Denmark
98426839 (tel)
Sells high-end home audio equipment, such as speakers, cassette decks, amplifiers, preamplifiers, etc.

Tape World
220 Spring St., Butler, PA 16003 USA
(800) 245-6000 (tel)
(412) 283-8298 (fax)
Sells more than 500 different types of blank recording cassettes, DATs, video tapes, MiniDiscs, etc. Offers a free catalog.

TAPR
8987-309 E. Tanque Verde Rd., #337, Tucson, AZ 85749 USA
(817) 383-0000 (tel)
(817) 566-2544 (fax)
Manufactures DSP kits for a wide variety of amateur radio applications.

Tartan Electronics
P.O. Box 36841, Tucson, AZ 85740 USA
(520) 577-1022 (tel)
Sells a variety of military surplus electronics.

Tascam
7733 Telegraph Rd., Monetbello, CA 90640 USA
(800) 827-2268 (tel)
(213) 726-0303 (tel)
340 Brunel Rd., Mississauga, ON L4Z 2C2 Canada
(905) 890-8008 (tel)
Manufactures recorders for professional studio recording and broadcasting applications, including multichannel open-reel recorders, reel-to-reel decks, and DAT decks.

Tasco Electronics
#101A, 627 South St., Honolulu, HI 96813 USA
(800) 435-7788 (tel)
(808) 524-7788 (tel/fax)
Visa, MasterCard
Manufactures color slow-scan television (SSTV) converters with digital signal processing for amateur radio applications.

Tatung
2850 El Presido St., CA 90810 USA
(800) 829-2850 (tel)
(310) 637-2105 (tel)
(213) 979-7055 (tel)
Manufactures computer video monitors and system platforms for a wide variety of personal and professional computer applications.

Tauber Electronics
10656 Roselle St., San Diego, CA 92121 USA
(800) 927-0004 (tel)
Manufactures batteries for a wide variety of electronics applications.

TC Computers
P.O. Box 10428, New Orleans, LA 70181 USA
(800) 723-8282 (tel)
(504) 733-2527 (tel)
(800) 723-9497 (Fortune 1000 sales tel)
(800) 723-9489 (educational sales tel)
(800) 723-9501 (dealer sales tel)
(800) 723-6380 (tech support tel)
(800) 723-9487 (government sales tel)
http://www.tccomputers.com
Visa, MasterCard, Discover, American Express
Sells system upgrades, floppy drives, hard drives, monitors, motherboards, video cards, modems, keyboards, etc.

t.c. electronic
705 Lakefield Rd., Ste. A, Westlake Village, CA 91361 USA
(800) 738-4546 (tel)
http://tcelectronic.com
Manufactures digital-audio mainframe effects processors for professional studio recording applications.

T.C. Electronics
(800) 775-0444 (tel)
Visa, MasterCard, Discover, American Express, COD
Sells cable TV descramblers and converters.

TC Sounds
6199 Cornerstone Ct. E., Ste. 105, San Diego, CA 92121 USA
(619) 622-1212 (tel)
(619) 622-9293 (fax)
Manufactures loudspeakers for high-end home audio and professional audio applications.

TDK
12 Haber Park Dr., Port Washington, NY 11050 USA
(800) 835-8273 (tel)
tdk@usa.pipeline.com
http://www.tdk.com
Manufactures CD-ROMs, audio cassettes, VHS video cassettes, QIC data cartridges, floppy disks, etc. for a wide variety of home and professional data and audio recording applications.

TDL Electronics
P.O. Box 2015, Las Cruces, NM 88004 USA
(505) 382-8175 (tel)
(505) 382-8810 (fax)
Manufactures over-/under-voltage fax protectors.

TD Systems
24 Payton St., Lowell, MA 01852 USA
(508) 937-9465 (tel)
(508) 458-1820 (fax)
Manufactures data storage accessories for a wide variety of professional computer network applications.

Teac
7733 Telegraph Rd., Montebello, CA 90640 USA
(800) 888-4XCD (tel)
(213) 726-0303 (tel)
(800) 366-8868 (fax)
(213) 727-7656 (fax)
Manufactures home stereo systems (including cassette decks, amplifiers, and CD decks) and CD-ROM drives for personal computer systems.

Tea Cellular Network Services
1301 Hightower Trail, #300, Atlanta, GA 30350 USA
(404) 992-7003 (tel)
(404) 992-8432 (fax)
Manufactures cellular networks for professional cellular microwave radio applications.

Team America
16104 Covello St., Van Nuys, CA 91406 USA
(818) 787-1920 (tel)
(818) 787-9118 (fax)
Visa, MasterCard
Sells modems, hard drives, controller cards, sound cards, cases, etc. for a wide variety of personal computer applications.

Team-Hold Technology
No. 532, Chung Yang Rd., Taipei, Taiwan
886-2-788-3393 (tel)
886-2-788-7788 (fax)
Manufactures backup power supplies for a wide variety of professional electronics applications.

Team Orion
23352-J Madero Rd., Mission Viejo, CA 92691 USA
(714) 707-4683 (tel)
363 Rue de Bernex CH-1233, Bernex Switzerland
Manufactures batteries for radio-control models.

Team Systems
2934 Corvin Dr., Santa Clara, CA 95051 USA
(408) 720-8877 (tel)
(408) 720-9643 (fax)
Manufactures digital video generators for professional electronics testing applications.

Tech Electronics
P.O. Box 5275, Gainesville, FL 32602 USA
(352) 376-8080 (tel)
Sells high-end home audio equipment, such as speakers, cassette decks, amplifiers, preamplifiers, etc.

Tec-Hill
7/F, Edward Wong Tower, 910 Cheung Sha Wan Rd., Kowloon, Hong Kong
852-2370-8818 (tel)
852-2370-8829 (fax)
Distributes memory and CPU chips for a wide variety of computer applications.

Techmatix
8309 Linden Oaks Ct., Lorton, VA 22079 USA
(800) 777-4932 (tel)
(703) 912-6649 (fax)
Visa, MasterCard, Discover
Sells personal desktop computers and printers.

Technics
1 Panasonic Way, Secaucus, NJ 07094 USA
(201) 348-7000 (tel)
1707 N. Randall Rd., Elgin, IL 60123 USA
(708) 468-4308 (tel)
6550 Katella Ave., Cypress, CA 90630 USA
San Gabriel Industrial Park, Avenue 65 de Infanteria, Km. 9.5, Carolina, PR 00984 Puerto Rico
(809) 750-4300 (tel)
Manufactures receivers, CD players, loudspeakers, cassette decks, headphones, equalizers, turntables, etc. for home stereo applications.

Technological Arts
1644 Bayview Ave., Box 1704, Toronto, ON M4G 3CS Canada
(416) 963-8996 (tel/fax)
309 Aragona Blvd., #102, Box 418, Virginia Beach, VA 23462 USA
http://www.io.org/~techart
Visa, MasterCard
Manufactures solderless breadboard adapter modules for a wide variety of electronics applications.

Technologic Systems
16610 E. Laser Dr., #10, Fountain Hills, AZ 85268
(800) 967-0101 (tel)
(614) 777-8598 (info fax)
(602) 837-5300 (fax)
Visa, MasterCard
Manufactures computer printer buffers.

Technologie MDB
5445 de Lorimier, #108, Montreal, PQ H2H 2C3 Canada
(514) 526-8851 (tel)
(514) 891-6265 (tel)
(514) 443-5485 (fax)
Sells a variety of hardware, kits, and sound systems for high-end home audio and professional audio applications.

Technology Distribution Network
1000 Young St., #270, Tonawanda, NY 14150 USA
(800) 420-3636 (tel)
(716) 746-0195 (tel)
(716) 743-0198 (fax)
Visa, MasterCard, Discover, American Express, COD
Sells personal computer systems; also sells computer parts and boards, such as floppy drives, hard drives, monitors, video cards, modems, mainboards, tape drives, keyboards, etc.

Technology Renaissance
2134 Plasters Bridge Rd., Atlanta, GA 30324 USA
(404) 876-4440 (tel)
(404) 876-0988 (fax)
Visa, MasterCard
Sells modems, terminals, ports, etc. for a wide variety of computer applications.

Technology Works
44-81-747-8666 (tel)
Manufactures accelerator boards for a wide variety of personal computer applications.

TechnoParts
5060 N. Royal Atlanta Dr., Tucker, GA 30084 USA
(770) 414-9105 (tel)
(770) 414-9115 (fax)
Sells IBM personal home computers and parts and accessories.

Tech-Source
442 S. North Lake Blvd., #1008, Altamonte Springs, FL 32701 USA
(800) 330-8301 (tel)
(407) 262-7100 (tel)
(407) 339-2554 (fax)
Manufactures terminals for a wide variety of professional computer network applications.

Techzam
185-E E. Easy St., Simi Valley, CA 93065 USA
(805) 520-9845 (tel)
(805) 520-9640 (fax)
Oriente 243-8, No. 3, Col. Agricola Oriental, 08500 Mexico D.F., Mexico
756-03-89 (tel/fax)
Sells hard drives for a wide variety of computer applications.

TECI
Rt. 3, Box 8C, Barton, VT 05822 USA
(802) 525-3458 (tel)
(802) 525-3451 (fax)
103006.612@compuserve.com
Manufactures microcontroller boards for educational computer applications.

Tee Vee Supply
407 R. Mystic Ave., P.O. Box 649, Medford, MA 02155 USA
(617) 395-9440 (tel)
(617) 391-8020 (fax)
Sells replacement parts and accessories for audio and video equipment.

Tekin Electronics
940 Calle Negocio, San Clemente, CA 92763 USA
(714) 498-9518 (tel)
(714) 498-6339 (fax)
Manufactures speed controls for radio-control boats.

Teknika Electronics
353 Rt. 46W, Fairfield, NJ 07004 USA
(201) 575-0380 (tel)
Manufactures televisions for home video applications.

Teknor Microsystems
5 Sussex Place., Downington, PA 19335 USA
(610) 873-8444 (tel)
(610) 873-8447 (fax)
616 Cure Boivin, Boisbriand, PQ J7G 2A7 Canada
(514) 437-5682 (tel)
(514) 437-8053 (fax)
Birkenstr. 8, D-84095 Furth, Germany
+49-8704-9213-0 (tel)
+49-8704-9213-33 (fax)
Unit 2/35 Hartington St., Kew, Victoria 3101 Australia
(613) 853-9927 (tel/fax)
Manufactures single-board computers for professional and industrial computer applications.

Tekserve
9400-14 Ransdell Rd., Raleigh, NC 27603 USA
(919) 557-6242 (tel)
(919) 557-6236 (fax)
Visa, MasterCard
Buys and sells IBM personal desktop computers and parts.

Tektronix
P.O. Box 1520, Pittsfield, MA 01202 USA
(800) 426-2200 (tel)
(800) 479-4490 (tel)
http://www.tek.com/mbd/w441
http://www.tek.com/measurement
Manufactures video measurement sets, signal level meters, etc. for professional video test applications.

Telamon
492 Ninth St., #310, Oakland, CA 94607 USA
(800) 622-0630 (tel)
(916) 622-0630 (tel)
(916) 622-0738 (fax)
Manufactures network engines for professional computer applications.

Telco Services
2609 Greenleaf Ave., Elk Grove Village, IL 60007 USA
(708) 439-9030 (tel)
(708) 439-9394 (fax)
Manufactures enclosures for cellular microwave radio repeater installations.

Telebit
1 Executive Dr., Chelmsford, MA 01824 USA
(800) TELEBIT (tel)
(408) 734-4333 (tel)
(408) 745-3872 (fax)
(408) 745-3848 (tech support fax)
(408) 745-3310 (fax back)
(408) 745-3861 (BBS)
info@telebit.com
modems@telebit.com
support@telebit.com
http://www.telebit.com
Manufactures Internet routers and modems for professional network computer applications.

Telecommunications Power Systems
2 Emery Ave., Randolph, NJ 07869 USA
(201) 361-1313 (tel)
Manufactures dc backup power systems for professional PCS and remote-cell site applications.

Telecom Solutions
85 W. Tasman Dr., San Jose, CA 95134 USA
(408) 433-0910 (tel)
(408) 428-7895 (fax)
secure7@telecom.com
Manufactures data networks for professional cable TV and telephony applications.

Teledyne Solid State Products
12525 Daphne Ave, Hawthorne, CA 90250 USA
(213) 777-0077 (tel)
(213) 779-9161 (fax)
49-0-611-763-6147 (Germany tel)
44-0-181-571-9586 (UK tel)
32-0-2-721-5252 (Belgium tel)
81-03-3423-1051 (Japan tel)
Manufactures semiconductors for a wide variety of electronics applications.

Telefunken Semiconductors
P.O. Box 54951, Santa Clara, CA 95056 USA
(408) 988-8000 (tel)
(408) 970-3950 (fax)
Manufactures integrated circuits for a wide variety of electronics applications.

Telefunken Sendertechnik
Sickingenstrasse 20-28, D-1000 Berlin 21, Germany
+49-30-3463-0 (tel)
+49-30-3463-2733 (fax)
Manufactures and sells a variety of transmitters, data communication links, and antennas for professional broadcasting applications.

TeleVideo
550 E. Brokaw Rd., San Jose, CA 95112 USA
(800) 835-3228 (tel)
(408) 954-0622 (fax)
Manufactures sound cards for a wide variety of personal computer applications.

Teleview Distributors
P.O. Box 71465, Las Vegas, NV 89170 USA
(702) 253-1852 (tel)
(702) 253-1854 (fax)
Visa, MasterCard
Sells cable TV decoders.

TeleWire Supply
2100-A Nancy Hank Dr., Norcross, GA 30071 USA
(800) 433-3765 (tel)
(800) 643-2288 (TX tel)
(800) 458-4524 (NJ tel)
(800) 428-7596 (IL tel)
(800) 227-2869 (CA tel)
Manufactures equipment for professional cable TV applications.

Telex
9600 Aldrich Ave. S., Minneapolis, MN 55420 USA
(612) 884-4051 (tel)
(612) 884-0043 (fax)
Manufactures wireless microphone systems for professional studio recording and live music performances; also manufactures LCD projection panels for computer multimedia applications. Offers a three-year warranty.

Telex Communications
8601 E. Cornhusker Hwy., Lincoln, NE 68505 USA
(402) 467-5321 (tel)
(402) 467-3279 (fax)
Manufactures vertical and beam antennas for the HF amateur radio bands. Offers a two-year warranty.

TelGuard
P.O. Box 192, Dorking, Surrey RH4 3YJ UK
01306-877889 (tel)
Manufactures telephone video link equipment for law enforcement applications.

Telos Zephyr
2101 Superior Ave., Cleveland, OH 44114 USA
(216) 241-7225 (tel)
(216) 241-4103 (fax)
info@zephyr.com

Manufactures a digital network audio transceiver for broadcast stations.

TelTec
7890 12th Ave. S., Minneapolis, MN 55425 USA
(800) 233-2001 (tel)
(612) 854-9177 (tel)
(612) 854-8601 (fax)
Manufactures dc and ac fans, heatsinks, cables, and connectors for a wide variety of computer and electronics applications.

TEM Antennas
P.O. Box 381, Milford, NH 03055 USA
Manufactures beam antennas for amateur radio applications.

Temic
2201 Laurelwood Rd., P.O. Box 54951, Santa Clara, CA 95056 USA
(408) 988-8000 (tel)
Manufactures semiconductors for a wide variety of electronics applications.

Tempest Micro
375 N. Citrus Ave., #611, Azusa, CA 91702 USA
(800) 818-5163 (tel)
(800) 848-5167 (customer service tel)
(818) 858-5163 (international tel)
(818) 858-5166 (fax)
Visa, MasterCard, Discover
Manufactures and sells personal desktop computers; also sells upgrade computer parts, such as memory, video cards, modems, etc.

Ten Lab
2217 Malcolm Ave., Los Angeles, CA 90064 USA
(310) 474-6909 (tel)
(310) 441-1807 (fax)
http://www.tenlab.com
Manufactures digital video converters, rear projectors, etc. for home theater applications.

Tennadyne
P.O. Box 1894, Rockport, TX 78381 USA
(512) 790-7745 (tel/fax)
Manufactures and sells broadband log-periodic antennas for amateur radio applications. Offers free specifications.specifications.

Ten-Tec
1185 Dolly Parton Pkwy., Seiverville, TN 37862 USA
(800) 833-7373 (tel)
(615) 428-0364 (repair tel)
(615) 453-7172 (kits tel)
(615) 428-4483 (fax)
Visa, MasterCard, Discover
Manufactures and sells amateur radio transceivers for the shortwave bands and electronics kits, including receivers, etc. Offers a VHS video "tour of the factory" for $10 and a 30-day money-back guarantee.

Tentronix
3605 Broken Arrow, Coeur d'Alene, ID 83814 USA
(208) 664-2312 (tel)
Manufactures and sells a Part 15 stereo FM broadcast transmitter kit.

Ten X Technology
4807 Spicewood Springs Rd., #3200, Austin, TX 78759 USA
(800) 922-9050 (tel)
(512) 346-8360 (tel)
(512) 346-9580 (fax)
Manufactures data storage systems for a wide variety of professional computer network applications.

Teradyne
30801 Agoura Rd., Agoura Hills, CA 91301 USA
(818) 991-2900 (tel)
info@std.teradyne.com
Manufactures integrated circuits for computer applications.

Terk Technologies
65 E. Bethpage Rd., Plainview, NY 11803 USA
(516) 756-6000 (tel)
Manufactures amplified AM, FM, and TV antennas for home audio and home video applications.

Terminals, Etc.
1225 E. Crosby Rd., Bldg. A, Unit 14, Carrollton, TX 75006 USA
(800) FIX-WYSE (tel)
(214) 242-1979 (tel)
(214) 245-4646 (fax)
Visa, MasterCard
Sells refurbished terminals for professional computer applications.

Tern
216 F St., #104, Davis, CA 95616 USA
(916) 758-0180 (tel)
(916) 758-0181 (fax)
tern@netcom.com
ftp://ftp.netcom.com/pub/te/tern (Internet ftp site)
Manufactures data acquisition cards for a wide variety of professional computer applications.

Test Equipment Sales
P.O. Box 986, Londonderry, NH 03503 USA
(603) 434-2544 (tel)
(603) 425-2945 (fax)
Visa, MasterCard
Sells test equipment, including oscilloscopes, pulse generators, SWR meters, distortion analyzers, etc. for a wide variety of electronics testing applications.

Testronics
1320 Millwood Rd., McKinney, TX 75069 USA
(214) 542-3111 (tel)
(214) 542-2131 (fax)
Manufactures manufacturing defects analyzers for professional industrial applications.

Texas Hardware
12029 Denton Dr., Dallas, TX 75234 USA
(214) 247-6506 (tel)
(214) 247-1236 (fax)
Sells printers, terminals, monitors, disk drives, etc. for a wide variety of computer applications.

Texas Instruments
6620 Chase Oaks Blvd., Plano, TX 75023 USA
(800) 336-5236 (tel)
(214) 995-2067 (tel)
(214) 995-6611 (fax)
P.O. Box 172228, Denver, CO 80217 USA (literature response center)
(800) TI-TEXAS (tel)
(800) 477-8924 (tel)
(713) 274-2323 (BBS)
+33-1-3070-1199 (Europe BBS)
2ti@msg.ti.com
http://www.ti.com
Manufactures and sells integrated circuits and personal laptop computers.

Texas ISA
14825 St. Mary's Ln., #120, Houston, TX 77079 USA
(713) 493-9925 (tel)
(713) 493-2724 (fax)
Manufactures data storage systems for a wide variety of professional computer network applications.

Texas Microsystems
5959 Corporate Dr., Houston, TX 77036
(800) 627-8700 (tel)
(713) 541-8200 (tel)
(713) 541-8226 (fax)
webmaster@texmicro.com
http://www.texmicro.com
Manufactures computers, PCs, peripherals, and laptops for professional and industrial computer applications.

Texas Timers
3317 Pine Timbers Dr., Johnson City, TN 37604 USA
Manufactures timers for model airplanes.

Texas Towers
1108 Summit Ave., Ste., #4, Plano, TX 75074 USA
(800) 272-3467 (tel)
(214) 422-7306 (tel)
(214) 881-0776 (fax)
Sells a wide variety of amateur radio equipment and accessories, including amateur radio transceivers, receivers, antenna tuners, antennas, amplifiers, etc.

Thaler
2015 N. Forbes Blvd., Tucson, AZ 85737 USA
(800) 827-6006 (tel)
(602) 742-9826 (fax)
Manufactures integrated circuits for a wide variety of electronics applications.

Themis Computer
3185 Laurelview Ct., Fremont, CA 94538 USA
(510) 252-0870 (tel)
(510) 490-5529 (fax)
+33-1-46-23-22-57 (Germany tel)
+44-734-258-080 (UK tel)
+81-78-261-6443 (Japan tel)
Manufactures processor boards for professional and industrial computer applications.

Thiel
1026 Nandino Blvd., Lexington, KY 40511 USA
(606) 254-9427 (tel)
(606) 254-0075 (fax)
Manufactures loudspeakers for high-end home audio and home theater applications.

Thomas Distributing
128 Eastwood, Paris, IL 61944 USA
(800) 821-2769 (tel)
Sells transceivers, microphones, scanners, antennas, etc. for citizens band radio applications.

Thomas Higgins
Capiol Bldgs., 10 Seaview Rd., Wallasey L45 4TH UK
+44-051-691-2817 (tel)
+44-051-630-2824 (fax)
Manufactures file servers for a wide variety of professional computer network applications.

Thomcast
1, rue de Hautil, BP 150, 78702 Conflans-Ste-Honorine Cedex, France
33-1-34-90-31-00 (tel)
33-1-34-90-30-00 (fax)
Bahnhofstrasse, CH-5300 Turgi, Switzerland
41-56-79-44-44 (tel)
41-56-28-11-25 (fax)
Ohmweg 11-15, D-68197 Mannheim, Germany
49-0621-8101-270 (tel)
49-0621-8101-290 (fax)
Manufactures and sells a wide variety of broadcasting transmitters.

Thorens Iberica
Gerona, 61, E-08009 Barcelona, Spain
Sells head units, loudspeakers, amplifiers, receivers, cassette decks, CD players, etc. for home audio and car audio applications.

Thoroughbred Music
P.O. Box 8009, Tampa, FL 33674 USA
(800) 800-4654 (tel)
(813) 885-9644 (tel)
(813) 881-1896 (fax)
jarata@ix.netcom.com
http://www.tbred-music.com/network
Visa, MasterCard, Discover, American Express
Sells electric guitars, amplifiers, special effects, keyboard synthesizers, lighting equipment, and professional audio equipment for a wide variety of professional and personal audio and musical applications. Offers a free catalog.

3Com
(800) NET-3COM (tel)
Level 7, 99 Walker St., North Sydney, NSW 2060 New Zealand
09-358-0322 (tel)
09-414-0100 (tel)
02-956-6247 (fax)
http://www.3com.com
Manufactures ISDN modems and adapters for professional computer network applications.

3M
Visual Systems Division
6801 River Place Blvd., Austin, TX 78726 USA
(800) 328-1371 (tel)
(800) 328-0016 (tel)
Manufactures connectors for a wide variety of applications, and disks, cassettes, etc. for magnetic recording applications.

3S Navigation
23141 Plaza Pointe Dr., Laguna Hills, CA 92653 USA
(714) 830-3777 (tel)
(714) 830-8411 (fax)
nav3s@aol.com
Manufactures GPS/GLONASS receiving systems for professional aviation applications.

Thrifty Distributors
641 W. Lancaster Ave., Frazer, PA 19355 USA
(800) 342-3610 (tel)
(610) 647-6289 (fax)
Visa, MasterCard, Discover
Sells home and professional audio and video equipment, such as wireless microphones, battery packs, magnetic recording media, etc.

Throttle Up!
P.O. Box 33924, Denver, CO 80223 USA
(303) 465-4435 (tel)
(303) 465-4191 (fax)
Manufactures digital sound systems for model railroads.

Thrustmaster
10150 SW Nimbus Ave., Portland, OR 97223 USA
(503) 639-3200 (tel)
(503) 620-8094 (fax)
Manufactures and sells joystick/game controllers for computer video games.

TIC General
P.O. Box 1, 302 3rd St E., Thief River Falls, MN 56701 USA
(800) 842-7464 (tel)
(218) 681-1119 (tel)
(218) 681-8509 (fax)
Manufactures and sells antenna rotators and controllers for amateur radio applications.

Tigertronics
400 Daily Ln., P.O. Box 5210, Grants Pass, OR 97527 USA
(800) 822-9722 (tel)
(503) 474-6700 (tel)
(503) 474-6703 (fax)
Visa, MasterCard
Manufactures and sells a packet modem for amateur radio applications.

Timeless Products
(218) 346-6665 (tel)
Manufactures and sells cable TV converters.

Timeline
2401 Dogwood Way, Vista, CA 92083 USA
(619) 727-3300 (tel)
(619) 727-3620 (fax)
Manufactures digital audio workstation and edit controller for professional studio recording applications.

Timeline
23605 Telo Ave., Torrance, CA 90505 USA
(800) 872-8878 (tel)
(800) 223-9977 (CA tel)
(310) 784-5488 (tech tel)
(310) 784-7690 (fax)
Visa, MasterCard
Sells surplus laser products, computer network boards, bar code readers, power supplies, monitors, computer systems, etc. Mininum order of $20.

Times Fiber Communications
358 Hall Ave., Wallingford, CT 06492 USA
(800) 677-CATV (tel)
(203) 265-8422 (tel)
+44-1279-600677 (UK tel)
Manufactures fiberoptic cable for professional cable TV applications.

Timewave Technology
2401 Pilot Knob Rd., St. Paul, MN 55120 USA
(612) 452-5939 (tel)
(612) 452-4571 (fax)
Manufactures DSP noise filters for shortwave radio and amateur radio-listening applications.

T.J. Antenna
1055 N. 1st Place 3E, Hermiston, OR 97838 USA

(800) 443-0966 (tel)
(503) 567-7885 (tel)
Visa, Master Card, COD
Manufactures and sells motorized, remotely tuned shortwave/VHF antennas for amateur radio applications. Offers a 30-day money-back guarantee.

TL Audio
Letchworth, SG6 1AN UK
+44 01462 490600 (tel)
+44 01462 490700 (fax)
(905) 469-8080 (US tel)
(905) 469-1129 (US fax)
Manufactures a variety of hollow state mixers, microphone preamplifiers, compressors, parametric equalizers, etc. for professional studio recording applications.

TM Brokers
5402 Hwy. 95, Cocolalla, ID 83813 USA
(208) 683-2797 (tel)
(209) 683-3998 (tel)
(208) 683-2019 (tel)
(208) 683-2374 (fax)
Buys and sells new and used professional cable TV equipment.

TM Rush
2705 Lee Blvd., Lehigh Acres, FL 33971 USA
(941) 369-5252 (tel)
(941) 368-6688 (fax)
Manufactures loudspeakers for car stereo applications.

TNC Electronics
2 White's Ln., Woodstock, NY 12498 USA
(914) 679-8549 (tel)
(914) 679-5542 (fax)
Manufactures digital optical tachometers for radio-control model airplanes.

TNR Technical
279 Douglas Ave., Altamonte Springs, FL 32714 USA
(800) 346-0601 (tel)
(407) 682-4469 (fax)
Sells transmitter and receiver packs, batteries, and servos for radio-control model airplanes.

Todays Computers
3000 E. Chambers, Phoenix, AZ 85040 USA
(602) 232-1876 (tel)
(602) 276-3210 (fax)
Visa, MasterCard, Discover, American Express
Sells used and refurbished personal home and laptop computers.

Todd Products
50 Emjay Blvd., Brentwood, NY 11717 USA
(516) 231-3366 (tel)
Manufactures multiple-output power supplies for a wide variety of electronics applications.

Toko America
1250 Feehanville Dr., Mt. Prospect, IL 60056 USA
(708) 297-0070 (tel)
Manufactures inductors for a wide variety of electronics applications.

Top Data
1105 N. Fair Oaks Ave., Sunnyvale, CA 94089 USA
(800) 888-3318 (tel)
(408) 734-9100 (tel)
(408) 734-9343 (tech support tel)
(408) 734-9340 (fax)
Visa, MasterCard, Discover, American Express
Manufactures and sells personal desktop and laptop computers; also sells computer parts and boards, such as monitors, controllers, video cards, multimedia upgrades, etc.

Top Shelf Music
1232 Hertel Ave., Buffalo, NY 14216 USA
(716) 876-6544 (tel)
Buys, sells, and trades used electric guitars for professional and personal musical applications. Offers a catalog for $1.

Toray Marketing and Sales
1875 S. Grant St., #720, San Mateo, CA 94402 USA
(415) 341-7152 (tel)
(415) 341-0845 (fax)
(212) 697-8150 (NY tel)
(212) 972-4279 (NY fax)

Toray Marketing and Sales

Manufactures oxygen analyzers for a wide variety of professional industrial applications; also manufactures data storage systems for a wide variety of professional computer network applications.

Toroid Corp. of Maryland
2020 Northwood Dr., Salisbury, MD 21801 USA
(800) 274-5793 (tel)
(410) 860-0300 (tel)
(410) 860-0302 (fax)
Manufactures specialty toroid transformers--especially for high-end solid-state and hollow-state audio amplifiers.

Toronto Microelectronics (TME)
5149 Bradco Blvd., Mississauga, ON L4W 2A6 Canada
(905) 625-3203 (tel)
(905) 625-3717 (fax)
Manufactures single-board computers for professional and industrial computer applications.

Toshiba America
Comsumer Electronics division
82 Totowa Rd., Wayne, NJ 07470 USA
(800) 457-7777 (tel)
(201) 628-8000 (tel)
Electronics Components division
9775 Toledo Way, Irvine, CA 92718 USA
(800) 879-4963 (tel)
National Parts center
1420 Toshiba Dr., Lebanon, TN 37087 USA
(800) 345-9785 (tel)
(615) 444-7481 (fax)
Manufactures computer memory chips and personal laptop computers.

Total Control Products
2001 N. Janice Ave., Melrose Park, IL 60160 USA
(800) 293-3568 (tel)
(708) 345-5670 (fax)
Manufactures computer touch screens for a wide variety of professional applications.

Touche Touch Pad
4640 W. 77th St., Ste. 311, Edina, MN 55435 USA
(800) 223-6433 (tel)
Manufactures computer pointing devices for a wide variety of personal computer applications.

Tower Electronics
(800) 662-3422 (tel)
Visa, MasterCard
Sells connectors, adapters, and dc wire, particularly for amateur radio applications.

Tower Electronics
281 S. Commerce Circle, Minneapolis, MN 55432 USA
(612) 571-3737 (tel)
(612) 571-5605 (fax)
towerinc@primenet.com
Manufactures power supplies for a wide variety of electronics applications.

Tower Hobbies
P.O. Box 9078, Champaign, IL 61826 USA
(800) 637-4989 (order tel)
(800) 637-6050 (info tel)
(217) 398-3636 (tel)
(800) 637-7303 (fax)
(217) 356-6608 (fax)
orders@towerhobbies.com (Internet order)
info@towerhobbies.com (Internet info)
Visa, MasterCard, Discover, COD ($5.99)
Sells radios, servos, power panels, batteries, and battery chargers for radio-control model airplanes and boats. Offers a catalog for $3.

Toys for Big Boys
1600 S. Anaheim Blvd., #G, Anaheim, CA 92805 USA
(800) 20-RC-FUN (tel)
(714) 758-0059 (tel)
(714) 758-0150 (fax)
rcboat4u@aol.com
Visa, MasterCard, Discover, American Express, COD ($5.50)
Sells sport and computer radios, batteries, battery chargers, and servos for radio-control model airplanes. Offers a catalog for $4.

The Toy Specialists
619 W. 54th, New York, NY 10019 USA
(212) 333-2206 (tel)
Sells used tape machines, mixing consoles, midi gear, computers, etc. for professional studio recording and live music performances.

Transcrypt International
4800 NW 1st St., Lincoln, NE 68521 USA
(800) 644-3277 (tel)
(800) 228-0226 (tel)
(402) 474-4800 (tel)
(402) 474-4858 (fax)
Manufactures encoder/decoder boards and security systems for cellular telephone applications.

Transducer Techniques
43178 Business Park Dr. B101, Temecula, CA 92590 USA
(909) 676-3965 (tel)
(909) 676-1200 (fax)
Manufactures load, force, and torque sensors for a wide variety of professional industrial applications.

Trans Electric
P.O. Box 18, Changhua, Taiwan
886-4-7633361 (tel)
886-4-7627018 (fax)
(908) 566-3232 (US tel)
(908) 566-3777 (US fax)
852-23318731 (Hong Kong tel)
852-23318716 (Hong Kong fax)
Manufactures converters, line splitters, distribution amplifiers, line extenders, combining networks, etc. for professional CATV applications.

Transel Technologies
123 E. South St., Harveysburg, OH 45032 USA
(800) 829-8321 (tel)
(513) 897-0738 (fax)
Visa, MasterCard
Sells antennas, mounts, antenna tuners, and accessories for cellular phones, amateur radios, and scanners.

TransInstruments (Jem Sensors)
No. 1 Cowell's Rd., Plainville, CT 06062 USA
(203) 793-4516 (tel)
(203) 793-4514 (fax)
Manufactures pressure transducers for a wide variety of professional industrial applications.

Transitional Technology
5401 E. La Palma Ave., Anaheim, CA 92807 USA
(714) 693-1133 (tel)
(714) 693-0255 (fax)
+44-01295-269000 (UK tel)
info@ttech.com
Manufactures data storage systems for a wide variety of professional computer network applications.

TransLogic
5641 Engineer Dr., Huntington Beach, CA 92649 USA
(714) 890-0058 (tel)
(714) 890-0788 (fax)
Manufactures temperature sensors for a wide variety of professional industrial applications.

Transmation
977 Mt. Reed Blvd., Rochester, NY 14606 USA
(800) 826-1292 (tel)
info@transmation.com
Manufactures control and communications systems for a wide variety of industrial applications.

Transparent Devices (CyberTouch)
853 Lawrence Dr., Newbury Park, CA 91320 USA
(805) 499-5000 (tel)
http://www.pacrain.com/-touchtone
touchtone@pacrain.com
Manufactures touch screens for a wide variety of electronics applications.

TranSys
5010 E. Shea Blvd., #C226, Scottsdale, AZ 85254 USA
(602) 483-7924 (tel)
Manufactures links for computer networks applications.

Travan Technology
P.O. Box 140407, Austin, TX 78754 USA
(800) 888-1889 (tel)
datastorage@mmm.com
Manufactures miniature tape cartridges for computer data storage applications.

Treasure Chest Peripherals
P.O. Box 10428, New Orleans, LA 70181 USA
(800) 677-9781 (tel)
(800) 723-9497 (corp tel)
(504) 733-2527 (tel)
http://www.tccomputers.com
Visa, MasterCard, Discover, American Express
Sells computer parts and boards, such as motherboards, hard drives, video cards, CD-ROM drives, modems, monitors, etc.

Tredex
5306 Beethoven Ave., Los Angeles, CA 90066 USA
(800) 899-6800 (tel)
(800) 899-4100 (corporate sales tel)
(310) 301-0300 (tel)
(800) 899-4556 (Canada tel)
(310) 301-0386 (tech support tel)
(310) 301-0313 (fax)
salesinfo@tredex.com
http://www.tredex.com
Visa, MasterCard, Discover, American Express
Sells personal desktop and laptop computers. Offers a 10-day money-back guarantee.

Triad Speakers
9106 NE Marx Dr., Portland, OR 97220 USA
(503) 256-2600 (tel)
Manufactures loudspeakers for car stereo applications.

Triad Transformers
90 E. Union St., Goodland, IN 47948 USA
(219) 297-3111 (tel)
(219) 297-3554 (fax)
Manufactures transformers for a wide variety of electronics applications.

Triangle Product Distributors USA
5750 Lakeshore Dr., Holland, MI 49424 USA
(616) 399-6390 (tel)
Sells decoders for satellite TV receiving applications.

Tricord Systems
2800 Northwest Blvd., Plymouth, MN 55441 USA
(800) TRICORD (tel)
(612) 557-9005 (tel)
(612) 557-8403 (fax)
Manufactures file servers for a wide variety of professional computer network applications.

Trident Computer
277 Park St., Troy, MI 48083 USA
(810) 585-8100 (tel)
(909) 784-4445 (California tel)
(810) 585-8108 (fax)
Buys and sells Texas Instruments computer systems, printers, terminals, etc. for a wide variety of computer applications.

Tri-Ex Tower
7182 Rasmussen Ave., Visalia, CA 93291 USA
(800) 328-2393 (tel)
(209) 651-7859 (tech support tel)
(209) 651-5157 (fax)
Visa, MasterCard
Manufactures and sells antenna towers for amateur radio applications.

Trilithic
9202 E. 33rd St., Indianapolis, IN 46236 USA
(800) 344-2412 (tel)
(317) 895-3600 (tel)
(317) 895-3613 (fax)
http://www.trilithic.com
Manufactures leakage detectors, preamplifiers, frequency counters, alignment systems, etc. for professional cable TV applications.

Trilogy Communications
2910 Hwy. 80 E., Pearl, MS 39208 USA
(800) 874-5649 (tel)
(601) 932-4461 (tel)

(601) 939-6637 (fax)
Manufactures coaxial cable for professional CATV applications.

Trilogy Magnetics
424 N. Mill Creek Rd., Quincy, CA 95971 USA
(800) 873-4323 (tel)
(916) 283-3736 (tel)
(916) 283-3122 (fax)
Sells cartridge drive parts for a wide variety of computer applications.

Trimble
2105 Donley, Austin, TX 78758 USA
(512) 873-9100 (tel)
(512) 836-9413 (fax)
Manufactures GPS systems for professional aviation applications.

Trimm Technologies
350 Pilot Rd., Las Vegas, NV 89115 USA
(800) 423-2024 (tel)
http://www.trim.com
Manufactures enclosures for a wide variety of computer equipment.

Trinity Products
1901 E. Linden Ave., #8, Linden, NJ 07036 USA
(908) 862-1705 (tel)
(908) 862-6875 (fax)
Sells batteries, battery packs, and accessories for radio-control model applications.

Tripp Lite
500 N. Orleans, Chicago, IL 60610 USA
(312) 329-1601 (tel)
(312) 644-6505 (fax)
Manufactures uninterruptible power supplies, surge supressors, and power inverters for a wide variety of personal computer applications.

Tri State Camera & Video
650 6th Ave., New York, NY 10011 USA
(800) 537-4441 (tel)
(212) 633-2290 (tel)

(212) 633-7718 (fax)
Visa, MasterCard, Discover, American Express, Canon, COD
Sells high-end home, amateur, and professional audio, video, computer, and office equipment, such as camcorders, VCRs, video editors, monitors, TVs, fax machines, etc.

Tritronics
1306 Continental Dr., Abingdon, MD 21009 USA
(800) 638-3328 (tel)
(410) 676-7300 (tel)
(800) 888-3293 (fax)
1015 Northwest 52nd St., Ft. Lauderdale, FL 33309 USA
(800) 365-8030 (tel)
(305) 938-8030 (tel)
(800) 999-3292 (fax)
Sells replacement parts and accessories for audio equipment.

Tri Valley Technology
2468 Armstrong St., Livermore, CA 94550 USA
(510) 447-2030 (tel)
(510) 447-4559 (fax)
Manufactures ruggedized rack-mount computer equipment for professional and industrial computer applications.

Trompeter Electronics
31186 LaBaya Dr., Westlake Village, CA 91362 USA
(800) 982-COAX (tel)
(818) 707-2020 (tel)
(818) 706-1040 (fax)
Manufactures jacks for a wide variety of electronics applications.

Troniks
Via N. Tommaseo 15, 1-35131 Padova, Italy
049-654220 (tel)
Sells equipment for shortwave radio receiving applications.

Tropic Aero
1090 NW 53rd St., Ft. Lauderdale, FL 33309 USA
(800) 351-9272 (tel)

Tropic Aero

(305) 491-6355 (tel)
(305) 772-3327 (fax)
Visa, MasterCard, Discover
Sells GPS and Loran for professional aviation applications.

Tropic Electronics

(800) 500-9758 (tel)
Sells cable TV converters and descramblers.

TrueData Products

775 Quaker Hwy., Rt. 146A, P.O. Box 347, Uxbridge, MA 01569 USA
(800) 635-0300 (tel)
(508) 278-6555 (fax)
Visa, MasterCard, Discover, American Express
Sells a variety of surplus personal home and laptop computers.

Trylon Manufacturing

P.O. Box 186, 21 Howard Ave., Elmira, ON N3B 2Z6 Canada
(519) 669-5421 (tel)
Manufactures and sells self-supporting and guyed towers, particularly for amateur radio applications.

Tryonics

8 Merrill Industrial Dr., Hampton, NH 03842 USA
(800) 551-6236 (tel)
Sells remanufactured UNIX workstations for professional computer applications.

Tubes

555 Seminole Woods Blvd., Geneva, FL 32732 USA
Sells a variety of electron tubes. Offers a tube list for a SASE.

Tucker Electronics & Computers

1717 Reserve St., Garland, TX 75042 USA
(800) 527-4642 (tel)
(214) 340-5460 (fax)
(214) 348-0367 (fax)
Sells a variety of test equipment, computers, shortwave receivers, amateur radio equipment, electronics kits, RTTY and fax equipment, accessories, and literature.

Tulsat

P.O. Box 2227, Broken Arrow, OK 74013 USA
(800) 331-5997 (tel)
Sells new and used traps, modulators, demodulators, satellite receivers, etc. for professional cable TV applications.

Tumbler Technologies

19201 Phil Ln., Cupertino, CA 95014 USA
(408) 996-8276 (tel/fax)
Manufactures dc/ac power inverters for a wide variety of electronics applications.

Turck

3000 Campus Dr., Minneapolis, MN 55441 USA
(800) 544-7769 (tel)
(612) 553-7300 (tel)
(612) 553-0708 (fax)
Manufactures miniature connectors for a wide variety of electronics applications. Offers free samples and a free 180-page catalog.

Turner Electronics

16701 Main St., #121, Hesperia, CA 92345 USA
Sells electron tubes for receiving applications. Offers a list for a first-class stamp.

Tucson Computer

6043-103 St., Edmonton, AB Canada
(403) 437-9797 (tel)
(403) 435-9400 (fax)
(403) 461-6903 (fax back)
tuscon@oanet.com
Visa, MasterCard
Manufactures memory adapter boards for computer applications.

TVS Holding

+40-991-35319 (Romania tel/fax)
Sells a wide variety of equipment for professional cable TV applications.

T & W Computer

361 Oak Place, #B, Brea, CA 92621 USA
(800) 965-3683 (tel)

(714) 990-3891 (tel)
(714) 990-3893 (fax)
Visa, MasterCard, COD
Sells personal computer parts and boards, such as motherboards, memory chips, hard drives, speakers, etc.

TW Enterprises
1099 Wyatt Dr., #6, Santa Clara, 95054 USA
(800) 407-4009 (tel)
(408) 727-7081 (tel)
(408) 727-7484 (fax)
Sells network and multimedia components for a wide variety of computer applications.

Twin Cities Digital
12217 Wood Lake Dr., Burnsville, MN 55337 USA
(612) 895-0522 (tel)
(612) 895-0521 (fax)
Buys and sells personal desktop computer systems; also sells computer parts and boards, such as memory chips, processors, network equipment, etc.

Twinhead
1537 Centre Pointe Dr., Milpitas, CA 95035 USA
(800) 995-8946 (tel)
(408) 945-0808 (tel)
(408) 945-1080 (fax)
Manufactures and sells personal laptop computers.

Two-Bit Score
4418 Pack Saddle Pass, Austin, TX 78745 USA
(512) 447-8888 (tel)
twobit@io.com
http://www.io.com/_7Etwobit
Manufactures keyboard emulators for computer testing applications.

Two Technologies
419 Sargon Way, Horsham, PA 19044 USA
(215) 441-5305 (tel)
(215) 441-0423 (fax)
(215) 441-8219 (BBS)
Manufactures handheld terminals and computers for professional electronics applications.

TX RX Systems
8625 Industrial Pkwy., Angola, NY 14006 USA
(716) 549-4700 (tel)
(716) 549-4772 (fax)
Manufactures preamplifiers, multicouplers, duplexers, cavity filters, etc. for two-way radio applications.

Typetronics
P.O. Box 8873, Ft. Lauderdale, FL 33310 USA
(305) 583-1340 (tel)
(305) 791-8337 (fax)
Sells electron tubes for a wide variety of electronics applications.

Ultimate Sound
138 University Pkwy., Pomona, CA 91768 USA
(800) 829-8818 (tel)
(909) 594-2604 (tel)
(909) 594-0191 (fax)
Manufactures amplifiers and loudspeakers for mobile audio applications.

Ultralife Batteries
1350 Rt. 88 S., Newark, NY 14513 USA
(315) 332-7100 (tel)
(315) 331-7800 (fax)
Manufactures a variety of long-lasting lithium batteries.

Ultralinear
12119 SE 82nd Ave., Portland, OR 97266 USA
(503) 653-5276 (tel)
(503) 653-7502 (fax)
Manufactures loudspeakers for car stereo applications.

Ulveco
10401 Westoffice Dr., Houston, TX 77042 USA
(800) 527-7042 (tel)
(713) 977-2500 (tel)
(713) 977-5031 (fax)
Manufactures transformers for a wide variety of electronics applications.

Umax
3353 Gateway Blvd., Fremont, CA 94538 USA
(800) 562-0311 (tel)
(510) 651-8883 (tel)
(510) 651-8834 (fax)
(510) 651-3710 (fax back)
Manufactures image scanners for a wide variety of computerized image-processing applications.

Uncle's Stereo
581 Broadway, New York, NY 10012 USA
(800) 978-6253 (tel)
(800) 924-2083 (tel)
(212) 343-9111 (tel)
(212) 343-9142 (fax)
Sells a wide variety of high-end audio and video equipment, including DSPs, CD players, amplifiers, receivers, laser disc players, loudspeakers, VCRs, TVs, etc.

Under the Wire Electronics
235 E. Colorado Blvd., #211, Pasadena, CA 91101 USA
(818) 930-1121 (tel)
(818) 930-1123 (fax)
Sells CPU chips for a wide variety of computer applications.

Unex Headsets
(800) 345-UNEX (tel)
Manufactures headsets for a variety of professional telecommunications applications.

Ungermann-Bass
3990 Freedom Circle, P.O. Box 58030, Santa Clara, CA 95030 USA
(800) 777-4LAN (tel)
(408) 496-0111 (tel)
(408) 970-7337 (fax)
Manufactures routers for professional computer network applications.

Unholtz-Dickie
6 Brookside Dr., Wallingford, CT 06492 USA
(203) 265-3929 (tel)
(203) 265-2690 (fax)
Manufactures vibration test equipment for a variety of professional industrial applications.

UNICO
P.O. Box 2486, Papeete, Tahiti
Sells head units, loudspeakers, amplifiers, receivers, cassette decks, CD players, etc. for home audio and car audio applications.

Unicom Paging Network of Texas
(713) 337-6556 (tel)
Sells pagers for cellular telephone applications.

Unicore Software
1538 Turnpike St., N. Andover, MA 01845 USA
(800) 800-2467 (tel)
(508) 686-6468 (tel)
(508) 683-1630 (fax)
Visa, MasterCard, Discover, American Express
Sells memory chips for a wide variety of computer applications.

Unicorn Electronics
1142 State Rt. 18, Aliquippa, PA 15001 USA
(800) 824-3432 (tel)
(412) 495-1230 (tel)
(412) 495-7882 (fax)
Sells electronics components and equipment, including video camera modules, robotic arms, laser diodes, laser pointers, power supplies, EPROM erasers, etc. Offers a free catalog.

Uniden
4700 Amon Carter Blvd., Ft. Worth, TX 76155 USA
(817) 858-3300 (tel)
(317) 842-2483 (tech tel)
Parts & Service division
8707 North By Northeast Blvd., P.O. Box 501368, Indianapolis, IN 46250 USA
(317) 842-2483 (tel)
(800) 323-2641 (fax)
Manufactures cordless and cellular telephones, scanners, radar detectors, citizens band transceivers, satellite receivers, commercial two-way transceivers, marine transceivers, pagers, etc.

Uni-firm
No. 7, 1/F, Hope Sea Industrial Centre, 26 Lam Hing St., Kowloon Bay, Kowloon Hong Kong
852-2796-6686 (tel)
852-2796-0125 (fax)
Sells telephones and accessories for cellular telephone applications.

Union Electronics
16012 S. Cottage Grove, S. Holland, IL 60473 USA
(800) 245-2492 (tel)
(708) 333-4100 (tel)
(708) 339-2777 (fax)
Sells replacement parts and accessories for audio equipment.

Unipower
3900 Coral Ridge Dr., Coral Springs, FL 33065 USA
(305) 346-2442 (tel)
Manufactures dc-to-dc power converters for a wide variety of electronics applications.

Unison Information Systems
21 Walsh Way, Framingham, MA 01701 USA
(508) 879-3200 (tel)
(508) 879-0772 (fax)
Manufactures disk drive controllers for a wide variety of professional computer network applications.

Unisys
2400 N. First St., San Jose, CA 95134 USA
(800) 874-8647 (tel)
(800) 5-UNISYS (tel)
(408) 434-2108 (tel)
http://www.unisys.com/adv
Manufactures network servers for a wide variety of professional computer applications.

United Computer Group
3751 6th Ave., San Diego, CA 92103 USA
(800) 877-2545 (tel)
(619) 291-9748 (fax)
Sells network and laptop computers, terminals, printers, keyboards, etc.

United Electronics
(800) 526-1275 (tel)
(201) 751-2591 (tel)
(201) 481-1524 (fax)
Sells a variety of electron tubes, tube manufacturing equipment, and assorted tube bases.

United Electronic Supply
P.O. Box 1206, Elgin, IL 60121 USA
(708) 697-0600 (tel)
Sells cable TV converters and universal TV remote controls.

United Technologies Microelectronics Center
1575 Garden of the Gods Rd., Colorado Springs, CO 80907 USA
(800) 645-UTMC (tel)
(719) 594-3486 (tel)
Manufactures integrated circuits for a variety of electronics applications.

Unitrode Integrated Circuits
7 Continental Blvd., Merrimack, NH 03054 USA
(603) 429-8610 (tel)
(603) 424-2410 (tel)
(603) 424-3460 (fax)
Manufactures semiconductors for a wide variety of electronics applications.

Universal Cross-Assemblers
9 Westminster Dr., Quispamsis, NB E2E 2V4 Canada
(506) 849-8952 (tel)
(506) 847-0681 (fax)
Visa, MasterCard, American Express
Manufactures EPROM programmers and emulators for professional computer applications.

Universal Manufacturing
43900 Groesbeck Hwy., Clinton Township, MI 48036 USA
(800) 542-3450 (tel)
(810) 463-2560 (tel)
(810) 463-2964 (fax)
Manufactures and sells aluminum towers for amateur radio applications.

Universal Radio
6830 Americana Pkwy., Reynoldsburg, OH 43068 USA
(800) 431-3939 (tel)
(614) 866-4267 (tel)
(614) 866-2329 (fax)
dx@universal-radio.com
http://www.universal-radio.com
Sells a variety of shortwave receivers, amateur radio equipment, accessories, and literature and manufactures RTTY and radio fax decoders for shortwave radio applications.

Universal Sales Agency
230 Duffy Ave., Unit R-1, Hicksville, NY 11801 USA
(516) 932-1400 (tel)
(516) 932-1449 (fax)
Visa, MasterCard
Sells new and used IBM personal home and laptop computers and parts and accessories, including memory chips, system boards, hard drives, power supplies, controllers, etc. Offers a 30-day money-back guarantee.

Universal Video & Camera
1104 Chestnut St., Philadelphia, PA 19107 USA
(800) 477-1003 (tel)
(215) 889-7628 (tel)
(215) 647-5830 (fax)
Visa, MasterCard, Discover, American Express
Sells high-end home, amateur, and professional audio and video equipment, such as camcorders, titlers, VCRs, video lights, etc.

University Audio Shop
402 S. Park St., Madison, WI 53715 USA
(608) 284-0001 (tel)
Sells loudspeakers, cassette decks, TVs, receivers, amplifiers, etc. for home audio and home video applications.

Unplugged Communications
20526 Gramercy Pl., Torrance, CA 90501 USA
(800) 279-CELL (tel)
(310) 787-9400 (tel)
(800) 787-9498 (fax)
(310) 787-9444 (fax)
Sells telephones, and other related equipment for cellular telephone applications.

Uptown Automation Systems
6205 Lookout Rd., Unit C, Boulder, CO 80301 USA
(303) 581-0400 (tel)
(303) 581-0114 (fax)
(516) 249-1399 (New York tel)
(310) 306-8823 (Los Angeles tel)
Manufactures moving fader automation packages for professional studio recording applications.

Urban Audio Works
11400 Downey Ave., Downey, CA 90241 USA
(800) 283-4698 (tel)
(310) 923-6302 (fax)
Manufactures amplifiers and loudspeakers for car stereo applications.

URS Information Systems
36 Jonspin Rd., Wilmington, MA 01887 USA
(508) 657-6100 (tel)
(508) 694-1444 (tel)
Sells personal computer parts and accessories, including system boards, hard drives, power supplies, keyboards, etc.

USA Computer Connections
4986-3 Euclid Rd., Virginia Beach, VA 23462 USA
(800) USA-6588 (tel)
Visa, MasterCard, Discover
Sells computer parts and boards, such as hard drives, motherboards, memory chips, CD-ROM drives, keyboards, etc.

USA Direct
894 Green Place, Woodbury, NY 11958 USA
(800) 830-1515 (tel)
Sells high-end home, amateur, and professional audio and video equipment, such as camcorders, VCRs, CD players, TVs, home theater equipment, etc.

USA Flex
444 Scott Dr., Bloomingdale, IL 60108 USA

U.S. Robotics

(800) 723-2259 (tel)
(708) 582-6206 (tel)
(708) 351-7204 (fax)
Visa, MasterCard, Discover, American Express, Flex Tech, COD
Manufactures and sells personal desktop and laptop computers, printers, and scanners; also sells computer parts, such as sound cards, monitors, tape drives, hard drives, modems, etc.

U.S. Amps
7325-100 NW 13th Blvd., Gainesville, FL 32606 USA
(904) 338-1926 (tel)
(904) 371-4122 (fax)
usamps@usamps.com
http://www.usamps.com
Manufactures amplifiers for mobile audio applications.

U.S. Black Magic
18001 Mitchell S., Irvine, CA 92714 USA
(714) 250-0155 (tel)
(714) 250-0122 (fax)
Manufactures loudspeakers for car stereo applications.

U.S. Cable TV
4100 N. Powerline Rd. F-4, Pompano Beach, FL 33073 USA
(800) 772-6244 (tel)
Sells cable TV descramblers. Offers a free catalog and a 30-day money-back guarantee.

US Computer
14677 E. Easter Ave., #A, Englewood, CO 80112 USA
(800) 683-6616 (tel)
(303) 690-1093 (tel)
(303) 690-7563 (fax)
16291 Gothard St., Huntington Beach, CA 92647 USA
(800) 787-4947 (tel)
(714) 841-4081 (tel)
(714) 841-4262 (fax)
Manufactures and sells personal desktop computers; *also sells computer parts and boards, such as hard drives, monitors, speakers, video cards, modems, etc.*

USD Audio
1030 N. Main St., Orange, CA 92667 USA
(714) 997-2475 (tel)
(714) 997-7360 (fax)
+61-9316-4040 (Australia tel)
(204) 256-2082 (Canada tel)
+46-156-10523 (Sweden tel)
(714) 961-0495 (Thailand/USA tel)
+44-081-756-0078 (UK tel)
Manufactures loudspeakers for mobile audio applications.

U.S. Design
9075 Guilford Rd., Columbia, MD 21046 USA
(800) 622-8732 (tel)
(410) 381-3000 (tel)
(410) 381-3235 (fax)
Manufactures data storage systems for a wide variety of professional computer network applications.

US Digital
380 Rougeau Ave., Winnipeg, MB R2C 4A2 Canada
(204) 661-6859 (tel)
(204) 667-9894 (fax)
MasterCard
Manufactures and sells a pocket simplex repeater system for amateur radio applications.

USRadio
377 Plaza, Granbury, TX 76048 USA
(800) 433-SAVE (tel)
Sells Radio Shack scanners and radios.

U.S. Robotics
7770 N. Frontage Rd., Skokie, IL 60077 USA
(800) DIAL-USR (tel)
(800) 845-0908 (tel)
(708) 933-5552 (tech support fax)
(800) 762-6163 (fax back)
(708) 933-5092 (BBS)
011-33-2019-1959 (Europe tel)
011-33-2005-3240 (Europe fax)
sales@usr.com (Internet sales)

U.S. Robotics

support@usr.com (Internet tech support)
http://www.usr.com
Manufactures computer modems.

US Tower
1220 Marcin St., Visalia, CA 93291 USA
(209) 733-2438 (tel)
(209) 733-7194 (fax)
Manufactures and sells crank-up towers, tubular and lattice towers, and accessories, particularly for amateur radio applications.

UTI Microtree
6910 Hayvenhurst Ave., # 100, Van Nuys, CA 91406 USA
(800) 892-2313 (tel)
(818) 373-1111 (tel)
(818) 373-1128 (fax)
Sells personal desktop and laptop computers and printers; also sells parts and accessories, including keyboards, monitors, etc.

V88 Computer Systems
1550 Montague Expressway, San Jose, CA 95131 USA
(800) 888-4828 (tel)
(408) 321-9353 (tech support tel)
http://www.V88.com
Sells personal laptop computers; also sells computer parts and boards, such as floppy drives, hard drives, motherboards, modems, controller cards, etc. Offers a 30-day money-back guarantee.

VAC (Valve Amplification Co.)
P.O. Box 4609, Sarasota, FL 34230 USA
Manufactures tube audio amplifiers for high-end home audio and home theater applications.

Vacuum Tube Industries
P.O. Box 2009, Brockton, MA 02405 USA
(800) 528-5014 (tel)
(508) 584-4500 (tel)
Sells rebuilt power amplifier electron tubes, particularly for commercial broadcasting applications.

Valhalla Scientific
9955 Mesa Rim Rd., San Diego, CA 92121 USA
(619) 457-5576 (tel)
Manufactures harmonic energy analyzer and data loggers for a variety of professional industrial applications.

Valitek
100 University Dr., Amherst, MA 01002 USA
(800) 825-4835 (tel)
(413) 549-2700 (tel)
(413) 549-2900 (fax)
Manufactures tape drives for professional computer data storage applications.

Valley Forge Computer
1206 Joshua Dr., West Chester, PA 19380 USA
(610) 429-1776 (tel)
(610) 429-1777 (fax)
Buys and sells memory chips and network equipment for professional and personal computer applications.

Valley Radio Center
1522 N. 77 Sunshine Strip, Harlingen, TX 78550 USA
(800) 869-6439 (tel)
(210) 423-6407 (tel)
(210) 423-1705 (fax)
Sells a wide variety of amateur radio equipment and accessories, including amateur radio transceivers, receivers, antenna tuners, antennas, amplifiers, etc.

Valor Enterprises
1711 Commerce Dr., Piqua, OH 45356 USA
(800) 543-2197 (tel)
(513) 778-0074 (tel)
(513) 778-0259 (fax)
Manufactures mobile antennas and accessories for citizens band, cellular, and amateur radio applications.

Valpey-Fisher
75 South St., Hopkinton, MA 01748 USA
(800) 982-5737 (tel)
(508) 435-6831 (tel)

(508) 497-6377 (fax)
Manufactures crystal clock oscillator modules for SONET, SDH, and ATM applications.

Valtron Technologies
28309 Avenue Crocker, Valencia, CA 91355 USA
(800) 2-VALTRON (tel)
(805) 257-0113 (fax)
Sells data storage systems, including optical drives, hard drives, floppy drives, tapes drives, etc., for a wide variety of computer applications.

Vandersteen Audio
116 W. 4th St., Hanford, CA 93230 USA
(209) 582-0324 (tel)
Manufactures high-end speakers for audio and home theater applications.

Van Gorden Engineering
P.O. Box 21305, S. Euclid, OH 44121 USA
(216) 481-6590 (tel)
(216) 481-8329 (fax)
Manufactures wire antennas, center insulators, baluns, etc. for amateur radio applications.

Vanguard
7202 Huron River Dr., Dexter, MI 48130 USA
(800) 875-3322 (tel)
Manufactures lighting systems for amateur and professional video production.

Vann Draper Electronics
Alexander House, Brampton Close, Wigston, Leicester LE18 2RZ UK
0116-2813091 (tel)
0116-2570893 (fax)
Sells test equipment, including function generators, multimeters, continuity testers, etc., for a wide variety of electronics testing applications.

Vantac Manufacturing
P.O. Box 469, Oak Lawn, IL 06453 USA
Sells electronic accessories, speed controls, and servos for radio-control model airplanes.

Vantage
345 E. 800 S., Orem, UT 84058 USA
(801) 229-2800 (tel)
http://www.transera.com
Manufactures control stations for home theater applications.

Vantec
460 Casa Real Pl., Nipomo, CA 93444 USA
(800) 882-6832 (tel)
(805) 929-5055 (tel)
Visa, MasterCard, COD
Manufactures channel expanders, speed controllers, electronics throttles for radio-control model airplanes and boats.

H.C. Van Valzah
1140 Hickory Trail, Downers Grove, IL 60515 USA
(800) HAM-0073 (tel)
(708) 852-1469 (tech tel)
(708) 852-1469 (fax)
Sells amateur radio accessories, including antenna tuners and antennas, etc.

Variant Microsystems
46560 Fremont Blvd., #105, Fremont, CA 94538 USA
(800) VARIANT (tel)
(510) 440-2870 (tel)
(510) 440-2873 (fax)
Manufactures bar code scanners for a wide variety of computerized business applications.

VDO-Pak
413 Oak Pl., #3-M., Port Orange, FL 32127 USA
(800) 767-9771 (tel)
Manufactures lighting systems for amateur and professional video production.

Vecmar International
990 Erie Rd., Eastlake, OH 44095 USA
(216) 953-1119 (tel)
Manufactures data storage systems for a wide variety of professional computer network applications.

Vector Technology
9318 N. 95th Ave., Ste. 201, Scottsdale, AZ 85258 USA
(602) 966-2770 (tel)
(602) 966-3710 (fax)
Buys, sells, and trades new and used UNISYS equipment for professional computer applications.

Vectron Technologies
267 Lowell Rd., Hudson, NH 03051 USA
(800) NEED-FCP (tel)
(603) 598-0070 (tel)
(603) 598-0075 (fax)
Manufactures clock, VCXO, oscillator, and timing-recovery unit chips for a wide variety of electronics applications.

Vega Signaling Products Group
9900 E. Baldwin Pl., El Monte, CA 91731 USA
(800) 877-1771 (tel)
(818) 442-0782 (tel)
(818) 444-1342 (fax)
(818) 444-2017 (fax back)
(800) 274-2017 (fax back)
Manufactures radio controllers for a two-way radio applications.

Vektron International
2100 N. Hwy. 360, #1904, Grand Prarie, TX 75050 USA
(800) 725-0047 (tel)
(800) 725-0038 (government/education orders tel)
(800) 725-0078 (dealers tel)
(214) 606-0280 (tel)
(214) 606-1278 (fax)
(214) 606-0444 (BBS tech support tel)
Visa, MasterCard, Discover, American Express
Manufactures and sells personal desktop computers; also sells computer parts and boards, such as hard drives, monitors, motherboards, video cards, modems, etc.

Vela Research
P.O. Box 9090, Clearwater, FL 34618-9090 USA
(813) 572-1230 (tel)
Manufactures audio/video decoders for personal and professional applications.

Velodyne Acoustics
1070 Commercial St., #101, San Jose, CA 95112 USA
(408) 436-7270 (tel)
Sells loudspeakers for high-end home and car audio applications.

Vencon Technologies
5 Graymar Ave., Downsview, ON M3H 3B5 Canada
(416) 398-0261 (tel)
(416) 398-0625 (fax)
vencon@enterprise.ca
http://www.enterprise.ca/_7Evencon
Manufactures battery analyzers for radio-control model airplanes.

Vento Associates
88 Ford Rd., Denville, NJ 07834 USA
(201) 627-8500 (tel)
(201) 627-6070 (fax)
Visa, MasterCard
Sells hard drives for a wide variety of computer applications.

Venture Fourth
572 Charcot Ave., San Jose, CA 95131 USA
(408) 428-9030 (tel)
Visa, MasterCard, Discover
Sells new and refurbished computer systems and parts for personal and professional computer applications.

VersaTech Electronics
P.O. Box 18476, Boulder, CO 80308 USA
(800) 940-5590 (tel)
(303) 440-5590 (tel)
(303) 440-9467 (fax)
(303) 440-7828 (BBS)
Visa, MasterCard
Manufactures FBASIC microcontroller boards for experimental data logging and robotics applications.

VersaTel
(800) 456-5548 (tel)
(307) 266-3010 (fax)

Buys and sells transceivers, etc. for two-way radio communications.

Vertex Industries
23 Carol St., Clifton, NJ 07014 USA
(201) 777-3500 (tel)
(201) 472-0814 (fax)
Manufactures workstations for a wide variety of professional computer network applications.

Vertex Radio Communications
(see Yaesu)

Vestax
2870 Cordelia Rd., #100, Fairfield, CA 94585 USA
(707) 427-1920 (tel)
(707) 427-2023 (fax)
Manufactures an eight-track digital recorder for professional studio recording.

VHF Communications
280 Tiffany Ave., Jamestown, NY 14701 USA
(800) 752-8813 (tel)
(716) 664-6345 (tel)
Sells amateur radio equipment and accessories, including amateur radio transceivers, receivers, antenna tuners, antennas, amplifiers, etc.

Vic Hi-Tech
El Segundo, CA
(310) 643-5193 (tel)
Manufactures a full-motion video/audio capture, storage, and playback computer board for home, amateur, and professional video editing applications via computer.

Vicor
23 Frontage Rd., Andover, MA 01810 USA
(800) 735-6200 (tel)
(508) 470-2900 (tel)
(508) 475-6715 (fax)
(408) 522-5280 (CA tel)
(408) 774-5555 (CA fax)
Carl-von-Linde Strasse 15, D-85748 Garching-Hochbruck, Germany
+49-89-329-2763 (tel)
+49-89-329-2767 (fax)
10F, No. 7, Ho Ping E. Rd., Sec. 3, Taipei, Taiwan
+886-2-708-8020 (tel)
+886-2-755-5578 (fax)
Manufactures semiconductors for a wide variety of electronics applications.

Victoreen
6000 Cochran Rd., Cleveland, OH 44139 USA
(216) 248-9300 (tel)
Manufactures and sells resistors for a wide variety of electronics applications.

Victory Trading
209 E. Ben White Blvd., Ste. 111, Austin, TX 78704 USA
(512) 912-8980 (tel)
(512) 912-8998 (fax)
Buys and sells IBM computer systems; also sells computer parts and boards, such as hard drives, memory chips, tape drives, etc.

Video Communications Systems (VCS)
(800) 872-6585 (tel)
Sells video descramblers and converters.

Video Direct
116 Production Dr., Yorktown, VA 23693 USA
(800) 368-5020 (tel)
(804) 595-2574 (tel)
Sells home audio and video equipment, such as camcorders, VCRs, cables, etc.

Video Discount Warehouse
295 Greewich St., #360, New York, NY 10007 USA
(800) 301-0028 (tel)
(212) 432-6058 (tel)
(212) 432-6104 (fax)
Visa, MasterCard, Discover, American Express, COD
Sells high-end home, amateur, and professional video equipment, such as camcorders, VCRs, video editors, etc. Offers a 7-day money-back guarantee.

Video Innovators
P.O. Box 4130, Frisco, CO 80443 USA

Video Innovators

(800) 832-6840 (tel)
Visa, MasterCard
Manufactures and sells shoulder rests, tripod shelves, vehicle camcorder brackets, and copy stands for amateur and professional video recording applications.

VideoLabs
10925 Bren Rd. E., Minneapolis, MN 55343 USA
(612) 988-0055 (tel)
(612) 988-0066 (fax)
Manufactures external video frame grabbers for a wide variety of personal computer applications.

VideoLogic
1001 Bayhill Dr., #310, San Bruno, CA 94066 USA
(800) 494-4903 (tel)
(800) 578-5644 (tel)
(415) 875-0606 (tel)
Manufactures graphics cards for a wide variety of personal computer applications.

Videomedia
175 Lewis Rd. #23, San Jose, CA 95111 USA
(408) 227-9977 (tel)
75026.1317@compuserve.com
Manufactures interfaces, databases, and recorders for computerized digital audio editing and recording.

Video Necessities
P.O. Box 411, Homecrest Stn., Brooklyn, NY 11229 USA
(800) 228-8480 (tel)
(718) 692-4512 (tel)
(718) 951-0843 (customer service tel)
(718) 951-9606 (fax)
Visa, MasterCard, Discover, American Express
Sells high-end home, amateur, and professional audio and video equipment, such as camcorders, VCRs, video editors, monitors, TVs, home theater equipment, etc.

Videonics
1370 Dell Ave., Campbell, CA 95008 USA
(800) 338-EDIT (tel)

(408) 866-8300 (tel)
Manufactures editing suites for home video editing.

Videospectra
P.O. Box 755, Agoura, CA 91301 USA
(800) 835-8335 (tel)
Visa, MasterCard, Discover, American Express
Manufactures and sells a synthesized function generator.

Videotique
1375 Coney Island Ave., #422, Brooklyn, NY 11230 USA
(800) 950-2122 (tel)
(718) 437-1777 (tel)
Buys and sells amateur, broadcast, and professional audio and video equipment, such as camcorders, video editors, switchers, etc. Offers a free catalog.

Videx
1105 NE Circle Blvd., Corvallis, OR 97330 USA
(503) 758-0521 (tel)
(503) 752-5285 (fax)
Manufactures portable, programmable data collectors for professional inventory and warehousing applications.

Vidicomp Distributors
10998 Wilcrest, Houston, TX 77099 USA
(800) 263-8211 (tel)
(800) 833-1960 (fax)
6303 Blue Lagoon Dr., Miami, FL 33126 USA
(305) 265-9339 (tel)
(305) 265-1164 (fax)
Visa, MasterCard, Discover, American Express
Sells amateur, broadcast, and professional audio and video equipment, such as camcorders, video editors, computer-based editing systems, monitors, etc.

Vidikron
150 Bay St., Jersey City, NJ 07302
(201) 420-6666 (tel)
Manufactures projection TVs for high-end home theater applications. Offers a free catalog.

ViewSonic
20480 Business Pkwy., Walnut, CA 91789 USA
(800) 888-8583 (tel)
(909) 869-7976 (tel)
(909) 869-7958 (fax)
(909) 869-7318 (fax on demand)
73374.514@compuserve.com
http://www.viewsonic.com
Manufactures video monitors and uninterruptible power supplies for computer applications.

Viewsonics
6454 E. Rogers Circle, Boca Raton, FL 33487 USA
(800) 645-7600 (tel)
(407) 998-9594 (tel)
(407) 998-3712 (fax)
Manufactures amplifiers, splitters, etc. for professional cable TV applications.

Viewstar
18 Dufflaw Rd., Toronto, ON M6A 2W1 Canada
(416) 789-1285 (tel)
(416) 784-9968 (fax)
Manufactures tuners for cable TV applications.

Vigra
10052 Mesa Ridge Ct., San Diego, CA 92121 USA
(619) 597-7080 (tel)
sales@vigra.com
http://www.vigra.com
Manufactures video and audio boards for a wide variety of computer applications.

Viking International
150 Executive Park Blvd., #4600, San Francisco, CA 94134 USA
(415) 468-2066 (tel)
(415) 468-2067 (fax)
CODs but no credit cards
Sells a professional 10-hour cassette recorder. Offers a free 32-page catalog.

Vintage Radio Co.
457 St., Manchester, NH 03102 USA
(603) 625-1165 (tel)
Manufactures and sells hollow-state receiver, transmitter, and power supply kits for amateur radio applications.

VIP Cable Services
2255 NW 102 Place, Miami, FL 33172 USA
(305) 994-7606 (tel)
(305) 994-7608 (fax)
Sells and remanufactures a wide variety of equipment for professional cable TV applications.

Viper
(800) 283-1302 (tel)
(800) 361-1312 (tel)
(619) 599-1394 (tel)
(800) 361-7271 (Canada tel)
525-531-6131 (Mexico tel)
809-751-7836 (Caribbean tel)
+56-2235-2328 (Chile tel)
+31-10262-1744 (Netherlands tel)
+61-7849-4006 (Australia tel)
+551-92553722 (Brazil tel)
095-156-2936 (Russia tel)
Manufactures systems for car security applications.

The Virtual Group
2950 Waterview Dr., Rochester Hills, MI 48309 USA
(810) 853-6000 (tel)
http://www.virtualgrp.com
Manufactures data storage systems for a wide variety of professional computer network applications.

Virtual Open Network Environment (V-ONE)
12300 Twinbrook Pkwy., Rockville, MD 20852 USA
(800) 881-7090 (tel)
(301) 838-8900 (tel)
(301) 838-8909 (fax)
http://www.v-one.com
Manufactures user authentication card equipment for professional computerized security systems.

Virtual Vision Sport
4132 E. Speedway Blvd., Tuscon, AZ 05712 USA
(800) 729-1020 (tel)

Virtual Vision Sport
Manufactures a projection TV built into tinted goggles for home audio and video applications.

VIS
P.O. Box 17377, Hattiesburg, MS 39404 USA
(800) OKK-HAMS (tel)
(601) 261-2601 (tel)
Visa, MasterCard, Discover, American Express
Sells mobile vertical antennas for amateur radio applications. Offers a free brochure.

Vision Electronics
2125 S. 156th Circle, Omaha, NE 68130 USA
(800) 562-2252 (tel)
Visa, MasterCard, Discover, American Express
Sells video stabilizers to eliminate copy protection. Offers a two-year warranty and a money-back guarantee.

Vision Microsystems
5501 East Rd., Bellingham, WA 98226 USA
(360) 398-1833 (tel)
(360) 398-1663 (fax)
Manufactures GPS and Loran systems, transceivers, batteries, etc. for professional aviation applications.

Vision & Motion Surplus
241 E. Main St., Westboro, MA 01581 USA
(508) 366-8370 (tel)
Visa, MasterCard
Sells a wide variety of electronics surplus, including components, miniature video cameras, bar code scanners, electro-pneumatic components, etc.

Vision Specialties
P.O. Box 2782, Rancho Cucamonga, CA 91729 USA
(800) 233-8431 (tel)
(909) 483-3848 (tel)
Visa, MasterCard, Discover, American Express
Sells cosmetic replacement parts, components, converters, specialty tools, remote controls, transformers, batteries, and hardware for professional cable TV applications.

Visiontek
(800) 360-7185 (tel)
Manufactures memory modules for a wide variety of computer applications.

Visitect
P.O. Box 14156, Fremont, CA 94539 USA
(510) 651-1425 (tel)
(510) 651-8454 (fax)
Visa, MasterCard
Manufactures miniature transmitters and receivers for a wide variety of electronics applications.

Visual Circuits
3989 Central Ave NE, Ste. 630, Minneapolis, MN 55421 USA
(612) 781-2186 (tel)
http://www.vcircuits.com
Manufactures and sells MPEG playback boards for amateur and professional video production.

Visual Communications
615 Main St., #206, Stroudsburg, PA 18360 USA
(800) GO-CABLE (tel)
(717) 620-4363 (tel)
Visa, MasterCard, American Express
Sells converter diagnostic test modules. Offers a 30-day money-back guarantee.

Visual Information Technologies
3460 Lotus Dr., #100, Plano, TX 75075 USA
(800) 325-6467 (tel)
(214) 596-5600 (tel)
(214) 867-4489 (fax)
Manufactures video accelerator boards for a wide variety of computer applications.

Vitrek
7576 Trade St., San Diego, CA 92121 USA
(619) 586-1695 (tel)
(619) 586-0802 (fax)
info@vitrek.com
Manufactures test equipment for a variety of professional industrial applications.

Vive Synergies
30 W. Beaver Creek Rd., Unit 101, Richmond Hill, ON L4B 3K1 Canada
(800) 567-4954 (tel)
(905) 882-6107 (tel)
(905) 882-6238 (fax)
info@vive.com
http://www.vive.com
Manufactures switches and professional, business, and personal Caller ID telephone applications.

Vivid Technology
4168 Avenida de la Plata, Oceanside, CA 92056 USA
(619) 631-7122 (tel)
Manufactures touch screens for a wide variety of computer applications.

VL Products
7871 Alabama Ave., #16, Canoga Park, CA 91304 USA
Sells batteries, battery chargers, radio-control systems, etc. for radio-control model airplanes. Offers a catalog for $1.

VLSI Technology
1109 McKay Dr., San Jose, CA 95131 USA
(408) 434-3000 (tel)
(408) 263-2511 (fax)
Manufactures integrated circuits for a wide variety of electronics applications.

VMPS Audio Products
3429 Morningside Dr., El Sobrante, CA 94803 USA
(510) 222-4276 (tel)
(510) 232-3837 (fax)
Manufactures high-end loudspeakers for home hi-fi systems.

VoCom Products
731 W. Lunt Ave., Schaumburg, IL 60193 USA
(800) 872-6233 (tel)
(708) 924-9078 (fax)
Manufactures power amplifiers for VHF and UHF radio applications.

Voltech
200 Butterfield Dr., Ashland, MA 01721 USA
(508) 881-7329 (tel)
(508) 879-8669 (fax)
Manufactures power analyzers for a variety of professional industrial applications.

VOVOX
Stationsgasse 2, CH-8155 Niederhasli, Switzerland
01-8506374 (tel)
Sells equipment for shortwave radio receiving applications.

VRA Batteries
16541 Redmond Way 251, Redmond, WA 98052 USA
(800) 747-8876 (tel)
(206) 881-8874 (fax)
Visa, MasterCard
Sells batteries for a wide variety of electronics applications, including laptop computers, camcorders, cellular phones, etc.

W9INN Antennas
P.O. Box 393, Mt. Prospect, IL 60056 USA
(847) 394-3414 (tel)
Manufactures wire antennas for amateur radio applications. Offers a catalog for a 52-cent SASE.

Wacom Technology
501 S.E. Columbia Shores Blvd., #300 Vancouver, WA 98661 USA
(800) 922-6620 X304 (tel)
(206) 750-8882 (tel)
(206) 750-8924 (fax)
(408) 982-2737 (BBS)
Manufactures computerized drawing boards for artistic applications.

Waddington Electronics
25 Webb St., Cranston, RI 02920 USA
(401) 781-3904 (tel)
(401) 781-1650 (fax)
Manufactures analog function modules for industrial machinery control applications.

Wall Industries
5 Watson Brook Rd., Exeter, NH 03833 USA
(603) 778-2300 (tel)
(603) 778-9797 (fax)
Manufactures dc-to-dc power converters for a wide variety of electronics applications.

Wandel & Goltermann
3000 Aerial Center, #110, Morrisville, NC 27560 USA
(919) 460-3000 (tel)
(919) 460-3030 (fax)
P.O. Box 1262, D-72795 Eningen uA, Germany
Manufactures optical power meters, optical test systems, backplane emulators, PCM measuring sets, etc. for professional industrial applications.

Wangrow Electronics
1500 W. Laverne, #K, Park Ridge, IL 60068 USA
(708) 696-3294 (tel)
(708) 696-3797 (fax)
Manufactures and sells digital command controls for model railroading systems. Offers more information for a large SASE.

The Warehouse
2071-20 Emerson St., Jacksonville, FL 32207 USA
(800) 483-8273 (tel)
(904) 399-0424 (tel)
Visa, MasterCard, Discover
Sells blank recording audio and video cassettes.

Washburn & Company
3800 Monroe Ave., Pittsford, NY 14534 USA
(800) 836-8028 (fax back)
(800) 836-8026 (tel)
(800) 836-8027 (motherboards tel)
(716) 385-5200 (tel)
(716) 381-7549 (fax)
Visa, MasterCard, Discover, American Express
Sells motherboards and raid systems for a wide variety of personal and professional computer applications.

Washburn Computer Group
4206 Park Glen Rd., St. Louis Park, MN 55416 USA
(612) 925-3220 (tel)
(612) 925-2422 (fax)
Buys and sells IBM computer systems.

Watkins-Johnson
Electronic Equipment Division
700 Quince Orchard Rd., Gaithersburg, MD 20878 USA
(800) 954-3577 (tel)
(301) 948-7550 (tel)
Piazza G. Marconi, 25, 00144 Roma, Italy
39-6-591-2515 (tel)
39-6-591-7342 (tel)
Manufactures receivers for amateur, commercial, military, and professional monitoring of shortwave, VHF, and UHF radio-monitoring applications

Watlow Anafaze
334 Westridge Dr., Watsonville, CA 95076 USA
(408) 724-3800 (tel)
(408) 724-0320 (fax)
Manufactures PID controllers for a wide variety of professional industrial applications.

Watlow Electric
12001 Lackland Rd., St. Louis, MO 63146 USA
Manufactures low-temperature infrared sensors for a wide variety of professional industrial applications. Offers a free, 16-page guide.

Wavetek
5808 Churchman Bypass, Indianapolis, IN 46203-6109 USA
(800) 622-5515 (tel)
(800) 392-8100 (tel)
128 Wheeler Rd., Burlington, MA 01803 USA
(617) 229-7585 (tel)
(617) 273-5943 (fax)
9045 Balboa Ave., San Diego, CA 92123 USA
(619) 279-2200 (tel)
(619) 565-9558 (fax)
(317) 788-9351 (tel)
852-2788-6221 (Pacific tel)
861-500-2255 (China tel)
65-356-2522 (Southeast Asia tel)
81-427-57-3444 (Japan tel)

431-813-5628 (Eastern Europe tel)
49-89-996410 (Western Europe tel)
44-1603-404824 (UK tel)
http://www.wavetek.com
Manufactures multichannel testers, reverse sweep testers, two-way radio test sets, etc. for a wide variety of professional testing applications.

Wawasee Electronics
P.O. Box 36, 400 S. Sycamore St., Syracuse, IN 46567 USA
(219) 457-3191 (tel)
Manufactures and sells wattmeters, frequency monitors, accessories, etc. for citizens band radio applications.

Wayne Communications
P.O. Box 23223, Rochester, NY 14692 USA
(716) 427-0830 (tel)
(716) 427-7163 (fax)
Sells used transceivers, chargers, pagers, etc. for cellular telephone applications.

Webco Sales
2112 S. Memorial Pkwy., Huntsville, AL 35801 USA
(205) 534-7356 (tel)
(205) 534-7357 (fax)
Sells surplus components, including semiconductors, transformers, TV/VCR tuners, relays, video splitters, connectors, etc. for a wide variety of electronics applications.

Weber-Sterling
130 Greeves Rd., P.O. Box 384, New Hampton, NY 10958 USA
(914) 355-1910 (tel)
(914) 355-6300 (fax)
Buys and sells ATMs, encoders, endorsers, teller terminals, etc. for a wide variety of professional banking applications.

Weeder Technologies
P.O. Box 421, Batavia, OH 45103 USA
(513) 752-0279 (tel)
Manufactures and sells a variety of electronics kits, including telephone scramblers, caller IDs, vocal filters, frequency counters, caller blocks, etc.

Weed Instrument
707 Jeffrey Way, Round Rock, TX 78664 USA
(800) 880-WEED (tel)
(512) 255-7043 (tel)
(512) 388-4362 (fax)
Visa, MasterCard
Manufactures temperature sensors and catalog probes for professional industrial applications.

Welbourne Labs
P.O. Box 260198, Littleton, CO 80126 USA
(303) 470-6585 (tel)
(303) 791-5783 (fax)
Visa, MasterCard
Sells a variety of high-end audio kits (including amplifiers, phono drives, line conditioners, crossovers, etc.) and components.

Wells-Gardner Electronics
2701 N. Kildare Ave., Chicago, IL 60639 USA
(312) 252-8220 (tel)
Manufactures monitors for a wide variety of computer applications.

Weltronics
P.O. Box 80584, San Marino, CA 91108 USA
(818) 799-6396 (tel)
(818) 799-6541 (fax)
Manufactures digital-to-analog converters for professional audio applications.

WestCoast Computer Parts
7746 Sedan Ave., West Hills, CA 91304 USA
(818) 992-6116 (tel)
(818) 992-6188 (fax)
Sells hard drives, controller cards, motherboards, CPUs, etc. for a wide variety of computer applications.

West Coast Radio
1355 Westwood Blvd., Ste. 200, Los Angeles, CA 90024 USA

West Coast Radio

(800) 881-6722 (tel)
Sells transceivers, etc. for cellular microwave telephone and two-way radio applications.

Western Data Entry Systems
20 W. Stone Rd., Unit 8, Marlton, NJ 08053 USA
(800) 778-4864 (tel)
western@jersey.net
Sells computers and printers for a wide variety of professional computer network applications.

Western Digital
(800) 832-4778 (tel)
(714) 932-4300 (fax)
http://www.wdc.com
Manufactures hard drives for personal desktop computers.

Western Multiplex
300 Harbor Blvd., Belmont, CA 94002 USA
(415) 592-8832 (tel)
(415) 592-4249 (fax)
Manufactures transmitters for professional cellular microwave radio applications.

Western Test Systems
530 Compton St., Unit #C, Broomfield, CO 80020 USA
(800) 538-1493 (tel)
(303) 438-9662 (tel)
(303) 438-9685 (fax)
Visa, MasterCard, American Express
Sells test equipment, including oscilloscopes, voltmeters, waveform generators, LCR meters, power supplies, frequency counters, oscillators, etc. for a wide variety of electronics testing applications.

Westlake Audio
2696 Lavery Ct., Unit 18, Newbury Park, CA USA
(805) 499-3686 (tel)
(805) 498-2571 (fax)
Manufactures loudspeakers for high-end home audio and home theater applications and reference monitors for professional studio recording.

West L.A. Music
11345 Santa Monica Blvd., Los Angeles, CA 90025 USA
(310) 477-1945 (tel)
(310) 477-2476 (fax)
Visa, MasterCard, Discover, American Express
Sells a wide variety of equipment for live music performances and professional studio recording, including multitrack tape recorders, DATs, processors, equalizers, microphones, monitors, computers, keyboards, guitar amplifiers, etc.

West-Tech
P.O. Box 1415, Sebring, FL 33871 USA
(941) 471-0922 (tel)
Sells record player parts, needles, cartridges, etc. for home audio applications.

Wetec (West Tennessee Electronic) Electronics
1295A Hwy. 51 Bypass, Starsburg, TN 38024 USA
(800) 249-1250 (tel)
Visa
Sells equipment for professional cellular microwave radio applications.

Wetex International
624 S. Hambledon Ave., City of Industries, CA 91744 USA
(800) 759-3839 (tel)
(818) 854-6600 (tel)
(818) 854-6611 (fax)
Sells computer parts and boards, such as cases, power supplies, removable storage modules, etc.

Wheatstone MicroSystems
3 Marine Ave., Clinton, CT 06413 USA
(203) 669-0401 (tel)
(203) 669-2838 (fax)
Manufactures microcontroller boards for a wide variety of professional computer applications.

Wheels & Wings, etc.
7620 Lyndale Ave. S., Minneapolis, MN 55423 USA
(800) 896-2582 (tel)

(612) 861-6261 (tel)
Visa, MasterCard, COD ($5.95)
Sells batteries, battery chargers, radio systems, etc. for radio-control models.

Whirlwind Music
99 Ling Rd., Rochester, NY 14612 USA
(800) 733-9473 (tel)
(716) 663-8820 (tel)
(716) 865-8930 (fax)
Manufactures mixing consoles for professional broadcasting, professional studio recording, and live music performances.

W. Whitaker & Associates
120 Professional Park #1, P.O. Box 6327, Lafayette, IN 47903 USA
(800) 433-8775 (tel)
(317) 447-5064 (fax)
Sells equipment for professional cable TV applications.

White Electronics
P.O. Box 403, Stanfod, IN 47463 USA
(812) 825-3355 (fax)
Sells loudspeakers for high-end home audio applications. Offers a free catalog.

Whiterock Products
309 S. Brookshire, Ventura, CA 93003 USA
(805) 339-0702 (tel)
Manufactures and sells binary clocks, LED clocks, and the "Time Beacon"--either assembled or in kit form. Offers a free catalog.

White Sands
335 W. Melinda Dr., Phoenix, AZ 85027 USA
(800) JUMPERS (tel)
(602) 582-2915 (tel)
(602) 581-0331 (fax)
Manufactures custom cable assemblies, particularly for professional cable TV applications.

White-Star Electronics
(405) 631-5153 (tel)
(405) 631-4788 (fax)
Sells cable TV converters and descramblers. Offers a free catalog.

J.C. Whitney
1917-19 Archer Ave., P.O. Box 8410, Chicago, IL 60680 USA
Sells auto parts, but also plenty of car stereo components, including head units, amplifiers, and speakers.

Wholesale Cable
(718) 262-0900 (tel)
(718) 297-9221 (tel)
(718) 657-4015 (fax)
Sells cable TV converters, descramblers, and universal remote controls.

Wholesale Cellular USA
5732 W. 71st St., Indianapolis, IN 46278 USA
(800) 243-1227 (tel)
(317) 297-6114 (fax)
Sells antennas, telephones, batteries etc. for cellular microwave radio applications.

WIBU Systems
Rueppurrer Strasse 54, D-76137 Karlsruhe, Germany
+49-721-93172-0 (tel)
+49-721-93172-22 (fax)
Manufactures and sells copy protectors for a wide variety of computer applications.

Wilderness Radio
P.O. Box 734, Los Altos, CA 94023 USA
(415) 494-3806 (tel)
Manufactures CW transceivers for amateur radio applications.

Will-Burt
P.O. Box 900, Orrville, OH 44667 USA
(216) 682-7015 (tel)
(216) 684-1190 (fax)
Manufactures and sells pneumatic telescoping antenna masts.

Williams Radio Sales
600 Lakedale Rd., Colfax, NC 27235 USA
(919) 993-5881 (tel)
Sells amateur radio accessories, including antenna tuners, antennas, etc.

Wilson Antenna
1181 Grier Dr., #A, Las Vegas, NV 89119 USA
(800) 541-6116 (tel)
Manufactures 10-meter antennas for citizens band and amateur radio applications.

WinBook Computer
1160 Steel Wood Rd., Columbus, OH 43212 USA
(800) 468-3341 (tel)
(800) 468-1037 (tel)
Visa, MasterCard, Discover
http://www.winbookcorp.com
Manufactures laptop personal computers. Offers a 30-day money-back guarantee.

Windows Memory
920 Kline St., #100, La Jolla, CA 92037 USA
(800) 454-9701 (tel)
(619) 454-9703 (fax)
Visa, MasterCard, Discover, American Express, COD
Buys and sells memory chips for a wide variety of computer applications.

Windward Products
P.O. Box 378, Moffett Field, CA 94035 USA
(800) 470-9463 (tel)
(408) 987-7735 (fax)
Visa, MasterCard
Sells several DMMs. Offers free shipping.

Win Systems
715 Stadium Dr., Arlington, TX 76011 USA
(817) 274-7553 (tel)
(817) 548-1358 (fax)
Manufactures motherboards for a wide variety of professional and industrial applications.

Wintenna
911 Amity Rd., Anderson, SC 29621 USA
(800) 845-9724 (tel)
(803) 261-3965 (tech tel)
Manufactures antennas and accessories for citizens band, amateur radio, scanner, cellular, and cordless phone applications.

Hank Winter & Associates
1870 W. Prince Rd., #38, Tucson, AZ 85705 USA
(520) 888-2040 (tel)
(520) 888-1970 (fax)
Sells IBM computer systems and parts.

The Wireman
261 Pittman Rd., Landrum, SC 29356 USA
(800) 433-9473 (tel)
(803) 895-4195 (tel)
(803) 895-5811 (fax)
Sells a variety of coaxial cable, twin lead, open wire transmission line, and antenna wire for amateur radio applications.

Wiscom Auto International
Block 261, Waterloo St., #02-25, Waterloo Centre, Singapore 0718
Sells head units, loudspeakers, amplifiers, receivers, cassette decks, CD players, etc. for home audio and car audio applications.

C.W. Wolfe Communications
1113 Central, Billings, MT 59102 USA
(406) 252-9220 (tel)
(406) 252-9617 (fax)
Sells transceivers, etc. for two-way radio communications.

Wollerman Guitars
P.O. Box 457, 118 S. Washington St., Sheffield, IL 61361 USA
(815) 454-2775 (tel)
(815) 454-2700 (fax)
Visa, MasterCard
Manufactures and sells electric guitars and parts for professional and personal musical applications. Offers a 100-page catalog for $5.

Wonderex
825 S. Lemon Ave., Walnut, CA 91789 USA
(800) 838-7988 (tel)
(800) 570-1337 (tech support tel)
(909) 595-1811 (tel)
(909) 598-9108 (fax)
Visa, MasterCard, American Express, COD
Manufactures and sells personal desktop computers. Offers a 30-day money-back guarantee.

Daniel Woodhead
3411 Woodhead Dr., Northbrook, IL 60062 USA
(800) 225-6243 (tel)
http://www.danielwoodhead.com
Manufactures connectors for a wide variety of professional industrial applications. Offers a 148-page designer's guide.

Workshop Records
P.O. Box 49507, Austin, TX 78765 USA
(800) 543-6125 (tel)
Sells half-speed recorders, portable studios, practice amplifiers, etc. for a wide variety of professional audio and musical applications.

Workstations International
601 Lakeshore Pkwy., #1160, Minnetonka, MN 55305 USA
(800) 842-4781 (tel)
(612) 449-9000 (tel)
(612) 449-3198 (fax)
Sells new and refurbished Hewlett-Packard workstations, parts, and upgrades for a wide variety of computer applications.

Workstation Solutions
1 Overlook Dr., Amherst, NH 03031 USA
(603) 880-0080 (tel)
(603) 880-0696 (fax)
Manufactures data storage systems for a wide variety of professional computer network applications.

Worldata
563 Main St., Bolton, MA 01740 USA
(800) WLDESK (tel)

(508) 779-8383 (tel)
(508) 779-6259 (fax)
Manufactures terminals for a wide variety of professional computer network applications.

World Audio Design
64 Castellain Rd., Maida Vale, London W9 1EX UK
+44-171-289-3533 (tel)
+44-171-289-5620 (fax)
Manufactures and sells a variety of tube audio amplifiers and loudspeaker kits.

World Class Batteries
1430 Florida Ave., #217, Longmont, CO 80501 USA
(303) 684-9450 (tel)
(303) 684-9613 (fax)
Sells batteries for radio-control models.

Worldcom Technology
P.O. Box 3364, Ft. Pierce, FL 34948 USA
(407) 466-4640 (tel)
Visa, MasterCard
Manufactures and sells accessories (including an indoor active antenna, a radio protector, and filters) for radio hobbyists. Offers a free catalog with an order or for postage.

World Trade Video
7 Broadway #1048, New York, NY 10004 USA
(800) 253-2639 (tel)
(212) 432-3156 (tel)
Visa, MasterCard, American Express, COD
Sells home, amateur, and professional video equipment, such as camcorders, VCRs, video editors, monitors, etc.

Worldwide Discount Computers
800 Bussey Ct., Streamwood, IL 60107 USA
(800) 700-4776 (tel)
(708) 830-3344 (shipping tel)
(708) 830-3694 (fax)
Visa, MasterCard, Discover, American Express
Sells computer accessories and software, including printers, modems, etc. Offers a free catalog.

Worldwide Memory
3900 E. Miratoma Ave., #C, Anaheim, CA 92806 USA
(800) 666-6117 (tel)
(714) 666-8877 (fax)
Visa, MasterCard, American Express
Sells memory chips and computers for personal computer applications.

Worldwide Technologies
437 Chestnut St., Philadelphia, PA 19106 USA
(800) 457-6937 (tel)
(215) 928-9407 (tel)
(215) 928-9406 (tel)
(215) 922-0116 (fax)
Sells computer parts and boards, such as processor upgrades, floppy drives, hard drives, monitors, video cards, modems, motherboards, memory chips, keyboards, etc.

Worthington Data Solutions
3004 Mission St., Santa Cruz, CA 95060 USA
(800) 345-4220 (tel)
(408) 458-9938 (tel)
(408) 458-9964 (fax)
353-1-6614-566 (Ireland tel)
353-1-6614-622 (Ireland fax)
Visa, MasterCard, American Express
Sells UPC scanners, radio-frequency terminals, etc. for computerized business applications. Offers a free catalog and a 30-day money-back guarantee.

WPI
23 Front St., Salem, NJ 08079 USA
(609) 935-7560 (tel)
Manufactures wire-net cable assemblies for a wide variety of electronics applications.

Wray Music House
P.O. Box 419, Lemoyne, PA 17043 USA
(717) 761-8222 (tel)
wrays@net-works.net
http://www.winways.com/wrays.html
Sells instruments for professional and personal musical applications.

WR Consultant Associates
5 Lucas Ln., Freehold, NJ 07728 USA
(908) 462-0062 (tel)
(908) 462-8455 (fax)
Buys and sells modems, controllers, etc. for a wide variety of computer applications.

The Wright Marketing Group
3498 Delaware Ct., Pennsauken, NJ 08109 USA
(800) 933-8550 (tel)
(609) 486-7973 (tel)
(609) 486-9084 (fax)
Visa, MasterCard
Buys and sells memory chips for a wide variety of computer applications.

Wright Microphones & Monitors
2091 Faulkner Rd. NE, Atlanta, GA 30324 USA
(800) 487-3886 (tel)
Manufactures microphones and monitors for broadcast stations and for professional studio recording.

WS Battery
50 Tannery Rd., Unit 2, North Branch, NJ 08876 USA
(800) 653-8294 (tel)
(908) 534-4630 (tel)
(908) 534-1614 (fax)
Visa, MasterCard
Sells batteries for laptop computers.

E.U. Wurlitzer Music
65 Bent St., Cambridge, MA 02141 USA
(800) 886-5397 (tel)
(617) 738-6203 (fax)
keyboard@wurlitzer.com
guitarworld@wurlitzer.com
http://www.wurlitzer.com
Sells electric guitars, keyboard synthesizers, audio mixers, recorders, etc. for professional and personal audio and musical applications.

W&W Associates
800 S. Broadway, Hicksville, NY 11801 USA
(800) 221-0732 (tel)
(516) 942-0011 (tel)

(516) 942-1944 (fax)
Visa, MasterCard, Discover
Manufactures and sells a wide variety of batteries and chargers. Offers a free catalog and a price list.

Wyman Research Laboratory
8339 S. 850 W., Waldron, IN 46182 USA
(317) 525-6452 (tel)
Visa, MasterCard
Manufactures antennas, amplifiers, transmitters, transceivers, receivers, and converters for two-way amateur TV (ATV) communications.

WYSE Guys
4747 Pioneers Blvd., Lincoln, NE 68506 USA
(800) 279-7312 (tel)
(402) 489-2717 (tel)
(402) 489-2370 (fax)
Visa, MasterCard
Sells refurbished terminals for professional computer applications.

Wyse Technology
3471 N. First St., San Jose, CA 95134 USA
(800) GET-WYSE (tel)
(408) 473-1200 (tel)
(408) 473-1222 (fax)
Manufactures workstations, file servers, etc. for a wide variety of professional computer network applications.

Xandi Electronics
P.O. Box 25647, Tempe, AZ 85285 USA
(800) 336-7389 (tel)
(602) 731-4748 (fax)
(602) 894-0992 (tech support tel)
Visa, MasterCard, COD
Manufactures and sells a variety of kits, including FM transmitters, bugs, bug detectors, voice changers, scanner converters, etc. Offers a free catalog.

Xantrex
8587 Baxter Pl., Bernaby, BC V5A 4V7 Canada
(800) 667-8422 (tel)
Manufactures dc power supplies for a wide variety of electronics applications. Offers a free catalog.

Xceed Technology
48700 Structural Dr., Chesterfield, MI 48051 USA
(800) 642-7661 (tel)
(810) 598-8008 (fax)
Manufactures graphics cards and memory products for personal computer applications.

Xecom
374 Turquoise, Milpitas, CA 95035 USA
(408) 945-6640 (tel)
(408) 942-1346 (fax)
info@xecom.com
Manufactures modem chips for a wide variety of computer applications.

Xicor
1511 Buckeye Dr., Milpitas, CA 95035 USA
(408) 432-8888 (tel)
(408) 432-0640 (fax)
Manufactures integrated circuits for a wide variety of computer applications.

Xilink
2100 Logic Dr., San Jose, CA 95124 USA
(408) 559-7778 (tel)
(408) 559-7114 (fax)
Manufactures integrated circuits for a wide variety of computer applications.

Xionics Document Technologies
70 Blanchard Rd., Burlington, MA 01803 USA
(800) 864-6243 (tel)
(617) 229-7000 (tel)
http://www.xionics.com
Manufactures and sells image accelerator boards for computer laser printing applications.

Xircom
2300 Corporate Center Dr., Thousand Oaks, CA 91320 USA
(800) 438-4526 (tel)
(805) 376-9020 (automated tel)
(800) 775-0400 (fax back)
(805) 376-9100 (tech support fax)
(805) 376-9311 (fax)

Xircom
(805) 376-9130 (BBS)
http://www.xircom.com
cs@xircom.com
Manufactures network adapters for professional computer network applications.

Xitron Technologies
6295 Ferris Square, Ste. D, San Diego, CA 92121 USA
(619) 458-9852 (tel)
Manufactures power analyzers for a variety of professional industrial applications.

XLO Electric
9480 Utica Ave., #612, Rancho Cucamonga, CA 91730 USA
(909) 466-0382 (tel)
(909) 466-3662 (fax)
Manufactures cables for audio/video applications.

Xtant
7901 E. Pierce St., Scottsdale, AZ 85257 USA
(602) 970-9900 (tel/fax)
Manufactures amplifiers for car stereo applications.

Xtreme Electronics
(908) 969-2088 (tel)
(908) 541-6348 (fax)
xtreme@paranoia.com
Manufactures test modules for cable TV converters.

XXera Technologies
9665 E. Las Tunas Dr., Temple City, CA 91780 USA
(818) 286-5569 (tel)
(818) 286-5228 (fax)
Visa, MasterCard, Discover, American Express
Sells computer parts and boards, such as motherboards, cases, floppy drives, monitors, video cards, modems, keyboards, etc.

Xylan
26679 W. Agoura Rd., Calabasas, CA 91302 USA
(800) 789-9526 (tel)
(818) 880-3500 (tel)
(818) 880-3505 (fax)
Manufactures access switches for professional computer network applications.

Xylogics
53 Third Ave., Burlington, MA 01803 USA
(800) 89-ANNEX (tel)
(617) 272-8140 (tel)
(617) 273-5392 (fax)
Manufactures terminal servers for professional computer network applications.

Xytek Industries
19431 W. Davison, Detroit, MI 48223 USA
(313) 838-6961 (tel)
(313) 838-3960 (fax)
Buys and sells used test equipment for professional industrial and electronics applications.

Yaesu
17210 Edwards Rd., Cerritos, CA 90701 USA
(310) 404-2700 (tel)
Manufactures amateur and shortwave radio transceivers, receivers, amplifiers, and accessories.

Yamaha Electronics
P.O. Box 6600, Buena Park, CA 90622 USA
(800) 492-6242 (tel)
(800) 937-7171 (tel)
(714) 522-9011 (tel)
Yamaha Computer Related Products
6600 Orangethrope Ave., Buena Park, CA 90620 USA
(800) 333-4442 (tel)
(800) 832-6414 (tel)
(714) 228-3913 (fax)
Yamaha Systems Technology
100 Century Center Ct., San Jose, CA 95112 USA
(408) 467-2300 (tel)
(408) 437-8791 (fax)
135 Milner Ave., Scarborough, ON M1S 3R1 Canada
(416) 298-1311 (Canada tel)
Manufactures a variety of semiconductor chips, stereo equipment, portable mixers, professional special effects units, computer multimedia speakers, and musical electronics, including electric guitars.

YHC Cassette
75 Saintsbury Sq., Scarborough, ON M1V 3K1 Canada
(416) 321-1179 (tel)
(416) 321-8451 (fax)
Manufactures floppy diskettes for a wide variety of computer applications.

Yokogawa of America
2 Dart Rd., Newnan, GA 30265 USA
(800) 258-2552 (tel)
(404) 251-2088 (fax)
Manufactures test equipment, including benchtop digital multimeters, for a wide variety of professional electronics testing applications.

Yokohama Telecom
1251-401 Manassero St., Anaheim, CA 92807 USA
(800) 965-6426 (tel)
(714) 693-1870 (tel)
(714) 693-1871 (fax)
Visa, MasterCard, Discover, American Express
Sells personal desktop computers; also sells computer accessories, such as modems, voice mail systems, etc.

Yokomo
25-2 Senju Motomachi Adachi-Ku, Tokyo 120, Japan
011-81-3-3882-8872 (tel)
Manufactures and sells battery chargers, etc. for radio-control models.

Yorkshire Computers
11515 W. Carmen, Milwaukee, WI 53225 USA
(800) 375-1667 (tel)
Visa, MasterCard, Discover, American Express
Manufactures and sells personal desktop computers; also sells computer parts and boards, such as floppy drives, hard drives, monitors, modems, keyboards, controllers, etc. Offers a 30-day money-back guarantee.

Yorkville Sound
4625 Witmer Industrial Estate, Niagra Falls, NY 14305 USA
550 Granite Ct., Pickering, ON L1W 3Y8 Canada
Manufactures studio reference amplifiers for professional studio recording.

Yorx Electronics
405 Minnisink Rd., Totowa, NJ 07512 USA
(201) 256-0500 (tel)
(201) 523-4012 (fax)
Manufactures a wide variety of consumer electronics products, including personal cassette players, portable stereos, etc.

E.H. Yost
2211D Parview Rd., Middleton, WI 53562 USA
(608) 831-3443 (tel)
Manufactures and sells NiCd, sealed lead, gel cells, mercury, alkaline, lithium, etc. batteries for a wide variety of electronics applications.

Young Minds
1910 Orange Tree Ln., #300, Redlands, CA 92374 USA
(800) YMI-4-YMI (tel)
(909) 335-1350 (tel)
(909) 798-0488 (fax)
Manufactures data storage systems for a wide variety of computer applications.

Yuasa-Exide
P.O. Box 14145, Reading, PA 19612 USA
(800) 538-3627 (tel)
(610) 372-8613 (fax)
Manufactures batteries and UPS backup power supplies for a wide variety of computer and electronics applications.

Zalytron Industries
469 Jericho Turnpike, Mineola, NY 11501 USA
Sells a variety of loudspeakers for high-end home audio and professional audio applications. Mininum order of $50.

ZAPCO
413 S. Riverside Dr., Modesto, CA 95354 USA
(800) 775-3557 (tel)
(209) 577-4268 (tel)

ZAPCO

(209) 577-8548 (fax)
zapco@zapco.com
http://s2.sonnet.com/zapco
Manufactures amplifiers, crossovers, and preamplifiers for mobile audio applications.

Zenith
1900 Austin Ave., Chicago, IL 60639 USA
(800) 808-4468 (tel)
(800) 788-7244 (tel)
(800) 227-3360 (computer info tel)
(800) 374-4890 (tel)
(800) 239-0900 (tel)
(312) 745-5152 (tel)
(312) 745-2000 (tel)
(708) 808-4468 (tech support fax)
Z Data Systems
2150 E. Lake Cook Rd., Buffalo Grove, IL 60089 USA
(708) 808-5000 (tel)
(708) 808-4434 (fax)
(708) 808-4942 (BBS)
http://www.zds.com
Manufactures televisions and personal computer systems. Also analog systems and equipment for professional cable TV applications.

Zenon Technology
18343 Gale Ave., Industry, CA 91748 USA
(800) 899-6119 (tel)
(800) 229-7898 (tech support tel)
(818) 935-1828 (tel)
(818) 935-1826 (fax)
Manufactures and sells personal desktop computers. Offers a 30-day money-back guarantee.

Zeos
1301 Industrial Blvd., Minneapolis, MN 55413 USA
(800) 272-8993 (tel)
(800) 228-5390 (tech support tel)
(800) 245-2449 (government orders tel)
(612) 362-1222 (tech support tel)
(800) 362-1205 (fax)
(612) 362-1205 (fax)
(612) 633-4607 (tech support fax)
(612) 362-1219 (BBS)
support@zeos
Manufactures and sells personal desktop and laptop computers.

Zero Surge
944 State Rt. 12, Frenchtown, NJ 08825 USA
(800) 996-6696 (tel)
(908) 996-7773 (fax)
Manufactures powerline surge protectors for a variety of electronics applications.

ZeTeK PLC
Fields, New Rd., Chadderton, Oldham OL9 8NP UK
0161-627-5105 (tel)
0161-627-5467 (fax)
Manufactures integrated circuits for a wide variety of electronics applications.

Zetron
12034 134th Ct. NE, P.O. Box 97004, Redmond, WA 98052 USA
(206) 820-6363 (tel)
(206) 820-7031 (fax)
Manufactures wireless links, recall recorders, etc. for professional telephone applications.

Ziatech
1050 Southwood Dr., San Luis Obispo, CA 93401 USA
(805) 541-0488 (tel)
Manufactures power systems for professional electronics applications.

Zinwell
3/F., No. 506-4, Yuan-Shan Rd., Chung-Ho City, Taipei Hsien, Taiwan
886-2-2251929 (tel)
886-2-2251447 (fax)
Manufactures headend systems, amplifiers, splitters, satellite multiswitches, etc. for professional cable TV applications.

ZK Celltest
2320 Walsh Ave., Bldg. H, Santa Clara, CA 95051 USA
(408) 986-8080 (tel)

(408) 986-8178 (fax)
Manufactures monitors for professional cellular microwave radio applications.

Zoom Telephonics
(617) 423-1076 (tel)
(617) 423-9231 (fax)
(617) 423-3733 (BBS)
76711.770@compuserve.com
Manufactures modems for personal computer communications applications.

Zorin
Unit B, P.O. Box 30547, Seattle, WA 98103 USA
(206) 286-6061 (tel)
http://zorinco.com
info@zorinco.com
Sells a variety of microcontroller products. Offers a free brochure.

Z-Systems Audio Engineering
4641F NW 6th St., Gainesville, FL 32609 USA
(904) 371-0990 (tel)
(904) 371-0093 (fax)
Manufactures sample rate converters, router remote controllers, distribution amplifiers, etc. for professional studio recording and live music performances.

Z-World Engineering
1724 Picasso Ave., Davis, CA 95616 USA
(916) 757-3737 (tel)
(916) 753-5141 (fax)
(916) 753-0618 (fax data)
Manufactures programmable miniature controllers for a wide variety of professional industrial applications.

ZyPCsom
2301 Industrial Pkwy W., Bldg. 7, Hayward, CA 94545 USA
(510) 783-2501 (tel)
(510) 783-2414 (fax)
Manufactures modems for personal computer communications applications.

Zzyzx Workstations and Peripherals
5893 Oberlin Dr., #104, San Diego, CA 92121 USA
(800) 876-7818 (tel)
(619) 588-7800 (tel)
(619) 588-8283 (fax)
Manufactures data storage systems for a wide variety of professional computer network applications.

Product Index

The following product index is a listing of the different companies in this book, organized by the different types of products that they manufacture or sell. If you want to maximize the usefulness of this book, you should look in this section first, find the product area that you are interested in, then look those companies up in the body of the book. Remember that it is impossible to pigeonhole thousands of companies; many dabble in numerous fields. The index is intended to help you narrow your search.

The index is divided into the following groups:

* **Amateur and shortwave radio dealers** This group is mostly comprised of businesses that either specialize in amateur radio or shortwave listening, although some of the companies sell both. A few of the companies sell a wide variety of consumer items, and also sell a few models of shortwave receivers.

* **Amateur and shortwave radio manufacturers** Many of these companies manufacture either amateur transceivers or shortwave radios and sell directly to the public. However, accessories (amplifiers, CW keyers, antenna tuners, etc.) manufacturers are also included in this list.

* **Amateur and shortwave radio antenna manufacturers** Maybe this sounds like a real niche, but a load of antenna companies are in operation!

* **Aviation electronics** You won't find any planes for sale here, but you will find panel meters, aviation radios, direction-finding systems, etc.

* **Bar code scanners** Perfect for the businessperson, this section lists companies that manufacture and/or sell bar code readers.

* **Batteries/power supplies** The companies in this section aren't the typical D cell manufacturers or dealers, which would include every convenience store in the U.S. Most of these companies manufacture or sell specialty batteries or power supplies, such as those for computers, amateur radio equipment, test equipment, etc.

* **Cable and connector manufacturers** This section lists companies that primarily manufacture wire, multiconductor line or cable, coaxial cable, etc. or connectors (BNC, UHF, jacks, etc.). For cable and connector dealers, see *Electronics parts dealers*.

* **Cable TV equipment** This section includes everything from splitter and amplifier manufacturers to descrambler dealers. Some of this equipment is illegal to use without proper authorization, so be careful

Product index

if you're a consumer who wants to descramble some cable programs!

* **Car and marine electronics dealers** These companies sell 12-V equipment for mobile applications. Most sell stereo systems and accessories, but radar detectors and other gadgets can also be found here.

* **Car and marine electronics manufacturers** Most of these companies manufacture car stereo head units or amplifiers, although radar detectors, etc. are also manufactured by some of them.

* **CB dealers** These companies deal in citizens band radio equipment, although many also sell a few models of shortwave radios and amateur radio transceivers.

* **CB radio manufacturers** The companies in this list primarily manufacture citizens band transceivers.

* **Cellular telephone** The companies in this list manufacture or sell cellular telephones, pagers, professional cellular equipment, and accessories.

* **Chip programmers** If you want to program ICs, such as EEPROMs, check out the companies in this list.

* **Computer power protection** Most of the companies in this list manufacture power-surge protection devices.

* **Computer audio/video** This category includes manufacturers of MIDI interface boards, multimedia speakers, video boards, multimedia projectors, TV scan converter boards, graphics adapters, etc.

* **Computer communications** The companies in this list primarily manufacture computer modems.

* **Computer drives** The companies here manufacture or sell computer hard disk drives, CD-ROM drives, tape drives in a wide variety of formats, optical disk drives, etc.

* **Computer memory** If your computer or printer needs more RAM, check here for memory chip dealers. Some of these companies also sell microprocessor chips, or other computer components.

* **Computer monitors** Most of the companies included in this list manufacture monitors, but a few specialize in monitor repair and sales.

* **Computer network** This broad category includes manufacturers and dealers of network servers, routers, multiplexers, disk controller boards, terminals, etc.

* **Computer parts and accessories** This very broad category includes companies that sell or manufacture motherboards, floppy drives, video cards, CPUs, monitors, cases, power supplies, mice, trackballs, etc. See also *Computer systems*.

* **Computer printers and scanners** Many of these companies are manufacturers of printers and the rest are dealers that specialize in printers and scanners.

* **Computer systems** These companies manufacture computer systems and often sell them directly to the public. Most of these companies only assemble the computers--the parts are purchased from other manufacturers. Most of these companies also sell parts and accessories (printers, scanners, data storage systems, etc.).

* **Crystals** These companies manufacture and sell crystals, which are commonly used in a wide variety of radio and computer applications.

* **Educational** It could be argued that every electronics- and computer-related product and kit is educational. The companies in this section sell products that are part of an instructional course, or something similar.

* **Electronics kits** These companies manufacture and sell electronics kits: every-

Product index

thing from color organs to hi-fi amplifiers to wireless microphones to amateur radio transceivers.

* Electronics parts dealers Some of these companies sell standard electronics parts and others sell just specific replacement parts for consumer equipment. See also *Surplus electronics*.

* Electronics parts manufacturers The companies in this section manufacture electronics parts. Most are very specialized and focus only on one topic, such as capacitors, transformers, or resistors.

* Fiberoptics These companies manufacture fiberoptic cable and connectors.

* Home stereo dealers If you want to buy or price home stereo equipment, look here. Many of these companies also sell home video equipment and magnetic recording media.

* Home stereo manufacturers Most of these companies manufacture a wide variety of consumer electronics, in addition to home stereo products.

* Home video dealers If you want to buy or price home video equipment, look here. Many of these companies also sell home stereo equipment and magnetic recording media.

* Home video manufacturers Most of these companies manufacture a wide variety of consumer electronics, in addition to home video products.

* Industrial computer and electronics This broad category of companies sells such things as: single-board computers, ruggedized computers, data acquisitions systems, industrial gauges, etc.

* Lasers Many of these companies are laser dealers. See also *Surplus*.

* Lighting systems The lights in this section are primarily those for live music performances

* Loudspeaker manufacturers It's amazing just how many companies are dedicated to manufacturing loudspeakers. A large number of these also manufacture car stereo amplifiers.

* Magnetic recording The companies in this listing either manufacture magnetic recording media or are dedicated to selling it. If you are looking for specialized media, check here; for consumer media, check here and also under *Home stereo dealers* and *Home video dealers*.

* Microphones A number of companies specialize in microphone production. Most of these companies sell microphones for professional recording applications or live music performances.

* Miscellaneous Here is the catch-all for unusual products. A few of the products include: timers, air ionizers, infrared product controllers, Van de Graaf generators, battery monitors, vehicular monitors, EMI filters, some DSP boxes (DSPs for specific applications are listed under that application, e.g., amateur/shortwave radio, computer audio/video, home stereo, etc.), pocket computer organizers, and much more.

* Musical instruments and equipment The companies in this category manufacture or sell such things as: electrical or electronic instruments (electric guitars, synthesizers, etc.), instrument amplifiers, live music processing equipment (limiters, compressors, distortion pedals, etc.), guitar pickups, and more. Although some of the same general types of equipment are used in professional broadcasting, there is no overlap here. See also *Studio recording dealers* and *Studio recording manufacturers*.

* Office equipment These companies manufacture or sell electronic office equipment, such as fax machines, photocopiers, paper shredders, etc.

* **Professional broadcasting** Most of the companies in this listing manufacture broadcast equipment that is sold directly to radio and TV stations.

* **Professional video dealers** In this case, "professional" means video editing equipment, high-end VCRs, video effects, etc. Although it might not all be broadcast-quality equipment, it is, at least, acceptable for weddings, etc.

* **Professional video manufacturers** In this case, "professional" means video editing equipment, high-end VCRs, video effects, etc. Although it might not all be broadcast-quality equipment, it is, at least, acceptable for weddings, etc.

* **Radio-control models and model trains** The companies in this listing sell or manufacture electronics products for models. The typical products here are: transmitters for radio-control models, electronic sound effects, etc.

* **Robotics** The companies listed here focus on selling robots and robot components.

* **Satellite TV/video** The companies here manufacture or sell satellite TV receivers, antennas, downconverters, etc. for consumer and professional applications.

* **Solar power** These companies primarily sell solar systems: solar panels, power inverters, and deep-cycle batteries.

* **Sports electronics** Although not a very large category, these companies sell radar guns for sporting applications.

* **Studio recording dealers** These companies sell equipment for recording studios. See *Professional broadcasting* and *Musical instruments and equipment*.

* **Studio recording manufacturers** These companies manufacture equipment to outfit recording studios; some sell direct. See also *Professional broadcasting* and *Musical instruments and equipment*.

* **Surplus electronics (parts and more)** If you need surplus electronics, these companies have it. See also *Electronics parts dealers* and *Vintage electronics*.

* **Surveillance and security** The companies here sell everything from wireless microphones to bug detectors to voice scramblers to mini infrared video cameras.

* **Telephone** These companies primarily manufacture telephones, answering machines, voice-mail systems, and large telephone "networks"--for everything from consumer to professional business applications. See also *Cellular telephone*.

* **Test equipment dealers** These companies sell new and used oscilloscopes, signal generators, frequency counters, audio generators, etc.

* **Test equipment manufacturers** These companies manufacture oscilloscopes, signal generators, frequency counters, audio generators, etc.

* **Two-way radio** The companies in this listing primarily manufacture or sell transceivers for VHF/UHF commercial or governmental communications. Most have nothing to do with amateur or CB radio.

* **VHF/UHF scanners** The companies in this listing manufacture or sell VHF/UHF scanning receivers for consumer monitoring. Many of these companies also manufacture or sell amateur, CB, or shortwave radio equipment.

* **Vintage electronics** This is your best bet for finding rare electron tubes, and classic radio and audio equipment. See also *Surplus electronics*.

Amateur and shortwave radio dealers

ABC Communications
ACE Communications
Ace Systems
Ack Radio Supply
Affordable Portables
Alden Electronics
Alexander Aeroplane Co.
Amateur Communications
Amateur Electronic Supply (AES)
Amateur Radio Supply
Amateur Radio Team of Spokane
ARE Communications
Associated Radio
A-TECH Electronics
Austin Amateur Radio Supply
Aventel OY
Barry Electronics
BC Communications
B&H Sales and Service
A. Böck
Boger-funk
Burk Electronics
Burqhardt Amateur Center
C-Comm
COMDAC (Communications Data Corp.)
Comm-Pute
Communication Headquarters
Communications Electronics
Com-West Radio Systems
Copper Electronics
Cotronics
C.P.I.
C. Crane
Dandys
Davis RF
De La Hunt Electronics
Deltron Communications International
Dentronics
Doc's Communications
Doeven Electronika
Dubberley's on Davie
Durham Radio Sales and Service
EDCO (Electronic Distributors)
EEB (Electronic Equipment Bank)
Electro-Comm
Electronic Center
Eli's Amateur Radio
El Original Electronics
Erickson Communications
Erie Aviation
First Call Communications (FCC)
F&M Electronics
Fortex Enterprises
Galaxy Electronics
Generale Electronique Services
GES
G and G Electronics of Maryland
Gilfer Shortwave
GMW
Grove Enterprises
Ham Buerger
The Ham Contact
Ham Radio Outlet
The Ham Station
Hamtronics Trevose
Hardin Electronics
Hatry Electronics
Harold Heaster
Henry Radio
Hialeah Communications
Hirsch Sales
Honolulu Electronics
Hooper Electronics
IGS
International Radio & Computer
IRI Amateur Electronics
Jade Products
Javiation
J-Com
Jeremy Communications
J.R.S. Distributors
Jun's Electronics
KComm
KDC Sound
Länggasse
LaRue Electronics
Lentini Communications
Longs Electronics
Lowe Electronics
Madison Electronics
Maha Communications
Maryland Radio
MCM Electronics
Memphis Amateur Electronics

Amateur and shortwave radio manufacturers

Miami Radio Center
Michigan Radio
Mike's Electronics
Miltronix
National Tower
Norad A/S
Norham Radio
North Atlantic Radio Service
N & G Electronics
Oklahoma Comm Center
Omar Electronics
Omni Electronics
Pearl Electronics
Portland Radio Supply
QRV Electronics
Radio Center
Radio Center USA
Radio City
Radio Communications
Radio Communications Center
Radio Communications Services
The Radio Place
Radio Shack
Radio Wholesale Marketing
Ray's Amateur Electronics
Reynolds Radio
Rivendell Electronics
R&L Electronics
Rogus Electronics
Rosen's Electronics
Rosewood
Ross Distributing
RT Systems Amateur Radio Supply
Sidney Aircraft Radio Industries
Slep Electronics
Soundnorth Electronics
South Essex Communications
SRS
Stabo Ricofunk
STG International
Support Measures
Surplus Sales of Nebraska
Texas Towers
Troniks
Tucker Electronics & Computers
Universal Radio
Valley Radio Center
H.C. Van Valzah
VHF Communications
VOVOX
Williams Radio Sales

Amateur and shortwave radio manufacturers

Advanced Receiver Research
Advanced RF Design
Advanced Specialties
AEA (Advanced Electronic Applications)
Alinco Electronics
Allen Telecom Group
Alpha Delta Communications
Altronic Research
Amateur and Advanced Communications
Ameritron
ANLI
AOR
Auskits
Autek Research
Azden
Batima
Baycom
B and B Enterprises
John Bell
Bencher
Byers Chassis Kits
Centaur Electronics
Chilton Pacific
C3I
Collins
Comer Communications
Command Technologies
Communication Concepts
Communications Specialists
DGM Electronics
Digital Communications
Doppler Systems
Dovetron
Down East Microwave
R.L. Drake
DWM Enterprises
Dynamic Electronics
Electronics Information Systems
Eska & Edvis
ETO (Ehrhorn Technological Operations)
David & Ervena Gillespie

Amateur and shortwave radio manufacturers

Grundig
HAL Communications
Doug Hall Electronics
The Ham Key Co.
Hamtronics
Heil Sound
ICOM
Index Laboratories
Industrial Communications Engineers
Japan Radio (JRC)
J Comm
JPS Communications
Kanga US
Kantronics
K-COM
KD4YBC
Kenwood
Kilo-Tec
Kiwa Electronics
K2AW's "Silicon Alley"
Link Communications
Lowe Electronics
Maxcom
MFJ
Micro Control Specialties
Microcraft
Microtek
Midland International
Mirage Communications
Mirage Communications Equipment
North Country Radio
NYE Engineering
OFS WeatherFAX
PacComm Packet Radio Systems
Palomar Engineers
Panasonic
Pan-Com International
P.C. Electronics
PCRC
Premier Communications
Procomm
Psitech Plus
Quorum Communications
Radio Accessories
Radio Adventures Corp. (RAC)
Radio Control Systems (RCSI)
Radio-Tech
Ramsey Electronics

RCK
RF Applications
RF Concepts
RF Limited
ROBOT Research
Rohde & Schwarz
Sangean America
SGC
Sherwood Engineering
Siemens
Somerset Electronics
Sony Electronics
Spi-Ro Manufacturing.
SSB Electronic
S&S Engineering
Standard Amateur Radio Products
TAPR
Tasco Electronics
Ten-Tec
Tigertronics
Timewave Technology
US Digital
VoCom Products
Watkins-Johnson
Wilderness Radio
Worldcom Technology
Wyman Research Laboratory
Yaesu

Amateur and shortwave radio antenna manufacturers

Alliance Rotators
Aluma Tower Co.
American Antenna
Am-Search Tower
Antennas Are Us
Antennas Etc.
Antenna Mart
Antenna Supermarket
Antennas West
ASA
Atlantic Ham Radio
Austin Antenna
Bilal
BMG Engineering
Bozak Antenna
Butternut Manufacturing

Aviation electronics

Cable X-Perts
CATS
ClearTek
ComTek Systems
Cushcraft/Signals
Datong Electronics
Delta Loop Antennas
Dressler
Electronic Switch
Gap Antenna Products
High Sierra Antennas
Hustler Antennas
IIX Equipment
Jo Gunn Enterprises
Lakeview
Larsen Electronics
LF Engineering
Maldol Antennas
Glen Martin Engineering
McClaran Sales
M_FD Enterprises
Mirage Communications
Mobile Mark Communications Antennas
Mosley Electronics
Multi-Band Antennas
Nott
Outbacker Antenna Sales
Pacific Avionics
Palomar Engineers
Radio Works
ROHN
Rutland Arrays
Spi-Ro Manufacturing
H. Stewart Designs
Telex Communications
TEM Antennas
Tennadyne
TIC General
T.J. Antenna
Transel Technologies
Tri-Ex Tower
Trylon Manufacturing
Universal Manufacturing
US Tower
Valor Enterprises
Van Gorden Engineering
VIS
W9INN Antennas

Will-Burt
Wilson Antenna
Wintenna
The Wireman

Aviation electronics

Aero Product Supply
Aircraft Spruce & Specialty
AlliedSignal Aerospace
APTEC
Arnav Systems
Artais
Ashtech
Aviation Surplus
Ballard Technology
Barfield
Bendix/King
Billingsley Magnetics
Chief Aircraft
Classic Air Supply
Collins
Datum
Eastern Avionics International
Edmo Distributors
Electronics International
Emergency Beacon
Erie Aviation
Flightcom
Flight Products International
Garmin
Grand Rapids Technologies
Warren Gregoire & Associates
Gulf Coast Avionics
Howell Instruments
II Morrow
Insight
Interstate Electronics
J.A. Air Center
JP Instruments
Magellan
Man Technologie
Maplin Electronics
Maxon Systems
MentorPlus
Millimeter Wave Technology
Narco Avionics
Next Destination

347

Aviation electronics

North Star Avionics
Allen Osborne Associates
Pilot Avioncs
Rocky Mountain Instrument
Sennheiser
Sextant Avionique
Shadin
Sidney Aircraft Radio Industries
Sigtronics
3S Navigation
Trimble
Tropic Aero
Vision Microsystems

Bar code scanners

American Microsystems
American Precision Instruments
ASI
August Mark
Auto-ID Products
Bar Code Discount Warehouse
Bear Rock Technologies
Cardservice International
Data Hunter
DataVision
Hand Held Products
Intelligent Barcode Systems (IBS)
Interconnections
PC America
P.O.S. International
Ryzex Re-Marketing
Syntax Retail Systems
Variant Microsystems
Worthington Data Solutions

Batteries/power supplies

Absopulse Electronics
Acopian
Advance Power
AER Energy Resources
AES
Alexander Batteries
Alliant Techsystems
Alpha Technologies
American Power Conversion (APC)
American Reliance

Amptech
ANLI
Applied Kilovolts
Argaph
Argus Technologies
Astec
AstroFlight
Astron
ATC Power Systems
Ault
Axiom
Battery-Biz
Battery Engineering
Battery Express
Battery-Tech
Bertan High Voltage
Best Power
Cascade Audio
Clary
Condor D.C. Power Supplies
Conversion Equipment
Cool-Lux
CUI Stack
Cunard Associates
Del Electronics
Deltec Electronics
Digital Power
Eagle Picher Industries
Ebbett Automation
Energizer Power Systems
Exide Electronics
Falkner Enterprises of America Inc.
Fortron/Source
Friwo North America
GlobTek
Gould Electronics
Hamamatsu
HC Power
Intellipower
Jade Products
JBRO Batteries
Kepco
Keystone Electronics
Lambda Electronics
Lectro Products
Less Buster's Electronics
Liebert
Lightning Audio

LZR Electronics
Magellan's
Matsushita (Panasonic)
MC_FD
Memocom Electronics
Microenergy
Minuteman
Mr. NICD (E.H. Yost)
The NiCad Lady
No. 1 Battery Specialists
Oneac
Operating Technical Electronics
Optima Batteries
Performance Cable TV Products
Periphex
Philips Power Systems
Plainview Batteries
Power and Environment International
Power-One Power Supplies
Power-Sonic
Power Trends
ProSource Power
Quality Power Products
Renu International
S.A.C.
SAFT America
Self-Reliance
SK America
Sola
Superior Electric
Sutton Designs
Suzalex Technologies
Tadiran Electronic Industries
Tauber Electronics
Team-Hold Technology
Telecommunications Power Systems
Todd Products
Tower Electronics
Tripp Lite
Tumbler Technologies
Ultralife Batteries
ViewSonic
VL Products
VRA Batteries
WS Battery
W&W Associates
Xantrex
E.H. Yost
Yuasa-Exide
Ziatech

Cable and connector manufacturers

Andrew
Amphenol
Audioquest
Belden Wire and Cable
Belkin Components
Cables America
Cables & Connector Technologies
Cables to Go
Cable X-Perts
Conec
Connectors Unlimited
Cord-Lox
Alcoa Fujikura
CS Electronics
DGS Pro-Audio
Ding Yih Industrial
Electronic Switch
Esoteric Audio USA
Gepco International
Harting Elektronik
Hirose Electric
International Electronic Wire & Cable
JSC Wire & Cable
Kimber Kable
KurlyTie
Lemo USA
Lindsay Electronics
The Live Wire Companies
Mahogany Sound
Majestic
Matejka Cable Recovery
Molex
Monster Cable
MWS Wire Industries
Nemal Electronics
NorComp Interconnect Devices
Optec Sales
Orca
Panduit
Phoenix Contact
Precision Interconnect
Radio Shack
RIA Electronic

Cable and connector manufacturers

The Rip-Tie Company
Siecor
Sonance
Sound Connections International
Stirling Connectors
3M
Trilogy Communications
Trompeter Electronics
Turck
Daniel Woodhead
WPI
XLO Electric

Cable TV equipment

3M
Abate Electronics
ABC Cable Products
Advantage Electronics
AEL Industries
All Phase Video-Security
Alpha Technologies
Alternative Marketing
AM Communications
American Trap
ANDO
ARCOM
Arena Services
Aria Technologies
ASKA Communication
ATx Telecom Systems
Axsys Comunicaciones
Basic Electrical Supply & Warehousing
Bear Mountain Audio
Blonder Tongue Laboratories
BLR Communications
B&S Sales
Cable AML
Cable Link
Cable Linx
Cable Source International
Cable Technologies International
Cable Warehouse
California Amplifier
CATV Services
C-COR Electronics
C & E Communications & Energy
CFS Electronics
Chainford Electronics
Channell
Communications Supply Group
ComSonics
Conifer
Converters
dB Communications
Decibel Products
Delphi Components
Direct Cable Supply
B.E. Duval
DX Communications
Eagle Comtronics
Eagle Electronics
Eastern Electronic
Erivision
EXFO
The Filter Company (TFC)
Finline Technologies
Forest Electronics
Foss Warehouse Distributors
Galow Cable
General Instrument
Glade Communication Equipment
Greenleaf Electronics
Harmonic Lightwaves
Hero Communications
High Tech Cable
Holland Electronics
Idea/onics
Inphone Electronics Enterprise
Ipitek
Jebsee Electronics
Jolly R. Electronics
Jones Surplus
J.V. Electronics
Kaul-Tronics (KTI)
KDE Electronics
Klungness Electronic Supply
Leader Instruments
MD Electronics
Mega Electronics
M & G Electronics
MicroCom Industries
MicroSat
Microwave Filter
Microwave Networks
Midwest Cable Services

Midwest Electronics
Modern Electronics
Monroe Electronics
Moore Diversified Products
Mo-Tech
Multicom
Multi-Vision
Multi-Vision Electronics
National Cable Brokers
NCS Industries
N.E. Engineering
Nortel (Northern Telecom)
Northeast Electronics
Novaplex
Nu-Tek Electronics
Philips Consumer Electronics
Phillips-Tech Electronics
Pico Macom
Pico Products
Polotec
Power Guard
PTL Cable Service
Pulsar Barco
Qintar
Quality Cable & Electronics
Reltec
Riser Bond Instruments
RMS Electronics
Rocky Mountain Jumper Cables
Rotal Telecommunication
RP Electronics
Sadelco
Scientific Atlanta
Southern Electronics
South World
Standard Communications
Standford Telecom
Star Circuits
Star Electronics
Starnet Development
Super Cable
Superior Electronics Group
Synchronous Marketing
Taiwan Cable Connection
T.C. Electronics
Telecom Solutions
Teleview Distributors
TeleWire Supply

Timeless Products
TM Brokers
Trans Electric
Trilithic
Tropic Electronics
Tulsat
TVS Holding
U.S. Cable TV
Vela Research
Video Communications Systems (VCS)
Viewsonics
Viewstar
VIP Cable Services
W. Whitaker & Associates
White Sands
White-Star Electronics
Wholesale Cable
Xtreme Electronics
Zinwell

Car and marine electronics dealers

Ace Car Audio
Al-Salah Trading
Authorized Audio/Video/Car
Autronic
AVG
B and B Enterprises
Canseda
Coast to Coast
Crutchfield
Damark
Digital Auto-Radio Sarl
Electronics Depot
Erie Aviation
Executive Photo and Electronics
Exhibo
Frontgate
Francisco Ganán
Njal Hansson
Jolly Sound
J & R Music World
Korbon Trading
Lyra
Marine Electronics
Marine Park Camera & Video
One Call West
Radio Shack

Car and marine electronics dealers

Samman's Electronics
Servis-Elekto-Praha
The Sound Approach
Sound City
Sound Ideas
Sound Source
Sunny Bright Electronic
Superior Audio
Thorens Iberica
UNICO
J.C. Whitney
Wiscom Auto International

Car and marine electronics manufacturers

AAMP of America
Accusound
A.C.I.
Adcom
Advanced Composite Audio (A.C.A.)
Advanced Radar Technologies
Aiwa
Alphasonik
Alpine
ALS
American Bass
American Recorder
American Terminal
Amptech
Amprotector
Amroh
Apogee
Audio Car
Audio Gods
AudioLink
Audio Logic
Audiovox
Audison USA
Autech
Authorized Parts Company
Autotek
Avalanche
Avital Auto Security
Beals Brothers
Becker Autoradio
Blade
Blaupunkt
Blitz-Safe
Blue Moon Gadgets
Boss/AVA
Boston Acoustics
Bostwick
Canton
Carver
Cascade Audio
Clarion
Clifford Electronics
Cobra Electronics
Comroad
Concord
Coustic
Craig Consumer Electronics
Crunch
Crystal-Line
Custom Autosound
Denon Electronics
Directed Electronics
Dominator Technologies
Dr. Crankenstein
Earthquake Sound
Eclipse
Electrocom Communication Systems
Gamber-Johnson
Graffiti Electronics
Harrison Laboratory
Harvard Radio
Hawk
Hellfire
HiFonics
Hitron
Infinity
Jensen
JVC Company of America
Kenford
Kenwood
K40 Electronics
Kortesis Marketing
Lanzar
Lazer Industries
Lightning Audio
Matsushita (Panasonic)
Maxon Systems
Maxxima Marine
McIntosh
Milbert
MMATS Professional Audio

Cellular telephone

Mobile Authority
MTX
Nakamichi America
Ocean Electronics
Orion
Paramount Audio
Phase Linear
Pioneer Electronics
Power Acoustic
Precision Power
Profile Consumer Electronics
Proton
Pyramid
Radio Shack
Rockford Fosgate
Rockwood
R&T Enterprises
Sansui
Sanyo Fisher USA
Sanyo Technica USA
Scientific Dimensions (SDI)
Seco-Larm
Sherwood
Silent Sam
SOAT
Sparkomatic
Stillwater Designs
Ultimate Sound
Uniden
United Electronic Supply
Urban Audio Works
U.S. Amps
Viper
Vision Specialties
Xtant
ZAPCO

CB radio dealers

Advanced Specialties
American Electronics
AXM
Base Station
Bob's CB
CB Shop
Communications Electronics
Copper Electronics
Damark
Danitas Radio
Discount Electronics CB Sales
D&R Electronics
Durham Radio Sales and Service
Electric City Electronics
Eleven Meter Communications
F&M Electronics
Galaxy Electronics
Highway 155 CB Shop
Hobby Radio Stop
HPR
KDC Sound
Loveland CB
Maxtech
Medicine Man CB
Miami Radio Center
Radio Shack
R & R Communications
Scanner World, USA
SolarCon
Stewart Electronics
Thomas Distributing

CB radio manufacturers

C.B. City International (CBCI)
Charles Radio
Cobra Electronics
Firestik Antenna
Hustler Antennas
Jo Gunn Antennas
K40 Electronics
MaCo Manufacturing
Marvel Communications
Maxon Systems
Midland International
RF Limited
Shakespeare
Signal Engineering & Communications
Uniden
Valor Enterprises
Wawasee Electronics
Wilson Antenna
Wintenna

Cellular telephone

Accutech International

Cellular telephone

Argus Technologies
Aries Manufacturing
Audiovox
Axiom
Bonigor USA
BPB Associates
Buy-Comm
California Microwave
Cell-Tel International
Cellular Products
Cellular Security Group
Cellular Wholesalers
Celwave
Centurion
Charge Guard
Cobra Electronics
Coherent Communications Systems
Communications Associates
Communications Consultants
Com-Rad Industries
CSS Power
Curtis Electro Devices
Damark
D&L Communications
Doppler Systems
DSC Communications
DuraComm
Eagle
El Paso Communication Systems (Epcom)
Ericsson
Fourth Dimension Industry
FWT
Gardiner Communications
General Communications
GL Communications
Haven Industries
IDA
International Cellular Products
Japan Radio (JRC)
Kern Pager Repair
Larsen Electronics
Loral Microwave-Narda
Loral Test & Information Systems
Marvel Communications
MCI
MDM Radio
Mechem Electronics
Microwave Data Systems

M & K
Mobilex
Motorola
MoTron Electronics
Multiplier Industries
Nokia Mobile Phones
Novacomm
ORA Electronics
PageTap
Pana-Pacific
Pipo Communications
Plexsys International
Polaris Industries
Ptek
Radio Central
Radio One
Radio Shack
Reach Electronics
RF Industries
RMS Communications Group
Santa Fe Distributing
Scala
Schaefer Radio
Scott Communications
Second Source Cellular
Shakespeare
Sharp Communication
S & N Cellular
Spectrum Cellular
Sterling Communications
Superior Cellular Products (SCP)
Sutter Buttes 2-Way
SWS Security
Tea Cellular Network Services
Telco Services
Transcrypt International
Transel Technologies
Unicom Paging Network of Texas
Uniden
Uni-firm
Unplugged Communications
Valor Enterprises
Wayne Communications
West Coast Radio
Western Multiplex
Wetec (West Tennessee Electronic) Electronics
Wholesale Cellular USA
Wintenna

Computer audio/video

ZK Celltest

Chip programmers
Andromeda Research
Data I/O
Dataman Programmers
General Device Instruments
Intronics
M_FDL Electronics
MQP Electronics
Needham's Electronics
Universal Cross-Assemblers

Computer audio/video
Advent
AITech
Altec Lansing
Ampro Computers
Antares Microsystems
Antonio Precise Products Manufactory
Apollo Presentation Products
ASK Technology
ATI Technologies
Aura
AVerMedia
Axionix
Boxlight
Bright Creation International
Colorgraphic Communications
Connectix
Creative Labs
Data Translation
Diamond Multimedia
Digital Processing Systems
Dobbs-Stanford
Eastman Kodak
Electrohome Limited
Eletech Electronics
Envisions Solutions Technology
Fast Electronic US
Futuretouch
Future Video Products
Genoa Systems
HDSS Computer Products
Imaging Technology
InFocus Systems
Infotec International
Infotronic America
IntelliMedia
i Sight
Labtech
Lance Technology
Laser Graphics
Leigh's Computers
Logitech
Maxtek Components
Media Management
Megatronics
MicroSolutions
MIE Systems
Minnesota Western Visual Presentation Systems
Miro Computer Products
New Voice
Nichi Standard Enterprises
NuReality
nView
Optibase
Orchid Technology
Panelight Display Systems
PC Video Conversion
Polaroid
Proxima
Publishing Perfection
Reveal Computer Products
Revered Technology
Rhetorex
Select Solutions
Sharp
Sony Electronics
STB Systems
STD Entertainment
Swiss Global Informatica
Teac
TeleVideo
Telex
TW Enterprises
Vic Hi-Tech
VideoLabs
VideoLogic
Videomedia
Vigra
Visual Circuits
Visual Information Technologies
Xceed Technology

Computer audio/video

Yamaha Electronics

Computer communications

Apex Data
Best Data Products
BOCA Research
Brooktrout Technology
Devcom Mid America
Global Village Communications
GMM Research
Hayes Microcomputer Products
Information Data Products
Insite Technology
Management Products
MegaDyne Communications
Megahertz
National Data Mux
Omni Data Communications
Orion Telecom
Ositech Communications
Practical Peripherals
Prarie Digital
Promethus Products
Signal Enterprises
Spectrum Cellular
Supra
U.S. Robotics
Zoom Telephonics
ZyPCsom

Computer drives

ABC Drives
Advanced Digital Information
APS Technologies
ARC
Bason Hard Drive Warehouse
Better Business Systems
Carrot Computer
CD-R Solutions
Cherokee Data
CHI
Chinon America
Computer Disk Service
Computer Innovations
CompuTrend (CTI)
Concorde Technologies

Conner Peripherals
Cybernetics
Data Rite
Data Systems Service (D.S.S.C.)
Digitape Systems
Direct Connections
Dirt Cheap Drives
Drive Outlet Center
Essential Data
EZ Systems
F & F Electronics
Hard Drives & Accessories (HDA)
HDSS Computer Products
Herman Elig Computers
Hitachi
Intacta
K.D.S.
Key Solutions
Kingston Technology
Laguna Data
Magnatek
MBI
Mega Haus Hard Drives
Micropolis
Micro Sense
MPS
MPS Multimedia
Multimedia Effects
NESI
Noble House
Nordisk Systems
Optical Media International
Overland Data
PC Peripherals
Philips Consumer Electronics
Pinnacle Micro
Plextor
PRC
ProCom Technology
Pro-Disk Service
QSTAR Technologies
Qualstar
SAG Electronics
Seagate
Silicon Back-Up Solutions
Sony Electronics
Sound Electro Flight (SEF)
SpringBoard Technology

Storage USA
Sun Moon Star
Sunstar
Techzam
Valtron Technologies
Vento Associates
Western Digital
Young Minds
Zzyzx Workstations and Peripherals

Computer memory

Advanced Recovery
Autotime
Cameleon Technology
The Chip Merchant
Chrislin Industries
The Computer Memory Outlet
Data Memory Systems (DMS)
Eastern 21
Eritech
International Computer Purchasing (ICP)
Internet Memory Exchange
L.A. Trade
Memory Conversion Products
Memory and CPU Warehouse
Memory Depot
Memory, Etc.
The Memory Exchange Wholesalers
Memory Express
Memory 4 Less
Memory Source
Memory World
Micro Memory Bank
MicroModule Systems
Microprocessors Unlimited
Mosel Vitelic
North East Memory
RAM International Computer
Samsung
Smart Modular Technologies
Songtech International
Tec-Hill
Toshiba America
Unicore Software
Valley Forge Computer
Visiontek
Windows Memory

Worldwide Memory
The Wright Marketing Group
Xceed Technology

Computer monitors

Action Electronics
ADI Systems
Anbonn
AOC International
Apollo Presentation Products
ASK LCD
AVT
Clinton Electronics
Computer Dynamics
Computer Monitor Service
Crestron
CTX International
Daewoo Electronics
Daytek Electronics
Electrohome Limited
Harz
Hitachi
Hyundai Electronics
MAG InnoVision
Magnavox
MGC Technologies
Microtouch Systems
Nanao
Nokia Display Products
NSA
Pikul & Associates
Pixel Express
Portrait Display Labs
Princeton Graphic Systems
Radius
Samtron
Sharp
Sony Electronics
Tatung
Transparent Devices (CyberTouch)
ViewSonic
Vivid Technology
Wells-Gardner Electronics

Computer network

ABLE Communications

Computer network

Acropolis Systems
AC Technology
ADAX
A.D.C.S.
Addtron Technology
ADS International
Advanced Digital Information
Advanced System Products
AFIC Technologies
Alantec
Alis Technologies
Allied Telesyn International
Alphatronix
Alpine Computer Sales
AmberWave Systems
AMCC MarCom
American Computer Hardware
Ancot
Andataco
Applied Digital Systems
Artecon
Asanté Technologies
Ashton Systems
ATL Products
Audax Direct
Auspex Systems
Automated Control Concepts
Axil Computers
Axis Communications
Bay Networks
Benchmark Computer Systems
Bering Technology
Blue Line Communications Warehouse
Boffin
Box Hill Systems
Business Information Zygoun (BIZI)
Cabletron Systems
Cambex
Campbell Technical
CastleRock
CCNS Group
CD-Rom Direct
Central Data
Chase Research
CHI
Chipcom
Ciprico
Cirris Systems

CMD Technology
Cogent Data Technologies
Combinet
Comco DatNet
Compatible Systems
Component Distributors
Computer Network Integration
Computer Ventures
Computer Wholesalers
Computone
Comp View
Comstar
Comtec Automated Solutions
Comtrol
Conley
Corollary
Covia Technologies
Cranel
C.S.U. Industries
CT2M
Cubix
Cybex
Cyclades
Data General
Dataram
Data Research & Applications
DataSouth
Dayna Communications
Develcon
Development Concepts
Diamond Data
DigiBoard
Digital Equipment (DEC)
Digital Repair
Digital Technology
Disk Emulation Systems
D-Link
Eclipse Technologies
Eicon Technology
EMASS Storage Systems Solutions
Emulex
Encore Technology Group
Engineering Design Team
Exabyte
Extended Systems
Falco Data Products
Falcon Systems
Federal Computer Exchange (Fedcom USA)

Computer network

Frontier Technologies
Gao Systems
Helios Systems
Hostlink
HT Communications
Human Designed Systems
Huntsville Microsystems
IBM
IBM Credit
ICL
IEM
IGEL
IKON
Imagraph
Interface Data
Interlink Computer Sciences
Interphase
I/O Concepts
I/O Controls
Iomega
IOtech
Ironwood Electronics
JCC
JDS Microprocessing
J&L Information Systems
J&M Microtek
JVC Company of America
Keltronix
Kerotec
Knaus Systems
Lago Systems
Lantronix
Legato Systems
Leunig Communications
Liberty Electronic
Liberty Systems
Linco Computer
Link Technologies
L&L Systems
Longshine Microsystems
LSC
Madge Networks
Maximum Strategy
Maxpeed
Media Logic ADL
Merisel
MetaStor
Metrum Peripheral Products

Meyer Hill Lynch
Mian Data Systems
Micom
Micro Design International
Microplex Systems
Micro Technologies International
Microtest
Microtext Systems
Migration Solutions
Milan Technology
Mobius Computer
Morning Star Technologies
Most
MountainGate Data Systems
Multi-Tech Systems
Mylex
National LAN Exchange
National Peripherals
Network Appliance
Network Computing Devices
Network Equipment Technologies
Network Express
Newbridge Microsystems
Northwest DevTech
Nupon Computing
Open Connect Systems
Optical Data Systems
Optimum Data
Opus Systems
Pagine
PairGain Technologies
The Panda Project
Paralan
Parity Systems
The Peak Technologies Group
Penril Datability Networks
Performance Technology
Peripheral Systems (PSI)
Persoft
Phase X Systems
Pika Technologies
Pinnacle Data Systems
Pixelworks
PMC-Sierra
Printec Enterprises (PEI)
Proteon
Pyramid Technology
Qualix Group

Computer network

Quantum
Qume
Racal-Datacom
Radlinx
Radstone Technology
Raritan Computer (RCI)
R-Cubed
R & D Computers
Relisys
Retix
Rockwell International
Rose Electronics
Sales Systems
Sanar Systems
Sangoma Technologies
SBE
SciNet
Sci Tran Products
Sea Change
SEEK Systems
Semaphore Communications
Shaffstall
Sherwood Technologies
A.T. Shindler Communications
Signatec
Silicon Graphics
R.E. Smith
SMS Data Products
The Software Group
Solflower Computer
Solunet
Specialix
Spectra Logic
Stallion Technologies
StarTech Computer Products
Star Technologies
Storage Computer
Storage Concepts
Storage Dimensions
Storage Solutions
Storage Solutions (SSI)
StorageTek
Sun Microsystems
SunRiver
Sycard Technology
Systech
Systems Beyond
Systran
Tandem Computers
Tatung
TD Systems
Tech-Source
Telamon
Telebit
Ten X Technology
Terminals, Etc.
Texas ISA
Thomas Higgins
3Com
Toray Marketing and Sales
Transitional Technology
TranSys
Tricord Systems
Tryonics
Tucson Computer
TW Enterprises
Ungermann-Bass
Unison Information Systems
Unisys
United Computer Group
U.S. Design
Valitek
Valley Forge Computer
Vecmar International
Vector Technology
Venture Fourth
Vertex Industries
The Virtual Group
Washburn & Company
Western Data Entry Systems
Wheatstone MicroSystems
Workstation Solutions
Worldata
WYSE Guys
Wyse Technology
Xircom
Xylan
Xylogics

Computer parts and accessories

168 Club
Acecad
ACE Micro
AcerOpen
Ackerman Computer Sciences (ACS)

Computer parts and accessories

ACS
Adaptec
Advanced Computer Services
Advanced Digital Industrial
Advanced Educational Systems (AES)
Advanced Micro Solutions
Advanced Portable Technologies
All Electronics
Allen Systems
Alltech Electronics
Alphanumeric Keyboards
American Eltec
American Megatrends
American Micro Products Technology
AmRam
Antec
Apex PC Solutions
Apex Technology
Arcom Control Systems
Arista
ASK Technology
Associates Computer Supply
Atlanta Signal Processors (ASPI)
A-Trend Technology
ATTO Technology
Aurora Computech
Avalon Micro Liquidators
Big Blue Products
Bisme Computer
Blue Star Computer
BMC Communications
Bogen Communications
BWC
Cactus Logic
California PC Products
Cambridge Microprocessor Systems
Capital Resource Recovery
The Case Depot
Centurian Surplus
The Cerplex Group
The Chip Ship
Circo Technology
Cirrus Logic
Coactive Aesthetics
Columbia Data Products
Compass Design Automation
Comptek Computer
Compu America

Compudoc
Comp-U-Plus
Compustar Computers
Computek Systems
ComputerBoards
Computer Classics
Computer Component Source
Computer Gate International
Computer Junction
Computer Parts & Pieces
Computer Products
Computer Recyclers
Computer Terminal Repair
Computer Warehouse
Connectware
R.J. Cooper & Associates
Custom Computers
DakTech
Dalco Electronics
Data Communications Service
Data Exchange
DCC Computer Parts Outlet
DC Industries
Dexon Systems
DFA
Digi-Rule
Easytech Computer
Electronics Design Specialists
ELSA
EMJ Data Systems
Envoy Data
Equinox Systems
Expen Tech Electronics
FarPoint Communications
First Source International
Fittec Electronics
FOB Tech-Specialties
Focus Computer Center
Focus Enhancements
FTG Data Systems
Full-Ten International
Future Domain
Future Micro
Gage Applied Sciences
General Service Computers (GSCI)
Global Computer Sources (GCS)
Globetek
Granite Digital

361

Computer parts and accessories

Hagstrom Electronics
H. Co. Computer Products
HD Computer Products
Hi-Tech Component Distributors
Hostlink
HSB Computer Laboratories
IDER (International Digital Equipment Research)
Image Computers
InBus Engineering
Indus Group
Infogrip
Innovation West
International Electronic
International Parts & Systems (IPSI)
ITAC Systems
Jameco Electronics
Kahlon
Kaitek Engineering
Key Computers
Kloot Computer
Jeffrey A. Krinsky
La Paz Electronics International
Lyben Computer Systems
TA Macalister
Maplin Electronics
Marrick
Matrix Portable Computer Services
Maxpoint Computers
MDL
Megabytes
Megatech
MEI
Merchant Data Systems
Metrocom
Micro-Assist
MicroCache
The Micro Group
Micro Media Computers
Micro Parts & Supplies
MicroTech
Micro Trade
Midwest MAC
Midwest Micro-Tek
Milford Instruments
Mitsumi Electronics
Modern Gold Enterprises
Moneke
Motherboard Discount Center

Motherboards International
Motorola
MultiMedia Direct
National Peripheral Services
Nationwide Computer Distributor
NetFRAME Systems
Net Works
Newport Computer Services
Nichi Standard Enterprises
NMB Marketing
Northeast Minicomputer
Oasis Technologies
Overcity International
Pacific Coast Micro
Pacific Computer Products
PartStock
PC Connection
PC Service Source
Pre-Owned Electronics
Prime Electronic Components
Priority Computer Parts
Professional and Leisure Computers
Quanta Micro
Rigel
Seewhy Computer Systems
S.E.G. Technologies
Shreve Systems
Sierra Systems
SimmSaver Technology
Sky1 Technologies
Softhard Systems Enterprises
Source Innovations
Southern Computer Repair
Spectrum Computer
S.S.R.
StarKom
Suncom Technologies
SunDisk
Sunrise Micro Systems
Symbios Logic
Synapse Micro
Synctronics
TC Computers
Team America
Technology Renaissance
Technology Works
Texas Hardware
Total Control Products

Computer printers and scanners

Touche Touch Pad
Treasure Chest Peripherals
Trilogy Magnetics
Trimm Technologies
T & W Computer
Under the Wire Electronics
URS Information Systems
USA Computer Connections
WestCoast Computer Parts
Wetex International
Worldwide Technologies
WR Consultant Associates
Xecom
XXera Technologies

Computer power protection

Citel
Curtis Manufacturing
Cylix
Delta Warranty
Harris Computer Systems
Panamax
Power Pros
Recoton
RF Engineering and Systems
Spi-Ro Manufacturing
TDL Electronics
Zero Surge

Computer printers and scanners

AC Enterprises
American Ribbon & Toner
Atlantic Logic South
Automated Tech Tools
Brother International
Buffalo
CAD & Graphics
Canon
Chinon America
Computer Automated Technology
Computer Friends
Computer Service Center
Corporate Raider
CoStar
Dataco
Data Reliance
dpi Electronic Imaging
Dytec Technologies

Epson
Genicom
C. Hoelzle Associates
Item
Kroy
Kyocera Electronics
Laser Impact
Laser Master
Laser Products
Laserstar
Lexmark International
Mandeno Electronic Equipment
Market Point
MicroLogic Systems
Microtek
Minolta
Mustek
North American CAD
O'Gara Associates
Okidata
Optronics
Peripherals Unlimited
P.I.E. Engineering
Primax Electronics
The Printer Connection
Printer Plus
The Printer Works
Publishing Perfection
Océ-Bruning
O.S. Computers
QMS
Results Computer
RMSE
SC Engineering
Seiko Instruments
Sentex
Sermax
Shenandoah Equipment
Source Graphics
Southern Technical
Spectron Systems
Storm Software
Synergistic Solutions
Technologic Systems
Umax
Wacom Technology
Xionics Document Technologies

Computer systems

Computer systems

111 Systems
AA Computertech
AB Computer
Aberdeen
ABS Computer Technologies
ABTECH
Acer
Acma
Acom
ACS Equipment
Advance
Advanced Equipment Products
Advanced Logic Research
Advantage Microsystems International
A.I. Credit
Alpha International Business
Alternative Computer Products (ACPC)
Alternative Computer Solutions
A Matter of Fax
American Micro Computer Center (AMCC)
American Wholesale Center
Americomp
AmeriData
AMEX Computers
AMPAC Computers & Software
AMS Direct
A-Plus Computer
Apple Computer
The ARC
Arlington Computer Products
ARM Computer
ASAPc Direct
Aserton
Ashtek
Aspen Computer Products
AST Research
A2Z Computers
AT
Atari
Atlantic Select
AT&T
Austin Direct
Barnett's Computers
BIT Computer
Bits & Byte Technologies
Blackrock Hardware Store
BNFE
Boca Computer Technology (BCT)
Brant Computer
BSI
Bulldog Computer Products
Business System Solutions
CAD Warehouse
CalTex Computer
Capital Computer Group
Capricorn Capital Group
Carlson Computer International
Carolina Computer Concepts
C & A Systems
C.B.S. of Tidewater
Centari Technologies
Centurion Technology Systems (CTS)
CMO
Compaq Computer
Compaq Works
Comp Express
Complete Systems-N-More
Compucon Computers
Compu.D International
CompUSA Direct
CompuShack
Computer Add-Ons
Computer and Software Sales
Computer Brokerage Service (CBS)
Computer Commodity (CCI)
Computer Connection
Computer Discounters
Computer Discount Warehouse
Computer Goldmine
Computerlane
Computer Leasing Exchange
Computer Marketing International
Computer Marketing Investments
Computer Network Systems
The Computer Outlet
Computer Parts Unlimited
Computer Reset
Computer Salvage Discount
Computers Direct
Computers, Parts, and Commodities (CPAC)
Computer Trade Exchange
Computerware Plus
CompuWorld
Comtrade

Computer systems

Connecticut Microcomputer
Continental Computers
C.S. Source
CyberMax Computer
Dallas Computer Parts
Damark
Ted Dasher & Associates
Data Express Systems
Datalux
Data Station
Dell
Delta Technologies
DFI (Diamond Flower)
Digilog Electronics
Digital Data Systems and Communications
Digital Equipment (DEC)
Digital Integrated Systems (DIS) Computers
Direct Ware
DirectWave
Dolch Computer Systems
Dollar Computer
DTI Factory Outlet
DTK Computer
Eagle Resource Marketing
Eclectic Computer Services
Electrified Discounters
Electrocom Communication Systems
Encore Computer Systems
EPS Technologies
Ergo Computing
Erin Multimedia
ESS Computer
Eurocom
Eurocomp
Everex
Expert Computers
Expotech Computer
Express Computer Systems
Fabrini Technology
Fairfax
Far Electronics
Fifth Ave. Computers
1st CompuChoice
First Computer Systems
First International Computer
5D Technology
Forz
Fujitsu Microelectronics

Future Micro
Galaxy Computers
Gao Systems
Gateway 2000
Global Computing
GTSI (Government Technology Services)
Harmony Computers
Hartford Computer Group
Haven Industries
Heartwood Computer
Hewlett Packard
Hi-Tech USA
H & J Electronics
HyperData
IBIP
IBM
IBM Credit
Image Systems (ISI)
Infomatic
Integrix
Intellicomp Technologies
Intergraph Computer Systems
Interlynx Computer Systems (ICS)
International Business Systems
InterPro Microsystems
IPC Business Centre
IPC Technologies
Island Micro Systems
Izzy's PC Systems
Jade Computer
JDR Micro Devices
Jinco Computers
J & R Music World
J.S. Computer
J.T. Computer Marketing
Kenosha Computer Center
Lake Erie Systems and Services
Laurel Computer Systems
Legend Micro
LNJ Enterprises
Lynnhaven Custom Computers
Magitronic
Magnum Computer Products
Marathon Computer
Market Value Programs
MasPar Computer
Master Lease
Maxim Technology

Computer systems

Maximus
MBS
McCormick Computer Resale
MD & I Computers
MEC Computer Express
MegaBytes
Megacomp
Micom
Micro City
Micro Computer Center
Micro Exchange
Microintel
Micro International
Micro Mart
Micron Electronics
Micro Pro
MicroProfessionals
Microsource
Micro Time
MicroXperts
Micro X-Press
Midland ComputerMart
Midwest Computer Brokers
Midwest Computer Electronics
Midwest Electronics
Midwest Micro
Minnesota Computers
MIS
Mitra
Modular Communication Systems
Monterey Bay Communications
Monument Computer
MPC Technologies
Multimedia Enterprises
Multiwave Technology
NAi (National Advantages)
National Computer Resources
Nationwide Computers Direct
NCA Computer Products
NEC
NECX
Net Computers International
Next Generation
Next International
Nico Systems
NICS (New Interactive Concept Systems)
NIE International
Novacomm
Nova Computers
NYCE
Omni Data Systems
One Stop Micro
Optima
Optimal Computing Systems
Pacific Computer Exchange
Pacific Electronics
Packard Bell
Page Computer
Panasonic
Paramount Computer
PC Computer Solutions
PC Direct
PC Importers
PC Interactive
PC International
PC Parts
PC Power
PCs Compleat
The PC Zone
Peak Computing
Peninsula Computer
J.C. Penney
Percomp Microsystems
Polywell Computers
Premio Express
Professional Technologies
ProGen Systems
ProStar
Proven Technology Group
Publisher's Toolbox
Quantex Microsystems
Quark Technology
Radio Shack
RCR
Reckon Data Systems
Reliable Computer Parts
Robotech
Royal Computer
Sager
Saguaro Computer Services
Samsung
Sears
Sequent Computer Systems
Sequoia Systems
Shoppers Market
Signature Computer

Electronics kits

Southwest Computers
Spectra
SRS
Star Computer
Stargate Microsystems
Starlite Electronic
Starquest Computers
Sun Data
Sunset Computer Services
Sunshine Computers
Surplus Supply
Swan Technologies
Tagram System
Techmatix
Technology Distribution Network
TechnoParts
Tekserve
Tempest Micro
Texas Instruments
Texas Microsystems
Todays Computers
Top Data
Toshiba America
Tredex
Trident Computer
TrueData Products
Tucker Electronics & Computers
Twin Cities Digital
Twinhead
United Computer Group
Universal Sales Agency
USA Flex
US Computer
UTI Microtree
V88 Computer Systems
Vektron International
Venture Fourth
Victory Trading
Washburn Computer Group
WinBook Computer
Hank Winter & Associates
Wonderex
Workstations International
Worldwide Discount Computers
Yokohama Telecom
Yorkshire Computers
Zenith
Zenon Technology

Zeos

Crystals

CAL Crystal Lab
CEH/Kirby Enterprises
Crystek Crystals
Crystronics Pager Parts Unlimited
CW Crystals
Frequency Management Crystals
Global Electronics International
Hy-Q International
International Crystal
JAN Crystals
NKX
Pagecorp Industries (PCI)
Sentry Manufacturing

Educational

Educational Insights
G-Vox
Heath
M & G Electronics
The Saelig Company
Square 1 Electronics
TECI

Electronics kits

Agrelo Engineering
AMC Sales
Anderson Electronics
Audio by Van Alstine
Audio Crafters Guild
Auskits
Bagnall Electronics
Bear Labs
Berger Electronics
Big Briar
Bull Electrical
C & A Electronics
Communication Concepts
Electrokraft
The Electronic Goldmine
Electronic Rainbow
Elenco Electronics
Gats Electronics
Gleason Research

Electronics kits

Hart Electronic Kits
Jameco Electronics
JCG Electronic Projects
J Comm
Kanga US
KD4YBC
Kilo-Tec
Magenta Electronics
Maplin Electronics
Marchand Electronics
Mark V Electronics
North Country Radio
Northeast Electronics
Old Colony Sound Lab
Pan-Com International
Progressive Concepts
Quasar Electronics
Radio Adventures Corp. (RAC)
Ramsey Electronics
RTVC
Sheffield Electronics
S&S Engineering
Suma Designs
SVS
Ten-Tec
Tentronix
Vintage Radio Co.
Weeder Technologies
Welbourne Labs
Whiterock Products
Xandi Electronics
Zorin

Electronics parts dealers

Allied Electronics
Alltronics
Andrew Electronics
APEM
Appleseed Electronics
Arrow Electronics
Audio Video Parts
AVEL Transformers
AVX
N. R. Bardwell
Bergquist
BHA Trading
Capital Electronics

Circuit Specialists
Cirkit Distribution
Cititronix
Compelec
Component Technology
The CR Supply
C & S Electronics
Peter Dahl
Dee Electronics
Del Electronics
DH Distributors
Digi-key
Diversified Parts
Eagle Distributors
Electronics Plus
E-Mark
ESR Electronic Components
Fertik's Electronics
Flame Enterprises
John Ford
Fox International
GMB Sales
Herman Electronics
IME
Insight Electronics
Jabbour Electronic Suppliers
Jameco Electronics
JCPenney
J&J International
Johnson Shop Products
K&D Electric
Lucas Transformer
Maplin Electronics
MCM Electronics
Milo Associates
Mis-Cyn Electronics
Mouser Electronics
Newport Digital
Omni Electronics
Pacific Coast Parts Distributor
Panson Electronics
The Parts Connection
Parts Express
Michael Percy
Purchase Radio Supply
Quality RF Services
Radio Component Specialists
Radio Shack

Electronics parts manufacturers

Red Castle Services
The RF Connection
Service Trading
Solen
Tee Vee Supply
Tower Electronics
Tritronics
Union Electronics

Electronics parts manufacturers

AAVID Thermal Technologies
Abbott Electronics
ACK Technology
A/D Electronics
Advanced Computer Products
Advanced Linear Devices
Advanced Micro Devices
AJ's Power Source
Alpha Semiconductor
Altera
AMD (Advanced Micro Devices)
Amega Technology
AMI (American Microsystems) Semiconductors
AMP
Amphenol
Amptech
Amtel
Analog Devices
Apollo Display Technologies
ARK Logic
Asahi Kasei Microsystems
ASCOR
AT&T
Authorized Parts Company
Babcock Display Products Group
Barker & Williamson
Floyd Bell
F.W. Bell
Benchmarq Microelectronics
Bliley Electric
Bourns
Broadcom
Burr-Brown
B/W Controls
Byers Chassis Kits
Caddock Electronics
Calex Manufacturing
Cambridge AccuSense
Carlingswitch
Cedko Electronics
Central Semiconductor
Centronic
Champion Technologies
Chelco Electronics
Cherry Electrical Products
Cherry Semiconductor
C&K Components
Coilcraft
Comet North America
Comlinear
Computer Products
Conversion Devices
Copley Controls
Cornell Dubilier
Crystaloid
C Sys Labs
Cytec
Dallas Semiconductor
Data Display Products
DATEL
Densitron
Dialight
Digital Semiconductor
Dinexcom
EAO Switch
Earth Computer Technologies
East Coast Transistor Parts
Scott Edwards Electronics
Elpac
Emco High Voltage
Endicott Research Group
Epson
E-Switch
E-T-A Circuit Breakers
ETI Systems
Eurom Flashware Solution
EXAR
GBC
Gilway Technical Lamp
Gould Shawmut
Grayhill
Hantronix
Harris Allied
Hobbs
Hoyt Electrical Instrument Works

Electronics parts manufacturers

Hybrids International
ILC Technology
Illinois Capacitor
Industrial Electronic Engineers
Integrated Circuit Systems (ICS)
Integrated Device Technology (IDT)
International Instruments
International Power Sources
Interpoint
IQD
IRC
Ishizuka Electronics
ITT Schadow
ITU Technologies
ITW Paktron
KIC
Kilovac
Lamp Technology
Lansing Instrument
Lattice Semiconductor
LDG Electronics
Leader Tech
Ledtronics
LG Semiconductor
Linear Systems
Linear Technology
Littelfuse
Logical Systems
Lumberg
Lumex Opto/Components
Lumitex
Marktech International
Maxim Integrated Products
Melcher
Melcor
MF Electronics
MGR Industries
Micrel Semiconductor
Micro Linear
Milwaukee Resistor
Mini-Circuits
Mitsubishi Cable America
Mitsubishi Electronics America
Motorola
Murata Electronics
Murrietta Circuits
National Semiconductor
Newport Technology

NKK Switches
Noritake
North American Capacitor
N.T.E. Electronics
OKI Semiconductor
Orbit Semiconductor
Oslo Switch
Parallax
Parker Hannefin
Pico Electronics
PolyPhaser
Portage Electric Products
Potter & Brumfield
Power Convertibles
Power-Lab
Prem Magnetics
Purdy Electronics
QT Optoelectronics
Quality Semiconductor
QuickLogic
Raytheon Electronic Systems
Renco Electronics
RF Monolithics (RFM)
RF Parts
RO Associates
Rolyn Optics
Rose Enclosures
Samtec
Schott
Schurter
Seiko Instruments
Semtech
Senix
Sescom
SGS-Thomson Microelectronics
Shogyo International
Shurter
Siemens
Signal Transformer
Siliconix
Siliconrax
Silicon Systems
Simpson Electric
Sipex
S-MOS
Sony Electronics
Sound Quest
Standard Microsystems

Home stereo dealers

Starled
Synergy Semiconductor
Talema Electronic
Technological Arts
Teledyne Solid State Products
Telefunken Semiconductors
TelTec
Temic
Teradyne
Texas Instruments
Thaler
Toko America
Toroid Corp. of Maryland
Triad Transformers
Ulveco
Unipower
United Technologies Microelectronics Center
Unitrode Integrated Circuits
Valpey-Fisher
Vectron Technologies
Vicor
Victoreen
VLSI Technology
Wall Industries
Xicor
Xilink
Yamaha Electronics
ZeTeK PLC

Fiberoptics

Alcoa Fujikura
Corning
Dolan-Jenner Industries
Fibertron
Gould Fiber Optics
Lightline Engineering
Methode Electronics
Moore Diversified Products
Ortel
Philtec
Pirelli Cable
Sceon Lighting Systems
SoLiCo (Sorenson Lighted Controls)
Times Fiber Communications

Home stereo dealers

ABC (American Buyer's Club)
Abe's of Maine
AC Components
ACE Audio
Accurate Audio Video
ACPEAM
Adcom
A&D Electronics
Alpha Stereo
American Audio & Video
Amkotron Electronics
Klay Anderson Audio
Arnes Radio
Audio Classics
Audio Excellence
Audio Nexus
Audio Outlet
Audioscan
Audio Solutions
Audio Sound
audio studio
Audio Unlimited
Autech
Authorized Audio/Video/Car
Autronic
AVG
B and B Enterprises
Beach
Borbely Audio
Bri-Tech A/V
C & A Audio
Cambridge Soundworks
Camera Sound
Camera World of Oregon
Canseda
Cititronix
Coast to Coast
Colonel Video & Audio
ComputAbility Consumer Electronics
COSMOSIC
Hal Cox
Damark
DataVision
DAVES (Direct Audio Video Entertainment Systems)
Definition Audio Video
Denon Electronics
Disco & Lydteknik (DLT)
Eastern Audio

Home stereo dealers

Electronics Depot
Executive Photo and Electronics
Express Computer & Video
Factory Direct
Family Photo & Video
John Ford
Free Trade Video
Frydendahl Hi-Fi
Gabriel Video & Camera
Francisco Ganán
Global Video
Gre-Trading
Njal Hansson
The Happy Medium
Haven Industries
HCM Audio
Hi-Fi Entusiasten
Hi Fi Exchange
Hi-Fi Farm
Home Theatre Systems
Ideal Acoustics
Image Communications
Independent Audio
Innovation Specialties
JCPenney
Jolly Sound
J & R Music World
Kinovox
Korbon Trading
KT Radio
LAT International
Legacy Audio
David Lewis Audio
Luxman Electronics
Lyle Cartridges
Lyra
Marine Park Camera & Video
Megasound
Meniscus
Metronome Music
The Microphone Company
MusiCraft
Nady Systems
Needle Doctor
Ohm Acoustics
One Call West
Paramount Video & Audio
The Parts Connection
Parts Express
Peach State Photo
J.C. Penney
Philly's Camera
Powervideo
PrimeTime Video and Cameras
Profeel Marketing
Radio Shack
Reed Brothers
Reference Audio Video
Rotel of America
Samman's Electronics
Savant Audio
Scientific Stereo
Sears
Servis-Elekto-Praha
Sessions Music
6th Avenue Electronics
Smile Photo & Video
Soborghus Hi-Fi
The Sound Approach
Sound City
Sound Ideas
The Sound Seller
Stepp Audio Technologies
Stereo Classics
The Stereo Trading Outlet
Stereo World
Superior Audio
Supreme
S&W Computers & Electronics
Sweetwater Sound
Systems Design Group
Tape Connection
Tech Electronics
Thorens Iberica
Thrifty Distributors
Tri State Camera & Video
Uncle's Stereo
UNICO
Universal Video & Camera
University Audio Shop
USA Direct
Video Direct
Video Necessities
Vidicomp Distributors
West-Tech
Wiscom Auto International

Home stereo manufacturers

Acoustic Research
Advent Electronics
Aiwa
Akai
Allpass Technologies
Yakov Aronov Audio Lab
Audio Crafters Guild
audio research
AudioSource
Bang & Olufsen
BBE Sound Inc.
Bear Labs
B&K Components
Bose
Bryston
C & A Electronics
Canton
Carver
Chase Technologies
Conrad-Johnson Design
Dale Pro Audio
DB Systems
Electronics Hospital
Emerson Radio
Fisher
G&D Transforms
GE
Harmon/Kardon
Harrison Laboratory
Hart Electronic Kits
HHB Communications
Intensitronics
JVC Company of America
Kenwood
Kinergetics Research
Legacy Audio
Linn Hi-Fi
Madrigal Audio Labratories
Magnavox
Marantz
Marchand Electronics
Mark V Electronics
Matsushita (Panasonic)
McIntosh
Meridian America
Monarchy Audio
Mondial Design Limited
Musical Concepts
NAD
Nagra
Nakamichi America
North Country Radio
Old Colony Sound Lab
Onkyo
Orca
Ortofon
Panasonic
Parasound Products
Philips Consumer Electronics
Picotronic
Pioneer Electronics
Polyfusion Audio
RTVC
Sansui
Sanyo Fisher USA
SDI Technologies
Sennheiser
Sharp
Sherwood
Siemens
Sonic Frontiers
Sony Electronics
Soundstream Technologies
Storm Power Sound
Sunfire
Teac
Technics
Technologie MDB
Tentronix
VAC (Valve Amplification Co.)
Welbourne Labs
World Audio Design
Yamaha Electronics
Yorx Electronics

Home video dealers

ABC (American Buyer's Club)
Abe's of Maine
Accurate Audio Video
A&D Electronics
AlTech
Aljon Technologies
Alpha Stereo

Home video dealers

American Audio & Video
Klay Anderson Audio
A.S. & S.
Audio by Van Alstine
audio studio
Authorized Audio/Video/Car
B and B Enterprises
Beach
Berger Brothers
Bri-Tech A/V
Camera Sound
Camera World of Oregon
Coast to Coast
Colonel Video & Audio
ComputAbility Consumer Electronics
Crutchfield
Damark
DataVision
DAVES (Direct Audio Video Entertainment Systems)
Definition Audio Video
Electronic Mailbox
Emerald Vision & Sound
Executive Photo and Electronics
Express Computer & Video
Factory Direct
Family Photo & Video
Free Trade Video
Frontgate
Gabriel Video & Camera
Genesis Camera
G&G Technologies
Global Video
Halcyon Group
Haven Industries
Home Theatre Systems
Innovation Specialties
J & R Music World
Marine Park Camera & Video
Markertek Video Supply
Marketing by Design
Megasound
MusiCraft
Nady Systems
One Call West
Paramount Video & Audio
Peach State Photo
J.C. Penney
Philly's Camera
Powervideo
PrimeTime Video and Cameras
Profeel Marketing
PRV Sales
Radio Shack
Reference Audio Video
Samman's Electronics
Savant Audio
Sears
Sessions Music
6th Avenue Electronics
Smile Photo & Video
Sound City
Supreme
S&W Computers & Electronics
Thrifty Distributors
Tri State Camera & Video
Uncle's Stereo
Universal Video & Camera
University Audio Shop
USA Direct
Video Direct
Video Discount Warehouse
Video Necessities
Videotique
Vidicomp Distributors
World Trade Video

Home video manufacturers

Advent Electronics
Alliance Rotators
AmPro
Angstrom
Black Feather Electronics
Brandess-Kalt
Brookline Technologies
Carver
Channel Master
Chyron
Citizen
Comtrad Industries
Da-Lite Screen Company
Dolgin Engineering
D-Vision Systems
Funai USA
Future Video Products
GE

Industrial computer and electronics

Generad
Glidecam Industries
Goldstar Electronics International
Go Video
Hotronic
JVC Company of America
Laserdisken
Magnavox
Matsushita (Panasonic)
Micro Thinc
Minolta
Mitsubishi Electronics America
North Country Radio
Panasonic
Philips Consumer Electronics
Phillips-Tech Electronics
Picotronic
Pioneer Electronics
Planet Electronics
Quasar
Runco
Samsung
Sharp
Sony Electronics
Stewart Filmscreen
Supercircuits
Surround Sound (SSI)
Teknika Electronics
Ten Lab
Terk Technologies
Vantage
Vic Hi-Tech
VideoLabs
Videonics
Vidikron
Virtual Vision Sport
Visual Circuits
Zenith

Industrial computer and electronics

Acqutek
Acromag
ACR Systems
Action Instruments
Adastra Systems
Advantech
AEROTECH
Alligator Technologies
American Advantech
Amot Controls
Amplifier Research
Anritsu America
Antenna Research
APPRO International
Aries Electronics
Arnet
Ashcroft
Axiom Technology
Azonix
Barco Chromatics
Barksdale
Baumer Electric
BEI Sensors & Systems
Bi-Link Computer
Bittware Research Systems
Bustronic
Cablemaster
California Instruments
Camintonn
Campbell Scientific
Capteur Sensors & Analysers
Cincinnati Electronics
Com-Power
Cooper Instruments & Systems
Crydom
Crystal Group
Dawn VME Products
DC Micros
Decibel Instruments
Directed Energy
Diversified Technology
Dolan-Jenner Industries
Dranetz Technologies
Eastman Kodak
EG & G IC Sensors
Eldre
Electro
Electromatic Equipment
Elma Electronics
EMAC
EPIX
FieldWorks
Figaro USA
Four Pi Systems
Gemco

Industrial computer and electronics

Gems Sensors Division
Gould Instrument Systems
Green Spring Computers
Gulton Statham
Hyde Park Electronics
IBI Systems
ICS Electronics
ILC Data Device
ImageNation
IMO TransInstruments
Inframetrics
Instek
Intec Inoventures
Integrand Research
Integrated Solutions
Intel
Intelligent Instrumentation
Interface Technology
International Computers
ITT Barton
ITT Pomona
JK Microsystems
Johnson Yokogawa
Kessler-Ellis Products
KIC Thermal Profiling
K-Tec
Lake Monitors
Leica
Linear Laboratories
Logan Enterprises
Logical Design Group
Macrolink
Magtrol
Marshall Industries
Max Machinery
McMillan
M/D Totco Instrumentation
Megatel
Microbus
Micro Pneumatic Logic
Microstar Laboratories
Micro/sys
Minco Products
Mizar
Moore Products
Morehouse Instrument
M-Systems
Multi-Micro Systems

MVS
National Instruments
NEFF Instrument
New Micros (NMI)
Omega Technologies
OmniTester
Onset Computer
OPTIM Electronics
Pacific Avionics
Pacific Power Source
PCB Piezotronics
PEP Modular Computers
Philtec
PMC
PMC/BETA Limited Partnership
PowerTronics
Princo Instruments
Pro-Log
Prologic Designs
Proteus Industries
Qualimetrics
Racal Recorders
RCC Electronics
RDP Electrosense
Recortec
Rustrak
Sciemetric
Scientific Technologies (STI)
Sejus
SensorData Technologies
Sensotec
Silicon Designs
Sky Computers
Sony Mangnescale America
Spectron
Stanford Research Systems
Summit Instruments
Suncoast Technologies
Synergy Microsystems
Teknor Microsystems
Tern
Testronics
Themis Computer
Toray Marketing and Sales
Toronto Microelectronics (TME)
Transducer Techniques
TransInstruments
TransLogic

Transmation
Tri Valley Technology
Two Technologies
Valhalla Scientific
Vitrek
Voltech
Waddington Electronics
Wandel & Goltermann
Watlow Anafaze
Watlow Electric
Weed Instrument
Win Systems
Xitron Technologies
Z-World Engineering

Lasers

Hi-Q Products
MWK Industries
SVS
Timeline

Lighting systems

Frezzolini Electronics
Ikelight
Jasco
Kalimar
Lenmar Enterprises
Lowel-Light Manufacturing
Proflex Lighting
Pro Sound & Stage Lighting
Rolls Music Center
Sam Ash Music Stores
Satter
Silicon Sound
Smith-Victor
Sunpak
Thoroughbred Music
Vanguard
VDO-Pak

Loudspeaker manufacturers

Acoustic Research
a/d/s
Advent
A&S Speakers
Allison Acoustics

Loudspeaker manufacturers

Altec Lansing
Ambiance Acoustics
Apogee Acoustics
Atlantic Technology
Atlas/Soundolier
Audio Art
Audio Artistry
Audio Car
Audio Design Innovations
Audio Gods
Audio Influx
audiophile
Audio Products
AudioSource
Audison USA
Avalanche
Bang & Olufsen
Barbetta
BIC America
B.K. Electronics
Bose
Boss/AVA
B&W Loudspeakers of America
Cadence
Canon
Canton
Celestion
Cerwin-Vega
Clif Designs
Cobalt Industrial
Collins Sun Electronics
Concord
Coustic
Craig Consumer Electronics
Critical Mass
Crunch
Dansk Audio Teknik
dB Speakerworks
Definitive Technology
Design Acoustics
Diablo Acoustics
Diamond Audio
Digital Phase
Directed Electronics
Dr. Crankenstein
Dynamic Control
Dynaudio
dynaudio acoustics

Loudspeaker manufacturers

Eclipse
Eminent Technology
Excalibur
Focal America
German Physiks
Goetz/Vanderschaaf
Gold Sound
Hales Design Group
Halladay Acoustics
Harrison Laboratory
HiFonics
Hitron
HSU Research
HTP
Infinity
Jamo Hi-Fi
JBL
Jensen
JL Audio
KEF Electronics of America
Kenford
Kenwood
KLH
Klipsch
Koss
KRK Monitoring Systems
Lanzar
Legacy Audio
Macrom
Madisound Speaker Components
Martin Logan
MB Quart
McIntosh
Miller & Kreisel Sound
Mirage
MMATS Professional Audio
Mobile Authority
Morel Acoustics USA
MTX
NHT (Now Hear This)
Nichi Standard Enterprises
Niles Audio
North Creek Music Systems
NSM
Orion
Paradigm Electronics
Paramount Audio
Petras

Philips Consumer Electronics
Phoenix Gold
Polk Audio
Polydax Speaker
Power Acoustic
Proac USA
ProSystems
Proton
PSB Speakers
P&W International
Pyle
Pyramid
Q Components
RBH Sound
Recoton
Rockford Fosgate
Rockwood
Ross Systems
R&T Enterprises
SC&T International
SEA
SOAT
Sonance
Sound Dynamics
SoundTech
Soundwave
Southern Audio Services Inc. (SAS)
Speaker City, USA
Speakers Etc.
Speakerworks
Speaker World
Spica
STD Entertainment
Stillwater Designs
Surround Sound (SSI)
Tannoy/TGI
TC Sounds
Thiel
TM Rush
Triad Speakers
Ultimate Sound
Ultralinear
Urban Audio Works
U.S. Black Magic
USD Audio
Vandersteen Audio
Velodyne Acoustics
VMPS Audio Products

Miscellaneous

Westlake Audio
White Electronics
World Audio Design
Zalytron Industries

Magnetic recording

3M
American Video Tape Warehouse
Ampex
Andol Audio Products
ARCAL
AVI
BASF
Burlington A/V Recording Media
Cassette House
Diskette Connection
Diskettes Unlimited
Diversified Systems Group
Excel Technology
EXXUS
Greencorp USA
Inmark Industrial
Interface Systems
JAC Industrial
Mann Endless Cassette Industries
Maxell
Media Factory
Media Source
Mediastore
Midwestern Diskette
National Recording Supplies
Nichi Standard Enterprises
Now! Recording Systems
Project One AV
ProMedia Computer Supplies
RDL Acoustics
Recording Systems
Recoton
Snell Acoustics
Sound Investment
SoundSpace
Steadi Systems
Tape World
TDK
Travan Technology
The Warehouse
YHC Cassette

Microphones

Bill Bradley Microphones
AKG Acoustics
Audio-Technica
beyerdynamic
Conneaut Audio Devices (CAD)
Earthworks
Electro-Voice
Gentex Electro-Acoustics
GT Audio
ILIADIS
Manley Laboratories
Mic Heaven
Neumann
Recoton
Shure
Stedman
Telex
Wright Microphones & Monitors

Miscellaneous

BMI (Benjamin Michael Industries)
CIMCO
Compac
CompCo Engineering
Component Parts
Comroad
Concept Seating
Creative Micro Electronics (CME)
Cruising Equipment
Davis
Deco Industries
Delta Products
Desco Industries
Diversified Electronics
Domain Technologies
Dynamic Systems
Edmund Scientific
Electrogadgets
Electroman
Enterprise Radio Applications
Franklin Electronic Publishers
Gemeni Industries
GETTECH
Gordon Products
JMS

Miscellaneous

Laser Sensor Technology
Lasiris
Magellan's
Nikon
M.M. Newman
Ohio Automation
Outdoor Outfitters
Paramax
Performance Motion Devices
Pulse Metric
Quantum Composers
Radio Astronomy Supplies
Radio Engineers
Riegl
Science First/Morris & Lee
R. Scott Communications
SkyMall
Smart Choice
Thrustmaster
Videx
Weber-Sterling

Musical instruments and equipment

8th St. Music
Ampeg
Analogics
Analog Modular Systems
Anatek
AOS
Applied Research & Technology
Aria
Armadillo Enterprises
ART
Arteq Sound & Light
Ashly Audio
Audio Village
Audio World
AVR Systems
Barcus-Berry
Bartolini
Basslines
Bedrock Amplification
Bellari
Big Briar
Black Cat Products
Boomerang Musical Products
Buchla and Associates

Caruso Music
Carvin
Casio
Charlie Stringer's Strings
CLE
Cort
Costello's Music
Crate
Daddy's Junky Music
Dave's Guitar Shop
DiscoTronix
Discount Music Supply
DSL Electronics
Echo Park Music
Eclair Engineering Services
Edirol
Elderly Instruments
EMG
Epiphone
Factory Music
Fernandez Guitars
GB Labs
Gibson
G&L
GMP Precision Products
Godin
Goff Professional
Griblin Engineering
Gruhn Guitars
GSF Agency
GT Audio
The Guitar Broker
Guitar Center
Guitar Emporium
Guitar Network
G-Vox
Harper's Guitars
Hi-MU Amplifiers
Ibanez
Interactive Light
Interstate Musician Supply
Jackson Guitar
Kendrick Amplifiers
Korg
Kramer
Kurzweil
Jim Laabs Music Superstore
Leigh's Computers

Ligon Guitars
L&M Music
Los' Music Center
LT Sound
Manny's Mailbox Music
Marshall
Media Mation
The Midi Station
Morley
Mountain International
Mr. Mike's Music
Music Industries
Nadine's Music & Pro Audio
New Sensor
New York Music
Numerous Complaints Music
Opcode
Ovation
Parker Guitars
Paul Reed Smith Guitars
Peavey Electronics
Pro-Rec Synth Sounds
Pro Sound & Stage Lighting
Randall Amplifiers
Resurrection Electronics
Rich Music
Rickenbacker International
Rock'n Rhythm
Rocktek America
Rogue Music
Roland
Rolls
Rolls Music Center
Rondo Music
Roxy Music
Sabine
Sam Ash Music Stores
Scharff Weisberg
Secondhand Sounds
Seymour Duncan
Shivelbine Music & Sound
Soldano Custom Amplification
Sound Deals
Sound Ethix
SoundTech
Southern California Pro Audio
Speir Music
The Starving Musician

Stick Enterprises
Stoneman Guitars
Studio Electronics
Thoroughbred Music
Top Shelf Music
West L.A. Music
Wollerman Guitars
Workshop Records
Wray Music House
E.U. Wurlitzer Music
Yamaha Electronics

Office equipment

ACCO USA
Allegheny Paper Shredders
A Matter of Fax
AMDEV Communications
Ameri-Shred
Camera World of Oregon
Central Technologies
Computerwise
Corporate Raider
Cummins-Allison
Damark
Danka
Duplo
ECCO Business Systems
Executive Photo and Electronics
Fairfax
GE
Gestetner
Image Systems (ISI)
International Electronic
JetFax
J & R Music World
Lanier Information Systems
MBM
Minolta
Neopost
Northern Exposure Office Products
Omation
Ricoh
Royal Copystar
Sears
Shredex
Smith Corona
SNI Innovation

Office equipment

Sony Electronics
Sparton Technology
S&W Computers & Electronics
Tri State Camera & Video

Professional broadcasting

Ace Music Center
AEQ America
Akai
Alcatel Network Systems
Altronic Research
Klay Anderson Audio
Aphex
Arrakis Systems
Audioarts Engineering
Audio Broadcast Group (ABG)
Audio Technologies
Audio Upgrades
Audio Village
Auditronics
Autogram
BAF Communications
Barco Communications Systems
Benchmark Media Systems
BEXT
B&H Audio-Video
Broadcast Devices
Broadcast Electronics
Burk Technology
CCA Electronics
CED
Cellcast
Circuit Research Labs
Comet North America
Comrex
Continental Electronics
Control Signal (CSC)
Crown International
Cutting Edge
Dale Pro Audio
Dan Discolight
DB Elettronica Telecomunicazioni S.p.A.
Dielectric Communications
Digicomm International
Disco & Lydteknik (DLT)
Di-Tech
Dolby Laboratories

EAR Professional Audio/Video
Econco
Electric Works
Electronics Research
Elite Video
E-mu Systems
Energy-Onix
ENI
ESE
Eventide
Fairlight
Fostex
Gentner Communications
Hall Electronics
Hallikaainen & Friends
Harris Allied
Harrison
Henry Engineering
Holaday Industries
Inovonics
International Datacasting
Jampro Antennas & RF Systems
JT Communications
KinTronic Labs
Microelectronics Technology
Moseley Associates
MPR Taltech
MYAT
Nagra
National Sound and Video
Nott
NSM
Orban
Otari
Pacific Recorders & Engineering
Potomac Instruments
Progressive Concepts
Psitech Plus
QEI
Radio Design Labs (RDL)
Reel-Talk
Satellite Systems
Sennheiser
Silicon Valley Power Amplifiers
STS Electronics
Studio Audio Digital Equipment (SADiE)
Studio Consultants
Systems With Reliability (SWR)

Radio-control models and model trains

Tascam
Telefunken Sendertechnik
Telos Zephyr
Thomcast
Videomedia
Whirlwind Music

Professional video dealers

Abe's of Maine
American Audio & Video
Beach
Better Reception
B&H Audio-Video
Coast to Coast
Colonel Video & Audio
ComputAbility Consumer Electronics
Electronic Mailbox
Elite Video
Executive Photo and Electronics
Express Computer & Video
Factory Direct
Family Photo & Video
Free Trade Video
Gabriel Video & Camera
Genesis Camera
G&G Technologies
Global Video
Markertek Video Supply
Marketing by Design
Nadel Enterprises
Nady Systems
NRG Research
Paramount Video & Audio
Peach State Photo
Philly's Camera
Powervideo
PrimeTime Video and Cameras
Profeel Marketing
Pro Sound & Stage Lighting
PRV Sales
Rader Video Products
Selectra
Smile Photo & Video
Sound City
STS Electronics
Studio 1 Productions
Supreme
Tri State Camera & Video
Universal Video & Camera
Video Discount Warehouse
Video Innovators
VideoLabs
Video Necessities
Videotique
Vidicomp Distributors
World Trade Video

Professional video manufacturers

Allegro
AmPro
Arriflex
Azden
Berger Brothers
Bogen Photo
Carlson-Strand
Chimera Lighting
Chyron
Citizen
Cool-Lux
Digicomm International
Digital Processing Systems
Dobbs-Stanford
Dolgin Engineering
D-Vision Systems
EAR Professional Audio/Video
Fairlight
Future Video Products
Generad
Glidecam Industries
Goldstar Electronics International
Go Video
Harrison
Hotronic
JVC Company of America
Mega Hertz
Planet Electronics
Sima Products
Stanton Video Services
Vic Hi-Tech
Videonics

Radio-control models and model trains

ACE Hardware Hobbies

Radio-control models and model trains

Ace R/C
Aero Scientific
Aero Spectra
Aero Sport
Aerotech Models
Airtronics
Ambrosia Microcomputer Products
American Hobby Center
America's Hobby Center
Associated Electrics
AstroFlight
Bodden Model Products
Bruckner Hobbies
B & T RC Products
Calculated Industries
Cannon R/C Systems
Central Model Marketing
Cermark
Charlie's
Charlie's Trains
C&M Engineering
Competition Electronics
Competition R/C Imports
C-Tronics
DAD
ElectroDynamics
Elite Speed Products
Futaba of America
Global Hobby Distributors
Hitec RCD
Hobby Club
Hobby Mart
Hobby Shack
Hobbytech
Hobby Warehouse of Sacramento
Horizon Hobby Distributors
I & G Hobbies
Irwin Performance Products
J & C Hobbies
JEM R/C Electronics Research
Just Hobbies
Kress Jets
Leisure Electronics
The Likes Line
Major Hobby
MC_FD
Measurement & Control Products
Micropace

Mondo-tronics
Morris Hobbies
Novak Electronics
OmniModels
Omnionics
Ott Machine Services
Pappy John's R/C Warehouse
Performance Hobby
Pro-Match
Radio South
Railway Depot
RAm
Ramfixx Technologies
The Ranch Pit Stop
R/C Skydivers
RC Systems
RT Aerospace
SavOn Hobbies
Serpent
S.G.
Sheldon's Hobbies
Smiley Antenna
Southside Hobbys
SR Batteries
Stormer Racing
SubTech
Sullivan Products
Team Orion
Tekin Electronics
Texas Timers
Throttle Up!
TNC Electronics
TNR Technical
Tower Hobbies
Toys for Big Boys
Trinity Products
Vantac Manufacturing
Vantec
Vencon Technologies
Wangrow Electronics
Wheels & Wings, etc.
World Class Batteries
Yokomo

Robotics

Graymark
Model Control Devices

Model Rectifier
Mondo-tronics
A.K. Peters
Precision Micro Dynamics
VersaTech Electronics

Satellite TV/video

BAF Communications
BLR Communications
Channel Master
Chaparral Communications
DBS Satellite Television
DH Satellite
R.L. Drake
Ecosphere
Home Theatre Systems
International Datacasting
Kathrein-Werke
MultiFAX
NBO Distributors
PTL Cable Service
RC Distributing
RMR Electronics
Satellite Systems
Satman
Skyvision
Swagur Enterprises
Sync-A-Link Communications
Triangle Product Distributors USA
Tulsat
Uniden

Solar power

Dallas Solar Power
Direct Power and Water
Hutton Communications
Photocomm
Self-Reliance
Solar Electric
Solar Electric Specialties

Sports electronics

AIS
Radar Sales

Studio recording dealers

Ace Music Center
American Pro Audio
Klay Anderson Audio
Berler Pro-Audio
Audio Intervisual Design
Audio Village
Audio World
AVR Systems
Caruso Music
Echo Park Music
Grandma's Music & Sound
Green Dot Audio
Harborsound & Video
HAVE
Independent Audio
Lexicon
L&M Music
The Means of Production
The Microphone Company
The Midi Station
Milam Audio
National Sound and Video
Lee Pepper Sound Studios
Professional Audio Design
Pure Audio
Rhythm City
Rich Music
Sam Ash Music Stores
Southern California Pro Audio
Speir Music
Stewart Electronics
Strings & Things
Studio Consultants
Sweetwater Sound
Systems Design Group
The Toy Specialists
West L.A. Music
Workshop Records

Studio recording manufacturers

Acoustic Sciences Corp. (ASC)
Adams-Smith
AD Systems
AET
Akai
Alesis
AMS Neve
Anatek

Studio recording manufacturers

api
Ashly Audio
AudioControl
Audio Intervisual Design
Audio Precision
Audio Toys (ATI)
Audio Upgrades
Audio Visual Assistance
Avalon Design
BBE Sound Inc.
Bellari
Boomerang Musical Products
Brent Averill Enterprises
Bryston
CCS Audio Products
Communication Task Group
Crest Audio
Crown International
Curcio Audio Engineering
Anthony DeMaria Labs
Demeter Amplification
Digidesign
DigiTech
East Coast Music Mall
Eclair Engineering Services
E-mu Systems
Euphonix
Fostex
Furman Sound
Gefen Systems
Gold Line
Hafler Professional
John Hardy
HHB Communications
Hydra Audio
The Kaba Group
Mackie
Manley Laboratories
Media Management
Micro Technology
Mytek Digital
Odyssey Pro Sound
Orca
Otari
Pacific Recorders & Engineering
Peavey Electronics
Popless Voice Screens
Premier Technologies
PreSonus Audio Electronics
QSC Audio
Raindirk
Rane
Reamp
Rocksonics
Roland
Rorke Data
RSP Technologies
Russian Dragon
Samson Technologies
Solid State Logic
Sonic Solutions
Sony Electronics
Soundscape
Spectral
Spirit
StarrSwitch
Studer Editech
Studio Audio Digital Equipment (SADiE)
Studio Techniques
Tascam
t.c. electronic
Timeline
TL Audio
Uptown Automation Systems
Vestax
Videomedia
Weltronics
Whirlwind Music
Yamaha Electronics
Yorkville Sound
Z-Systems Audio Engineering

Surplus electronics (parts and more)

All Electronics
Alltech Electronics
Alltronics
American Science and Surplus
A.R.E. Electronic Surplus
Brigar Electronics
C & H
Compac Electronics
Cricklewood Electronics
Davilyn Corp.
DC Electronics
Debco Electronics

The Electronic Goldmine
Fair Radio Sales
Falkner Enterprises of America Inc.
Gateway Electronics
Greenweld Electronic Components
Halted Specialties
International Microelectronics
Javanco
J & N Factors
Johnson Shop Products
Johns Radio
JPG Electronics
Marymac
MECI
MEM Electronics
M.O.O.
Ontario Surplus
Pembleton Electronics
RA Enterprises
Roger's Systems Specialist
R & S Surplus
Skycraft Parts & Surplus
Special Economies
Star-Tronics
Surplus Center
Surplus Sales of Nebraska
Surplus Supply
Surplus Traders
Surplus Trading
Synergetics Surplus
Tartan Electronics
Timeline
Vision & Motion Surplus
Webco Sales

Surveillance and security

A&D Electronics
All Phase Video-Security
AMC Sales
American Innovations
Ar_FD Communications Products
A.S. & S.
Bull Electrical
Cimarron Technologies
Closed Circuit Products
Communications Systems Laboratories
Detection Dynamics
Direct CCTV
EDE
Electrim
Great Southern Security
Information Unlimited
Knight Industries
Lintel Security
McCallie
Novacomm
Otto
Pan-Com International
Personal Security
Quasar Electronics
Security Call
Security Dynamics
Security Trade
Semaphore Communications
Sheffield Electronics
Sonic Communications
Spy Emporium
Spy Outlet
Spy Supply
Suma Designs
Supercircuits
TelGuard
Transcrypt International
Viking International
Virtual Open Network Environment (V-ONE)
Visitect
WIBU Systems
Xandi Electronics

Telephone

ADC Video Systems
Allied Resources
AT&T
Automation & Electronics Engineering (AEE)
Avtec
Bang & Olufsen
Casio
Cobra Electronics
Connect Systems (CSI)
Damark
DataCom
Dexis
Dictaphone
Digital Electronic Systems

Telephone

Fabtronics
Frontgate
GE
JC Marketing
Jescom International
Logical Design Tools
Alex Mastin
Midian Electronics
NCI
Nortel (Northern Telecom)
Omnicron Electronics
Orion Telecom
Pan-Com International
PhoneMate
Primax Electronics
Radio Shack
Scanner World, USA
SNI Innovation
Synergetics Surplus
Telecom Solutions
Uniden
Vive Synergies
Weeder Technologies
Zetron

Test equipment dealers

Beyond Oil
BMC
A-Comm Electronics
All Phase Video-Security
Alpha Electronics
Bartek Electronics
B and B Enterprises
Cooke International
C&S Sales
Danbar Sales
Dexis
Electronics Center
Electro Tool
Feitek
Fordham Radio Supply
Fotronic
G and G Electronics of Maryland
International Components Marketing (ICM)
ISA
J.B. Electronics
Lehman Scientific

LG Precision
Maplin Electronics
MCM Electronics
Metric Equipment Sales
MHz Electronics
Milo Associates
Mueller
Penstock
Phelps Instruments
Processor Technology
Products International
RAG Electronics
Specialized Products
Spectra Test Equipment
Stewart of Reading
Test Equipment Sales
Tucker Electronics & Computers
Vann Draper Electronics
Visual Communications
Western Test Systems
Xytek Industries

Test equipment manufacturers

A&A Engineering
Accurite Technologies
Advanced Testing Technologies (ATTI)
Advantage Instrument Corp.
Advin Systems
AEMC
Alfa Electronics
Allied Electronics
Allison Technology
AllMicro
AlphaLab
Alta Engineering
Amprobe Instrument
Applied Microsystems
Aristo Computers
Avcom of Virginia
Bagnall Electronics
Barfield
Bel Merit
B+K Precision
Bird Electronic
Cablescan
CED
Charleswater

Two-way radio

C & H Technologies
Clarke-Hess Communication Research
Columbia Research Laboratories
Com-Power
CST
Darkhorse Systems
Digital Products
Electronics Design Specialists
Electrotech Systems
Fieldpiece Instruments
Fluke
GN Nettest
Grayson Electronics
Hart Scientific
Hewlett Packard
Richard Hirschman
Hitachi
Hitachi Denchi America
IC Engineering
IFR Systems
Jensen Tools
Keithley Instruments
Kelvin Electronics
LeCroy
Marconi Instruments
Metrix Instruments
Micro 2000 Europe
MicroTools
Multi-Contact USA
Neutrik
Noise Com
North Atlantic Instruments
Optoelectronics
Oxley
Pico Technology
ProBoard Circuits
Programmable Designs
Protek
PWI/MobileFax
Quantum Data
Ramsey Electronics
Sencore
Sigmatest
Signum Systems
Slaughter
Smith Design
A.W. Sperry Instruments
Startek International

Team Systems
Tektronix
Two-Bit Score
Unholtz-Dickie
Videospectra
Wavetek
Windward Products
Yokogawa of America

Two-way radio

ACE Communications
Air Comm
Allen Telecom Group
Allied Resources
American Electronics
Antenna Research
Antennas America
Astatic
Astron
Azden
Brix
Cadex Electronics
David Clark
CLM Technologies
Comm-Net
Comtelco Industries
Connect Systems
CPI
DataCom
Delta Communications
Electro Automatic
Electronics Center
El Paso Communication Systems (Epcom)
ETrunk Systems
Gamber-Johnson
GMRS Radio Sales
Haewa Communications
JPS Communications
Marty's Emergency Products
MDM Radio
M_FD Enterprises
Metroplex Mobile Data
Midian Electronics
Midland International
Modular Communication Systems
Motorola
Mx-Com

Two-way radio

Naval Electronics
NCG Companies
Newmar
Norcomm
Nott
NS Electronics Service
Oar
Omnicron Electronics
Orbacom Systems
Otto
Pana-Pacific
Pyramid Communications
Radio Communications
Radio Communications of Charleston
Radio Communications Wholesalers
Radio Express
RadioMate
Radio Wholesale Marketing
Railcom
Relm Communications
Reynolds Radio
Safari Radio
Scanner World, USA
SGC
Shakespeare
Sharp Communication
SIRIO Antenna
Sports-Communications Dist.
Stanlite
STI-CO Industries
SWS Security
Tad Radio of Canada
Tait Electronics
TX RX Systems
Unex Headsets
Uniden
Vega Signaling Products Group
VersaTel
VoCom Products
West Coast Radio
C.W. Wolfe Communications

VHF/UHF radio scanners

ACE Communications
American Electronics
AOR
Communications Electronics
Computer Aided Technologies
Cotronics
C. Crane
CTP
Discount Electronics CB Sales
EDCO (Electronic Distributors)
EEB (Electronic Equipment Bank)
Emkay Enterprises
Galaxy Electronics
GRE America
Grove Enterprises
Hobby Radio Stop
HPR
Javiation
G.E. Jones Electronics
Marvel Communications
MAX System Antennas
MetroWest
Miami Radio Center
Par Electronics
Pioneer Data
Quement Communications
Radio Shack
Radioware
Railcom
Scanner World, USA
Spectronics
Stewart Electronics
Systems & Software International
Transel Technologies
USRadio
Wintenna

Vintage electronics

Antique Audio
Antique Electronic Supply
Antique Radio Components
Antique Triode
Arlen Supply
Associated Radio
Thomas J. Bruckner
Carolina Tubeworks
C & N Electronics
Custom Autosound
Peter Dahl
Daily Electronics
Dovetron

Vintage electronics

Electronics Plus
Electron Tube Enterprises
ESCR
Fair Radio Sales
International Components
ISA
Kirby
Kurluff Enterprises
Miltronix
Old Tyme Radio
Purchase Radio Supply
Quest Electronics
Radio Electric Supply
Sidney Aircraft Radio Industries
Slep Electronics
Steinmetz
Surplus Sales of Nebraska
Svetlana Electron Devices
A.G. Tannenbaum
Tubes
Turner Electronics
Typetronics
United Electronics
Vacuum Tube Industries
Vintage Radio Co.

Other books by Andrew Yoder

Pirate Radio Stations: Tuning in to Underground Broadcasts
Build Your Own Shortwave Antennas--2nd edition
Auto Audio: Choosing, Installing, Maintaining, and Repairing Car Stereo Systems
Shortwave Listening on the Road: The World Traveler's Guide
Pirate Radio: The Incredible Saga of America's Underground, Illegal Broadcasters
Pirate Radio Operations
The Complete Shortwave Listener's Handbook--5th Edition
Home Audio
Home Video

About the Author

Andrew Yoder is the author of more than 10 technical books on **communications and various** audio topics. He has also written a number of technical articles for such magazines as *Popular Electronics*, *Radio!*, *New Jersey Monthly*, and *Popular Communications*. Mr. Yoder was an electronics editor and associate managing editor at TAB Books from 1990 to 1996.

Get Up to Speed!

With your *Free Subscription* to:

Computer Service & Repair Magazine

... the only publication devoted to technical computer support professionals.

Qualified individuals can receive a complimentary one year subscription to *Computer Service & Repair Magazine* by calling (303) 797-8522 or visiting us on the World Wide Web at www.csrmagazine.usa.net.

VALUABLE DISCOUNT COUPON
FREE COMPUTER PROGRAM

Secondhand Sounds

Free copy of "Home Inventory" with any purchase

Secondhand Sounds: 3867 W. Market St. #188, Akron, OH 44333
330-665-4755 (tel) 330-666-2903 (fax)

VALUABLE DISCOUNT COUPON
$10 OFF ANY PURCHASE OVER $100

Ham Radio Outlet

Limit 1 HRO coupon per customer. This coupon may not be used with any other promotional discount/coupon

Order on-line at http://www.hamradio.com or visit/call our stores.
See pages 139 and 140 for the location of our stores nearest you

---

VALUABLE DISCOUNT COUPON
5% OFF MIDI AND DIGITAL RECORDING HARDWARE AND SOFTWARE

Leigh's Computers

Leigh's Computers: 129 W. Eagle Rd., Havertown, PA 19083
610-896-4414 (fax) 800-321-6434 (tel)

VALUABLE DISCOUNT COUPON
10% OFF CUSTOMER'S FIRST PURCHASE

Cassette House

800-321-5738 (tel)
artmuns@tape.com (e-mail) http://www.tape.com (Web page)

VALUABLE DISCOUNT COUPON
GOOD FOR 15% OFF YOUR FIRST ORDER

Magellan's
CATALOG OF TRAVEL ESSENTIALS

Subtract 15% from sub-total of first order and attach coupon,
or phone 1-800-962-4943 to place your first order

Magellan's, P.O. Box 5485, Santa Barbara, CA 93150

A Yellow Pages 15% Off Coupon

VALUABLE DISCOUNT COUPON
$5 OFF ANY ONE TELEPHONE-RELATED ELECTRONIC KIT

Digital Products Co.

One coupon per order. May not be used with other discounts.

Digital Products Co., 134 Windstar Circle, Folsom, CA 95630
916-985-7219 (tel)   916-985-8460 (fax)   digprod@aol.com (e-mail)

A Yellow Pages $5 Off Coupon

VALUABLE DISCOUNT COUPON

10% OFF LIST PRICE OF INFRARED KEYBOARDS

Two-Bit Computing

Two-Bit Computing, 4418 Pack Saddle Pass, Austin, TX 78745
sales@embed.com (e-mail)   http://www.embed.com (WWW)
512-447-8888 (tel)   512-447-8895 (fax)

A Yellow Pages 10% Off Coupon

VALUABLE DISCOUNT COUPON
10% OFF NICAD BATTERY PACKS
FOR MILITARY, AEROSPACE, GENERAL AVIATION, & CONSUMER APPLICATIONS

SR Batteries, Inc.

SR Batteries, Inc., P.O. Box 287, Bellport, NY 11713
516-286-0079 (tel)   74167.751@compuserve.com   516-286-0901 (fax)

A Yellow Pages 10% Off Coupon

VALUABLE DISCOUNT COUPON
10% OFF CUSTOMER'S FIRST PURCHASE

Zorin

Control your world with Zorin's Microcontroller products!

Zorin: Unit B, P.O. Box 30547, Seattle, WA 98103
206-286-6061 (tel)
http://zorinco.com   info@zorinco.com

VALUABLE DISCOUNT COUPON
FREE SHORT-FORM PRODUCT CATALOG

Information on networking solutions for ATM, T1/E1, and SONET/SDH applications

PMC-Sierra, Inc.

For free catalog, please request MAR-950200 to: pmc-sierra@akabs.com and include your full name, company, telephone number, and mailing information.

VALUABLE DISCOUNT COUPON
$2 Off a one-year subscription

Hobby Broadcasting

Dedicated to low-power radio and TV: cable-access TV, and Internet, college, international shortwave, pirate, Part-15, cable, and carrier-current radio

Hobby Broadcasting: P.O. Box 642, Mont Alto, PA 17237

VALUABLE DISCOUNT COUPON
10% OFF OPTICAL AND TAPE DRIVES

EZ Systems

EZ Systems: 5122 Bolsa Ave., #109, Huntington Beach, CA 92649
714-379-8383 (tel)   800-392-6962 (tel)   714-379-8391 (fax)
http://www.ezsystems.com   sales@ezsystems.com

VALUABLE DISCOUNT COUPON
NMIY-0031 board & software: only $39 with coupon
New Micros, Inc.

Supports Small C, Forth, Basic, and an Assembler 51
This board is 8051-based with RS-232

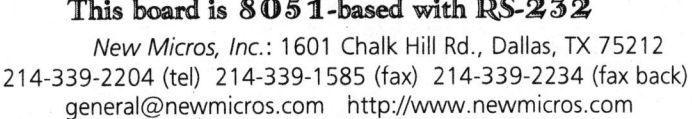

New Micros, Inc.: 1601 Chalk Hill Rd., Dallas, TX 75212
214-339-2204 (tel)   214-339-1585 (fax)   214-339-2234 (fax back)
general@newmicros.com   http://www.newmicros.com

VALUABLE DISCOUNT COUPON
10% DISCOUNT ON STORAGE DEVICES
MBI USA

Hard drives, tape drives, optical drives, CD players

MBI USA: 507 Highview St., Newbury Park, CA 91320
805-499-4993 (tel)   805-499-9142 (fax)
mbiusa@mbiusa.com

Tired of fixing your PC "in the dark"?

Now you don't have to! Subscribe to:

The PC Toolbox™

Hints, Tips, and Fixes for Every PC User

A publication of Dynamic Learning Systems, P.O. Box 282, Jefferson, MA 01522-0282
Tel: 508-829-6744 Fax: 508-829-6819 BBS: 508-829-6706 Internet: sbigelow@cerfnet.com

Finally, there is a newsletter that brings computer technology into perspective for PC users and electronics enthusiasts. Each issue of *The PC Toolbox*™ is packed with valuable information that you can use to keep your system running better and longer:

- Step-by-step PC troubleshooting information that is easy to read and inexpensive to follow. Troubleshooting articles cover all aspects of the PC and its peripherals including monitors, modems, and printers.

- Learn how to upgrade and configure your PC with hard drives, video boards, memory, OverDrive™ microprocessors. scanners, and more.

- Learn to understand the latest terminology and computer concepts. Shop for new and used systems with confidence.

- Improve system performance by optimizing your MS-DOS® and Windows™ operating systems.

- Learn about the latest PC-oriented web sites, ftp locations, and other online resources. Check out our web site at http://www.dlspubs.com/

- Get access to over 2000 PC utilities and diagnostic shareware programs through the Dynamic Learning Systems BBS.

There is no advertising, no product comparisons, and no hype - just practical tips and techniques cover to cover that you can put to work right away. If you use PCs or repair them, try a subscription to *The PC Toolbox*™. If you are not completely satisfied within 90 days, just cancel your subscription and receive a full refund - *no questions asked!* Use the accompanying order form, or contact:

Dynamic Learning Systems, P.O. Box 282, Jefferson, MA 01522-0282 USA.

MS-DOS and Windows are trademarks of Microsoft Corporation. The PC Toolbox is a trademark of Dynamic Learning Systems.
The PC Toolbox logo is Copyright©1995 Corel Corporation. Used under license.

The PC Toolbox™

Use this form when ordering *The PC Toolbox*™. You may tear out or photocopy this order form.

YES! I'm tired of fixing computers in the dark! Please accept my order as shown below: (check any one)

_____ Please start my *one year subscription* (6 issues) for $39 (USD)

_____ Please start my *two year subscription* (12 issues) for $69 (USD)

PRINT YOUR MAILING INFORMATION HERE:

Name: Company:

Address:

City, State, Zip:

Country:

Telephone: (   )   Fax: (   )

PLACING YOUR ORDER:

By FAX: Fax this completed order form (24 hrs/day, 7 days/week) to 508-829-6819

By Phone: Phone in your order (Mon-Fri; 9am-4pm EST) to 508-829-6744

By Web: Complete the online subscription form at http://www.dlspubs.com/

___ MasterCard   Card: __ __ __ __ __ __ __ __ __ __ __ __ __ __ __ __

___ VISA   Exp: ___/___   Sig: _____

Or by Mail: Mail this completed form, along with your check, money order, PO, or credit card info to:

Dynamic Learning Systems, P.O. Box 282, Jefferson, MA 01522-0282 USA

Make check payable to Dynamic Learning Systems. Please allow 2-4 weeks for order processing. Returned checks are subject to a $15 charge. There is a 90 day unconditional money-back guarantee on your subscription.

Have a Product You'd Like to Sell? Promote it on the World Wide Web!

Internet Sales Solutions

How can the Internet Work for You? Here are the Facts:

- Potential audience of 40 million people
- Available in every market, 24 hours/day
- Excellent demographic statistics
- Colorful promotions at low rates
- Instant customer response

Selling Your Products on the Web

Grove Enterprises, a successful pioneer in Internet sales and marketing, will help you develop web pages to promote your products, increase company recognition, and boost sales—all at a surprisingly low cost. Please call or e-mail us to learn more!

Where will Customers Find Your World Wide Web Pages?

Grove is pleased to offer you retail space in our newest addition to the World Wide Web, the attractive **Blue Marble Mall**. Retailers will profit from the thousands of hits the Grove site receives each day. Check out the Blue Marble Mall at **http://www.grove.net/bluemarble.html**, then call us to find out how to begin your successful Internet marketing program.

What is Sales Facilitation?

After we create attractive, effective advertising and promote your product to millions of potential customers on the Internet, then, we can handle your sales!

As your distributor, we can provide a toll-free (800) order line and accept VISA, Mastercard, and Discover credit cards.

Our friendly, highly trained staff can also maintain inventory as well as package and ship your product to the customers.

Our sales facilitation program can be custom tailored to fit many large or small business needs. Call today or e-mail our Internet Services Coordinator: melody@grove.net

One of our current facilitation customers is Clarity Hearing Instruments. Our client furnishes the products, and Grove handles promotion, packaging, and all sales transactions.

Grove's Other Industries

You may already know Grove for our sales and service of shortwave receivers and scanner equipment and accessories. We are also publishers of *Monitoring Times* and *Satellite Times* magazines. Call today for a free catalog. And visit our web site at **http://www.grove.net**.

GROVE

Grove Enterprises
P.O. Box 98, 7540 Hwy. 64 W.
Brasstown, NC 28902
1-800-438-8155
(704) 837-9200, (704) 837-2216 Fax